New Nanomaterials and Techniques for Tumor-targeted Systems

Rongqin Huang • Yi Wang
Editors

New Nanomaterials and Techniques for Tumor-targeted Systems

Springer

Editors
Rongqin Huang
Department of Pharmaceutics,
School of Pharmacy, Key Laboratory
of Smart Drug Delivery
Ministry of Education, Fudan University
Shanghai, China

Yi Wang
Center for Advanced Low-dimension Materials
Donghua University
Shanghai, China

ISBN 978-981-15-5161-1 ISBN 978-981-15-5159-8 (eBook)
https://doi.org/10.1007/978-981-15-5159-8

This Springer imprint is published by the registered company Springer Nature Singapore Pte Ltd.
The registered company address is: 152 Beach Road, #21-01/04 Gateway East, Singapore 189721, Singapore

Contents

1 **Introduction** 1
Rongqin Huang

2 **Barriers for Tumor Drug Delivery** 5
Qiuyue Huang and Jinzhi Du

3 **Tumor-targeted Strategies** 27
Min Liu and Weiyue Lu

4 **Tumor-Responsive Drug Release Strategies** 57
Zhaoqing Shi, Yun Zhou, and Lin Mei

5 **Tumor Diagnosis Patterns** 87
Xinwei Li and Cong Li

6 **Tumor Therapeutic Modes** 135
Yu Zhong Peng, Li Jun Yang, Hang Hong Lo, Betty Yuen Kwan Law, and Vincent Kam Wai Wong

7 **Carbon-Based Tumour-targeted Systems** 231
Smriti Sri, Shweta Panwar, and Pratima R. Solanki

8 **Silica-Based Tumor-targeted Systems** 271
Wei Guo, Min Qian, Xiaoyi Zhang, and Yi Wang

9 **Lipid-Based Tumor-targeted Systems** 293
Yaxi Li, Chen Zhang, Tianliang Min, Yuan Ping, and Kai Li

10 **Dendrimer-Based Tumor-targeted Systems** 337
Zhijun Ouyang, Du Li, Mingwu Shen, and Xiangyang Shi

11 **Polymer-Based Tumor-targeted Nanosystems** 371
Teoman Benli-Hoppe and Ernst Wagner

12 Other New Tumor-targeted Systems 413
Yibo Xie, Min Qian, Xiaoyi Zhang, and Rongqin Huang

13 Clinical Applications of Tumor-targeted Systems 437
Xinxin Zhang

14 Challenges and Perspectives of Tumor-targeted Systems 457
Yi Wang

Chapter 1
Introduction

Rongqin Huang

Due to the complex and fast-changing tumor microenvironment, abnormal metabolic behavior, and uncontrolled growth and metastasis, cancers become more and more invasive and also the main cause of disease-associated death [1]. This situation leads to almost no improvement to overall patient survival rates based on the current treatment modalities for tumors such as surgery, chemotherapy, and radiotherapy [2, 3]. How to treat tumors effectively and safely has become a great challenge that can't be ignored in today's world development.

In recent decades, nanotechnology has already infiltrated into all areas of biomedical science. Nanoscaled materials have emerged as indispensable elements for preparation of novel and multifunctional drug delivery systems (DDS), providing a new avenue for tumor therapy, especially for tumor-targeted therapy [4]. By controlling particle size [3, 5], molecular mass [6], and charge [7] or adding targeting moieties such as antibodies or their fragments, peptides, proteins, or other ligands [8, 9], nanocarriers by passive or active targeting delivery can increase the accumulation of the cargos to tumor cells or tissues while simultaneously limit their accumulation in the healthy cells and tissues, enhancing the efficacy of the clinical diagnosis and treatment. In addition to solubilize poorly soluble drugs and protect drugs from degradation in vivo, the nanocarriers could prolong the circulation time of free drugs [10]. Moreover, the release of the cargos from the nanomedicines can be controlled in a spatiotemporal manner [11], making the cargos exert their pharmacological activities only at the target sites. To date, several classes of nanocarriers have been exploited for tumor-targeted therapy, including liposomes [12, 13], polymeric micelles [14], polymer-drug conjugates [9, 15], nanogels [16–18], mesoporous silica nanoparticles [11, 19], carbon nanodots [20], metal organic framework (MOF) [21–23], and so on. It is rational to believe that nanotechnology

R. Huang (✉)
Department of Pharmaceutics, School of Pharmacy, Key Laboratory of Smart Drug Delivery, Ministry of Education, Fudan University, Shanghai, China
e-mail: rqhuang@fudan.edu.cn

R. Huang, Y. Wang (eds.), *New Nanomaterials and Techniques for Tumor-targeted Systems*, https://doi.org/10.1007/978-981-15-5159-8_1

could provide cancer treatment with great promise and expect to offer superior therapeutic outcomes.

In this book, we focus on new nanomaterials and techniques for tumor-targeted systems to tackle the complexity of cancer treatment. Firstly, we gave a brief mention of the advances of nanomaterials in cancer diagnosis and therapy. Secondly, the physiological barriers affecting tumor treatment in the circulatory system and the tumor microenvironment were introduced. Thirdly, tumor-targeted strategies and tumor-responsive drug release strategies based on tumor microenvironment were discussed. And on this basis, the diagnosis patterns and therapeutic modes about tumor were described. In this part, based on the previous introduction to tumor development and biological barriers, we summarized the current strategies and methods that might be used in tumor-targeted therapy. After all of the above, we systematically summarized the latest development of nanomaterials and techniques in the field of tumor-targeted diagnosis and therapy, providing valuable scientific references for beginners and senior researchers. In this part, the preparation methods, physical and chemical properties of the materials, as well as their unique advantages as tumor-targeted carriers and their applications in tumor-targeted therapy were introduced in classification. Meanwhile, we compared the nanomaterials discussed in this book so as to show their unique advantages in tumor-targeted therapy more precisely. In addition, the development of some new tumor-targeted systems which are not included in the above mentioned was reviewed. In this section, we discussed their characteristics and highlights as tumor-targeted nanomaterials. And then, in the next canto, we reviewed the current clinical applications of tumor-targeted systems. Eventually, the challenges and future perspectives of tumor-targeted systems for cancer diagnosis and therapy were prospected.

We believe that the publication of this book will contribute to the renewal and development of the vision of oncotherapy and the development of tumor-targeted delivery systems and highly effective tumor therapy. Furthermore, this book will help graduate students and researchers to get an overall picture of tumor-targeted diagnosis and therapy and may also yield benefits for pharmaceutical companies with regard to drug discovery based on new nanomaterials and techniques. Meanwhile, the chief editors express the sincere gratitude to the editorial committee unit's leadership, each expert editorial committee, and the publishing house's editorial staffs of this book. It is your hard work that gives this book even more valuable meaning.

Of course, even wise men sometimes miss things. Due to the little talent and less learning, omissions and imperfections are inevitable. We also look forward to your comments and suggestions, which are very valuable for making a better reprint and contributing to the development of diagnostic, therapeutic, and theranostic tools of cancers.

References

1. Hanahan D, Weinberg RA (2011) Hallmarks of cancer: the next generation. Cell 144 (5):646–674
2. Phuengkham H, Ren L, Shin IW, Lim YT (2019) Nanoengineered immune niches for reprogramming the immunosuppressive tumor microenvironment and enhancing cancer immunotherapy. Adv Mater 31:1803322
3. Nguyen TL, Nguyen Y, Kim J (2019) Mesoporous silica as a versatile platform for cancer immunotherapy. Adv Mater 31:1803953
4. Liu R, Yu MN, Yang XT, Umeshappa CS, Hu C, Yu WQ, Qin L, Huang Y, Gao HL (2019) Linear chimeric triblock molecules self-assembled micelles with controllably transformable property to enhance tumor retention for chemo-photodynamic therapy of breast cancer. Adv Funct Mater 29:1808462
5. Majumder J, Taratula O, Minko T (2019) Nanocarrier-based systems for targeted and site specific therapeutic delivery. Adv Drug Deliv Rev 144:57–77
6. Maeda H (2017) Polymer therapeutics and the EPR effect. J Drug Target 25:781–785
7. Zhou FY, Feng B, Yu HJ, Wang DG, Wang TT, Ma YT, Wang SL, Li YP (2019) Tumor microenvironment-activatable prodrug vesicles for nanoenabled cancer chemoimmunotherapy combining immunogenic cell death induction and CD47 blockade. Adv Mater 31:1805888
8. Wang SS, Li CY, Qian M, Jiang HL, Shi W, Chen J, Lachelt U, Wagner E, Lu WY, Wang Y, Huang RQ (2017) Augmented glioma-targeted theranostics using multifunctional polymer-coated carbon nanodots. Biomaterials 141:29–39
9. Li CY, Qian M, Wang SS, Jiang HL, Du YL, Wang JX, Lu WY, Murthy N, Huang RQ (2017) Aptavalve-gated mesoporous carbon nanospheres image cellular mucin and provide on-demand targeted drug delivery. Theranostics 7(13):3319–3325
10. Feng B, Niu ZF, Hou B, Zhou L, Li YP, Yu HJ (2019) Enhancing triple negative breast cancer immunotherapy by ICG-templated self-assembly of paclitaxel nanoparticles. Adv Funct Mater 30:1906605
11. Yang XT, Hu C, Tong F, Liu R, Zhou Y, Qin L, Yang LO, Gao HL (2019) Tumor microenvironment-responsive dual drug dimer-loaded PEGylated bilirubin nanoparticles for improved drug delivery and enhanced immune-chemotherapy of breast cancer. Adv Funct Mater 29:1901896
12. Zylberberg C, Matosevic S (2016) Pharmaceutical liposomal drug delivery: a review of new delivery systems and a look at the regulatory landscape. Drug Deliv 23:3319–3329
13. Deng C, Zhang Q, Jia M, Zhao J, Sun X, Gong T, Zhang Z (2019) Tumors and their microenvironment dual-targeting chemotherapy with local immune adjuvant therapy for effective antitumor immunity against breast cancer. Adv Sci 6:1801868
14. Li CY, Du YL, Qian M, Jiang HL, Wang JX, Murthy N, Wang Y, Huang RQ (2018) Side effects-avoided theranostics achieved by biodegradable magnetic silica-sealed mesoporous polymer-drug with ultralow leakage. Biomaterials 186:1–7
15. Li X, Schumann C, Albarqi HA, Lee CJ, Alani WG, Bracha S, Milovancev M, Taratula O (2018) A tumor-activatable theranostic nanomedicine platform for NIR fluorescence-guided surgery and combinatorial phototherapy. Theranostics 8:767–784
16. Chen Q, Wang C, Zhang XD, Chen GJ, Hu QY, Li HJ, Wang JQ, Wen D, Zhang YQ, Lu YF, Yang G, Jiang C, Wang J, Dotti G, Gu Z (2019) In situ sprayed bioresponsive immunotherapeutic gel for post-surgical cancer treatment. Nat Nanotechnol 14:89–97
17. Yang L, Lu ZZ, Xu SS, Li M, Wang XH, Zhang ZR, He Q (2019) Self-delivery micellar nanoparticles prevent premetastatic niche formation by interfering with the early recruitment and vascular destruction of granulocytic myeloid-derived suppressorcells. Nano Lett 20 (4):2219–2229. https://doi.org/10.1021/acs.nanolett.9b03883
18. Ruan HT, Hu QY, Wen D, Chen Q, Chen GJ, Lu YF, Wang JQ, Cheng H, Lu WY, Gu Z (2019) A dual-bioresponsive drug-delivery depot for combination of epigenetic modulation and immune checkpoint blockade. Adv Mater 31:1806957

19. Wang Y, Wang KY, Zhang R, Liu XG, Yan XY, Wang JX, Wagner E, Huang RQ (2014) Synthesis of core-shell graphitic carbon@silica nanospheres with dual-ordered mesopores for cancer-targeted photothermo chemotherapy. ACS Nano 8(8):7870–7879
20. Qian M, Du YL, Jiang HL, Huo TT, Yang YF, Guo W, Wang Y, Huang RQ (2019) Biodegradable mesoporous silica achieved via carbon nanodots-incorporated framework swelling for debris-mediated photothermal synergistic immunotherapy. Nano Lett 19(12):8409–8418
21. Chen LL, Qian M, Jiang HL, Zhou YW, Du YL, Yang YF, Huo TT, Huang RQ, Wang Y (2020) Multifunctional mesoporous black phosphorus-based nanosheet for enhanced tumor-targeted combined therapy with biodegradation-mediated metastasis inhibition. Biomaterials 236:119770
22. Zhou J, Wang B (2017) Emerging crystalline porous materials as a multifunctional platform for electrochemical energy storage. Chem Soc Rev 46(22):6927–6945
23. Yang Q, Xu Q, Jiang HL (2017) Metal-organic frameworks meet metal nanoparticles: synergistic effect for enhanced catalysis. Chem Soc Rev 46(15):4774

Chapter 2
Barriers for Tumor Drug Delivery

Qiuyue Huang and Jinzhi Du

Abstract The ultimate goal of cancer nanomedicine is to specifically increase the drug accumulation in tumor and reduce side effects. Although nanomedicine has made great progress in recent decades, its delivery efficacy is still suboptimal, which hinders its clinical translation. Currently, most drug delivery vehicles are delivered by systemic administration, and physiological barriers prevent effective accumulation of nanomedicine at the tumor site. These nanocarriers must reach the tumor tissue after the long journey in blood circulation, then access to tumor cells deep inside the tumor by overcoming the tumor tissue barrier, and finally kill the tumor cells after entering the cell. Successful design of nanocarriers must fully consider the existing delivery barriers in order to achieve better therapeutic efficacy. This chapter summarizes obstacles experienced by nanomedicine from the site of injection and the site of action as well as exemplifies typical strategies for overcoming these barriers, hopefully providing insights into the design of more effective nanomedicine.

Keywords Nanomedicine · Drug delivery · Delivery barriers

2.1 Introduction

Traditional chemotherapeutic drugs are usually highly toxic and poorly soluble in water and can be rapidly cleared from the body. When they are injected directly into the body, they will diffuse and distribute throughout the body, causing unexpected side effects and limiting the effective dose at the tumor site [1]. In comparison, nanoscaled drug delivery systems have shown many advantages, including improving drug solubility, prolonging blood circulation, and enhancing drug accumulation in tumor tissues [2]. Due to these advantages, many nanoparticle (NP)-based

Q. Huang · J. Du (✉)
Institutes for Life Sciences, School of Medicine, South China University of Technology, Guangzhou, Guangdong, China
e-mail: djzhi@scut.edu.cn

R. Huang, Y. Wang (eds.), *New Nanomaterials and Techniques for Tumor-targeted Systems*, https://doi.org/10.1007/978-981-15-5159-8_2

formulations have been approved for clinical use, such as Doxil® and Abraxane®. However, these nanomedicines only show suboptimal benefits, especially in patients with aggressive solid tumors. Taking Doxil® as an example, it shows lower cardiotoxicity and higher safety profile; however, its therapeutic efficacy is not significantly improved compared with free doxorubicin [3].

Although nanomedicine increases the drug concentration at the tumor site by a factor of 5 or more compared to free drugs, there are only $<5\%$ of the total amount of drugs could be delivered to the tumor [4]. A recent meta-analysis of published literature showed that in preclinical animal models, only 0.7% (median) of injected doses of NPs were delivered to solid tumors [5]. They then studied the targeting efficiency of herceptin and folate-coated gold NPs and silica NPs to cancer cells and found that only 2% of tumor cells interacted with the NPs. More than 88% of intratumor NPs were trapped in the extracellular matrix (ECM) or ingested by tumor-associated macrophages around blood vessels [6]. Although these data are debatable, they reflect to some degree that the delivery efficacy of nanomedicine is still not satisfactory. Accordingly, how to reduce the loss of NPs during delivery to increase their accumulation in tumor tissues and promote efficient uptake by tumor cells are urgent problems to be solved.

One prerequisite to solve these problems is to understand the delivery barriers in a thorough manner. After intravenous injection, blood is the first stop on the journey of the drug carriers in the body. NPs in blood will interact with proteins, which facilitate the elimination of NPs by mononuclear phagocyte system (MPS). During circulating in the blood, NPs accumulate preferentially in tumor tissue through enhanced permeability and retention (EPR) effect. After entering the tumor tissue, heterogeneous tumor blood vessel network, dense extracellular matrix, high tumor interstitial fluid pressure, and tumor interstitial cells are all huge obstacles to the penetration of NPs, resulting in the heterogeneous distribution of NPs within the tumor. Most of NPs were trapped by the extracellular matrix or internalized by tumor stromal cells, and only a few of NPs could interact with tumor cells. But these particles still face barriers from tumor cells (membrane barriers, endosomal/lysosomal escape, nuclear membrane barriers, drug efflux pumps, etc.). Each physiological and pathological barrier will affect the eventual therapeutic effect. It can be imagined that if any individual step is delivered inefficiently, the overall delivery efficacy will be greatly reduced [7]. Therefore, it is necessary to understand tumor drug delivery barriers for the design of drug carriers. The main purpose of this chapter is to understand these delivery barriers and how they affect the delivery efficacy. In addition, design strategies which are capable of overcoming these barriers will also be included.

2.2 Blood Barriers for Tumor Drug Delivery

Sequestration by Mononuclear Phagocyte System (MPS) After entering the blood circulation via intravenous administration, NPs first face the blood barrier, which mainly includes the phagocytosis and clearance of mononuclear phagocyte system (MPS) from the liver and spleen. MPS is mainly composed of macrophages, dendritic cells, and monocytes located in the blood, lungs, liver, lymph nodes, and spleen. It is the first line of defense for the NPs after system administration [8]. The sequestration process of NPs by MPS can be divided into two steps, opsonization and phagocytosis. Once NPs are injected, the plasma proteins (including serum albumin, apolipoprotein, complement components, immunoglobulin, and so on) will attach onto the surface of the circulating NPs, forming the protein corona. Once the protein corona is formed, macrophages will recognize NPs as foreign substances through opsonin and other pattern recognition receptors, rapidly triggering the process of internalization, isolation, and subsequent clearance of NPs from circulation system [1].

Besides protein adsorption, the size, surface charge, and shape also affect the circulation of the NPs in blood. Take micelle as an example, micelle size has a great influence on the blood clearance rate. When the micelle was less than 100 nm, the blood clearance rate was negatively correlated with the particle size. When the micelle size increased to 100 ~ 160 nm, the blood clearance rate was very similar. Further enlargement of the micelles speeded up their blood clearance [9]. On the other hand, nanoparticles less than 5.5 nm were quickly eliminated by the kidneys [10], while larger particles (> 200 nm) accumulated in the liver and spleen [1]. Positively charged NPs have a shorter blood circulation half-life, while neutral and slightly negatively charged NPs, by contrast, have a longer blood circulation time [11, 12]. Non-spherical particles, such as discoidal particles, have a larger surface area than traditional spherical particles and a greater tendency to contact with the blood vessel wall, because they have more contact/binding points with the blood vessel and are more likely to marginate to vessel walls. However, the spherical particles showed minimal lateral drift to vessel walls and were unlikely to marginate vascular walls and establish contact/binding points with the endothelial cells [13–15]. Other studies also showed that elongated particles could attenuate phagocytic uptake and circulate longer than spherical particles in vivo [16]. Moreover, compared with hard NPs, soft NPs prolonged blood circulation in the body [17].

Renal Clearance The kidney plays a key role in the transport and clearance of NPs in the body. Renal clearance is mainly carried out in the glomeruli, and the size, charge, and shape of materials will affect the filtration efficiency. Spherical nanoparticles with a diameter of less than 5.5 nm are easily cleared by the kidney and have a half-life of less than 4 h in vivo, while nanoparticles with a diameter of 8 nm are difficult to pass the glomerular filtration barrier [10, 18]. Because of the negatively charged glomerular basement membrane, cationic nanoparticles have stronger renal clearance ability than neutral or negatively charged nanoparticles [19].

2.3 Strategies to Overcome the Blood Barriers

PEGylation of the Delivery Systems PEGylation refers to the modification of NPs with poly(ethylene glycol) (PEG). The hydration layer of PEG chain can effectively prevent the recognition of opsonin protein and the binding of plasma protein, thus reducing the phagocytosis of MPS [20]. Generally, longer and denser PEG chains or branched PEG chains on NPs surface can result in longer blood circulation [21]. Dispersed PEG brush of varying length is more effective in reducing protein adsorption than single dispersed PEG layer [22]. Zhou et al. prepared PEG-modified NPs with dual physical effects (Fig. 2.1) [23]. It had a dense PEG inner layer that could block the protein from approaching the NPs through steric repulsion and a sparse PEG outer layer (a density close to the mushroom-to-brush transition) that could interfere with protein binding by conformational changes. Compared to conventional PEGylated NPs, the outer dynamic PEG layer significantly prolonged their blood circulation by reducing the uptake of NPs by hepatocytes, particularly liver sinusoidal endothelial cells (LSECs) rather than macrophages. However, repeated injection of PEGylated NPs can induce immune responses in vivo, resulting in accelerated blood clearance phenomenon, the so-called ABC phenomenon [24].

Biomimetic Delivery Systems Another strategy to prolong particle circulation in bloodstream is using the biomimetic delivery systems. By simulating the structure and function of some endogenous proteins in mammalian cells and pathogens, the immunogenicity of NPs can be reduced, the retention time of drugs in blood can be prolonged, and the targeting ability of NPs can be improved. Endogenous proteins such as serum albumin [25] and lipoprotein [26] have been used as drug delivery vectors or as coating layers. For most NPs, the protein corona produced on the surface of NPs is dominated by albumin, the most abundant protein in the serum

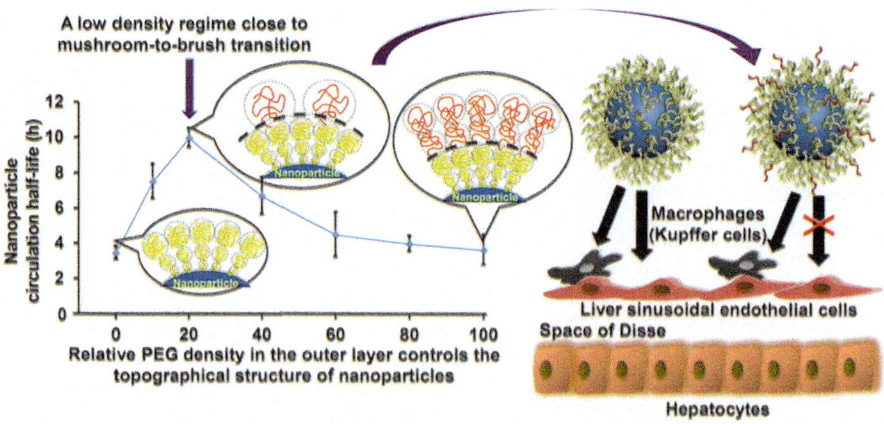

Fig. 2.1 Polyethylene glycol polymer with a dynamic outer structure can effectively prolong the blood circulation time of NPs through reducing the clearance of NPs by liver vascular endothelial cells. (Reprinted with permission from Ref. [23]. Copyright 2018 American Chemical Society)

(55%) [27]. Albumin-based delivery systems can extend the blood circulation time and increase NPs uptake in tumor tissue [28]. Chemically attach the cellular components of endogenous cells such as CD47 to the carrier surface is another option. CD47 self peptide can inhibit phagocytic clearance by binding with CD172a (also known as signal regulatory protein-α, SIRPα) on phagocytes and prolong the circulation [29, 30].

Coating the carrier with intact cell membrane is a robust way to overcome the blood barriers [31–33]. Zhang et al. developed a tumor microenvironment-responsive macrophage-membrane coated NP, which was a sustained-release delivery system for tumor-targeted chemotherapy. Macrophage membrane-coated NPs not only evaded MPS and immune surveillance but also enhanced tumor accumulation because of their tumor homing properties. Under weakly acidic tumor conditions, these pH-responsive NPs removed the membrane coating and exposed the targeted ligands. After being internalized by tumor cells, the drug was rapidly released from the NPs under the acidic condition of endosome [34].

2.4 Tumor Tissue Barriers for Tumor Drug Delivery

Heterogeneous Tumor Vascular Network The distribution of drugs in the tumor site can be strongly affected by the heterogeneous tumor vascular network [35]. Drugs travel through the bloodstream by convection and then permeate blood vessels by diffusion or convection to enter the tumor tissue. However, tumor vascular network is heterogeneous, which is not only reflected in its structure and function but also in its spatial and temporal distribution. Tumor vascular network varies with tumor type, also varies among individuals with the same tumor type [35, 36]. The peripheral vascular density of the tumor is high, while the vascular density of the interior of the tumor is low [37]. Such uneven spatial and temporal distribution will lead to the uneven penetration and diffusion of drugs, which limits the distribution of drugs in the tumor. In addition, the growth rate of tumor vessels cannot keep pace with the rapid proliferation of tumor cells, so tumor cells are generally far away from blood vessels [38].

High Interstitial Fluid Pressure (IFP) The interstitial pressure is high in most solid tumors, due to the defective blood vessels, poor lymphatic drainage, as well as the dense extracellular matrix [39]. The IFP value of normal tissue is about 0–3 mmHg, while the IFP value of solid tumor is generally 5–40 mmHg. The IFP value of fibrous proliferative pancreatic tumor can even be increased to 75–130 mmHg [40]. The soluble protein macromolecules and NPs were mainly transported in the interstitium by convection, while the low-molecular weight compounds such as glucose and oxygen were mainly transported by diffusion [41]. High IFP limits convection, which prevents NPs from entering the tumor, but paradoxically promotes passive diffusion. Nevertheless, diffusion is a much slower process across blood vessels than convection, especially for large entities [42, 43]. In addition, stromal cells can compress the blood and lymphatic vessels, further inhibiting the penetration of NPs in the tumor tissue [44]. Studies have shown that increased IFP in tumors limited the entry of antibodies into target cells. And the penetration of antibodies into tumors was limited to the area surrounding blood vessels [45]. Therefore, EPR effect can promote the perfusion of NPs at tumor sites, but vascular network heterogeneity and increased IFP hinder the penetration of NPs to sites far away from tumor blood vessels [46]. Normalization of tumor vascular abnormalities is an effective method to reduce IFP in tumor tissues [47, 48].

Dense Extracellular Matrix (ECM) ECM is mainly composed of collagen network, microfibrin elastin, glycosaminoglycan (GAG), proteoglycan, and other polysaccharides with different physical and biochemical properties. ECM is a highly dynamic and complex molecular network in tumors [49]. Tumor ECM is denser than normal ECM due to high density collagen, elevated lysyl oxidase (LOX) levels, and the presence of enhanced integrin receptors [50]. Dense ECM is the major obstacle to NP penetration in solid tumors, which will be manifested in the following aspects. Firstly, ECM is a cross-linked gel-like structure, and its naturally viscous property limits the convection and diffusion of NPs in tumor tissues [51]. Secondly, high cell density in the tumor compresses the ECM collagen network and increases solid stress and IFP, thus impeding the movement of NPs within the tumor [52]. Thirdly, the pores in the ECM are generally less than 40 nm [46], and NPs larger than 60 nm are difficult to diffuse in the ECM matrix [51].

Stromal Cells Tumor stromal consists of tumor cells, tumor-associated fibroblasts (CAFs), angiogenesis related cells (endothelial cells and pericytes), and tumor-associated immune cells (T cells and B cells, natural killer T cells, dendritic cells, tumor-associated macrophages (TAMs), and myeloid-derived suppressor cells (MDSCs)) [53]. CAFs and TAMs can not only promote the proliferation and metastasis of tumor cells but also affect the uptake of NPs and the distribution of NPs in tumor tissues [52, 54]. CAFs can produce a large amount of collagen and proteoglycan, which are involved in the formation of ECM. Receptors expressed in tumor cells may also be expressed in CAFs, which will form a binding site barrier near tumor blood vessels [55]. That is, NPs extravasated from tumor blood vessels will be absorbed by CAFs, resulting in off-target distribution [52, 56]. TAMs can be actively infiltrated in tumor tissue, and its phagocytosis of NPs has a great impact on the accumulation degree of NPs in tumor [53, 57]. The HT1080 xenograft tumor model, in which the number of tumor cells far exceeds TAMs, was adopted to study the uptake capacity of ~100 nm NPs in tumor tissues at the cellular level. Although TAMs only accounted for 4% in tumor, TAMs internalized 30% of NPs compared with tumor cells. Moreover, the intake of NPs in TAMs was also the highest at the single cell level [58]. Chan's research group designed a trastuzumab encapsulated NPs targeting tumor cells to study the targeting effect of NPs on tumor cells. Studies have shown that up to 90% of NPs in tumors were ingested by macrophages. In comparison, the possibility of interaction between NPs and TAMs was 7–38 times higher than that of tumor cells, and TAMs accounted for 1.6–2.1 more NPs in a single cell level than that of tumor cells [54].

2.5 Strategies to Overcome the Tumor Tissue Barriers

Tumor tissue permeability is an important problem to be solved during drug delivery and a major obstacle to improving therapeutic efficacy. At present, these obstacles are mainly solved from two aspects: reasonable design of NPs and remodeling the tumor stromal microenvironment.

Rational Design of NPs Physical and chemical properties of NPs, such as particle size, shape, and surface modification (ligand modification or surface charge), are the main factors affecting the penetration ability of NPs in tumors [46]. NPs of small size have a fast clearance rate in the blood and a limited accumulation in the tumor tissue during the circulation period, but their diffusion barriers in the tumor tissue are small, and their penetration ability is strong. Although the carrier with large particle size is easy to overflow from tumor blood vessels, its permeability in tumor tissue is poor [59–61]. When designing the size of nanocarrier, it is necessary to find a balance between tumor tissue accumulation and tumor tissue permeability. In other words, the carrier needs to have different particle sizes at different delivery stages. During the blood circulation, particles with larger size can achieve effective accumulation in tumor tissues through EPR effect, while particles with smaller size can meet the deep penetration ability of NPs in tumor tissues [9, 18, 62]. Based on this principle, different size switch systems that sensitive to pH [63], MMPs [63], and light [64] have been developed. Li et al. designed intelligent nanocarrier iCluster, which could rapidly change the size and improve tumor penetration under the acidic tumor environment (Fig. 2.2). With an initial size of 100 nm, iCluster prolonged blood circulation and accumulated in tumor tissue. The weakly acidic environment in the tumor triggered the release of dendritic prodrugs (~5 nm) from iCluster. Small-sized dendrimer prodrugs further diffused throughout the tumor tissue, effectively delivering anticancer drugs to tumor cells. The internalized dendrimer prodrugs then decomposed in the cell to release cisplatin and kill cancer cells [63]

The effect of surface charge on tumor penetration should be considered when designing deep penetration NPs. Neutral or negative charge carriers have advantages in blood circulation, tumor accumulation, and penetration in tumor stroma but are not easy to be engulfed by tumor cells. Because of the high clearance rate in blood and enhanced cell absorption, the positive charge carriers often have poor accumulation effect in tumor tissue and penetration ability in tumor stroma. Besides, positive charge carriers are easy to be internalized by tumor cells [65, 66]. In other words, the positive carrier will be first consumed by tumor stromal cells or tumor cells surrounding tumor blood vessels, that is, binding site barrier, and only a few particles can enter the deep site of the tumor. While the neutral or negative charge carrier is consumed to a small extent, its penetration capacity is stronger than that of the positive carrier. Therefore, size/charge dual switch NPs have designed. Chen and his colleagues prepared a novel shell-stacked NP (SNP). In acidic tumor tissues, the size of these NPs was significantly reduced from about 145 to 40 nm, and the surface charge was reversed from −7.4 to 8.2 mV, thus enhancing tumor penetration and

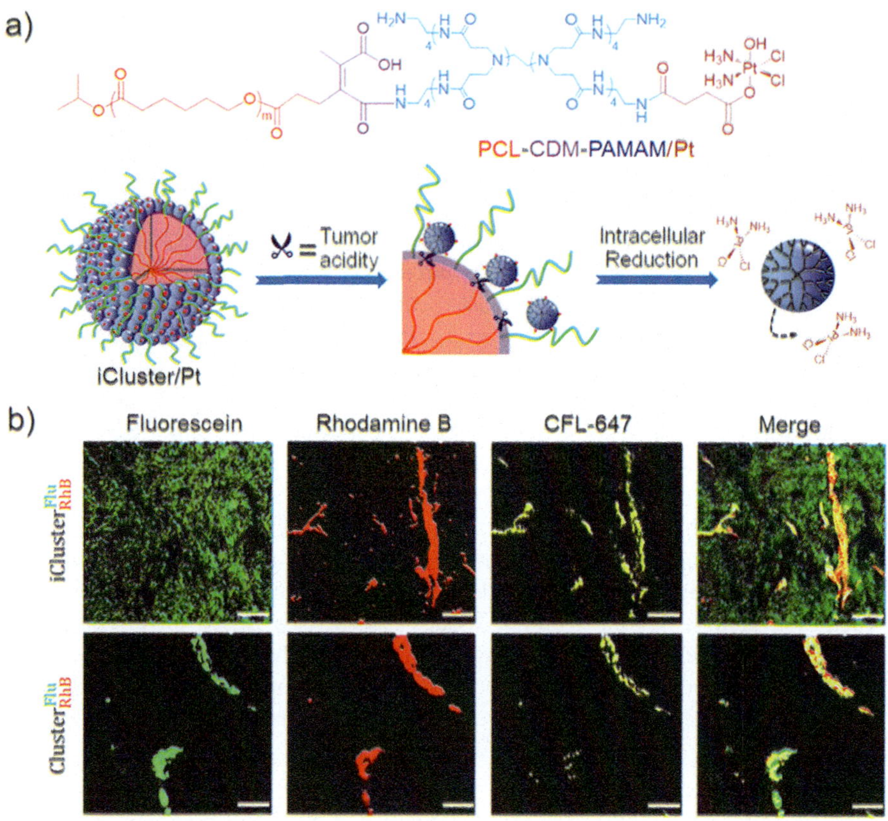

Fig. 2.2 (**a**) Structure of PCL-CDM-PAMAM/Pt, its self-assembly into iCluster/Pt, and iCluster/Pt's response to tumor acidity and the intracellular reductive environment. (**b**) Immunofluorescence images showing the microdistribution of iCluster in BxPC-3 xenograft tumor at 4 h postinjection. PAMAM, green; iCluster core, red; blood vessels, yellow. The scale bars represent 50 μm. (Reprinted with permission from Ref. [63]. Copyright 2016 National Academy of Sciences)

uptake of tumor cells. The penetration depth of SNPs in xenografted A549 lung carcinoma was about 1 mm, which was four times than that of normal NPs. In addition, the core formed by disulfide bond cross-linking rapidly disintegrated in the intracellular reduction environment and accelerates the release of drugs into the cytosol [67].

Remodeling the Tumor Microenvironment Remodeling the tumor microenvironment involves remodeling the tumor vascular system and the tumor stromal microenvironment. EPR effect is heterogeneous in different individuals, different tumor types, and even in different areas and progress stages of the same tumor [68]. Regulation of tumor vascular system is an effective strategy to deal with EPR heterogeneity. There are two main strategies to regulate the tumor vascular systems: one strategy is disrupting the blood vessels to improve its permeability. Enhanced vascular permeability can improve the potential for intratumoral transport of NPs [69]. Physical methods such as radiation [70], ultrasound [71], and near-infrared laser irradiation [72] or disrupting agents such as NO [73] and antiangiogenic agents [74] can achieve vascular damage, improve tumor vascular permeability, and promote the penetration of NPs in tumor tissues. Platelets are capable of maintaining tumor vascular integrity, and tumor-associated platelets depletion is a novel therapeutic strategy for tumor vascular regulation [75]. The other one is on the opposite, which is normalizing the blood vessels, thus reducing IFP and promoting the convection of drug delivery vectors [46, 76, 77]. Vascular normalization can enhance blood perfusion, reduce IFP, and promote the convection and diffusion of NPs in tumor [76, 78]. Currently, tumor vascular normalization strategies mainly combine antiangiogenic drugs such as VEGF receptor-2 (VEGFR2), sinomenine, and bevacizumab with NPs [77]. However, this normalized vascular system results in decreased pore size of the neovasculature. Therefore, vascular normalization can only enhance tumor penetration of small NPs, but not promote the delivery of large NPs [47, 79].

Disruption of the tumor ECM is another strategy to facilitate particle distribution in the tumor tissues. There are three main strategies for disrupting tumor ECM. First, physical methods like photothermal, hyperthermia, and ultrasound [80] have been widely used to disrupt tumor ECM [76], but it is also possible to destroy normal tissues [76]. Second, biochemical enzymes such as collagenase [81] and hyaluronidase [82] can degrade ECM and reduce the diffusion barrier in tumor, but it is usually difficult to find the appropriate formulation for loading protein NPs [83]. Moreover, this strategy may also increase tumor proliferation and metastasis, limiting the clinical application of these proteins [46]. Third, chemical agents such as cyclopamine [84] and pirfenidone [85] can also be utilized to facilitate the penetration of NPs by disrupting the tumor ECM. Zhang et al. demonstrated that cyclopamine, a drug with strong anti-fibrotic activity, promoted the penetration and efficacy of nanotherapeutics in pancreatic cancer. Cyclopamine could disrupt the extracellular fibrinoprotein, reduce tumor vascular pressure, and improve tumor perfusion. Thus, cyclopamine improved the tumor accumulation, intratumoral distribution, and the tumor growth inhibiting effect of NPs [84].

Selective targeting of cancer-associated fibroblasts (CAFs) and tumor-associated macrophages (TAMs) has been designed to eliminate or deplete the stromal cell in tumor to overcome the tumor tissue barrier. Recently, some researchers have proposed using stromal cells to suppress metastasis. Considering that most of NPs are delivered to CAFs and the depletion of CAFs may increase the risk of tumor metastasis, Lang et al. proposed a novel strategy to inhibit tumor metastasis and enhance the therapeutic effect by regulating the function of CAFs rather than directly consuming CAFs. CXCL12 released by activated CAFs not only promotes tumor metastasis but also maintains the tumor-promoting phenotype of CAFs. Therefore, in order to inactivate CAFs, they selected C-X-C motif chemokine ligand 12 (CXCL12) gene as the target to specifically deliver siRNA to CAFs through PNP/si*CXCL12*/mAb nanosystem. This nanosystem could be recognized and

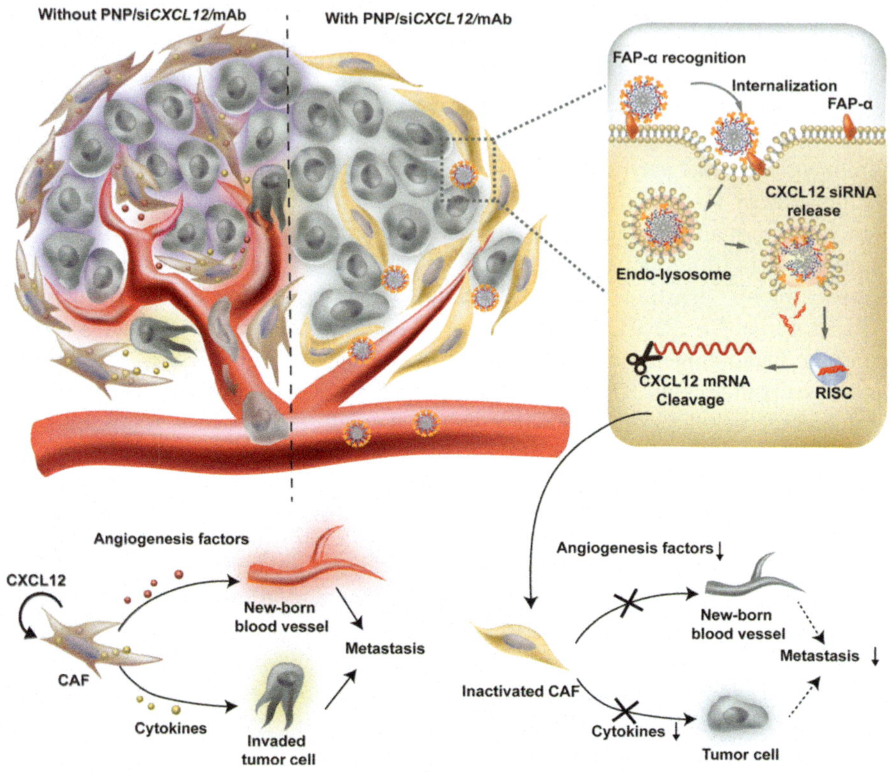

Fig. 2.3 Proposed mechanism of PNP/si*CXCL12*/mAb-mediated metastasis inhibition and CPP-mediated transfection of CXCL12 siRNA in CAFs. CAF-secreted cytokines contribute to tumor metastasis by promoting migration and invasion of tumor cells and angiogenesis. PNP/si*CXCL12*/mAb nanosystem recognizes FAP-α on CAFs and is internalized into CAFs, and then the released siRNA leads to CXCL12 gene silencing. As a consequence, CAFs are inactivated, and CAF-related prostate tumor metastasis is inhibited. (Reprinted with permission from Ref. [86]. Copyright 2019 American Chemical Society)

internalized by CAFs through fibroblast activation protein-α (FAP-α) antibody and then releases siRNA to silence CXCL12 gene, thus suppressing metastasis by inhibiting tumor cell migration and invasion and tumor angiogenesis [86] (Fig. 2.3).

2.6 Cellular Barriers for Tumor Drug Delivery

Even if NPs manage to overcome the blood barriers and tumor tissue barriers, most drugs still need to work inside tumor cells. NPs need to overcome the membrane barrier and enter the cell through endocytosis. Then NPs need to achieve endosomal/

lysosomal escape and release the drug into the cytoplasm. Finally, intracellular drug concentrations are also affected by drug efflux pumps.

Plasma Membrane Barrier The plasma membrane is the first barrier for NPs to interact with cells. Small hydrophobic molecules can easily diffuse into the membrane, while large or hydrophilic particles cannot passively diffuse through the lipid bilayer membrane and must be introduced through other mechanisms [87]. When NPs reach the membrane, they can interact with membrane components and enter the cell mainly through endocytosis [88]. Endocytosis can be divided into phagocytosis and non-phagocytosis (including pinocytosis and receptor-mediated endocytosis). Phagocytosis mainly occurs in specific phagocytes, such as macrophages, neutrophils, and dendritic cells [89]. Phagocytosis typically takes 30 min to several hours, depending on the cell type and the nature of the particle surface. Endocytosis includes different pathways including macropinocytosis, clathrin-dependent endocytosis, caveolae-dependent endocytosis, and clathrin/caveolae-independent endocytosis [90, 91].

Size affects the intracellular endocytosis process [92]. Chan's group studied the cell absorption of gold NPs at 14, 50, and 74 nm by HeLa cells. Results showed that the cell absorption kinetics and saturation concentration of gold NPs changed with the size of gold NPs, and it was found that gold NPs with a diameter of 50 nm could be internalized by cells most effectively [93, 94]. Subsequently, Fang Lu et al. also found that spherical silica NPs at 50 nm had the highest efficiency of cell internalization [95]. Similarly, restricted by the caveolae size, the uptake of 20 and 40 nm albumin-coated polymer NPs was 5–10 times greater than 100 nm particles [96]. These results suggest that there might be an optimal size for efficient cell uptake of nanomaterials. When the particle size is small, the contact area and the corresponding adhesion energy between the particle and the cell membrane are not enough to achieve the complete encapsulation of the particle. Larger particles require longer wrapping time because of slower diffusion kinetics [93, 97]. It has been confirmed that when the particle size is small, multiple particles can be wrapped by the cell membrane as a whole [98]. Jaskiewicz et al. used a vesicle-forming amphiphilic copolymer as the cell membrane model system and observed that several smaller particles enter the vesicles as a whole [99]. Similarly, Chithrani et al. observed that gold NPs at 14 nm require at least 6 NPs to form cluster to be endocytosed by cell [93]. Interestingly, computational predictions have shown similar results, suggesting that 30–50 nm particles may achieve better internalization efficiency [100].

Particle shape is another important factor affecting endocytosis of NPs [92]. Studies have shown that rod NPs have lower uptake than spherical NPs. For example, Chithrani et al. found that cellular uptake of spherical Au NPs at 74 nm and 14 nm was 5 and 3.75 times higher than that of rod Au NPs at 74 × 14 nm, respectively. Likewise, Qiu et al. reported that lower aspect ratio rod-shaped Au NPs were easier to be internalized by human breast cancer cells (MCF-7) than Au NPs with a higher aspect ratio [101]. The possible explanation is that the rod-shaped Au NPs has a larger surface area, so it takes longer to wrap the rod-shaped NPs than to wrap the spherical NPs [93]. However, this observation may also be affected by materials of the NPs. Gratton et al. prepared cubic block and cylindrical cationic cross-linked PEG hydrogel particles and observed the effect of HeLa cells on their internalization. Inversely, they found that HeLa cells internalized 150 × 450 nm rod-shaped (aspect ratio 3) NPs four times faster than 200 × 200 nm NPs [102].

The surface charge of NPs also has a certain effect on the uptake efficiency of NPs. Since cell membrane is electronegative, particles carrying positive charges can usually achieve higher uptake efficiency, whereas negative charged particles are usually liked by the cells [103, 104].

Endosomal/Lysosomal Degradation Most of the nanocarriers entered tumor cells by endocytosis. During endocytosis, a portion of the extracellular membrane is pinched out of the cell to form vesicles containing endocytosed materials, which will be fused with the early endosome and become part of it. In the early endosome, some NPs may be transported to the recycling endosomes and subsequently expelled from the cell, while most early endosomes mature and differentiate into late endosomes. The late endosome may fuse with the plasma membrane (releasing its contents outside the cell) or fuse with the lysosome to form endolysosomes, whose contents are exposed to various degrading enzymes, such as lysosomal hydrolases [105]. Lysosomes are highly acidic and full of proteases that can degrade organic NPs and even inactivate drugs [106]. Therefore, endosome/lysosome is the most important barrier for transporting carriers into cells.

Cell Nuclear Barriers Some small molecular drugs (e.g., doxorubicin) and large biomacromolecules (e.g., plasmid DNA) work inside nuclear, and thus they face cell nuclear barriers for effective delivery. The nuclear envelope consists of two parallel but discontinuous membranes. The nucleo-oriented layer is called inner nuclear membrane, and the cytoplasm-oriented layer is called outer nuclear membrane. The inner and outer nuclear membranes often fuse with each other to form annular openings called nuclear pores. The limit for passive transport through the nuclear pore is molecules less than 70 kDa or particles less than 10 nm.

Drug Efflux Pumps Most chemotherapeutic drugs can diffuse through the cell membrane, but these drugs can be repelled from intracellular environment due to multidrug resistance (MDR). There are three main mechanisms of multidrug resistance in cells: firstly, reducing water-soluble drugs such as folic acid antagonists, nucleoside analogs, and cisplatin into cells; secondly, affecting the ability of drugs to kill cells through the intracellular environment; and thirdly, increasing efflux of hydrophobic drugs. Among these mechanisms, the most common is the ABC transporter overexpressing the cell membrane, usually P-glycoprotein (P-gp) [107]. The ABC transporter relies on the energy released by ATP hydrolysis to pump intracellular drugs out of the cell, resulting in reduced intracellular drug concentrations and undesirable therapeutic effects [108]. ABC transporters include P-gp, MDR-associated protein 1, and breast cancer resistance protein. P-gp is a kind of drug efflux pumps capable of transporting hydrophobic chemotherapeutic drugs such as doxorubicin, vinblastine, and paclitaxel out of the cell [109, 110]. In fact,

P-gp molecules not only exist in the cell membrane but also in the Golgi apparatus and nucleus [111]. P-gp-induced drug resistance not only leads to lower intracellular drug concentration but also increases the likelihood that healthy cells will be exposed to chemotherapy drugs. To reach an effective drug concentration for killing tumor cells, the administered dose must be increased. But this often causes many side effects, leading to failure of the chemotherapy [107]. For nanocarriers that specifically release drugs in the tumor microenvironment, it utilizes the superior diffusion properties of small molecule drugs to enhance penetration of the drug-loaded NPs in tumor tissue. But this strategy tends to fail in drug-resistant cells [87].

2.7 Strategies to Overcome Cellular Barriers

Enhancing Cellular Internalization Effective uptake of NPs by tumor cells is essential for effective intracellular drug delivery. Overexpressed receptors on the surface of target cells such as folate receptor (FR), transferrin receptor (T_fR), epidermal growth factor receptor (EGFR), g-protein-coupled receptor (GPCR), prostate-specific membrane antigen (PSMA), and lectin can be used to improve cell uptake of NPs [112]. Ligand-conjugated NPs can enter cells through receptor-mediated endocytosis, improving therapeutic efficacy and reducing toxicity compared with untargeted NPs [113]. A typical example is a docetaxel-loaded targeted polymeric NPs (DTXL-TNP) modified with a small molecular ligand targeting to the type 2 integral membrane glycoprotein PSMA, which is overexpressed on prostate cancer cells as well as neovasculature of solid tumors. Compared with commercial solvent-based DTXL formulation (Taxotere), lower dose of DTXL-TNP can significantly inhibit tumor growth [114].

In addition, studies have shown that positively charged NPs have stronger interactions with cell membrane than negative or neutral ones and can be more easily internalized by cells. However, the positively charged NPs suffer from ready clearance from blood circulation [115]. Du et al. prepared a pH-responsive charge-conversional nanogel, which can transform from negative charge into positive charge in the weakly tumor acidic environment, enhancing the interaction between the nanogels and cells and the subsequent internalization [116].

Endosomal/Lysosomal Escape Endosome/lysosome is the most important barrier for intracellular transport, because acids and enzymes in endosome/lysosome can easily degrade the carrier and its loaded drug. In order to avoid the degradation of loaded drugs, nuclear acids, and proteins by lysosome, many types of nanocarriers that can realize endosomal/lysosomal escape have been developed. Currently, it is generally believed that nanocarriers can escape lysosomes through the following mechanisms: proton sponge effect, membrane fusion, pore formation, and membrane de-stabilization [117–119].

Proton sponge effect is the most commonly used method to achieve endosomal escape, which is usually mediated by polymers containing unsaturated amino groups such as polyethylenimine (PEI), poly(amidoamine) dendrimers, and poly (L-histidine). These polymers contain amines that can be protonated in the acidic environment of the endosome/lysosome. In the endosome/lysosome, these amines can absorb a great deal of protons, and the ATPase proton pumps will continuously transport protons into endosome/lysosome. To achieve the balance of charge and concentration, a large amount of counter ions and water influx from the cytoplasm. The endosomes/lysosomes swell and burst, releasing the contents into the cytoplasm, namely, the proton sponge effect [117, 120]. Cationic polymers with proton sponge effect are currently widely used in gene delivery. They can not only form polyelectrolyte complexes by electrostatic interaction with electronegative gene but also achieve endosomal/lysosomal escape [121]. However, studies have shown that it is a relatively inefficient escape process, and only a limited number of internalized polyplexes can effectively escape from the endosomes [122].

When the encapsulated nanocarriers contact with an already unstable membrane, the fusion between carriers and the inner endosomal membrane can lead to drug release in cytosol [117]. Lipids and peptides are the commonly used membrane fusion materials. 1,2-Dioleyl-sn-glycerol-3-phosphatidylethanolamine (DOPE) is the most commonly used membrane fusion lipid. Liposomes containing DOPE are stable under neutral conditions and can fuse with plasma membranes under acidic conditions [118]. Fusogenic peptides can destabilize the endosomal membrane and promote endosomal escape by changing their conformation when pH decreases. Hemagglutinin, for example, is a polypeptide that helps influenza viruses destroy lysosomes and enter the cytoplasm [123].

Pore formation induced by cationic amphiphilic peptides (AMPs) is another method of endosomal/lysosomal escape. Such polypeptides have a high affinity to phospholipid bilayers. The peptide preferentially binds to the rim of the pore, causing changes in membrane tension to form barrel-shaped or toroidal pores [117]. Cationic lipid complexes can destabilize the endosomal membrane. After entering the endosome, there is an electrostatic interaction between the cationic lipid complex and the cytoplasm-oriented anionic lipids. The anionic lipids of endosomal membrane diffuse laterally into the lipid complex and form charge neutralizing ion pairs with the cation lipids of the lipid complex. Finally, drugs are released into the cytoplasm [124].

Nuclear Delivery Active nuclear transport can be achieved by specific binding of oligopeptide sequences of NPs to nuclear localization signal (NLSs) receptors. NLS peptides have been shown to contribute to nuclear targeting of DNA. Even a simple mixture of negatively charged DNA and purified NLS peptides can significantly enhance gene transfection efficiency [125]. Direct binding of NLS peptides to DNA by chemical coupling has also been used to enhance nuclear localization of DNA. This strategy has also been extended to anticancer drug delivery. To more effectively deliver doxorubicin (DOX) directly to the nucleus, Misra et al. conjugated NLS peptide onto DOX-loaded poly(D,L -lactide-co-glycolide) (PLGA) NP. After

MCF-7 cells were treated with NLS conjugated DOX-loaded NP, the DOX concentration of in nucleus was 3.1 times and 1.7 times higher than that treated with natural DOX and unconjugated NPs, respectively [126].

A more sophisticated delivery system was developed by Ashley and his colleagues to overcome multiple biological barriers for nuclear delivery. Multiple modifications of lipid bilayers enabled the NPs to function effectively in cells of interest. PEG modification increased its circulation time by reducing the interaction with serum protein and decreasing the absorption of phagocytes. SP94 targeted peptide modification enhanced the affinity between NPs and human hepatocellular carcinoma (HCC). Histidine-rich fusogenic peptide (H5WYG) modification promoted the endosomal escape of NPs through proton sponge effect and then released NLS-modified drugs into the cytoplasm. Because NLS promotes the active transport of NPC, NLS-modified drugs (including DOX, calcein, and dsDNA) achieved better nuclear localization and effectively killed cancer cells [127]. In addition, the amphipathic alcohol trans-cyclohexane-1,2-diol (TCHD) is an alternative method to enhance the transfection efficacy by reversibly collapsing the permeability barrier of the NPCs at nontoxic concentrations [128].

Inhibition of Drug Efflux Pumps The key to overcoming MDR is to ensure the effective drug accumulation in MDR cells. The final intracellular drug concentration depends on the balance between drug uptake and drug efflux. Although free chemotherapeutic drugs can achieve high intracellular drug concentrations in a short period of time through passive diffusion, these rapidly internalized drugs will be quickly detected and excluded by the drug efflux pump in MDR cells. In comparison, NPs are mainly internalized by cells through endocytosis. This cell entry pathway allows NPs to enter deeper regions of the cell with endosomes/lysosomes, enabling them to bypass the drug efflux pump and reduce drug efflux [129]. Therefore, nanomedicine is a very promising strategy to overcome tumor MDR. Interestingly, some nanomaterials, especially Pluronic, have inherent anti-MDR properties, first discovered by Kabanov et al. They found that Pluronic overcome MDR through a variety of mechanisms, including reducing the membrane microviscosity, inducing ATP depletion, and inhibiting drug efflux transporters [130].

In addition, nanocarriers in combination with MDR inhibitors, either co-administration or co-delivery, is also effectively increase intracellular drug concentration [131]. Patel et al. found that tariquidar-paclitaxel-co-encapsulated liposomes can effectively reduce the drug resistance in MDR ovarian cancer cells. In non-resistant SKOV-3 and resistant SKOV-3TR cells, paclitaxel had initial IC50 values of 27 nM and 2743 nM, respectively. Then, after treatment with tariquidar-paclitaxel-co-loaded liposomes, the IC_{50} values of paclitaxel in SKOV-3 cells and SKOV-3TR cells were reduced to 18 nM and 34 nM [132]. Wong and his colleagues prepared polymer-lipid hybrid NPs (PLN) that can co-deliver doxorubicin and GG918, the third generation of P-gp inhibitors. They compared the anticancer effects of different formulations containing doxorubicin (Dox) and/or GG918 on P-gp overexpressing human breast cancer cells. It was found that compared to free

DOX or/and free GG918, PLN-loaded doxorubicin, and/or PLN-loaded GG918, dual-loaded DOX and GG918 PLN were not only more toxic to human breast cancer cells but also could effectively increase drug localization in the nucleus [132].

2.8 Conclusions

The goal of tumor drug delivery is to enhance the drug deposition in tumor cells and to improve drug bioavailability. However, there is a series of biological barriers from the site of injection to the site of action, which severely impeded therapeutic effectiveness of nanoparticulate drug delivery systems. In this chapter, we comprehensively discussed these barriers and exemplified typical design strategies to overcome these barriers. We hope our description could be helpful to the readers and researchers in this area and finally contribute to better design of the nanomedicine for clinical translation. However, it should be noted that each individual part of these barriers can be an independent topic, and this chapter cannot cover every details in such a limited space.

References

1. Blanco E, Shen H, Ferrari M (2015) Principles of nanoparticle design for overcoming biological barriers to drug delivery. Nat Biotechnol 33:941–951
2. Dawidczyk CM, Kim C, Park JH, Russell LM, Lee KH, Pomper MG et al (2014) State-of-the-art in design rules for drug delivery platforms: lessons learned from FDA-approved nanomedicines. J Control Release 187:133–144
3. Barenholz Y (2012) Doxil (R) – the first FDA-approved nano-drug: lessons learned. J Control Release 160:117–134
4. Park K, Bae YH, Mrsny RJ (2013) The missing components today and the new treatments tomorrow. In: Bae YH, Mrsny RJ, Park K (eds) Cancer targeted drug delivery: an elusive dream. Springer, New York, pp 689–707
5. Wilhelm S, Tavares AJ, Dai Q, Ohta S, Audet J, Dvorak HF et al (2016) Analysis of nanoparticle delivery to tumours. Nat Rev Mater 1:1–12
6. Dai Q, Wilhelm S, Ding D, Syed AM, Sindhwani S, Zhang YW et al (2018) Quantifying the ligand-coated nanoparticle delivery to cancer cells in solid tumors. ACS Nano 12:8423–8435
7. Sun Q, Zhou Z, Qiu N, Shen Y (2017) Rational design of cancer nanomedicine: nanoproperty integration and synchronization. Adv Mater 29:1606628
8. Von RC, Jiang W, Chan CK, Weissman IL, Kim BY (2017) Breaking down the barriers to precision cancer nanomedicine. Trends Biotechnol 35:159–171
9. Jinqiang W, Weiwei M, Lye Lin L, Jianbin T, Meihua S, Weilin S et al (2015) The role of micellesize in tumor accumulation, penetration, and treatment. ACS Nano 9:7195
10. Hak Soo C, Wenhao L, Preeti M, Eiichi T, Zimmer JP, Binil II et al (2007) Renal clearance of quantum dots. Nat Biotechnol 25:1165–1170
11. Arvizo RR, Miranda OR, Moyano DF, Walden CA, Giri K, Bhattacharya R et al (2011) Modulating pharmacokinetics, tumor uptake and biodistribution by engineered nanoparticles. PLoS One 6:e24374

12. Lee JS, Ankone M, Pieters E, Schiffelers RM, Hennink WE, Feijen J (2011) Circulation kinetics and biodistribution of dual-labeled polymersomes with modulated surface charge in tumor-bearing mice: comparison with stealth liposomes. J Control Release 155:282–288
13. Decuzzi P, Lee S, Bhushan B, Ferrari M (2005) A theoretical model for the margination of particles within blood vessels. Ann Biomed Eng 33:179–190
14. Gentile F, Chiappini C, Fine D, Bhavane RC, Peluccio MS, Cheng MMC et al (2008) The effect of shape on the margination dynamics of non-neutrally buoyant particles in two-dimensional shear flows. J Biomech 41:2312–2318
15. Ennio T, Xuewu L, Rohan B, Kevin P, Leonard AD, Katherine Price B et al (2008) Mesoporous silicon particles as a multistage delivery system for imaging and therapeutic applications. Nat Nanotechnol 3:151–157
16. Geng Y, Dalhaimer P, Cai SS, Tsai R, Tewari M, Minko T et al (2007) Shape effects of filaments versus spherical particles in flow and drug delivery. Nat Nanotechnol 2:249–255
17. Anselmo AC, Zhang MW, Kumar S, Vogus DR, Menegatti S, Helgeson ME et al (2015) Elasticity of nanoparticles influences their blood circulation, phagocytosis, endocytosis, and targeting. ACS Nano 9:3169–3177
18. Hak Soo C, Wenhao L, Fangbing L, Khaled N, Preeti M, Bawendi MG et al (2010) Design considerations for tumour-targeted nanoparticles. Nat Nanotechnol 5:42–47
19. Liu JB, Yu MX, Zhou C, Zheng J (2013) Renal clearable inorganic nanoparticles: a new frontier of bionanotechnology. Mater Today 16:477–486
20. Li SD, Huang L (2010) Stealth nanoparticles: high density but sheddable PEG is a key for tumor targeted. J Control Release 145:178–181
21. Jokerst JV, Lobovkina T, Zare RN, Gambhir SS (2011) Nanoparticle PEGylation for imaging and therapy. Nanomedicine 6:715–728
22. Matsumoto M, Matsusaki M, Akashi M (2014) Preparation of biodegradable peptide nanospheres with hetero PEG brush surfaces. Macromol Biosci 14:142–150
23. Zhou H, Fan ZY, Li PY, Deng JJ, Arhontoulis DC, Li CY et al (2018) Dense and dynamic polyethylene glycol shells cloak nanoparticles from uptake by liver endothelial cells for long blood circulation. ACS Nano 12:10130–10141
24. Ishida T, Ichihara M, Wang X, Yamamoto K, Kimura J, Majima E et al (2006) Injection of PEGylated liposomes in rats elicits PEG-specific IgM, which is responsible for rapid elimination of a second dose of PEGylated liposomes. J Control Release 112:15–25
25. Gradishar WJ, Sergei T, Neville D, Heather S, Neil D, Paul B et al (2005) Phase III trial of nanoparticle albumin-bound paclitaxel compared with polyethylated castor oil-based paclitaxel in women with breast cancer. J Clin Oncol 23:7794–7803
26. Kadiyala P, Li D, Nuñez FM, Altshuler D, Doherty R, Kuai R et al (2019) High-density lipoprotein-mimicking nanodiscs for chemo-immunotherapy against glioblastoma ultiforme. ACS Nano 13:1365–1384
27. Fleischer CC, Payne CK (2014) Nanoparticle-cell interactions: molecular structure of the protein corona and cellular outcomes. Acc Chem Res 47:2651–2659
28. Kratz F (2008) Albumin as a drug carrier: design of prodrugs, drug conjugates and nanoparticles. J Control Release 132:171–183
29. Rodriguez PL, Takamasa H, Christian DA, Pantano DA, Tsai RK, Discher DE (2013) Minimal "Self" peptides that inhibit phagocytic clearance and enhance delivery of nanoparticles. Science 339:971–975
30. Tsai RK, Rodriguez PL, Discher DE (2010) Self inhibition of phagocytosis: the affinity of 'marker of self' CD47 for SIRP alpha dictates potency of inhibition but only at low expression levels. Blood Cell Mol Dis 45:67–74
31. Hu CMJ, Li Z, Santosh A, Connie C, Fang RH, Liangfang Z (2011) Erythrocyte membrane-camouflaged polymeric nanoparticles as a biomimetic delivery platform. Proc Natl Acad Sci U S A 108:10980–10985

32. Alessandro P, Nicoletta Q, Ven AL, Van De CC, Michael E, Martinez JO et al (2013) Synthetic nanoparticles functionalized with biomimetic leukocyte membranes possess cell-like functions. Nat Nanotechnol 8:61–68
33. Kang T, Zhu QQ, Wei D, Feng JX, Yao JH, Jiang TZ et al (2017) Nanoparticles coated with neutrophil membranes can effectively treat cancer metastasis. ACS Nano 11:1397–1411
34. Zhang Y, Gai KM, Li C, Guo Q, Chen QJ, He X et al (2018) Macrophage-membrane-coated nanoparticles for tumor-targeted chemotherapy. Nano Lett 18:1908–1915
35. Dagogo-Jack I, Shaw AT (2018) Tumour heterogeneity and resistance to cancer therapies. Nat Rev Clin Oncol 15:81–94
36. Maeda H (2015) Toward a full understanding of the EPR effect in primary and metastatic tumors as well as issues related to its heterogeneity. Adv Drug Deliv Rev 91:3–6
37. Holash J, Maisonpierre PC, Compton D, Boland P, Alexander CR, Zagzag D et al (1999) Vessel cooption, regression, and growth in tumors mediated by angiopoietins and VEGF. Science 284:1994–1998
38. Minchinton AI, Tannock IF (2006) Drug penetration in solid tumours. Nat Rev Cancer 6:583–592
39. Jain RK (2012) Delivery of molecular and cellular medicine to solid tumors. Adv Drug Deliv Rev 64:353–365
40. Khawar IA, Kim JH, Kuh HJ (2014) Improving drug delivery to solid tumors: priming the tumor microenvironment. J Control Release 201:78–89
41. Heldin CH, Rubin K, Pietras K, Ostman A (2004) High interstitial fluid pressure – an obstacle in cancer therapy. Nat Rev Cancer 4:806–813
42. Nakamura Y, Mochida A, Choyke PL, Kobayashi H (2016) Nanodrug delivery: is the enhanced permeability and retention effect sufficient for curing cancer? Bioconjug Chem 27:2225–2238
43. Jain RK, Triantafyllos S (2010) Delivering nanomedicine to solid tumors. Nat Rev Clin Oncol 7:653–664
44. Padera TP, Stoll BR, Tooredman JB, Capen D, di Tomaso E, Jain RK (2004) Cancer cells compress intratumour vessels. Nature 427:695–695
45. Heine M, Freund B, Nielsen P, Jung C, Reimer R, Hohenberg H et al (2012) High interstitial fluid pressure is associated with low tumour penetration of diagnostic monoclonal antibodies applied for molecular imaging purposes. PLoS One 7:e36258
46. Zhang Z, Hong W, Tao T, Jie L, Wang Z, Li Y (2018) Rational design of nanoparticles with deep tumor penetration for effective treatment of tumor metastasis. Adv Funct Mater 28:1–13
47. Chauhan VP, Triantafyllos S, Martin JD, Zoran P, Ou C, Kamoun WS et al (2012) Normalization of tumour blood vessels improves the delivery of nanomedicines in a size-dependent manner. Nat Nanotechnol 7:383
48. Stylianopoulos T, Jain RK (2013) Combining two strategies to improve perfusion and drug delivery in solid tumors. Proc Natl Acad Sci U S A 110:18632–18637
49. Pengfei L, Weaver VM, Zena W (2012) The extracellular matrix: a dynamic niche in cancer progression. J Cell Biol 196:395
50. Kolacna L, Bakesova J, Varga F, Kostakova E, Planka L, Necas A et al (2007) Biochemical and biophysical aspects of collagen nanostructure in the extracellular matrix. Physiol Res 56: S51–S60
51. Barua S, Mitragotri S (2014) Challenges associated with penetration of nanoparticles across cell and tissue barriers: a review of current status and future prospects. Nano Today 9:223–243
52. Miao L, Lin CM, Huang L (2015) Stromal barriers and strategies for the delivery of nanomedicine to desmoplastic tumors. J Control Release 219:192–204
53. Overchuk M, Zheng G (2018) Overcoming obstacles in the tumor microenvironment: recent advancements in nanoparticle delivery for cancer theranostics. Biomaterials 156:217–237
54. Qin D, Stefan W, Ding D, Muhammad SA, Shrey S, Yuwei Z et al (2018) Quantifying the ligand-coated nanoparticle delivery to cancer cells in solid tumors. ACS Nano 12:8423–8435

55. Sherman MH, Yu RT, Engle DD, Ding N, Atkins AR, Tiriac H et al (2014) Vitamin D receptor-mediated stromal reprogramming suppresses pancreatitis and enhances pancreatic cancer therapy. Cell 159:80–93
56. Miao L, Newby JM, Lin CM, Zhang L, Xu F, Kim WY et al (2016) The binding site barrier elicited by tumor-associated fibroblasts interferes disposition of nanoparticles in stroma-vessel type tumors. ACS Nano 10:9243–9258
57. Cuccarese MF, Dubach JM, Pfirschke C, Engblom C, Garris C, Miller MA et al (2017) Heterogeneity of macrophage infiltration and therapeutic response in lung carcinoma revealed by 3D organ imaging. Nat Commun 8:14293
58. Miller MA, Zheng YR, Gadde S, Pfirschke C, Zope H, Engblom C et al (2015) Tumor associated macrophages act as a slow-release reservoir of nano-therapeutic Pt(IV) pro-drug. Nat Commun 6:8692–8692
59. Sykes EA, Juan C, Gang Z, Chan WCW (2014) Investigating the impact of nanoparticle size on active and passive tumor targeted efficiency. ACS Nano 8:5696–5706
60. Dreher MR, Wenge L, Michelich CR, Dewhirst MW, Fan Y, Ashutosh C (2006) Tumor vascular permeability, accumulation, and penetration of macromolecular drug carriers. J Natl Cancer Inst 98:335
61. Cabral H, Matsumoto Y, Mizuno K, Chen Q, Murakami M, Kimura M et al (2011) Accumulation of sub-100 nm polymeric micelles in poorly permeable tumours depends on size. Nat Nanotechnol 6:815–823
62. Li T, Xujuan Y, Qian Y, Kaimin C, Hua W, Isthier C et al (2014) Investigating the optimal size of anticancer nanomedicine. Proc Natl Acad Sci U S A 111:15344–15349
63. Li HJ, Du JZ, Du XJ, Xu CF, Sun CY, Wang HX et al (2016) Stimuli-responsive clustered nanoparticles for improved tumor penetration and therapeutic efficacy. Proc Natl Acad Sci U S A 113:4164–4169
64. Rong T, Chiang HH, Kohane DS (2013) Photoswitchable nanoparticles for in vivo cancer chemotherapy. Proc Natl Acad Sci U S A 110:19048–19053
65. Wang HX, Zuo ZQ, Du JZ, Wang YC, Sun R, Cao ZT et al (2016) Surface charge critically affects tumor penetration and therapeutic efficacy of cancer nanomedicines. Nano Today 11:133–144
66. Wang GY, Chen YY, Wang P, Wang YF, Hong H, Li YL et al (2016) Preferential tumor accumulation and desirable interstitial penetration of poly(lactic-co-glycolic acid) nanoparticles with dual coating of chitosan oligosaccharide and polyethylene glycol-poly(D, L-lactic acid). Acta Biomater 29:248–260
67. Chen JJ, Ding JX, Wang YC, Cheng JJ, Ji SX, Zhuang XL et al (2017) Sequentially responsive shell-stacked nanoparticles for deep penetration into solid tumors. Adv Mater 29:170–177
68. Danhier F (2016) To exploit the tumor microenvironment: since the EPR effect fails in the clinic, what is the future of nanomedicine? J Control Release 244:108–121
69. Yu M, Nichols JW, Toh K, Nomoto T, Cabral H, Miura Y et al (2016) Vascular bursts enhance permeability of tumour blood vessels and improve nanoparticle delivery. Nat Nanotechnol 11:533–538
70. Miller MA, Chandra R, Cuccarese MF, Pfirschke C, Engblom C, Stapleton S et al (2017) Radiation therapy primes tumors for nanotherapeutic delivery via macrophage-mediated vascular bursts. Sci Transl Med 9:eaal0225
71. Ho YJ, Chang YC, Yeh CK (2015) Improving nanoparticle penetration in tumors by vascular disruption with acoustic droplet vaporization. Theranostics 6:392–403
72. Liu FH, Cong Y, Qi GB, Ji L, Qiao ZY, Wang H (2018) Near-infrared laser-driven in situ self-assembly as a general strategy for deep tumor therapy. Nano Lett 18:6577–6584
73. Yin M, Tan S, Bao Y, Zhang Z (2017) Enhanced tumor therapy via drug co-delivery and in situ vascular-promoting strategy. J Control Release 258:108–120
74. Song WT, Tang ZH, Zhang DW, Li MQ, Gu JK, Chen XS (2016) A cooperative polymeric platform for tumor-targeted drug delivery. Chem Sci 7:728–736

75. Li S, Zhang Y, Wang J, Zhao Y, Ji T, Zhao X et al (2017) Nanoparticle-mediated local depletion of tumour-associated platelets disrupts vascular barriers and augments drug accumulation in tumours. Nat Biomed Eng 1:667–679
76. Chen Q, Liu GX, Liu S, Su HY, Wang Y, Li JY et al (2018) Remodeling the tumor microenvironment with emerging nanotherapeutics. Trends Phamacol Sci 39:59–74
77. Yang S, Gao HL (2017) Nanoparticles for modulating tumor microenvironment to improve drug delivery and tumor therapy. Pharmacol Res 126:97–108
78. Jain RK (2013) Normalizing tumor microenvironment to treat cancer: bench to bedside to biomarkers. J Clin Oncol 31:2205–U2210
79. Wen J, Yuhui H, Yi A, Kim BYS (2015) Remodeling tumor vasculature to enhance delivery of intermediate-sized nanoparticles. ACS Nano 9:8689
80. Lee S, Han H, Koo H, Na JH, Yoon HY, Lee KE et al (2017) Extracellular matrix remodeling in vivo for enhancing tumor-targeted efficiency of nanoparticle drug carriers using the pulsed high intensity focused ultrasound. J Control Release 263:68–78
81. Kuhn SJ, Finch SK, Hallahan DE, Giorgio TD (2006) Proteolytic surface functionalization enhances in vitro magnetic nanoparticle mobility through extracellular matrix. Nano Lett 6:306–312
82. Zhou H, Fan Z, Deng J, Lemons PK, Arhontoulis DC, Bowne WB et al (2016) Hyaluronidase embedded in nanocarrier PEG shell for enhanced tumor penetration and highly efficient antitumor efficacy. Nano Lett 16:3268
83. Sun QX, Ojha T, Kiessling F, Lammers T, Shi Y (2017) Enhancing tumor penetration of nanomedicines. Biomacromolecules 18:1449–1459
84. Zhang B, Jiang T, Shen S, She XJ, Tuo YY, Hu Y et al (2016) Cyclopamine disrupts tumor extracellular matrix and improves the distribution and efficacy of nanotherapeutics in pancreatic cancer. Biomaterials 103:12–21
85. Ji T, Lang J, Wang J, Cai R, Zhang Y, Qi F et al (2017) Designing liposomes to suppress extracellular matrix expression to enhance drug penetration and pancreatic tumor therapy. ACS Nano 11:8668
86. Lang J, Zhao X, Qi Y, Zhang Y, Han X, Ding Y et al (2019) Reshaping prostate tumor microenvironment to suppress metastasis via cancer-associated fibroblast inactivation with peptide-assembly-based nanosystem. ACS Nano 13:12357–12371
87. Nichols JW, Bae YH (2012) Odyssey of a cancer nanoparticle: from injection site to site of action. Nano Today 7:606–618
88. Bareford LA, Swaan PW (2007) Endocytic mechanisms for targeted drug delivery. Adv Drug Deliv Rev 59:748–758
89. Swanson JA (2008) Shaping cups into phagosomes and macropinosomes. Nat Rev Mol Cell Bio 9:639–649
90. Iversen TG, Skotland T, Sandvig K (2011) Endocytosis and intracellular transport of nanoparticles: present knowledge and need for future studies. Nano Today 6:176–185
91. Yan Y, Such GK, Johnston APR, Best JP, Caruso F (2012) Engineering particles for therapeutic delivery: prospects and challenges. ACS Nano 6:3663–3669
92. Zhu MT, Nie GJ, Meng H, Xia T, Nel A, Zhao YL (2013) Physicochemical properties determine nanomaterial cellular uptake, transport, and fate. Acc Chem Res 46:622–631
93. Chithrani BD, Chan WCW (2007) Elucidating the mechanism of cellular uptake and removal of protein-coated gold nanoparticles of different sizes and shapes. Nano Lett 7:1542–1550
94. Chithrani BD, Ghazani AA, Chan WCW (2006) Determining the size and shape dependence of gold nanoparticle uptake into mammalian cells. Nano Lett 6:662–668
95. Lu F, Wu SH, Hung Y, Mou CY (2009) Size effect on cell uptake in well-suspended, uniform mesoporous silica nanoparticles. Small 5:1408–1413
96. Wang ZJ, Tiruppathi C, Minshall RD, Malik AB (2009) Size and dynamics of caveolae studied using nanoparticles in living endothelial cells. ACS Nano 3:4110–4116
97. Saric A, Cacciuto A (2013) Self-assembly of nanoparticles adsorbed on fluid and elastic membranes. Soft Matter 9:6677–6695

98. Lunov O, Zablotskii V, Syrovets T, Rocker C, Tron K, Nienhaus GU et al (2011) Modeling receptor-mediated endocytosis of polymer-functionalized iron oxide nanoparticles by human macrophages. Biomaterials 32:547–555
99. Jaskiewicz K, Larsen A, Lieberwirth I, Koynov K, Meier W, Fytas G et al (2012) Probing bioinspired transport of nanoparticles into polymersomes. Angew Chem Int Ed 51:4613–4617
100. Gao HJ, Shi WD, Freund LB (2005) Mechanics of receptor-mediated endocytosis. Proc Natl Acad Sci U S A 102:9469–9474
101. Qiu Y, Liu Y, Wang LM, Xu LG, Bai R, Ji YL et al (2010) Surface chemistry and aspect ratio mediated cellular uptake of Au nanorods. Biomaterials 31:7606–7619
102. Gratton SEA, Ropp PA, Pohlhaus PD, Luft JC, Madden VJ, Napier ME et al (2008) The effect of particle design on cellular internalization pathways. Proc Natl Acad Sci U S A 105:11613–11618
103. Verma A, Stellacci F (2010) Effect of surface properties on nanoparticle-cell interactions. Small 6:12–21
104. Cho EC, Xie JW, Wurm PA, Xia YN (2009) Understanding the role of surface charges in cellular adsorption versus internalization by selectively removing gold nanoparticles on the cell surface with a I-2/KI etchant. Nano Lett 9:1080–1084
105. Behzadi S, Serpooshan V, Tao W, Hamaly MA, Alkawareek MY, Dreaden EC et al (2017) Cellular uptake of nanoparticles: journey inside the cell. Chem Soc Rev 46:4218–4244
106. Eskelinen EL, Tanaka Y, Saftig P (2003) At the acidic edge: emerging functions for lysosomal membrane proteins. Trends Cell Biol 13:137–145
107. Gergely S, Jill KP, Joseph AL, Catherine B-G, Michael MG (2006) Targeting multidrug resistance in cancer. Nat Rev Drug Discov 5:219–234
108. Fletcher JI, Haber M, Henderson MJ, Norris MD (2010) ABC transporters in cancer: more than just drug efflux pumps. Nat Rev Cancer 10:147–156
109. Gottesman MM, Fojo T, Bates SE (2002) Multidrug resistance in cancer: role of ATP-dependent transporters. Nat Rev Cancer 2:48–58
110. Aller SG (2010) Structure of P-Glycoprotein reveals a molecular basis for Poly-Specific drug binding. Biophys J 98:755a
111. Calcabrini A, Meschini S, Stringaro A, Cianfriglia M, Arancia G, Molinari A (2000) Detection of P-glycoprotein in the nuclear envelope of multidrug resistant cells. Histochem J 32:599–606
112. Lee SM, Park H, Choi JW, Park YN, Yun CO, Yoo KH (2011) Multifunctional nanoparticles for targeted chemophotothermal treatment of cancer cells. Angew Chem Int Ed Engl 50:7581–7586
113. Du JZ, Sun TM, Song WJ, Wu J, Wang J (2010) A tumor-acidity-activated charge-conversional nanogel as an intelligent vehicle for promoted yumoral-cell uptake and drug delivery. Angew Chem Int Ed 49:3621–3626
114. Hrkach J, Von Hoff D, Ali MM, Andrianova E, Auer J, Campbell T et al (2012) Preclinical development and clinical translation of a PSMA-targeted Docetaxel nanoparticle with a differentiated pharmacological profile. Sci Transl Med 4:128ra39
115. Oupicky D, Ogris M, Howard KA, Dash PR, Ulbrich K, Seymour LW (2002) Importance of lateral and steric stabilization of polyelectrolyte gene delivery vectors for extended systemic circulation. Mol Ther 5:463–472
116. Hrkach J, Von Hoff D, Mukkaram Ali M, Andrianova E, Auer J, Campbell T et al (2012) Preclinical development and clinical translation of a PSMA-targeted docetaxel nanoparticle with a differentiated pharmacological profile. Sci Transl Med 4:128ra39
117. Varkouhi AK, Scholte M, Storm G, Haisma HJ (2011) Endosomal escape pathways for delivery of biologicals. J Control Release 151:220–228
118. Martens TF, Remaut K, Demeester J, Smedt SCD, Braeckmans K (2014) Intracellular delivery of nanomaterials: how to catch endosomal escape in the act. Nano Today 9:344–364
119. Brock DJ, Kondow-McConaghy HM, Hager EC, Pellois JP (2019) Endosomal escape and cytosolic penetration of macromolecules mediated by synthetic delivery agents. Bioconjug Chem 30:293–304

120. Pei DH, Buyanova M (2019) Overcoming endosomal entrapment in drug delivery. Bioconjug Chem 30:273–283
121. Won YY, Sharma R, Konieczny SF (2009) Missing pieces in understanding the intracellular trafficking of polycation/DNA complexes. J Control Release 139:88–93
122. Rehman ZU, Hoekstra D, Zuhorn IS (2013) Mechanism of polyplex- and lipoplex-mediated delivery of nucleic acids: real-time visualization of transient membrane destabilization without endosomal lysis. ACS Nano 7:3767–3777
123. And DCW, Skehel JJ (1987) The structure and function of the Hemagglutinin membrane glycoprotein of influenza virus. Annu Rev Biochem 56:365
124. Wasungu L, Hoekstra D (2006) Cationic lipids, lipoplexes and intracellular delivery of genes. J Control Release 116:255–264
125. Subramanian A, Ranganathan P, Diamond SL (1999) Nuclear targeting peptide scaffolds for lipofection of nondividing mammalian cells. Nat Biotechnol 17:873–877
126. Misra R, Sahoo SK (2010) Intracellular trafficking of nuclear localization signal conjugated nanoparticles for cancer therapy. Eur J Pharm Sci 39:152–163
127. Ashley CE, Carnes EC, Phillips GK, Padilla D, Durfee PN, Brown PA et al (2011) The targeted delivery of multicomponent cargos to cancer cells by nanoporous particle-supported lipid bilayers. Nat Mater 10:389–397
128. Vandenbroucke R, Lucas B, Demeester J, De Smedt S, Sanders N (2007) Nuclear accumulation of plasmid DNA can be enhanced by non-selective gating of the nuclear pore. Hum Gene Ther 18:973–973
129. Kunjachan S, Blauz A, Mockel D, Theek B, Kiessling F, Etrych T et al (2012) Overcoming cellular multidrug resistance using classical nanomedicine formulations. Eur J Pharm Sci 45:421–428
130. Batrakova EV, Kabanov AV (2008) Pluronic block copolymers: evolution of drug delivery concept from inert nanocarriers to biological response modifiers. J Control Release 130:98–106
131. Kunjachan S, Rychlik B, Storm G, Kiessling F, Lammers T (2013) Multidrug resistance: physiological principles and nanomedical solutions. Adv Drug Deliv Rev 65:1852–1865
132. Patel NR, Rathi A, Mongayt D, Torchilin VP (2011) Reversal of multidrug resistance by co-delivery of tariquidar (XR9576) and paclitaxel using long-circulating liposomes. Int J Pharm 416:296–299

Chapter 3
Tumor-targeted Strategies

Min Liu and Weiyue Lu

Abstract The rising prominence of cancer as a leading cause of death reflects therapeutic predicament over malignancy, leading to intensive efforts for tumor treatment. Anticancer precision medicine has shown great progress in recent years, holding promise to reduce side effects and improve treatment outcome. The preferential accumulation of nanomedicines at pathological sites relies on the enhanced permeability and retention effect, which lays the foundation for the passive tumor-targeted strategy. In tumors, the abnormally wide fenestrations in the blood vessels allow for the extravasation of materials with sizes up to several hundreds of nanometers. To this end, a class of fairly functional nanomedicines ranging from 5–200 nm have been pursued, including liposomes, polymeric micelles, nanoparticles, biomimetic nanoparticles encapsulated in cell membrane, and polymer-drug conjugates. PEGylation and innate cell membrane encapsulation have been exploited to reduce both opsonization and the uptake by the reticuloendothelial system. Thus, long-circulating nanomedicines could perform the targeting process and convey cargos to tumors. The targets specifically overexpressed on tumor cells and tumor neovasculature provide insights for the active tumor-targeted strategy. The targeting molecules, including antibody, ligand, aptamer, and endogenous substance, specifically recognize antigen, receptor, and transporters and are linked with drugs to form conjugates or modified onto the surface of nanomedicines. Because some innate cells show natural tropism to pathological sites, cell membrane-coated nanoparticles also have great potential for actively delivering therapeutic agents to targeting sites. Using targeting molecules guidance in vivo, precision medicine could identify targets with ideal safety and efficacy profiles. Targeted therapy for cancer has become a critical strategy to revolutionize the traditional treatment of tumors.

Keywords Passive tumor-targeted strategy · Active tumor-targeted strategy

---

M. Liu · W. Lu (✉)
Department of Pharmaceutics, School of Pharmacy, Key Laboratory of Smart Drug Delivery, Ministry of Education, Fudan University, Shanghai, China
e-mail: wylu@shmu.edu.cn

R. Huang, Y. Wang (eds.), *New Nanomaterials and Techniques for Tumor-targeted Systems*, https://doi.org/10.1007/978-981-15-5159-8_3

3.1 Introduction

Cancer incidence and mortality are rapidly increasing worldwide. The GLOBOCAN 2018 estimated 18.1 million new cases of cancer and 9.6 million deaths from cancer in 2018. Cancer is an important cause of morbidity and mortality worldwide and in every global region, irrespective of the level of human development [1]. Over the past decades, the development of anticancer therapies has been fruitful, and advances in nanotechnology have led to promising tumor-targeted strategies.

Tumor-targeted strategies can be classified into passive tumor-targeted and active tumor-targeted strategies. The passive tumor-targeted strategy is defined as the preferable accumulation and entrapment of nanomedicines in tumors by passive diffusion or convection after systemic administration [2]. Notably, the passive tumor-targeted strategy was based on the enhanced permeability and retention (EPR) effect and mainly subject to the issue of diameter and circulation time. The EPR effect relies on specific pathophysiological characteristics of tumors that distinguishes tumors from healthy tissues, which was first put forward by Matsumura and Maeda in 1986 [3]. When the diameter of solid tumors reach about 2 mm, angiogenesis factors are released to induce microangiogenesis, and new angiogenic blood vessels in the tumor bulk show high density with increased growth characteristics, giving rise to the lack of a smooth muscle layer and pericytes. This disordered vasculature together with the absence of lymphatic drainage in tumors ensured effective and selective extravasation and retention of nanomedicines [4] (ranging from 5 to 200 nm [5]). Only sufficient circulation of nanomedicines could promise fenestration on the tumor blood vessels through blood flow and random diffusional processes. Nanomedicines without PEGylation are opsonized by plasma proteins, quickly recognized as foreign bodies, and rapidly captured by the reticuloendothelial system (RES). Positive nanoparticles (NPs) are easily scavenged by the RES. PEGylation was the mostly exploited strategy to prevent opsonization and reduce the uptake by RES, and therefore PEGylated nanomedicines were intensely investigated. To tackle the challenge of nanomedicines being recognized as foreign material, NPs encapsulated in cell membrane were established to camouflage as endogenous cells, leading to a prolonged half-life in the circulation and minimal adverse effects. Thus, only nanomedicines with a suitable diameter with a prolonged circulation half-life have a significant chance of encountering the tumor vasculature. Accordingly, various nanoscopic medicines have been developed that are encompassed by liposomes, micelles, NPs, biomimetic NPs encapsulated in cell membrane, and polymer-drug conjugates. These nanomedicines can be tailored and meet the particle size criterion for the EPR effect and show a prolonged half-life in the circulation, allowing for passive targeting drug delivery.

Currently, the concept of the "active targeting" includes a coordinated behavior of three components: (A) the drug, (B) a targeting moiety, and (C) a pharmaceutical carrier. Most nanomedicines work depending on dual roles of active targeting and passive targeting strategies. The nanosize effect based on EPR and tumor-specific uptake based on molecular recognition has helped optimize and advance the active tumor-targeted strategy. Because of the heterogeneity presented by tumor cells and the tumor neovasculature, the active tumor-targeted strategy was developed to specifically convey cargos to tumor sites without harming normal cells. These strategies depend on the recognition between overexpressed receptor/antigen/transporter on cancerous cells and the targeting molecules tethered on the surface of nanomedicines. The surface antigens on diseased cells as well as overexpressed transporters and receptors have become intriguing targets for nanomedicines. The targeting molecules, such as antibodies and their fragments, nucleic acid aptamers, peptides, carbohydrates, and small molecules, are decorated onto nanocarriers via covalent or noncovalent binding, which can help promote nanomedicines to cross the physiopathological barrier membrane and target tumor cells, tumor stem cells, angiogenesis, and vasculogenic mimicry. Thus, several antibody-mediated, receptor-mediated, transporter-related, aptamer-mediated, and biomimetic membrane protein-mediated targeting drug delivery systems have been in development. In addition, the target molecules expressed on the surface of different tumor cells vary, and the expression level also differs across pathological stages. Hence, researchers should take into account the specificity and expression level of target molecules to specifically select and combine targeting molecules to maximize targeting efficiency and avoid off-target effects in normal tissues.

As mentioned above, the heterogenic or pathological substances in tumors provide the opportunity to customize effective medicines for an active tumor-targeted strategy.

3.2 Passive Tumor-targeted Strategy

The principles of passive tumor-targeted are mainly based on the EPR effect that was first reported by Matsumura and Maeda in 1986 [3]. The density of solid vascular tumors is higher than that of normal tissues and organs [6]. Angiogenesis in tumors is promoted by several cytokines (e.g., vascular endothelial growth factor) [7]. The newly formed blood vessels usually have abnormal architecture such as defective endothelial cells and irregular vascular alignment. Therefore, most solid tumor vessels exhibit increased permeability [8]. These architectural and anatomical features in tumor tissues constitute the foundation of the EPR effect, which allows for the accumulation of macromolecules (e.g., nanomedicines) in the tumor site [9]. Antitumor drugs used in intravenous administration usually encounter several drawbacks, such as poor water solubility, nonspecific biodistribution, and severe side effects [10]. Hence, nanomedicines provide an alternative strategy to improve the pharmacokinetic and pharmacodynamic properties of antitumor drugs. The EPR effect has become the "gold standard" for the design and targeting delivery of antitumor drugs [11]. Over the past decades, passive tumor-targeted has been widely used for antitumor drug delivery, gene therapy, and molecular imaging. Passive tumor-targeted strategies for nanomedicines mainly include liposomes, polymeric micelles, NPs, and polymer-drug conjugates. Below we discuss the various strategies in more detail.

3.2.1 Liposomes

Liposomes mainly consist of phospholipids that have a hydrophilic "head" and two hydrophobic fatty acid "tails." When dispersed in aqueous condition, phospholipids tend to form two lipid layers due to the amphiphilicity [12]. Like the cell membrane, the hydrophobic chains of each layer face each other and enclose an aqueous core. To improve liposome stability, cholesterol is often added [13]. Because of their versatile structure, liposomes can encapsulate both hydrophilic and hydrophobic agents. Over the past decades, liposomes have gained a great detail of attention because of their desirable biocompatibility, biodegradability, variable size, large drug loading capacity, low toxicity, and low costs [14]. Since the discovery of the EPR effect, liposomes with a size of around 100 nm have been widely used for passive tumor-targeted delivery of chemotherapeutic drugs, proteins, and genes [15, 16]. Liposomes can selectively extravasate from tumor leaky vessels, resulting in their accumulation in the tumor site. In contrast, antitumor drugs with low molecular weight cannot be retained in the tumor microenvironment as they easily reenter the circulation via diffusion. Thus, liposomes can improve the pharmacokinetic and pharmacodynamic profiles of antitumor drugs.

There are two types of liposomes, conventional liposomes and long-circulating liposomes, and each of these liposome types possesses different advantages. For example, conventional liposomes can be phagocytosed by cells in the RES after intravenous injection and are used to target organs such as liver and spleen, though the circulation time of these liposomes are short [17]. In contrast, long-circulating liposomes such as PEGylated liposomes are protected against uptake by RES because polyethylene glycol (PEG), also referred to as poly(ethylene oxide), forms a hydrophilic protective layer around the liposomes, preventing the liposomes from aggregating and interacting with the plasma proteins [18]. However, recent studies have shown that PEGylated liposomes can trigger the production of IgM antibodies, leading to accelerated clearance after repeated injection [19]. Despite the immunogenicity of PEGylated liposomes, they still have wide application in tumor imaging and gene delivery. For example, Feng and colleagues [20] designed a PEGylated liposome that encapsulated a commercially hydrophilic hypoxia-activated prodrug termed AQ4N into its aqueous cavity. The hydrophobic photosensitizer chlorin (hCe6) conjugated with ^{64}Cu (^{64}Cu-hCe6) was embedded in the liposome bilayer. After intravenous injection, the liposomes loaded with AQ4N and ^{64}Cu-hCe6 demonstrated an excellent tumor imaging ability under positron emission tomography as well as effective chemotherapeutic efficiency. For gene delivery using liposomes, the most commonly used non-viral delivery systems are cationic liposomes such as 1,2-dioleoyl-3-trimethylammonium-propane (DOTAP) and N-[1-(2,3-dioleoyloxy)propyl]-N,N,N-trimethyl-ammonium methyl sulfate

(DOTMA). The positive charged surface of cationic liposomes can condense negative charged DNA or siRNA by electrostatic interaction, resulting in high cellular uptake [21]. However, the cytotoxicity caused by the highly positive charge limits the clinical application. Recently, researchers synthesized cationic liposomes using DOTAP and cholesterol. To reduce the surface charge, the cationic liposomes were modified by hyaluronic acid (HA) [22]. HA-modified liposomes showed significantly less cytotoxicity than unmodified liposomes partly because of the blockage of their surface charge, which provided a strategy to reduce the unwanted cytotoxicity. However, liposomes that depend on the EPR effect may not be applicable for heterogeneous tumors, and multifunctional liposomes should be developed for wider application in the future.

3.2.2 Polymeric Micelles (PMs)

PMs consist of a hydrophobic inner core and a hydrophilic shell. The inner core is used to encapsulate hydrophobic drugs, while the outer shell is used to stabilize PMs in the aqueous environment [23]. PMs are nanosized (10–100 nm) and spherical. Most of the PMs are made up of amphiphilic copolymers with hydrophilic and hydrophobic block domains [24]. When exposed in the aqueous environment, PMs can form supramolecular structure self-assembly at or above the critical micelle concentration, which is determined by the physicochemical properties of the amphiphilic copolymers. The most common structural unit of the outer shell is PEG. Like with liposomes, PEG is also commonly used to coat PMs because PEG is hydrophilic, electrically neutral, nontoxic, and flexible [25]. Other hydrophilic block polymers such as chitosan, poly(N-vinyl pyrrolidone) (PVP), and poly (N-isopropylacrylamide) (pNIPAAm) are also used [26]. These hydrophilic polymers increase the stability of PM in water and help them escape from the RES. For the hydrophobic inner core, there are a variety of block-forming polymers, including poly(propylene oxide) (PPO), poly(lactideco-glycolic acid) (PLGA), poly(β-aminoesters), poly(L-histidine) (pHis), poly(L-aspartic acid) (pAsp), dioleoyl (phosphatidylethanolamine) (DOPE), and distearoyl(phosphatidylethanolamine) (DSPE) [27], among others. The inner core offers PMs the potential to solubilize water-insoluble drugs.

Since the reporting of PMs by Kataoka's group, PMs have shown promising application in the delivery of antitumor drugs, genes, and imaging agents [28]. PMs loaded with antitumor drugs show prolonged circulation time and selectively accumulate in the tumor site due to the EPR effect. Thus, they show improved pharmacokinetic properties and biodistribution. For example, by adding PEGylated phospholipid in the range of 4–10 wt%, Phei Er Saw and colleagues converted the disk-shaped liposomes to uniform spherical micelles, which were termed as hyper-cell-permeable micelles (HCPMi) [29]. When loaded with docetaxel, HCPMi not only showed superior cellular uptake but also improved serum stability compared with disk-shaped liposomes. Along with solubility improvement, docetaxel-loaded

HCPMi exhibited much better antitumor effects than docetaxel. Furthermore, PMs were easily eliminated by renal glomeruli when their diameters were 5 nm to 10 nm, whereas larger PMs with a diameter range from 50 nm to 100 nm were removed by the liver and spleen [30]. Thus, optimizing the PM size for drug delivery of tumor-targeted is critical. Recently, Ling Mei and coworkers designed a transformable micelle to treat the primary tumor and lymph node metastasis [31]. The transformable micelle loaded with paclitaxel was synthesized by DSPE-PEG and modified with azide/alkyne. When the transformable micelles reached tumor tissues, they aggregated through the click reaction after catalyst addition by intratumoral injection. The results showed that transformable micelles with a size around 25 nm exhibited better tumor-targeted efficiency and the size increased around 120 nm in the presence of catalyst, leading to long-term retention in tumor tissues. Another important application of PM is gene delivery. For example, cholesterol-conjugated polyamidoamine (PAM-Chol) was synthesized to combine the delivery of resveratrol and the HO-1 gene [32]. PAM-Chol self-assembled into micelles in an aqueous solution and showed efficient anti-inflammatory effects for acute lung injury treatment. In addition, PMs can be used to combine drug delivery with in vivo imaging. Nicolas Mackiewicz synthesized three types of micelles composed of polydiacetylene (PDA) amphiphiles with different surface coatings (nitrilotriacetic acids, PEG350 or PEG2000) to investigate their passive tumor-targeted ability [33]. The targeting efficiency in tumors was determined by co-localization with [18F]-fluorodeoxyglucose conjugated with the PDA micelles. The results demonstrated that paclitaxel-loaded PDA modified with PEG2000 micelles showed the most promising therapeutic and imaging effect.

3.2.3 Nanoparticles (NPs)

NPs are particles whose size ranges from 1 to 100 nm in at least one dimension. There are three types of NPs: lipid NPs, polymeric NPs, and inorganic NPs [34]. Lipid NPs (e.g., solid lipid NPs) exhibit good physical tolerability. Thus, they have been widely exploited for drug delivery [35]. For example, studies have shown that solid lipid NPs of cholesteryl butyrate could inhibit adhesion to human umbilical vein endothelial cells and migration of several cancer cell lines, resulting in improved antitumor effects compared with butyrate [36]. Polymeric NPs can be formed by biodegradable and non-biodegradable polymers. In general, the biodegradable polymeric nanomaterials are used for drug delivery. They can protect labile agents (e.g., DNA, RNA, and proteins) from degradation and offer controlled release. Recently, researchers designed a biodegradable polymeric NP using poly (β-amino esters) (PBAE) to deliver and protect DNA in vivo [37]. Treatment of an orthotopic glioma murine model by intratumoral injection of PBAE/DNA NPs led to more effective transfection than naked DNA as well as low cytotoxicity. Moreover, lyophilized PBAE/DNA NPs could be stored for at least 2 years without losing efficacy and thus showed excellent stability. Inorganic NPs including magnetic NPs, carbon NPs, and gold NPs also play important roles in passive tumor-targeted. Magnetic NPs are synthesized by ferromagnetic and ferrimagnetic materials and have shown excellent magnetic properties, stability, and low toxicity. The most commonly used magnetic materials are magnetite (Fe_3O_4) and maghemite (γ-Fe_2O_3). Both are widely used for fundamental investigations and biomedical applications [38]. Carbon NPs include graphene, carbon nanotubes, and carbon quantum dots. Among these, carbon nanotubes are widely explored for biomedical applications due to their ability to penetrate into all types of cells [39]. Carbon nanotubes are divided into single-walled carbon nanotubes (SWNTs) or multi-walled carbon nanotubes. SWNTs have recently demonstrated great potential for antitumor drug delivery, in vivo imaging, and diagnosis [40]. For example, Ali Razzazan and colleagues functionalized PEGylated SWNTs by gemcitabine (SWCNT-PEG-GEM), which was a highly hydrophilic antitumor drug [41]. SWCNT-PEG-GEM exhibited higher efficacy in suppressing tumor growth than gemcitabine in B6 nude mice. Gold NPs have efficient light-to-heat conversion ability that can be used for photothermal therapy. This property of gold NPs can be optimized by changing their structural dimensions, resulting in the absorption of near-infrared light. Near-infrared light can safely penetrate deeply through the healthy tissue to eventually reach the gold NPs remaining in the tumor site [42].

3.2.4 Polymer-Drug Conjugates

Polymer-drug conjugates are drugs in polymer molecules. The first polymer-drug conjugate was reported in 1955 by Jatzkewitz, who used a PVP polymer and dipeptide spacer conjugated to the mescaline drug [43]. Depending on the molecular weight of drugs, polymer-drug conjugates are classified as polymer-protein and polymer-small-molecule drug conjugates [44]. Proteins tend to show poor stability, a short half-life, and immunogenicity despite their excellent clinical values. PEG is one of the most commonly used polymers for polymer-protein conjugates as it is highly water-soluble, neutral and biocompatible. For small molecules, systemic administration is usually preferred when used for passive tumor-targeted. Similarly, polymers can offer numerous benefits to small-molecule drugs including enhanced drug solubilization, prolonged circulation, controlled delivery, reduced immunogenicity, and improved pharmacokinetics [44, 45]. So far, several polymer-small-molecule drug conjugates have been successfully translated into clinical trials such as poly(glutamic acid)-paclitaxel, N-(2-hydroxypropyl)methacrylamide-doxorubicin (HPMA-doxorubicin), and PEG-irinotecan. HPMA-doxorubicin was the first polymer-small-molecule drug conjugate translated into clinical trials and showed enhanced antitumor efficacy compared with doxorubicin [46]. Kopeček and coworkers demonstrated that the antitumor efficacy of the HPMA-doxorubicin conjugate was dependent on molecular weight [47]. The authors synthesized a series of water-soluble HPMA-doxorubicin conjugates with molecular weights of 22 kDa,

160 kDa, 895 kDa, or 1230 kDa. The HPMA-doxorubicin conjugate of 1230 kDa showed the highest half-life in blood and the lowest elimination rate from the tumor.

Long circulation polymer-drug conjugates can accumulate passively in solid tumor tissue because of the EPR effect, and polymer-drug conjugates provide controlled or sustained release of drugs, which is very important for highly toxic drugs. For example, Luis Rojo and colleagues designed a polymerizable derivative of dapsone [48]. The dapsone was functionalized with an acrylic moiety and co-polymerized with hydroxyethyl methacrylate (HEMA). The dapsone-HEMA conjugate exhibited high stability with slow release of dapsone from the polymeric backbone due to hydrolysis. Moreover, dendrimers with high molecular weight are also used for drug delivery. For instance, a poly(L-lysine) dendrimer-docetaxel conjugate exhibited improved safety, prolonged plasma half-life, lower peak blood concentrations and greater overall drug exposure compared with docetaxel [49]. Polyamidoamine (PAMAM) is another widely studied dendrimer, but it is rich in cationic amine and non-biodegradable. Hence, its application is limited by high toxicity and hemolysis. In addition to the high molecular weight polymer-drug conjugates, low molecular weight polymer-drug conjugates also exhibit promising antitumor effects. Although lower molecular weight polymers have faster clearance, they may facilitate deeper tumor penetration. Researchers recently conjugated poly (vinyl alcohol) (~10 kDa) with camptothecin and doxorubicin, which showed superior cancer cell inhibition [50]. This result shows the therapeutic potential of low molecular weight polymer-drug conjugates.

Although EPR effects are widely used in drug delivery, numerous problems still exist. Many studies have demonstrated the highly vascular permeability of tumors, but the EPR effect is heterogeneous and depends on the cancer cells [51]. Furthermore, some metastatic cancers do not exhibit the EPR effect. Larger tumors tend to have more necrotic tissues and highly hypovascular areas instead of neovascular regions. In addition, the mechanical stress in tumors increases due to the production of extracellular matrix and remodeling of the tumor margin, which along with the leaky vessels leads to increased interstitial fluid pressure [52]. Therefore, even though drug-loaded nanomedicines can reach the tumor site through the EPR effect, the high interstitial fluid pressure still prevents them from penetrating into tissues. Several efforts have been made to overcome these limitations, such as elevating blood pressure or using nitric oxide-releasing agents. However, the most effective strategy is to target the tumor cells through active strategies as described in a later section.

3.2.5 Biomimetic NPs

Although thousands of nanomedicines have been developed and tested preclinically, only a few of them have undergone clinical trial screening and have successfully been approved by the Food and Drug Administration (FDA) and translated into the clinic. This gap between scientific and clinical practice is mainly due to one reason:

the human body's ability to recognize and clear foreign material. In recent years, biomimetic drug delivery systems based on natural components in the body have attracted the attention of researchers. These systems use naturally occurring proteins or cells in the body to accurately deliver drugs to target sites, resulting in minimal adverse effects and optimal therapeutic results.

Cellular proteins have been proposed as natural counterparts for synthetic polymers for drug delivery systems, some of which have been regarded safe by the FDA. Albumin has been considered an ideal candidate for drug delivery because of its nontoxicity and biodegradability, as well as its abundance in nature. Albumin can be obtained from various sources, including egg white (ovalbumin), bovine serum albumin, and human serum albumin (HSA). As a major soluble protein in the circulatory system, albumin is involved in maintaining osmotic pressure and binding and transporting nutrients to cells [53]. Many drugs and endogenous molecules are known to bind to albumin. Albumin has a number of mesh-like voids with a special structure similar to a honeycomb, providing a favorable space for carrying drugs. Albumin-based NPs offer several advantages for the delivery of antitumor drugs: (1) it is the natural carrier of hydrophobic drugs; (2) they are rescued from systemic clearance and degradation by natural mechanisms; (3) they accumulate in vascular leaks based on the EPR effect; and (4) they are more highly taken up and metabolized by rapidly growing, nutrient-starved cancer cells compared with healthy cells [54]. In terms of preclinical research, several studies have already indicated encouraging results for albumin-based NPs both in vitro and in vivo. Sebak et al. [55] prepared noscapine-loaded HSA NPs. The NPs were optimized to achieve an average size from 150 to 300 nm with drug encapsulation efficiency of 85–96%. Lee JE et al. [56] introduced a facile and oil-free method to prepare albumin-based paclitaxel NPs using HSA and PEG conjugates. The spherical HSA-PEG NP with a hydrodynamic diameter of 280 nm mediated efficient cellular delivery, leading to comparable or even higher cytotoxicity in various breast cancer cells than that of the commercially available Abraxane®. When systemically administered in a mouse xenograft model for human breast cancer, the HSA-PEG-based NPs exhibited an extended systemic circulation for more than 96 h and enhanced intratumoral accumulation, resulting in a remarkable anticancer effect and prolonged survival of the animals.

Erythrocytes are a type of non-nucleated cell with a diameter of 7–8 μm, which is uniquely double-sided concave microdisk. Inspired by the long circulation of erythrocytes in the body, researchers developed an erythrocyte-membrane-camouflaged nanomedicine through a top-down method. Using this strategy, intact erythrocyte membranes were collected from red blood cells and coated onto the surface of synthesized NPs. The erythrocyte membrane-coated NPs had highly tunable physicochemical properties of synthetic nanomaterials as well as highly complex functions of host cell membranes. Based on lipid and functional proteins on the surface of erythrocytes, Dr. Zhang Liangfang prepared erythrocyte membrane-coated PLGA NPs by a co-extruding method, and these NPs could circulate in the body for a long time [57, 58]. The erythrocyte membrane-coated PLGA NPs had more significant antitumor effect in a mouse lymphoma model compared with doxorubicin.

3.3 Active Tumor-targeted Strategy

3.3.1 Antibody-Mediated Targeting Delivery

Active targeting can not only increase drug concentration in tumor tissues and improve therapeutic effect but also reduce side effects. The antibody-mediated targeting strategy uses the specific recognition mechanism between antigen-antibody. Humanized antibodies and antibody fragments are typically conjugated with drugs or modified on the surface of nanomedicines.

Antibody-drug conjugates (ADCs) are an emerging class of targeting therapies that can improve traditional chemotherapeutic indices. ADCs combine the advantages of antibody targeting and selectivity with the advantages of high cytotoxic drug activity, thereby effectively increasing the drug concentration in the diseased tissue and significantly improving the therapeutic effect. ADCs that have gained regulatory approval from the FDA include brentuximab vedotin for CD30-positive Hodgkin's lymphoma and trastuzumab emtansine for human epidermal growth factor receptor 2 (HER2)-positive breast cancer [59]. Several other agents are in advanced stages of clinical development, including sacituzumab govitecan for breast cancer, mirvetuximab soravtansine for ovarian cancer, rovalpituzumab tesirine for lung cancer, depatuxizumab mafodotin for glioblastoma, and oportuzumab monatox for bladder cancer. Tai et al. prepared J6M0-mc MMAF (GSK2857916), an antibody drug coupling agent for the treatment of multiple myeloma coupled by a non-degradable linker [60]. When J6M0-mc MMAF was added to myeloma cells and other normal cells, it could significantly inhibit the growth of myeloma cells without affecting other cells. In a multiple myeloma mouse model, J6M0-mc MMAF could rapidly eliminate myeloma cells in mice. These results indicated that J6M0-mc MMAF had highly efficient and selective activity against multiple myeloma cells, which also provided a basis for tumor immunotherapy.

Modification of antibodies on the surface of carriers can improve the targeting and therapeutic effects of nanomedicines. Epidermal growth factor receptor (EGFR) is overexpressed in many human cancers, including colorectal cancer, breast cancer, and lung cancer [61]. This suggests that anti-EGFR antibodies or fragments can be used as targeting devices for specific cancer cells. Altintas et al. [62] developed a novel EGFR-targeted albumin NP coated with PEG 3500 and a nanobody, the single variable domain of an antibody (Ega1) against EGFR. The authors coupled the multi-kinase inhibitor 17,864-Lx to the methionine residue of albumin and cross-linked it with glutaraldehyde to form particles of around 100 nm. Cellular uptake experiments showed that Ega1-PEG functionalized NPs showed 40-fold higher binding to EGFR-positive ^{14}C squamous head and neck cancer cells compared with PEGylated NPs. The intracellular routing of Ega1-targeted NPs led to a successful release of the kinase inhibitor in the cell and inhibition of proliferation, whereas the non-targeting formulations had no anti-proliferative effects on ^{14}C cells.

Double modification of the antibody on the surface of a carrier can further enhance the targeting and therapeutic effects of nanomedicines. B-chronic lymphocytic leukemia is a common type of adult leukemia for which current treatments are not curative. CD37 antigen has been recently identified as a potential target for therapy in B-cell malignancies [63, 64]. CD37, a 40–52 kDa glycoprotein, is highly expressed on B cells and has limited or no expression on other hematopoietic cells such as T cells and NK cells [63, 65]. Yu et al. [66] developed a new strategy for the treatment of B chronic lymphocytes, the dual-ligand immunoliposomes (dILs). The authors combined anti-CD19 or anti-CD20 with anti-CD37 to construct dILs. Anti-CD19/CD37 dILs showed significantly enhanced cell delivery efficiency and apoptosis induction compared with single antibody ILs, while anti-CD20/anti-CD37 dILs showed enhanced apoptosis induction in leukemia metastatic cell lines. Cirstoiu Hapca et al. [67] conjugated the anti-HER2 monoclonal antibody and the anti-CD20 monoclonal antibody to the NPs to obtain NPs of about 250 nm. The interaction between antibody NPs and SKOV-3 ovarian cancer cells, which overexpress HER2, was observed by laser confocal microscopy. Anti-CD20-labeled NPs bound to the cell surface, while anti-HER2-labeled NPs entered the cell interior.

Notably, due to the large volume of antibody molecules, the ability of antibody-coupled NPs to penetrate into tissues is weak, limiting their distribution within cells, especially in solid tumors. Furthermore, antibody-coupled NPs often show changes in the three-dimensional structure of the antibody [68]. Therefore, antibody fragments that bind only to antigens (such as Fab and scFv) are popular in antibody-mediated active targeting drug delivery systems. The functional NPs formed by coupling small antibody fragments have better localization of disease and better penetrability, which can improve bioavailability and increase efficacy. CD44 variant 6 (CD44v6) is overexpressed in a subset of epithelial-derived cancers (e.g., pancreatic cancer, hepatocellular carcinoma, colorectal cancer, and gastric cancer) and is directly involved in carcinogenesis, metastasis, and progression, suggesting that CD44v6 is an ideal antigen to mediate targeting delivery of anti-CD44v6-modified NPs [67, 68]. However, the high molecular weight (150 kDa) of the full-length antibody limits the number of ligands that can be linked to the surface of NPs. In addition, NPs bearing the full-length antibody have the potential to cause immunogenicity in vivo [69]. Qian et al. [70] selected anti-human CD44v6 scFv (scFvCD44v6, MW = 25 kDa) as a targeting ligand and encapsulated arsenic trioxide in scFvCD44v6-decorated PEG-PDLLA NPs (scFv-As-NPs). Compared with the full-length CD44v6 antibody, scFvCD44v6 had a much lower molecular weight but exhibited higher affinity, higher biophysical stability, and lower immunogenicity. In vitro results indicated that conjugation of scFvCD44v6 with mal-PEG-PDLLA (scFv-As-NPs) enabled more efficient delivery and exhibited higher cytotoxic activity than non-targeted NPs (As-NPs) in PANC-1human pancreatic cancer cells.

3.3.2 Ligand-Mediated Targeting Delivery

With the deepening of tumor biology and tumor signaling research, a series of potential tumor-targeted sites have been discovered. Receptor-mediated targeting delivery has been investigated as an attractive strategy, in which targets are receptors that are specific or overexpressed on the surface of tumor cells. Many specific receptors are overexpressed in tumor cells and in the tumor microenvironment. The receptor-mediated tumor-targeted strategy can improve the efficacy of antitumor drugs and reduce side effects. This chapter provides a brief analysis and summary of receptor-mediated tumor-targeted delivery strategies based on the selection of targeted sites (Table 3.1).

3.3.2.1 Tumor Cell Targeting

Many receptors that are specifically expressed in various tumor cells have been used as potential targets for tumor-targeted therapy. The current targets for nanomedicines research mainly include folate receptor (FR), transferrin receptor, CD44 receptor, low density lipoprotein receptor, and HER2, among others. With the development of phage display technology, a growing number of peptides have been screened out as targeting ligands.

FR overexpression is commonly associated with a broad variety of tumor types including ovarian, lung, brain, head and neck, renal cell, and breast cancers, while FR displays relatively low expression on non-tumorous tissues [89]. Han et al. [90] demonstrated that folate-modified polymer micelles loading 9-nitro-camptothecin (9-NC) had higher antitumor activity than both non-folate drug-loaded micelles and free anticancer agents. Lian et al. [71] reported a microvesicle-based drug delivery system endowed with high tumor-targeted toward breast cancer that was attributed to the modified folate onto the membrane. The antitumor effect could be improved up to 15% with the help of the improved tumor-targeted ability.

Transferrin-receptor (TfR) is another recent research hotspot. While TfR is expressed on both normal tissues and malignant tissues, overexpression of TfR on cancer cells can be up to 100-fold higher than normal cells. Hence, Tf (a hydrophilic b-1 glycopeptide) [91] has been used as a targeting ligand modified on NPs to deliver drugs into tumor cells [92]. Jain et al. developed Tf-modified methotrexate NPs that produced higher cytotoxicity to C6 glioma cells compared with non-modified methotrexate NPs. The authors demonstrated that Tf conjugation in targeted drug delivery could promote selective uptake of anticancer drugs into the tumor tissue [73].

EGFR upregulation is closely related to cancer progression and poor prognosis. Many studies have focused on using EGFR-binding ligands for modifying nanocarrier surfaces to achieve tumor-targeted drug delivery. In recent years, several EGFR-specific peptides have been used in tumor-targeted nanomedicines, including D4 and GE11 [93]. D4 (LARLLT) is a novel peptide ligand developed via

Table 3.1 A brief summary of targeted receptors and targeting ligands

Targeted cites	Receptors	Targeting Ligands	Ref
Tumor cell	Folate receptor	Folate	[71]
	CD44 receptor	Hyaluronic acid	[72]
	TfR	Transferrin	[73]
	EGFRs	D4 peptide (LARLLT); GE11 peptide (YHWYGYTPQNVI)	[74, 75]
	HER2	VIP peptide (HSDAYFTDNYTRLRKOMAYKKYLNSILN)	[76]
	Low density lipoprotein receptor	DWLKAFYDKVAEKLKEAFRLTRKRGLKLA	[77]
	Asialoglycoprotein receptor	Galactosamine	[78]
Tumor microenvironment	VEGFR-2	K237 peptide (HTMYYHHYQHHL)	[79]
	Tie2	PH-1 peptide (TMGFTAPRFPHY)	[80]
Tumor cell and tumor microenvironment	p32	LyP-1peptide (CGNKRTRGC)	[81, 82]
	Integrins	c(RGDyK)	[83]
	Platelet-derived growth factor-β	Ppb cylic peptide (CSRNLIDC)	[84]
	Neuropilin-1	$_{D}$A7R peptide (rpplwta), rgerppr	[85, 86]
	Nucleolin	F3 peptide (KDEPQRRSARLSAKPAPPKPEPKPKKAPAKK)	[87, 88]

computer-aided design [94]. D4-modified liposomes enter EGFR-overexpressing cancer cells (H1299) specifically and efficiently by endocytosis. The EGFR-specific peptide GE11 was developed by Li et al. via phage display technique. Song et al. modified GE11 to doxorubicin-loaded liposomes and found that the modified liposomes had similar pharmacological potency to free doxorubicin. In vivo fluorescence images of the doxorubicin-loaded liposomes also showed that the targeted liposomes were frequently internalized in the tumor tissues during the first 12 h, while non-targeted liposomes achieved lower levels of internalization [74].

3.3.2.2 Tumor Microenvironment Targeting

The neovascularization plays a very important role in the growth and metastasis of tumors. Tumor-associated neovascular endothelial cells grow rapidly and express some specific angiogenic markers. Compared with tumor cells, neovascular epithelial cells show relatively stable growth and have the same phenotype in different tumors. Therefore, treatment for tumor neovascularization results in lower drug resistance and can be applied to different tumors. In recent years, many drug delivery systems have been designed based on tumor neovascular-targeting to destroy the neovascularization and achieve the effect of starving tumor cells. The current targets for nanocarrier research mainly include integrins, vascular endothelial growth factor receptor-2 (VEGFR-2), and tyrosine kinase with immunoglobulin and epidermal growth factor homology-2 (Tie2).

Vascular endothelial growth factor (VEGF) is one of the most crucial mediators of tumor angiogenesis and is closely involved in tumor development and metastasis [95]. VEGFR-2 is a primary mediator of tumor angiogenesis. Recently, a peptide designated K237 (HTMYYHHYQHHL) isolated from a phage display peptide library was shown to bind VEGFR-2 with high affinity and specificity [96]. Yu et al. [79] prepared K237-modified paclitaxel-loaded NPs (K237-PTX-NP). The long-circulating property and the K237 ligand of K237-PTX-NPs resulted in rapid, long-term, and accurate in vivo tumor neovasculature targeting and the significant apoptosis of tumor neovasculature endothelial cells and necrosis of tumor tissues of breast tumor-bearing nude mice.

3.3.2.3 Tumor Cell and Tumor Microenvironment Targeting

Recent studies have identified specific receptors on the surface of tumor blood vessels that are different from normal blood vessels. These receptors, including integrins and nucleolin, are also highly expressed on the surface of tumor cells. Nano-delivery systems modified with ligands of these receptors have the function of simultaneously targeting tumor blood vessels and tumor cells. The targeting of tumor blood vessels can increase the retention of the drug delivery systems in the vicinity of tumor blood vessels, thereby increasing the amount of drug distribution into the tumor tissue and exerting antitumor growth effects mediated by receptors on the tumor cell surface. Among these ligands, peptides are widely used due to their simple and stable structure.

Integrins are heterodimeric transmembrane glycoproteins that are very important for cancer progression [97]. Among the 24 subtypes of integrins, $\alpha_v\beta_3$ and $\alpha_v\beta_5$ are overexpressed on glioma cells and neovasculature [98]. The RGD peptide motif was designed to be a pentacyclic peptide (cyclic RGDyK or RGDfK, cRGD) that could recognize integrin $\alpha_v\beta_3$ overexpressed on neovasculature and tumor cells [99]. RGD-modified drug delivery systems have advantages in delivering chemotherapeutics, peptides and proteins, nucleic acids, and irradiation targeting at tumor cells [100]. A c(RGDyK)-modified lipid disk was developed as a tumor-targeted drug delivery system for melittin. The biodistribution and tumor growth inhibitory experiments showed that c(RGDyK) modification increased the distribution of lipid disks in the tumor and the anticancer efficacy of melittin-loaded lipid disks, which was attributed to the specific binding of c(RGDyK) with integrin-overexpressed tumor cells [83].

In glioma treatment, the blood-brain barrier (BBB) and the blood-brain tumor barrier (BBTB) prevent most drugs from entering brain tumor tissues. Ying et al. [85] designed a peptide $_D$A7R with high binding affinity in vitro to VEGFR-2 and neuropilin-1 (NRP-1) overexpressed on glioma, glioma vasculogenic mimicry, and neovasculature. The $_D$A7R-modified doxorubicin-loading liposomes significantly inhibited tumor growth in a subcutaneous model to a greater extent compared with free doxorubicin.

The formation of tumor lymphatic vessels in tumor tissues greatly promotes lymphatic metastasis of tumors. Several studies have identified some common receptors specifically expressed on tumor lymphatic vessels and tumor cells. Thus, nanocarriers modified by their ligands have the effects of simultaneously killing tumor cells and destroying lymphatic vessels of tumors, thereby inhibiting tumor growth and inhibiting lymphatic metastasis of tumors. For example, the p32 receptor is overexpressed on a variety of tumor cells and tumor lymphatics [101]. LyP-1 (CGNKRTRGC), a cyclic nonapeptide, demonstrated specific binding ability with the p32 receptor [102, 103]. Yan et al. [81] prepared LyP-1-modified PEGylated liposomes loaded with doxorubicin. The LyP-1-modified liposomes exhibited good targeting ability to lymphatic metastatic tumors based on LyP-1 specific binding to tumor cells, tumor lymphatics, and tumor-associated macrophages in metastatic lymph nodes. LyP-1 modification significantly enhanced the inhibition effect of liposomal DOX on lymphatic metastatic tumors.

3.3.3 Aptamer-Mediated Targeting Delivery

Aptamers are single-stranded DNA or RNA with unique tertiary structures that are regarded as artificial nucleic acid ligands of biomolecules [104]. Aptamers are mainly screened via systematic evolution of ligands by exponential enrichment (SELEX), aiming at targets varying from small molecules to biomacromolecules, infected cells, stem cells, and cancer cells. Aptamers are also known as chemical antibodies because aptamers bind to target molecules in a manner resembling the antigen-antibody interaction. As a class of molecular ligands, aptamers have some advantages over other classes of ligands such as antibodies or peptides [105], including the following: (1) because of their chemical essence of DNA or RNA, aptamers are biocompatible and poorly immunogenic, which may address safety concerns; (2) aptamers can be applied to a wider array of molecular targets that are difficult to generate antibodies against or screen peptides by phage display; (3) aptamers have decent stabilities because of their relatively simple chemical structures; (4) current automated DNA/RNA synthesis technologies allow for easy, cost-effective, and scalable production of aptamers; and (5) aptamer manufacturing has higher reproducibility with low batch-to-batch variations compared with peptide and antibody synthesis. Aptamers have been applied in two main forms: aptamer-modified nanomedicines and aptamer-drug conjugates (ApDCs).

Aptamer-modified nanomedicines can realize targeted delivery of hydrophobic or hydrophilic drugs to tumors. For instance, aptamer-modified NPs were prepared for targeted co-delivery of doxorubicin hydrochloride (DOX) and paclitaxel (PTX) [106]. The aptamer sgc8, which can bind human protein tyrosine kinase 7 overexpressed on human T-cell acute lymphocytic leukemia cells, was linked to PEG-lipid. DOX was intercalated by preferential interaction with double-stranded GC/CG regions of aptamers. Hydrophobic poly(lactic-co-glycolic acid) (PLGA) with encapsulated PTX constituted the core structure and the DOX-loaded aptamer-PEG-lipid was inserted into PLGA cores having successfully co-delivered these two drugs with distinct solubility to target cancer cells. In addition, aptamer strengthened liposomes have also been reported. A 26-mer DNA aptamer showed high binding affinity to nucleolin (NCL), a bcl-2 mRNA binding protein involved in cell proliferation and linked to breast cancer. The NCL-aptamer was tagged by cholesterol and then inserted into the hydrophobic lipid membrane of cisplatin-loaded liposomes [107]. Aptamer-modified liposomes effectively delivered cisplatin in a cancer cell-specific manner, as evidenced by the significant killing of the target cancer cells but not the control cancer cells. Similarly, a therapeutic liposome drug delivery system was engineered by covalently linking the fsgc8 aptamer to the liposome through a PEG spacer [108]. The advantages of aptamers make the DNA-mediated liposome carrier an efficient strategy for drug transduction with high target cell specificity and sensitivity. Aptamers can also be decorated onto new nanostructures other than NPs and liposomes as a highly specific ligand. In a previous study, the multifunctional DNA nanostructures, termed as nanoflowers, was realized by long DNA building blocks of a designer template [109]. The concatemer aptamers in elongated single strand DNA decorated onto nanoflowers were expected to enhance the binding affinity to target cells. This strategy may provide insight for future construction of DNA/RNA nanostructures incorporated with targeting aptamers. Gold nanomaterials conjugated with aptamers hold great potential for photodynamic therapy and photothermal therapy. The aptamer-gold nanoparticle-hybridized graphene oxide facilitated targeted treatment of tumor cells by near-infrared light-activating photothermal therapy [110]. The type I transmembrane mucin glycoprotein binding aptamer attached to the surface of gold NPs enabled the nanocomposite to selectively target MUC1-positive MCF-7 human breast cancer cells and to exert therapeutic effects on MCF-7 cells at an ultralow concentration without inducing adverse effects in healthy cells. Together these studies indicate that aptamers as targeting molecules on the surface of nanomedicines including NPs, liposomes, and new nanostructures formed by DNA hybridization may be an effective active tumor-targeted drug delivery strategy.

In addition to aptamer-modified nanomedicines, ApDC represents a promising targeted drug delivery system for reducing toxicity while increasing the efficacy of chemotherapy. The DNA aptamer sgc8c was covalently linked with the antitumor agent DOX, and the resultant sgc8c-Dox conjugate specifically killed the target T-cell acute lymphoblastic leukemia but with minimal toxicity toward non-target cells [111]. Furthermore, a photocleavable linker was used to develop the ApDC sgc8-5FU to deliver 5-fluorouracil for targeting colon cancer, which was enabled by the aptamer sgc8 with photocontrollable drug release [112]. In addition to covalently linkages, embedding drugs through intermolecular forces into the three-dimensional structure of the adapter is a strategy to maximize drug therapeutic potency, because one aptamer could carry multiple copies of drugs. In one example, one aptamer was tethered with a long double-stranded DNA that was composed with almost 100% of drug-intercalating sites for targeted drug transport in cancer therapy [113]. Potent antitumor efficacy and reduced side effects of drugs were demonstrated in a mouse xenograft tumor model. The concept of ApDCs was also used to link macromolecular drugs such as antibodies and immunomodulatory molecules for cancer immunotherapy. Blocking of cytotoxic T lymphocyte-associated antigen-4 (CTLA-4) and B7 conjugation reshaped the host immune response and exerted a sustained antitumor effect in some cancer subpopulations. By the integration of two high-throughput platforms, SELEX and next-generation sequencing, a novel CTLA-4-antagonizing DNA aptamer was developed, and it functionally bound to CTLA-4 with high affinity to promote T-cell activity in the tumor microenvironment [114]. In another case, the model antigen ovalbumin was conjugated to selected aptamers that specifically bound the murine receptor, DEC205, a C-type lectin expressed predominantly on the surface of CD8α + dendritic cells; the resultant DEC205-targeted antigen cross-presentation was verified in vitro and in vivo by proliferation and cytokine production by primary murine CD8+ T cells [115].

All NPs using aptamers as target agents are still in the preclinical phase [116]. The current research indicates that aptamers can be successfully anchored to a myriad of different nanocarriers and conjugated with small molecules, small interfering RNA and antigens as a form of ApDCs for active tumor-targeted drug delivery. However, aptamers have some drawbacks, including in vivo nuclease instability, long length (typically 75–100 nucleotides), and lack of flexibility.

3.3.4 Biomimetic Membrane-Mediated Targeting Delivery

Natural cells such as red blood cells, stem cells, and dendritic cells have been studied as drug carriers [117, 118]. However, their micron size and the instability of these cell carriers largely limit their extravascular diffusion capacity and penetration at the tumor site. Cell membrane-coated NPs can effectively overcome the disadvantages of cell-based carriers (Table 3.2). Erythrocyte membranes have been used to recombine and encapsulate polymer NPs, and polymer NPs exhibit a better half-life than the PEGylated counterparts [124]. However, erythrocyte membranes do not contain any targeting ligands for cancer cells. Other cells of the human body (i.e., white blood cells, platelets, MSCs) have many recognition receptors on the membrane that can be used to distinguish between inflammation and diseased areas. Maintaining a natural cell membrane not only helps avoid phagocytosis by the mononuclear phagocytic system but also enhances the accumulation at the tumor site to improve the therapeutic effect. Cell membrane-coated NPs have great potential for delivering therapeutic agents for a variety of diseases.

3.3.4.1 Platelet Membrane-Coated NPs

One important cell type that has been explored is platelets, which are cytoplasmic fragments of mature macrophages. Some studies have also shown that platelets are crucial for canceration [125], and inflammation that occurs during tumor progression causes platelets to return to the tumor site, stimulating tumor angiogenesis. In addition, platelets maintain tumor cell extravasation and the survival of circulating tumor cells (CTCs) in the bloodstream [126], thereby facilitating metastatic spread.

CTCs form microthrombus with activated platelets and fibrin, which in turn facilitates the formation of metastases. Inspired by this observation, Li et al. produced silica (Si) NPs coated with membranes isolated from activated platelets along with surface conjugation of a tumor-specific apoptosis-inducing ligand cytokine, tumor necrosis factor-related apoptosis inducing ligand (TRAIL) [119]. Platelet membrane-coated Si-NPs were shown to express most of the platelet surface proteins (i.e., CD41, CD42b, and CD61) and glycans relevant for targeting CTCs and escaping phagocytosis. Evaluation of a variety of cancer-bearing murine models (i.e., human breast cancer, colon cancer, and a syngeneic metastatic colon cancer and

Table 3.2 Biomimetic membrane-mediated tumor-targeted strategy

Cell membranes	Targeting proteins on membrane	Ref
Platelet membrane	CD36, CD41, CD42d, CD61, CD40L, P-selectin	[119, 120]
Mesenchymal stem cell membrane	CD44, CD105	[121]
Cancer cell membrane	Cadherin	[122]
White blood cell membrane	C49d	[123]

melanoma mouse model) demonstrated that TRAIL-conjugated platelet membrane-Si particles were able to efficiently target CTCs in lung vasculature and dramatically decrease lung metastases compared with untreated mice or mice treated with empty platelet membrane-coated Si particles or soluble TRAIL.

Hu et al. [127] developed platelet-membrane (PM)-coated nanovesicles (PM-NVs) loaded with two anticancer components: TRAIL and DOX. Administration of PM-NVs in a mouse model of breast cancer showed accumulation of NPs at the tumor site and efficient delivery of TRAIL to the cancer cells. Based on the specific affinity between P-selectin on the surface of platelet membrane and CD44 overexpressed on tumor cells, Hu et al. [128] developed platelet membrane-coated glucan NPs loaded with the anticancer agent bortezomib and thrombolytic agent tPA. The nanomedicine could sequentially target the bone microenvironment and myeloma cells to enhance drug availability at the myeloma site and reduce off-target effects while inhibiting multiple myeloma growth and simultaneously eradicating thrombotic complications.

3.3.4.2 Mesenchymal Stem Cell (MSC)-Coated NPs

MSCs are pluripotent stem cells that have the ability to differentiate into osteoblasts, adipocytes, chondroblasts, fibroblasts, and pericytes. Stimulated by microenvironmental factors such as hypoxia and cytokine gradients in the extracellular matrix, MSCs migrate to damaged and inflamed sites. Since cancer is considered to be a chronic inflammatory disease, MSCs have been used in tumor therapy [129].

Timaner et al. [130] developed nanoghosts (NGs) derived from the plasma membrane of MSCs. These MSC-NGs retained the ability of MSCs to target inflammatory endothelium. In an orthotopic pancreatic tumor model, nanovesicles derived from MSC membranes and loaded with CXCR3 antagonists showed enhanced therapeutic results and delayed tumor regrowth when administered in combination with gemcitabine. Machluf and colleagues [131] used a similar method to prepare intratumoral MSC-derived nanovesicles. The surface-specific MSC integrin bound to and fused with the surface of tumor cells and destroyed the plasma membrane, resulting in cytotoxicity and tumor growth inhibition.

3.3.4.3 Cancer Cell Membrane-Coated NPs

The intrinsic homogenous aggregation properties of cancer cells and their ability to evade immune recognition make cancer cell-derived biomimetic strategies a promising candidate for surface functionalization of NPs with cell membranes. Unlike normal cells, cancer cells show rapid proliferation in vitro, making these cells suitable for cell membrane coating strategies.

Fang et al. [132] coated PLGA NPs with breast cancer cell membrane (CCNP) and demonstrated that the prepared NPs showed very similar protein expression to the source cancer cells. Flow cytometry results showed that CCNP showed a 40-fold and 20-fold increase in tumor uptake compared with RBC-coated NPs and uncoated NPs, respectively. In another study, the authors inserted monomeric phosphoryl lipid A as an immunoadjuvant into the membrane of CCNPs to enhance the immune response against low immunogenic antigens found on cancer membranes. The functionalized CCNPs showed the ability to deliver tumor antigens and induce denaturation of dendritic cells.

Sun et al. developed a new cancer cell biomimetic nanoparticle for targeted chemotherapy of homologous tumors [133]. A unique biomimetic drug delivery system composed of 4 T1 breast cancer cell membranes and paclitaxel-loaded polymeric NPs demonstrated superior interactions to the source tumor cells and extended half-life in blood circulation. Furthermore, the system displayed high cell-specific targeting of the homotypic primary tumor and metastases, with successful inhibition of the growth and metastasis of the breast cancer cells.

3.3.4.4 White Blood Cell (WBC) Membrane-Coated NPs

WBCs are composed of many different cells, including macrophages, dendritic cells, B cells, T cells, and neutrophils. WBCs exhibit unique functions to target specific sites, especially tumors or abnormal blood vessels.

Parodi A et al. [117] obtained a leukocyte-derived membrane by a sucrose density gradient purification method and coated it onto the surface of nanoporous silicon microparticles to generate the leukocyte-like carrier (LLV). The authors successfully transplanted more than 300 proteins on the NP surface and then identified proteins that promoted reduced macrophage uptake (i.e., clusterin [134]), immune tolerance-associated proteins (i.e., CD47 and CD45 [135]), and endothelial/tumor-targeted proteins (i.e., lymphocyte function-related antigen 1 and macrophage-1 antigen [117]). LLV preferentially accumulated in activated endothelial cells, leading to downstream phosphorylation of vascular endothelium-adhesin proteins [117]. This increase in phosphorylation resulted in a decrease in tight junctions between endothelial cells, thereby promoting penetration of the payload into the tumor microenvironment.

In addition to enhancing immune evasion and inflammatory targeting, WBC membranes can also target tumors through specific receptor-ligand binding. Krishnamurthy et al. [123] coated the mononuclear cell membrane around the PLGA NPs loaded with DOX. The presence of CD49d, a surface marker and the major ligand for vascular cell adhesion molecule 1, was demonstrated by western blotting, and its function as a targeting mechanism was then studied in vitro. More coated NPs were taken up by MCF-7 cancer cells than uncoated particles. Compared with uncoated NPs, membrane-coated NPs killed cancer cells better in vitro due to adhesion receptor-ligand interactions.

3.3.5 Transporter-Related Targeting Delivery

In addition to antibody, ligand, or aptamer modification used for nanomedicines, endogenous substances can also serve as targeting molecules for tumor therapy. On account of specific recognition and active transportation of transporters overexpressed on the targeting site, the targeting characteristics of corresponding substances unfold. Transporters are functional cell membrane proteins in various tissues throughout the human body that govern the movement of endogenous and exogenous substances in and out of cells. More than 400 transporters that mediate the movement of sugar, nucleosides, amino acids, and peptides across the cell membrane have been identified. According to the direction of substrate transmembrane transport, transporters can be divided into two categories: absorption and efflux transporters. Transporters mediating entry of substances into cells belong to the solute carrier group, which mainly includes L-type amino transporter, peptide transporters (PEPTs), sodium-dependent secondary active transporter, sodium-independent facilitated diffusion transporters (GLUTs), monocarboxylate transporters, organic anion transporters, and cationic transporters. The transporters mediating drug efflux mainly include P-glycoprotein (P-gp), multidrug resistance associated proteins, and breast cancer resistance protein (BCRP)1–5, which belong to the ATP binding cassette (ABC) family [136]. The ABC drug efflux pump uses the energy of ATP hydrolysis to transport drugs and endogenous substances.

Capitalizing on the transporter overexpression on target cells, the corresponding substances were conjugated to drugs or nanocarriers to tailor the transporter-mediated nanomedicines. For example, D-glucose can cross the BBB and be taken up by brain parenchymal cells via GLUT-mediated transcytosis. Approximately 6×10^6 GLUTs are expressed per brain capillary endothelial cell, and the transcytosis rate is considerably fast; the transcytosis ability of GLUT is 1420 nmol/(min·g), which is far higher than other transporters. In this line with this research, the heptapeptide ATWLPPR (A7R) bound specifically to VEGFR-2 and NRP-1 overexpressed in glioma cells, but A7R did not have the capacity to cross the BBB. Glycosylation of the peptide enabled its efficient traversing the BBB and led to uptake by glioma cells [137]. A novel glioma-targeting drug delivery system was constructed using glycopeptide as the targeting moiety and nanodisk as the carrier of paclitaxel. In addition, dehydroascorbic acid (DHA) as a targeting moiety has a high affinity for GLUT1 on the BBB. The surface of polymeric micelles (PEG-pLys-pPhe) was modified by DHA modified via a propynyl group to synthesize pPhe-pLys-PEG-DHA [138]. Compared with free itraconazole, the modified drug-containing micelles resulted in higher drug concentration in brain parenchyma and effective improvement of the storage time of drugs in blood. Another important membrane transporter is the sodium-dependent multivitamin transporter (SMVT), which is expressed along the small and large intestines. SMVT is responsible for the uptake of essential nutrients such as biotin, lipoate, and pantothenate. A previous study showed that biotinylated PAMAM dendrimer-FITC conjugate and biotinylated hydroxypropylmethacrylate (HPMA) polymer displayed substantially high uptake in HeLa human cervical carcinoma cells and ID8 murine cancer cells, respectively, via an interplay between biotin and SMVT [139].

In general, a substance is conjugated with a drug to form a prodrug to improve the bioavailability through influx mediated by the corresponding transporter. The peptide transporters PEPT1 is overexpressed in several cancer cells including the malignant ductal pancreatic cancer cell lines AsPc-1 and Capan-2 and the human fibrosarcoma cell line HT-1080 but not in normal cells. To improve the physiochemical properties and selectivity of the clinical anticancer agent floxuridine and to reduce its undesirable toxicity effects, prolyl and lysyl prodrugs were synthesized, and these exhibited enhanced uptake (two to eightfold) in PEPT1+ HeLa cells compared with HeLa cells [140]. Gemcitabine is a clinically approved anticancer agent for the treatment of breast, lung, ovarian, pancreatic, and bladder cancer. The amino acid ester conjugate of gemcitabine showed binding to one or both of the peptide transporters PEPT1 and PEPT2 using a radiolabeled substrate uptake assay [141].

In addition to influx transporters, efflux transporters also show potential utility for transporter-mediated tumor-targeted drug delivery systems. All the main ABC drug efflux transporters are primarily located in the plasma membrane and extrude a variety of structurally diverse drugs, drug conjugates and metabolites, and other compounds from the cell. ABC drug efflux pumps restrict the penetration of drugs into a range of important organs such as brain, testis, and fetus as well as specific cell and tissue compartments [142]. Pluronic block copolymers consist of ethylene oxide (EO) and propylene oxide (PO) segments arranged in the basic A-B-A structure: EOa-POb-EOa. These are known under their non-proprietary name "poloxamers" and can inhibit the P-gp drug efflux transport system that is overexpressed in multidrug resistant (MDR) cancer cells [143]. A mixed micellar system based mainly on the pH-sensitive copolymer of poly (L-histidine)-poly (D,L-lactide)-polyethyleneglycol-poly (D,L-lactide)-poly (L-histidine) (PHis-PLA-PEG-PLA-PHis) and Pluronic F127, some of which was conjugated with folate, was constructed to deliver the unimers of Pluronic P85 and doxorubicin to MDR cells [144]. The active targeting and bypassing of the efflux mediated by MDR-related proteins contributed to the MDR reversal effect of resultant micelles.

In conclusion, these studies suggest that substance-modified drugs and nanomedicines could potentiate drug absorption, penetration cross barrier membrane, and preferential retention in targeted tumor site actively in virtue of endogenous transport system. However, most transporters are distributed throughout the whole body, and thus specificity is not guaranteed. Transporter-mediated drug delivery increased drug retention in tumors, but the increased drug distribution in other organs may also be an issue of concern. Reasonable selection of substrates and accurate design of drug delivery system may consummate the transporter-mediated tumor-targeted drug delivery strategy.

References

1. Bray F, Ferlay J, Soerjomataram I, Siegel RL, Torre LA et al (2018) Global cancer statistics 2018: GLOBOCAN estimates of incidence and mortality worldwide for 36 cancers in 185 countries. CA Cancer J Clin 68:394–424
2. Haley B, Frenkel E (2008) Nanoparticles for drug delivery in cancer treatment. Urol Oncol 26:57–64
3. Matsumura Y, Maeda H (1986) A new concept for macromolecular therapeutics in cancer chemotherapy: mechanism of tumoritropic accumulation of proteins and the antitumor agent smancs. Cancer Res 46:6387–6392
4. Maeda H, Nakamura H, Fang J (2013) The EPR effect for macromolecular drug delivery to solid tumors: improvement of tumor uptake, lowering of systemic toxicity, and distinct tumor imaging in vivo. Adv Drug Deliv Rev 65:71–79
5. Greish K (2012) Enhanced permeability and retention effect for selective targeting of anticancer nanomedicine: are we there yet? Drug Discov Today Technol 9:e71–e174
6. Fang J, Nakamura H, Maeda H (2011) The EPR effect: unique features of tumor blood vessels for drug delivery, factors involved, and limitations and augmentation of the effect. Adv Drug Deliv Rev 63:136–151
7. Folkman J (1995) Angiogenesis in cancer, vascular, rheumatoid and other disease. Nat Med 1:27–31
8. Hashizume H, Baluk P, Morikawa S, McLean JW, Thurston G et al (2000) Openings between defective endothelial cells explain tumor vessel leakiness. Am J Pathol 156:1363–1380
9. Azzopardi EA, Ferguson EL, Thomas DW (2013) The enhanced permeability retention effect: a new paradigm for drug targeting in infection. J Antimicrob Chemother 68:257–274
10. Prabhakar U, Maeda H, Jain RK, Sevick-Muraca EM, Zamboni W (2013) Challenges and key considerations of the enhanced permeability and retention effect for nanomedicine drug delivery in oncology. Cancer Res 73:2412–2417
11. Torchilin V (2011) Tumor delivery of macromolecular drugs based on the EPR effect. Adv Drug Deliv Rev 63:131–135
12. Decker C, Schubert H, May S, Fahr A (2013) Phospholipid vesicles (liposomes) as models for biological membranes: their properties and interactions with cholesterol and proteins. J Control Release 66:277–285
13. Bawarski WE, Chidlowsky E, Bharali DJ, Mousa SA (2008) Emerging nanopharmaceuticals. Nanomedicine 4:273–282
14. Tiwari G, Tiwari R, Sriwastawa B, Bhati L, Pandey S, Pandey P et al (2012) Drug delivery systems: an updated review. Int J Pharm Investig 2:2–11
15. Etheridge ML, Campbell SA, Erdman AG, Haynes CL, Wolf SM, McCullough J (2013) The big picture on nanomedicine: the state of investigational and approved nanomedicine products. Nanomedicine 9:1–14
16. Ozpolat B, Sood AK, Lopez-Berestein G (2014) Liposomal siRNA nanocarriers for cancer therapy. Adv Drug Deliv Rev 66:110–116
17. Sharma US, Sharma A (1997) Liposomes in drug delivery: Progress and limitations. Int J Pharm 154:123–140
18. Malam Y, Loizidou M, Seifalian AM (2009) Liposomes and nanoparticles: nanosized vehicles for drug delivery in cancer. Trends Pharmacol Sci 30:592–599
19. Overchuk M, Zheng G (2018) Overcoming obstacles in the tumor microenvironment: recent advancements in nanoparticle delivery for cancer theranostics. Biomaterials 156:217–237
20. Feng L, Cheng L, Dong Z, Tao D, Barnhart TE et al (2017) Theranostic liposomes with hypoxia-activated prodrug to effectively destruct hypoxic tumors Post-photodynamic therapy. ACS Nano 11:927–937
21. Landen CN Jr, Chavez-Reyes A, Bucana C, Schmandt R, Deavers MT et al (2005) Therapeutic EphA2 gene targeting in vivo using neutral liposomal small interfering RNA delivery. Cancer Res 65:6910–6918

22. Qian Y, Liang X, Yang J, Zhao C, Nie W et al (2018) Hyaluronan reduces cationic liposome-induced toxicity and enhances the antitumor effect of targeted gene delivery in mice. ACS Appl Mater Interfaces 10:32006–32016
23. Movassaghian S, Merkel OM, Torchilin VP (2015) Applications of polymer micelles for imaging and drug delivery. Wiley Interdiscip Rev Nanomed Nanobiotechnol 7:691–707
24. Eetezadi S, Ekdawi SN, Allen C (2015) The challenges facing block copolymer micelles for cancer therapy: in vivo barriers and clinical translation. Adv Drug Deliv Rev 91:7–22
25. Wei J, Wang H, Zhu M, Ding D, Li D et al (2013) Janus nanogels of PEGylated Taxol and PLGA–PEG–PLGA copolymer for cancer therapy. Nanoscale 5:9902–9907
26. Biswas S, Kumari P, Lakhani PM, Ghosh B (2016) Recent advances in polymeric micelles for anti-cancer drug delivery. Eur J Pharm Sci 83:184–202
27. Tanbour R, Martins AM, Pitt WG, Husseini GA (2016) Advances in polymeric micelles for drug delivery and tumor targeting. Curr Pharm Des 22:2796–2807
28. Yokoyama M, Kwon GS, Okano T, Sakurai Y, Seto T et al (1992) Preparation of micelle-forming polymer-drug conjugates. Bioconjug Chem 3:295–301
29. Saw PE, Yu M, Choi M, Lee E, Jon S et al (2017) Hyper-cell-permeable micelles as a drug delivery carrier for effective cancer therapy. Biomaterials 123:118–126
30. Kabanov AV, Batrakova EV, Alakhov VY (2002) Pluronic block copolymers as novel polymer therapeutics for drug and gene delivery. J Control Release 82:189–212
31. Mei L, Rao J, Liu Y, Li M, Zhang Z et al (2018) Effective treatment of the primary tumor and lymph node metastasis by polymeric micelles with variable particle sizes. J Control Release 292:67–77
32. Kim G, Piao C, Oh J, Lee M (2018) Self-assembled polymeric micelles for combined delivery of anti-inflammatory gene and drug to the lungs by inhalation. Nanoscale 10:8503–8514
33. Mackiewicz N, Gravel E, Garofalakis A, Ogier J, John J et al (2011) Tumor-targeted Polydiacetylene micelles for in vivo imaging and drug delivery. Small 7:2786–2792
34. Kumari P, Ghosh B, Biswas S (2016) Nanocarriers for cancer-targeted drug delivery. J Drug Target 24:179–191
35. Wissing SA, Kayser O, Müller RH (2004) Solid lipid nanoparticles for parenteral drug delivery. Adv Drug Deliv Rev 56:1257–1272
36. Minelli R, Serpe L, Pettazzoni P, Minero V, Barrera G (2012) Cholesteryl butyrate solid lipid nanoparticles inhibit the adhesion and migration of colon cancer cells. Br J Pharmacol 166:587–601
37. Guerrero-Cázares H, Tzeng SY, Young NP, Abutaleb AO, Quiñones-Hinojosa A et al (2014) Biodegradable polymeric nanoparticles show high efficacy and specificity at DNA delivery to human glioblastoma in vitro and in vivo. ACS Nano 8:5141–5153
38. Behrens S, Appel I (2016) Magnetic nanocomposites. Curr Opin Biotechnol 39:89–96
39. Liu JH, Yang ST, Wang X, Wang H, Liu Y et al (2014) Carbon nanoparticles trapped in vivo—similar to carbon nanotubes in time-dependent biodistribution. ACS Appl Mater Interfaces 6:14672–14678
40. Sun H, Ren J, Qu X (2016) Carbon nanomaterials and DNA: from molecular recognition to applications. Acc Chem Res 49:461–470
41. Razzazan A, Atyabi F, Kazemi B, Dinarvand R (2016) In vivo drug delivery of gemcitabine with PEGylated single-walled carbon nanotubes. Mater Sci Eng C Mater Biol Appl 62:614–625
42. Riley RS, Day ES (2017) Gold nanoparticle-mediated photothermal therapy: applications and opportunities for multimodal cancer treatment. Wiley Interdiscip Rev Nanomed Nanobiotechnol 9:e1449
43. Pang X, Du HL, Zhang HQ, Zhai YJ, Zhai GX (2013) Polymer-drug conjugates: present state of play and future perspectives. Drug Discov Today 18:1316–1322
44. Ekladious I, Colson YL, Grinstaff MW (2019) Polymer-drug conjugate therapeutics: advances, insights and prospects. Nat Rev Drug Discov 18:273–294

45. Paramjot KNM, Kapahi H, Kumar S, Bhardwaj TR et al (2015) Role of polymer-drug conjugates in organ-specific delivery systems. J Drug Target 23:387–416
46. Kopecek J, Kopecková P, Minko T, Lu Z (2000) HPMA copolymer–anticancer drug conjugates: design, activity, and mechanism of action. Eur J Pharm Biopharm 50:61–81
47. Yang J, Kopeček J (2016) Design of smart HPMA copolymer-based nanomedicines. J Control Release 240:9–23
48. Rojo L, Fernandez-Gutierrez M, Deb S, Stevens MM, San RJ (2015) Designing dapsone polymer conjugates for controlled drug delivery. Acta Biomater 27:32–41
49. Zhou X, Zheng Q, Wang C, Xu J, Wu JP (2016) Star-shaped Amphiphilic Hyperbranched Polyglycerol conjugated with dendritic Poly(l-lysine) for the Codelivery of Docetaxel and MMP-9 siRNA in Cancer therapy. ACS Appl Mater Interfaces 8:12609–12619
50. Camacho KM, Menegatti S, Mitragotri S (2016) Low-molecular-weight polymer–drug conjugates for synergistic anticancer activity of camptothecin and doxorubicin combinations. Nanomedicine 11:1139–1151
51. Danhier F, Feron O, Préat V (2010) To exploit the tumor microenvironment: passive and active tumor targeting of nanocarriers for anti-cancer drug delivery. J Control Release 148:135–146
52. Swartz MA, Lund AW (2012) Lymphatic and interstitial flow in the tumour microenvironment linking mechanobiology with immunity. Nat Rev Cancer 12:210–219
53. Larsen MT, Kuhlmann M, Hvam ML, Howard KA (2016) Albumin-based drug delivery: harnessing nature to cure disease. Mol Cell Ther 4:3
54. Silvestris N, Gnoni A, Brunetti AE, Vincenti L, Santini D et al (2014) Target therapies in pancreatic carcinoma. Curr Med Chem 21:948–965
55. Sebak S, Mirzaei M, Malhotra M, Kulamarva A, Prakash S (2010) Human serum albumin nanoparticles as an efficient noscapine drug delivery system for potential use in breast cancer: preparation and in vitro analysis. Int J Nanomedicine 5:525–532
56. Lee JE, Kim MG, Jang YL, Lee MS, Kim NW, Yin Y (2018) Self-assembled PEGylated albumin nanoparticles (SPAN) as a platform for cancer chemotherapy and imaging. Drug Deliv 25:1570–1578
57. Hu CM, Zhang L, Aryal S, Cheung C, Fang RH et al (2011) Erythrocyte membrane-camouflaged polymeric nanoparticles as a biomimetic delivery platform. Proc Natl Acad Sci U S A 108:10980–10985
58. Luk BT, Fang RH, Hu CM, Copp JA, Thamphiwatana S et al (2016) Safe and Immunocompatible Nanocarriers Cloaked in RBC Membranes for Drug Delivery to Treat Solid Tumors. Theranostics 6:1004–1011
59. Younes A, Yasothan U, Kirkpatrick P (2012) Brentuximab vedotin. Nat Rev Drug Discov 11:19–20
60. Tai YT, Mayes PA, Acharya C, Zhong MY, Cea M et al (2014) Novel anti-B-cell maturation antigen antibody-drug conjugate (GSK2857916) selectively induces killing of multiple myeloma. Blood 123:3128–3138
61. Herbst RS, Shin DM (2002) Monoclonal antibodies to target epidermal growth factor receptor-positive tumors: a new paradigm for cancer therapy. Cancer 94:1593–1611
62. Altintas I, Heukers R, van der Meel R, Lacombe M, Amidi M et al (2013) Nanobody-albumin nanoparticles (NANAPs) for the delivery of a multikinase inhibitor 17864 to EGFR overexpressing tumor cells. J Control Release 165:110–118
63. Robak T, Robak P, Smolewski P (2009) TRU-016, a humanized anti-CD37 IgG fusion protein for the potential treatment of B-cell malignancies. Curr Opin Investig Drugs 10:1383–1390
64. Lapalombella R, Yeh YY, Wang L, Ramanunni A, Rafiq S et al (2012) Tetraspanin CD37 directly mediates transduction of survival and apoptotic signals. Cancer Cell 21:694–708
65. Zhao X, Lapalombella R, Joshi T, Cheney C, Gowda A et al (2007) Targeting CD37-positive lymphoid malignancies with a novel engineered small modular immunopharmaceutical. Blood 110:2569–2577

66. Yu B, Mao Y, Yuan Y, Yue C, Wang X et al (2013) Targeted drug delivery and cross-linking induced apoptosis with anti-CD37 based dual-ligand immunoliposomes in B chronic lymphocytic leukemia cells. Biomaterials 34:6185–6193
67. Avoranta ST, Korkeila EA, Syrjänen KJ, Pyrhönen SO, Sundström JT (2012) Lack of CD44 variant 6 expression in rectal cancer invasive front associates with early recurrence. World J Gastroenterol 18:4549–4556
68. Gun BD, Bahadir B, Bektas S, Barut F, Yurdakan G et al (2012) Clinicopathological significance of fascin and CD44v6 expression in endometrioid carcinoma. Diagn Pathol 7:80
69. Yang L, Mao H, Wang YA, Cao Z, Peng X et al (2009) Single chain epidermal growth factor receptor antibody conjugated nanoparticles for in vivo tumor targeting and imaging. Small 5:235–243
70. Qian C, Wang Y, Chen Y, Zeng L, Zhang Q et al (2013) Suppression of pancreatic tumor growth by targeted arsenic delivery with anti-CD44v6 single chain antibody conjugated nanoparticles. Biomaterials 34:6175–6184
71. Zhu L, Dong D, Yu ZL, Zhao YF, Pang DW et al (2017) Folate-engineered microvesicles for enhanced target and synergistic therapy toward breast Cancer. ACS Appl Mater Interfaces 9:5100–5108
72. Zhong Y, Goltsche K, Cheng L, Xie F, Meng F et al (2016) Hyaluronic acid-shelled acid-activatable paclitaxel prodrug micelles effectively target and treat CD44-overexpressing human breast tumor xenografts in vivo. Biomaterials 84:250–261
73. Jain A, Jain A, Garg NK, Tyagi RK, Singh B et al (2015) Surface engineered polymeric nanocarriers mediate the delivery of transferrin-methotrexate conjugates for an improved understanding of brain cancer. Acta Biomater 24:140–151
74. Song S, Liu D, Peng J, Sun Y, Li Z et al (2008) Peptide ligand-mediated liposome distribution and targeting to EGFR expressing tumor in vivo. Int J Pharm 363:155–161
75. Pi J, Jiang J, Cai H, Yang F, Jin H et al (2017) GE11 peptide conjugated selenium nanoparticles for EGFR targeted oridonin delivery to achieve enhanced anticancer efficacy by inhibiting EGFR-mediated PI3K/AKT and Ras/Raf/MEK/ERK pathways. Drug Deliv 24:1549–1564
76. Dagar S, Krishnadas A, Rubinstein I, Blend MJ, Onyüksel H (2003) VIP grafted sterically stabilized liposomes for targeted imaging of breast cancer: in vivo studies. J Control Release 91:123–133
77. Nikanjam M, Blakely EA, Bjornstad KA, Shu X, Budinger TF et al (2007) Synthetic nano-low density lipoprotein as targeted drug delivery vehicle for glioblastoma multiforme. Int J Pharm 328:86–94
78. Yousef S, Alsaab HO, Sau S, Iyer AK (2018) Development of asialoglycoprotein receptor directed nanoparticles for selective delivery of curcumin derivative to hepatocellular carcinoma. Heliyon 4:e01071
79. Yu DH, Lu Q, Xie J, Fang C, Chen HZ (2010) Peptide-conjugated biodegradable nanoparticles as a carrier to target paclitaxel to tumor neovasculature. Biomaterials 31:2278–2292
80. Mai J, Song S, Rui M, Liu D, Ding Q et al (2009) A synthetic peptide mediated active targeting of cisplatin liposomes to Tie2 expressing cells. J Control Release 139:174–181
81. Yan Z, Wang F, Wen Z, Zhan C, Feng L et al (2012) LyP-1-conjugated PEGylated liposomes: a carrier system for targeted therapy of lymphatic metastatic tumor. J Control Release 157:118–125
82. Yan Z, Zhan C, Wen Z, Feng L, Wang F et al (2011) LyP-1-conjugated doxorubicin-loaded liposomes suppress lymphatic metastasis by inhibiting lymph node metastases and destroying tumor lymphatics. Nanotechnology 22:415103
83. Gao J, Xie C, Zhang M, Wei X, Yan Z et al (2016) RGD-modified lipid disks as drug carriers for tumor targeted drug delivery. Nanoscale 8:7209–7216

84. Prakash J, de Jong E, Post E, Gouw AS, Beljaars L et al (2010) A novel approach to deliver anticancer drugs to key cell types in tumors using a PDGF receptor-binding cyclic peptide containing carrier. J Control Release 145:91–101
85. Ying M, Shen Q, Liu Y, Yan Z, Wei X et al (2016) Stabilized Heptapeptide A7R for enhanced multifunctional liposome-based tumor-targeted drug delivery. ACS Appl Mater Interfaces 8:13232–13241
86. Wang J, Lei Y, Xie C, Lu W, Wagner E et al (2014) Retro-inverso CendR peptide-mediated polyethyleneimine for intracranial glioblastoma-targeting gene therapy. Bioconjug Chem 25:414–423
87. Winer I, Wang S, Lee YE, Fan W, Gong Y et al (2010) F3-targeted cisplatin-hydrogel nanoparticles as an effective therapeutic that targets both murine and human ovarian tumor endothelial cells in vivo. Cancer Res 70:8674–8683
88. Gomes-da-Silva LC, Ramalho JS, Pedroso de Lima MC, Simões S et al (2013) Impact of anti-PLK1 siRNA-containing F3-targeted liposomes on the viability of both cancer and endothelial cells. Eur J Pharm Biopharm 85:356–364
89. Lu Y, Low PS (2002) Folate-mediated delivery of macromolecular anticancer therapeutic agents. Adv Drug Deliv Rev 54:675–693
90. Han X, Liu J, Liu M, Xie C, Zhan C et al (2009) 9-NC-loaded folate-conjugated polymer micelles as tumor targeted drug delivery system: preparation and evaluation in vitro. Int J Pharm 372:125–131
91. Leitner DF, Connor JR (1820) Functional roles of transferrin in the brain. Biochim Biophys Acta 2012:393–402
92. Sonali AP, Singh RP, Rajesh CV, Singh S et al (2016) Transferrin receptor-targeted vitamin E TPGS micelles for brain cancer therapy: preparation, characterization and brain distribution in rats. Drug Deliv 23:1788–1798
93. Kim SK, Huang L (2012) Nanoparticle delivery of a peptide targeting EGFR signaling. J Control Release 157:279–286
94. Yuan Y, Paunesku T, Arora H, Ward J, Vogt S et al (2011) Interrogation of EGFR targeted uptake of TiO2 Nanoconjugates by X-ray fluorescence microscopy. AIP Conf Proc 1365:423–426
95. Inai T, Mancuso M, Hashizume H, Baffert F, Haskell A et al (2004) Inhibition of vascular endothelial growth factor (VEGF) signaling in cancer causes loss of endothelial fenestrations, regression of tumor vessels, and appearance of basement membrane ghosts. Am J Pathol 165:35–52
96. Hetian L (2002) A novel peptide isolated from a phage display library inhibits tumor growth and metastasis by blocking the binding of vascular endothelial growth factor to its kinase domain receptor. J Biol Chem 277:43137–43142
97. Rechenmacher F, Neubauer S, Mas-Moruno C, Dorfner PM, Polleux J et al (2013) A molecular toolkit for the functionalization of titanium-based biomaterials that selectively control integrin-mediated cell adhesion. Chemistry 19:9218–9223
98. Bello L, Francolini M, Marthyn P, Zhang J, Carroll RS et al (2001) Alpha(v)beta3 and alpha(v) beta5 integrin expression in glioma periphery. Neurosurgery 49:380–389
99. Chen G, Xie Y, Peltier R, Lei H, Wang P et al (2016) Peptide-decorated gold nanoparticles as functional Nano-capping agent of Mesoporous silica container for targeting drug delivery. ACS Appl Mater Interfaces 8:11204–11209
100. Wang F, Chen L, Zhang R, Chen Z, Zhu L (2014) RGD peptide conjugated liposomal drug delivery system for enhance therapeutic efficacy in treating bone metastasis from prostate cancer. J Control Release 196:222–233
101. Fogal V, Zhang L, Krajewski S, Ruoslahti E (2008) Mitochondrial/cell-surface protein p32/gC1qR as a molecular target in tumor cells and tumor stroma. Cancer Res 68:7210–7218
102. Laakkonen P, Porkka K, Hoffman JA, Ruoslahti E (2002) A tumor-homing peptide with a targeting specificity related to lymphatic vessels. Nat Med 8:751–755

103. Laakkonen P, Akerman ME, Biliran H, Yang M, Ferrer F et al (2004) Antitumor activity of a homing peptide that targets tumor lymphatics and tumor cells. Proc Natl Acad Sci U S A 101:9381–9386
104. Ellington AD, Szostak JW (1990) In vitro selection of RNA molecules that bind specific ligands. Nature 346:818–822
105. Keefe AD, Pai S, Ellington A (2010) Aptamers as therapeutics. Nat Rev Drug Discov 9:537–550
106. Huang F, You M, Chen T, Zhu G, Liang H, Tan W (2014) Self-assembled hybrid nanoparticles for targeted co-delivery of two drugs into cancer cells. Chem Commun 50:3103–3105
107. Cao Z, Tong R, Mishra A, Xu W, Wong GC et al (2009) Reversible cell-specific drug delivery with aptamer-functionalized liposomes. Angew Chem Int Ed Engl 48:6494–6498
108. Kang H, O'Donoghue MB, Liu H, Tan W (2010) A liposome-based nanostructure for aptamer directed delivery. Chem Commun 46:249–251
109. Zhu G, Hu R, Zhao Z, Chen Z, Zhang X et al (2013) Noncanonical self-assembly of multifunctional DNA nanoflowers for biomedical applications. J Am Chem Soc 135:16438–16445
110. Yang L, Tseng YT, Suo G, Chen L, Yu J et al (2015) Photothermal therapeutic response of cancer cells to aptamer-gold nanoparticle-hybridized graphene oxide under NIR illumination. ACS Appl Mater Interfaces 7:5097–5106
111. Huang YF, Shangguan D, Liu H, Phillips JA, Zhang X et al (2009) Molecular assembly of an aptamer-drug conjugate for targeted drug delivery to tumor cells. Chembiochem 10:862–868
112. Wang R, Zhu G, Mei L, Xie Y, Ma H et al (2014) Automated modular synthesis of aptamer-drug conjugates for targeted drug delivery. J Am Chem Soc 136:2731–2734
113. Zhu G, Zheng J, Song E, Donovan M, Zhang K et al (2013) Self-assembled, aptamer-tethered DNA nanotrains for targeted transport of molecular drugs in cancer theranostics. Proc Natl Acad Sci U S A 110:7998–8003
114. Huang BT, Lai WY, Chang YC, Wang JW, Yeh SD et al (2017) A CTLA-4 antagonizing DNA Aptamer with antitumor effect. Mol Ther Nucleic Acids 8:520–528
115. Wengerter BC, Katakowski JA, Rosenberg JM, Park CG, Almo SC et al (2014) Aptamer-targeted antigen delivery. Mol Ther 22:1375–1387
116. Xiao Z, Farokhzad OC (2012) Aptamer-functionalized nanoparticles for medical applications: challenges and opportunities. ACS Nano 6:3670–3676
117. Parodi A, Quattrocchi N, van de Ven AL, Chiappini C, Evangelopoulos M et al (2013) Synthetic nanoparticles functionalized with biomimetic leukocyte membranes possess cell-like functions. Nat Nanotechnol 8:61–68
118. Cao B, Yang M, Zhu Y, Qu X, Mao C (2014) Stem cells loaded with nanoparticles as a drug carrier for in vivo breast cancer therapy. Adv Mater 26:4627–4631
119. Li J, Ai Y, Wang L, Bu P, Sharkey CC et al (2016) Targeted drug delivery to circulating tumor cells via platelet membrane-functionalized particles. Biomaterials 76:52–65
120. Hu Q, Sun W, Qian C, Bomba HN, Xin H et al (2017) Relay drug delivery for amplifying targeting signal and enhancing anticancer efficacy. Adv Mater 29:1605803
121. Pasto A, Giordano F, Evangelopoulos M, Amadori A, Tasciotti E (2019) Cell membrane protein functionalization of nanoparticles as a new tumor-targeting strategy. Clin Transl Med 8:8
122. Palomba R, Parodi A, Evangelopoulos M, Acciardo S, Corbo C et al (2016) Biomimetic carriers mimicking leukocyte plasma membrane to increase tumor vasculature permeability. Sci Rep 6:34422
123. Krishnamurthy S, Gnanasammandhan MK, Xie C, Huang K, Cui MY et al (2016) Monocyte cell membrane-derived nanoghosts for targeted cancer therapy. Nanoscale 8:6981–6985
124. Gao W, Hu CM, Fang RH, Luk BT, Su J et al (2013) Surface functionalization of gold nanoparticles with red blood cell membranes. Adv Mater 25:3549–3553

125. Schumacher D, Strilic B, Sivaraj KK, Wettschureck N, Offermanns S (2013) Platelet-derived nucleotides promote tumor-cell transendothelial migration and metastasis via P2Y2 receptor. Cancer Cell 24:130–137
126. Gay LJ, Felding-Habermann B (2011) Contribution of platelets to tumour metastasis. Nat Rev Cancer 11:123–134
127. Hu Q, Sun W, Qian C, Wang C, Bomba HN et al (2015) Anticancer platelet-mimicking nanovehicles. Adv Mater 27:7043–7050
128. Hu Q, Qian C, Sun W, Wang J, Chen Z et al (2016) Engineered nanoplatelets for enhanced treatment of multiple Myeloma and Thrombus. Adv Mater 28:9573–9580
129. Näkki S, Martinez JO, Evangelopoulos M, Xu W, Lehto VP et al (2017) Chlorin e6 functionalized Theranostic multistage nanovectors transported by stem cells for effective photodynamic therapy. ACS Appl Mater Interfaces 9:23441–23449
130. Timaner M, Letko-Khait N, Kotsofruk R, Benguigui M, Beyar-Katz O et al (2018) Therapy-educated mesenchymal stem cells enrich for tumor-initiating cells. Cancer Res 78:1253–1265
131. Toledano Furman NE, Lupu-Haber Y, Bronshtein T, Kaneti L, Letko N et al (2013) Reconstructed stem cell nanoghosts: a natural tumor targeting platform. Nano Lett 13:3248–3255
132. Fang RH, Hu CM, Luk BT, Gao W, Copp JA et al (2014) Cancer cell membrane-coated nanoparticles for anticancer vaccination and drug delivery. Nano Lett 14:2181–2188
133. Sun H, Su J, Meng Q, Yin Q, Chen L et al (2016) Cancer-cell-biomimetic nanoparticles for targeted therapy of Homotypic tumors. Adv Mater 28:9581–9588
134. Corbo C, Molinaro R, Taraballi F, Toledano Furman NE et al (2017) Unveiling the in vivo protein Corona of circulating leukocyte-like carriers. ACS Nano 11:3262–3273
135. Molinaro R, Corbo C, Martinez JO, Taraballi F, Evangelopoulos M et al (2016) Biomimetic proteolipid vesicles for targeting inflamed tissues. Nat Mater 15:1037–1046
136. Giacomini KM, Huang SM, Tweedie DJ, Benet LZ, Brouwer KLR et al (2010) Membrane transporters in drug development. Nat Rev Drug Discov 9:215–236
137. Wang H, Wang X, Xie C, Zhang M, Ruan H et al (2018) Nanodisk-based glioma-targeted drug delivery enabled by a stable glycopeptide. J Control Release 284:26–38
138. Shao K, Zhang Y, Ding N, Huang S, Wu J et al (2015) Functionalized nanoscale micelles with brain targeting ability and intercellular microenvironment biosensitivity for anti-intracranial infection applications. Adv Healthc Mater 4:291–300
139. Yang W, Cheng Y, Xu T, Wang X, Wen LP (2009) Targeting cancer cells with biotin-dendrimer conjugates. Eur J Med Chem 44:862–868
140. Landowski CP, Vig BS, Song X, Amidon GL (2005) Targeted delivery to PEPT1-overexpressing cells: acidic, basic, and secondary floxuridine amino acid ester prodrugs. Mol Cancer Ther 4:659–667
141. Gallop Mark A (2010) Gemcitabine prodrugs, pharmaceutical compositions and uses thereof. XenoPort, Inc., Santa Clare, pp 1–52
142. Borst P, Evers R, Kool M, Wijnholds J (2000) A family of drug transporters: the multidrug resistance-associated proteins. J Natl Cancer Inst 92:1295–1302
143. Alakhova DY, Kabanov AV (2014) Pluronics and MDR reversal: an update. Mol Pharm 11:2566–2578
144. Hong W, Chen D, Zhang X, Zeng J, Hu H et al (2013) Reversing multidrug resistance by intracellular delivery of Pluronic(R) P85 unimers. Biomaterials 34:9602–9614

Chapter 4
Tumor-Responsive Drug Release Strategies

Zhaoqing Shi, Yun Zhou, and Lin Mei

Abstract Generally, tumor is surgically removed or cured by chemotherapy in clinical treatment. However, due to the rapid proliferation of tumor cells and severe adverse effects, these traditional therapies cannot satisfy the needs of clinical situations. With the development of nanotechnology and deeper understanding of tumor in cellular and molecular levels, novel nanomaterial-based tumor-targeted systems are fabricated, which exhibit considerable effect compared to traditional therapies. In recent years, many versatile tumor-targeted systems that generally consist of carrier and active pharmaceutical ingredient (API) are fabricated to respond unique stimulation of the tumor cell or tumor microenvironment so as to achieve accurate drug release, which is inseparable from the various tumor-responsive drug release strategies. The tumor-responsive strategies are classified as vascular-responsive, physical-responsive strategies (photo-responsive, hyperthermia-responsive, magnetic-responsive, and ultrasonic-responsive), chemical-responsive strategies (pH-responsive, redox-responsive, and reactive oxygen species-responsive), biological-responsive strategies (enzyme-responsive and membrane-responsive), and other responsive strategies. These strategies own distinct mechanisms and work on different aspects in drug release; hence favorable antitumor effects can be obtained via employing suitable strategies and fabricating smarter delivery systems. Moreover, it is found that combination of different synergistic strategies exhibits enhanced therapy effects, which shows promising development in tumor-targeted systems. This chapter will mainly introduce various strategies and discuss their mechanisms, applications, and challenges.

Keywords Tumor-responsive release · Nanomaterial · Drug delivery

Z. Shi · Y. Zhou · L. Mei (✉)
School of Pharmaceutical Sciences (Shenzhen), Sun Yat-sen University, Guangzhou, China
e-mail: meilin7@mail.sysu.edu.cn

R. Huang, Y. Wang (eds.), *New Nanomaterials and Techniques for Tumor-targeted Systems*, https://doi.org/10.1007/978-981-15-5159-8_4

4.1 Introduction

Since the first antitumor molecular nitrogen mustard was found in the 1940s, many chemical molecules and natural products, e. g., cisplatin, paclitaxel (PTX), and doxorubicin (DOX), which possess antitumor activities, had been discovered for chemotherapy. Their discovery has propelled the evolution of tumor therapy. However, as a result of lacking tumor selectivity, most of these antitumor drugs not only attack tumor cells but also attack normal cells, which causes severe side effects and unbearable pains to patients. Moreover, the tumoral killing effect is also limited due to the imperfect properties of chemotherapy drugs, such as poor water solubility and instability, undesirable pharmacological properties, bad pharmacokinetic behaviors, and occurring multidrug resistance [1].

Owing to the revolutionary development in material science and technique, various versatile nanomaterials are employed as carriers to fabricate tumor-responsive drug delivery systems. Their developments provide hope for conquering the challenges of traditional chemotherapy. In the past decade, many novel nanomaterials have been investigated as carriers, such as nanocarbon, mesoporous silica, black phosphorus, metal nanoparticle, lipid, polymer, and dendrimer. Different from the molecular scale of traditional chemotherapy drugs, tumor-responsive drug delivery systems are fabricated on nanoscale. This chapter mainly introduces the developments, mechanisms, and the applications of different types of tumor-responsive drug release strategies.

Gene mutation is the origin of tumor; fortunately, the gene-mutated cells are eliminated by immune system so that a small number of mutations cannot eventually develop into tumor. Due to gene mutation, tumor cells are different from normal cells in cellular characteristics, and the hallmarks of tumor cells are sustaining proliferate signal, enabling replicate immortally, resisting apoptosis, inducing angiogenesis, and evading growth suppressions and metastasis [2].

The unique cell and tissue constitution and growth environment of solid tumor are called tumor microenvironment (TME), and TME is classified into primary TME, invasive TME, and metastatic TME according to their morphology, all of which play important roles in tumor generation, invasion, and metastasis. The TME not only consists of tumor cells, but also contains abnormal vasculature system and other cells such as cancer-associated fibroblasts, endothelial cells, pericytes, cancer stem cells, and even immune inflammatory cells [2]. Therefore, the complexity of TME is even regarded as an organ. The TME is supposed to be a hypoxia, acidic, and hyperthermia abnormal environment because of the rapid metabolism and relevant isolated structure. Thereby, tumor-responsive drug release strategies could be developed through responding the unique TME.

The novel nanomaterials employed in drug delivery systems mainly include inorganic nanomaterials (such as silica, carbon, metal oxide, and black phosphorus), organic nanomaterials (such as polymer and lipid), biomaterials (such as cellular membrane and virus), and other nanomaterials. These nanomaterials serve as carriers in drug delivery systems, which are responsible to various tumor environments to realize accurate drug release.

4.2 Applications of Different Tumor-Responsive Drug Release Strategies

Drug release strategies often involve carrier and drug, in which the carrier plays a role of car and the drug plays a role of bomb. The car (carrier) first reaches the tumor sites, and then the bomb (drug) unloaded from the car will kill the tumor cells. Moreover, as many outstanding materials possess multifunction, the preparation of a multi-responsive carrier becomes feasible, which provides the possibility to respond the complex TME. The cores of these strategies are how the carrier recognizes the tumor and how the drug can be released from the carrier. The responsive strategies are classified as physical-responsive strategies (hyperthermia, light, magnetic, and ultrasonic), chemical-responsive strategies (pH and redox), biological-responsive strategies (enzyme and receptor), and other responsive strategies (EPR effect and NanoEL effect) according to their responsive manners. The following part will discuss the responsive strategies in the unique mechanisms and the property of the nanomaterials. The most commonly used responsive strategies are shown in Fig. 4.1.

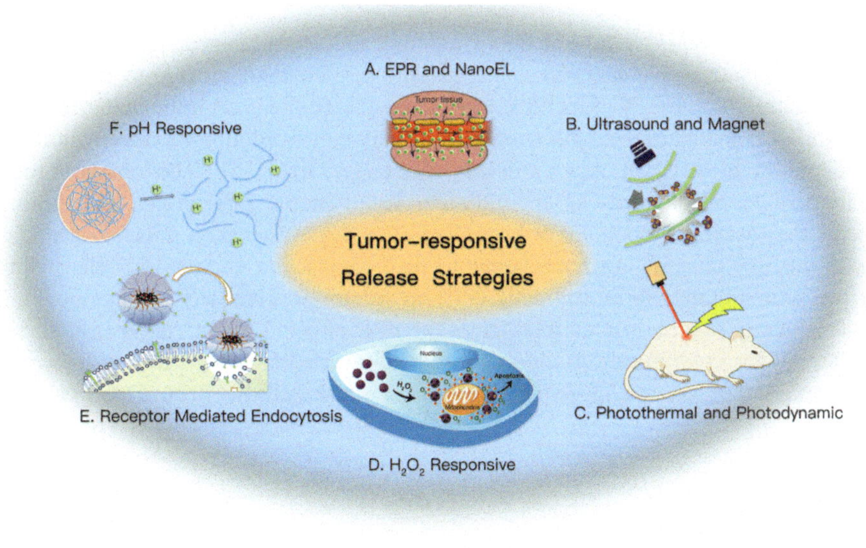

Fig. 4.1 The most commonly used responsive strategies. (**a**) EPR effect and NanoEL effect, the nanoparticles show enhanced accumulation in tumor sites due to highly leakiness vessels [3]; (**b**) ultrasound- and magnet-responsive drug release, nanoparticles can respond to applied ultrasound or magnetic field to release drug [4]; (**c**) Photothermal and photodynamic therapy, the nanoparticles can respond to specific laser and generate heat or ROS to realize antitumor effects; (**d**) H_2O_2-responsive drug release, ROS-sensitive nanoparticles can release drug in the presence of H_2O_2 [5]; (**e**) receptor-mediated endocytosis, receptors can bind to nanoparticles; thus the nanoparticles will enter the cell through receptor-mediated endocytosis pathway [6]; (**f**) pH-responsive drug release, pH-sensitive nanoparticles can respond to H^+ and release drug

4.2.1 EPR Effect and NanoEL Effect of Nanoparticles

The enhanced permeability and retention (EPR) effect was put forward by Matsumura in 1986 [7]; it was the first and most commonly employed tumor-responsive drug release strategy. Although EPR effect is a controversial mechanism now [8, 9], it has made a deep impact on tumor nanomedicine researches. This part will mainly introduce EPR effect.

The EPR effect mainly exists in solid tumor tissues, when the tumoral tissues are newly forming and the supplies of oxygen and nutrition are sufficient. However, the previous vascular vessel that supplies matter will be overloaded when the tumoral tissues are growing larger, which becomes the limitation of tumor proliferation. Hence, in order to obtain more matter and grow larger, the solid tumor cells could secrete factors such as VEGF to stimulate blood vessels and induce the collapse of blood vessels and lymphatic vessels in the occupied space [8, 10, 11]. Therefore, solid tumor tissues possess some characteristics: hypervascularity, irregular flow of blood, higher vascular permeability, and abnormal lymph drainage [1, 12]. Nanoparticles of suitable size are more likely to leak out from the tumor vasculature and accumulate in tumor tissues due to the unique vascular properties (as shown in Fig. 4.2a), which are vasculature leakage and impaired lymphatic drainage [14].

The EPR effect is a kind of passive targeting strategy influenced by many factors, including nanoparticle size and shape, particle surface material, charge, tumor type, and location [15, 16]. Nanoparticle size is considered the most influencing factor in EPR effect; nanoparticles with small size display enhanced tumor extravasation and penetration effect than nanoparticles with larger size [15]. The charge of the nanoparticles also serves as an important factor in the EPR effect. On the one hand, the vascular endothelial wall surfaces carry negative charge; thus nanoparticles with positive charges are more likely to bind to vascular endothelial cells, which cause unfavorable pharmacokinetic (PK) and pharmacokinetic (PD) behaviors and consequently reduce drug enrichment of the EPR effect [12]. On the other hand, the macrophages contributing to the major loss of injected nanoparticles are known as mononuclear phagocyte system (MPS), and it is reported that neutral nanoparticles

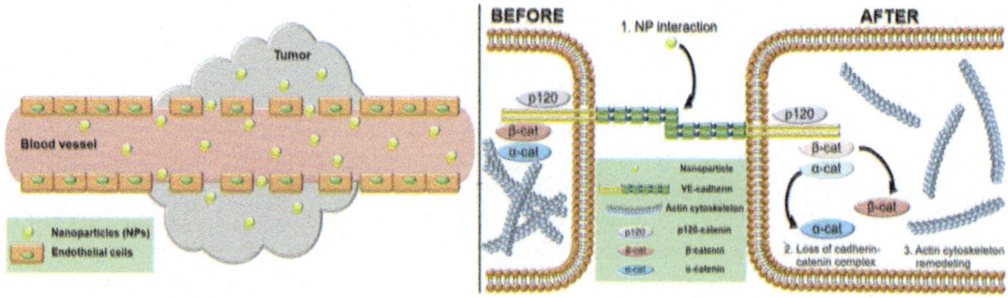

Fig. 4.2 Schematic illustration of (**a**) EPR effect and (**b**) NanoEL effect [13]

exhibit a decreased rate of MPS clearance [17]. Other factors can also influence the EPR effect, although the extent remains irregular.

To utilize the EPR effect, various novel multifunctional materials could be employed as carriers to fabricate nanoparticles with suitable size and shape. For example, nanoliposome, micelle, polymer particle, and mesoporous silica nanoparticle (MSN) are the most extensively investigated spherical structure nanoparticles, while graphene oxide (GO) and black phosphorus (BP) are popular two-dimensional (2D) nanomaterials in recent years. Other multifunctional nanomaterials such as nanotube and quantum dot (QD) have also been employed to develop drug delivery system for EPR effect.

Recently, researchers gradually turn their eyes to enhance the traditional EPR effect, which is generally called enhanced EPR effect. For instance, the EPR effect is enhanced by size-tunable nanoparticle or autonomous nanoparticle. Tan et al. have designed a kind of size-tunable nanoassembly system where the nanoparticle held the particle size of 80 nm under physiological pH (7.4) whereas transformed to much smaller particles (less than 10 nm) under tumoral acidic pH (6.0), which obviously enhanced the EPR effect by varying particle size [18]. Peng et al. have put forward a Janus nanomotor strategy to enhance the mobility of the nanoparticles. Due to the fact that the H_2O_2 level of tumor tissue was higher than that of normal tissue, and the Janus nanoparticle containing a Pt catalysis part which could catalyze H_2O_2 to produce O_2 gas so as to push forward itself [19]. Llopis-Lorente et al. also have fabricated an enzyme-powered MSN nanomotor which utilized urease to catalyze urea to produce CO_2 and NH_3 as pushing forces [20].

Nevertheless, such nano-based strategies exhibited some limitations in practical application. At first, the extent of vascular leakiness is hard to predict as the tumor type and growth condition are different from every patient [12, 21]; thus it is harder to kill the newly forming tumor tissues and metastatic tumor cells because of the limitation of the EPR effect as mentioned before. Secondly, the osmotic pressure in tumor tissues is higher than that of normal tissues and hence prevents the penetration of drugs, which is called tumoral hyper interstitial fluid pressure (IFP) [22]. Besides, the tumor cell density also appears important to the efficiency of the EPR effect, and with the presence of pericyte coverage that surrounds tumor tissues, the central part of tumor tissues might exhibit poor EPR effect [23]. Moreover, the tumor-to-body weight ratio is quite different between mice and human, i.e., the in vivo experimental results of EPR effect on mice could not be directly deduced to human [9].

To step over the limitations mentioned above, by utilizing the phenomenon that some inorganic nanoparticles could induce the vascular leakiness [24, 25], D.T. Leong and his coworkers had put out the concept of nanomaterial-induced endothelial leakiness (NanoEL) effect in 2013 [24].

NanoEL refers to a kind of phenomenon that nanomaterials could destroy the interaction of junctional protein and remodel the cytoskeleton, subsequently causing temporal gaps between vascular barriers that formed by cell-cell contact, which is also called endothelial leakiness [13]. The mechanism of NanoEL effect is shown in Fig. 4.2b. The primary interface between the tumor tissues and blood stream is considered as the vascular endothelial cells, which regulate the matter transportations; hence the nanoparticles encounter endothelial cells prior to underlying tissues [26]. The transportation of nanoparticles through vascular endothelial cells mainly depends on two pathways known as transcellular pathway and paracellular pathway [27], and the latter pathway provides more effective transportation ability for nanoparticles. Nevertheless, the proteins located among the endothelial cells, such as vascular endothelial-cadherin (VE-cadherin), are adherent and tight conjunct proteins which are likely attributed to prevent the nanoparticles transportation progress [28]. VE-cadherin is thought to present in most endothelial cell [28], which is a kind of transmembrane protein and consists of extracellular, transmembrane, and intracellular domain. In this context, through utilizing the characteristics of VE-cadherin [29], the penetration of nanoparticles could be realized by NanoEL effect artificially [13].

Apart from the physiological conditions mentioned before, the intrinsic properties of nanoparticles are essential for the NanoEL effect, such as shape, particle size, surface property, charge, and density [30]. For instance, long nanorods (NRs) are expected to have more vascular contacted part and less blood stream exposed part so as to apply forces on endothelial cells. Both computational model and experimental results show long NRs with smaller size exhibits better NanoEL effect [31]. As the VE-cadherin is positively charged, the negatively charged nanoparticles would create a "bouncing effect" which attracts and bounces the nanoparticles to the VE-cadherin site, while positively charged nanoparticles would encounter electrostatic repulsion [32].

Compared to the EPR effect, the NanoEL effect exhibits similarities in the enhanced penetration of nanoparticles. Nevertheless, they still have distinguished differences naturally. The EPR effect is generated by tumor itself, while the tumor cells need to expand. On the contrary, the NanoEL effect is artificially produced by nanoparticles, while the nanoparticles could cause temporal endothelial leakiness [13]. Moreover, the discovery of NanoEL effect has reversed the situation that the tumor-responsive strategies could only function by unpredictable passive EPR effect.

Various inorganic nanoparticles have been utilized for NanoEL effect as tumor-responsive drug release strategies, like TiO_2NPs, AuNPs, and SiO_2NPs [13].

4.2.2 Hyperthermia and Photothermal-Responsive Drug Release

Hyperthermia refers to local or systemic severe temperature increasing (up to around 43 °C), which can increase vascular permeability, damage tumor cell, induce cellar apoptosis, and even induce immune responses, such as immunogenic cell death (ICD) [33]. The fast proliferation and metabolism of tumor cells are usually thought to possess the characteristic of mild hyperthermia. However, the slight temperature difference between tumor tissues and normal tissues is hard to respond. Moreover, the hyperthermia in tumor tissue appears asymmetric, unstable, and hard to quantify, which becomes a nature obstacle to utilize this character of tumor itself. Therefore, external heat sources are chosen to achieve a hyperthermia environment. Traditionally, hot water sacks were applied to provide the heat. Now, applying near-infrared light (NIR) on photothermal-responsive materials such as thermo-responsive polymer and photothermal agent and applying alternating magnetic field on magnetic-responsive materials are the most widely investigated modern method to generate heat [34, 35]. This section will mainly introduce NIR-mediated photothermal-responsive strategies, and the magnetic-mediated magnetothermal-responsive strategies will be discussed in the latter section.

Photothermal therapy (PTT) refers to administration of photothermal agent, later applying light of specific wavelength locally, and then photothermal agent converting light energy to thermal energy. NIR is usually employed because it is capable of deep penetration (up to 10 cm in tissues) [35]. The advantages of PTT include being noninvasive to physiological structure and controllable of action site and power.

Photothermal agents could be classified as the following types: small molecular photothermal agents, organic photothermal nanoparticles, and inorganic photothermal nanoparticles [36]. Small molecular photothermal agent indocyanine green (ICG) had been found in the last century. However, the PK behaviors and toxicity of ICG were found to be not suitable for clinic therapy [37, 38]. With the rapid development of nanomaterials, gold nanoparticles (AuNPs) gradually aroused attentions of researchers due to its fine properties in 1990s, such as biocompatibility, safety, and inertness both in vivo and in vitro [35, 39]. Subsequently, in 2004, Geim et al. have prepared a kind of thin carbon atom layer called graphene dramatically by using tapes to peel graphite [40], and this remarkable work won the Nobel Prize in 2010. Graphene possesses various excellent properties including ultrabroad surface area and higher photothermal conversion efficiency [36]. More recently, in 2015, Zhang et al. have prepared an ultrasmall black phosphorus (BP) quantum dots (BPQDs), which also presented an excellent photothermal conversion efficiency [41, 42]. BPQDs showed excellent biocompatibility and biodegradation properties [42], which could be safely injected and degraded into nontoxic phosphate by the presence of water and oxygen. Except AuNPs, graphene, and BPQDs, other inorganic materials, such as Pd nanosheets (Pd NSs) [43], copper nanocrystals including CuS and Cu_2Se [44], Ag2S [45], and organic materials such as large π-conjugating polymer polyaniline [46] and polypyrrole [47], could also exhibit photothermal conversion ability. The photothermal agents could form different shapes to fabricate versatile photothermal systems. The shape of photothermal agents is shown in Fig. 4.3.

Photothermal effect is controllable by choosing proper therapy method and suitable nanomaterials. On the one hand, during treatment progress many influencing factors are actually adjustable, such as the dosage of drug [33], power of laser [35], and the treatment time gap after administrating photothermal agents [48], which can be adjusted when facing different kinds of tumor [36]. On the other hand, various photothermal agents could be designed to cater diverse situations, such

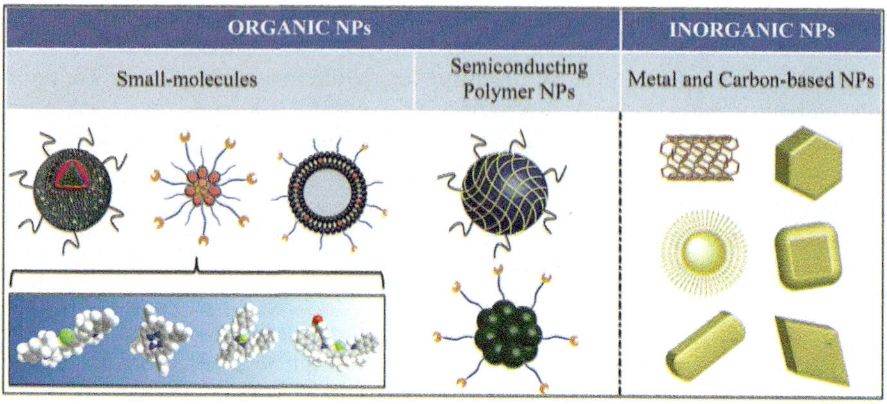

Fig. 4.3 Shape of different photothermal nanoparticles [36]

as designing suitable property of nanoparticle in size, shape, and material [49], developing photothermal agent that could be activated via wavelength of second NIR window range [50], and tuning surface or other chemical properties to accord with specific TME [51].

To utilize photothermal effect and attain hyperthermia-responsive drug release, many researchers had employed the materials discussed above to construct diverse photothermal drug delivery system on nanoscale in recent years. Li et al. have design a kind of PTT triggered hyperthermia-responsive gold nanocages coated via cancer cell membrane, which exhibited excellent antitumor effect on mice in vivo [33]. Zhang et al. have developed graphene oxide (GO)-based PTT antitumor strategies [52]. Mei et al. have fabricated a black phosphorus nanosheets constructed nano delivery platform to combine chemotherapy and PTT, showing excellent photothermal conversion efficiency and antitumor effect [53].

4.2.3 *Photodynamic Responsive Drug Release*

The photodynamic therapy (PDT) had undergone more than 100 years of history, and the first PDT clinical application could date back to 1901. In that time, Niels Finsen discovered that applying light to the smallpox and cutaneous tuberculosis could exert therapy effect, which was of epoch-making significance and won the Nobel Prize in 1903 [54, 55]. In 1961, Richard et al. opened up the modern era for the tumor diagnostic application of PDT, and they showed that hematoporphyrin derivative (HPD) could localized in tumor tissues and emit fluorescence under excitation [55]. In 1972, Diamond et al. employed the localization and phototoxic properties of porphyrins to kill tumor cells [55]. Thenceforth, small molecular photosensitizers had been extensively investigated, and eventually in the 1990s, various PDT drugs had been approved by FDA [56]. More recently, the talent for

PDT of many nanomaterials was found, which showed promising future of PDT [57].

PDT refers to administration of photosensitizer and later applying light of specific wavelength locally and the photosensitizer converting the light energy to photochemical or photobiological antitumor effects, such as generating reactive oxygen species (ROS) [56]. Compared to other strategies, PDT is also a noninvasive treatment method that causes low toxicity and less side effects, and it shows better selectivity under the accurate laser irradiation [42]. Both PTT and PDT are phototherapy, whereas each of therapy is based on different mechanisms.

The mechanism is shown in Fig. 4.4, after the nontoxic photosensitizer is administrated for a certain period of time, applying laser of specific wavelength on accumulated tissues which could excite the photosensitizer from ground state to excited state that could lead to two kinds of reactions (type I reaction and type II reaction). For type I reaction, the excited photosensitizer can react with the substrate directly, such as cell membrane or cellular molecule, and then transfer hydrogen atoms to form radicals, which will then react with oxygen to generate reactive oxygen species (ROS) to exhibit cytotoxicity to tumor cells. For type II reaction, the excited photosensitizer then transfers the absorbed energy to molecular oxygens that exist in tissues or cells and generates ROS [55, 57, 58]. Despite killing cells directly, the ROS can also cause vascular damage that leads to tumor tissue hypoxia. On the other hand, the PDT can trigger inflammatory progress that the inflammatory mediators release from treating part. Moreover, some released inflammatory mediators can in turn stimulate leucocyte and thus induce immune responses [58]. The combination of these mechanisms contributes to the considerable therapeutic effect of PDT.

The photosensitizer could be classified as three generations according to their chemical structures. Developed in the 1970s, the first-generation photosensitizers are porphyrin-based chemical structures, such as porphyrin, hematoporphyrin, and its derivative (HpD). The second-generation photosensitizers are porphyrin-based macrocyclic structures such as chlorins, Ce6, and m-THPC and the pro-drug of porphyrin such as 5-aminolevulinic acid (ALA) and its derivative [54, 57, 59]. Both the first and second generations of photosensitizers had been undergone clinical trials and had even been approved. However, the therapeutic effect of the first- and second-

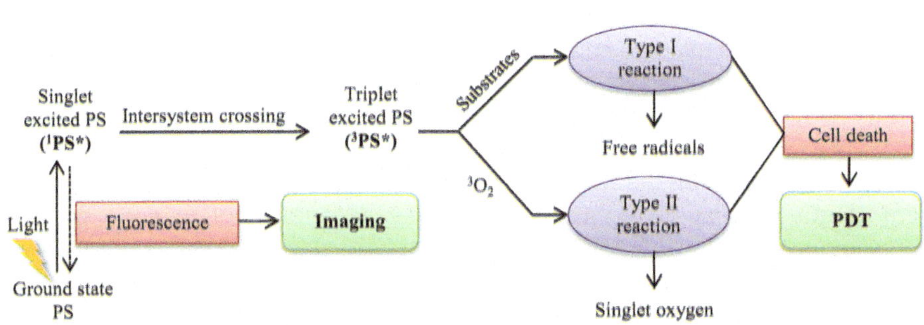

Fig. 4.4 Schematic illustration of photodynamic therapy mechanism [57]

generation photosensitizers could be influenced by various factors, and sometimes their PK behaviors and therapy effect in vivo are not outstanding enough. More importantly, they are more likely to be affected by many unpredictable factors, such as tumor selectivity and tissue penetration depth, which lead to different therapeutic effect between patient [54, 58]. Hence discovering a new generation of photosensitizer that possesses excellent in vivo properties is still an urgent task now.

While meeting with the nanotechnology, the PDT based on nanotechnology, also called the third-generation photosensitizers, shows immeasurable potential in the future [57]. In this context, nanoparticles could serve as carriers, downconverting photosensitizers, and even energy transducers [57, 60]. Nanomaterials such as polymers, two-dimensional materials, and inorganic nanoparticles are ideal carriers for small molecular photosensitizers because they possess many excellent properties. The PDT-based nano drug delivery system exhibits more accurate in tumor-targeted treatment and more effective in multidrug resistant tumor. Furthermore, it is feasible to combine PDT with chemotherapy known as chemophototherapy [56]. Wang et al. have prepared a kind of DTX-loaded star-shaping TAPP (a kind of photosensitizer) polymer nanoparticles, which showed synergistic effects between phototherapy and chemotherapy [61]. On the other hand, some nanomaterials could serve as photosensitizers naturally. Pan et al. have fabricated a kind of novel dual-responsive photosensitizers based on pyrite NSs, which not only could trigger PDT under the irradiation of 650 nm laser but also could trigger PTT under the irradiation of 808 nm laser. Lin et al. had firstly fabricated a biodegradable BP nanosheets Nd-sensitized photosensitizer [62].

4.2.4 Magnetic Localization and Magnetothermal-Responsive Drug Release

Magnetic-responsive strategies mainly include magnetic localization and magnetic heating. Magnetic localization refers to utilizing a magnet to locate magnetic nanoparticles in vivo or in vitro due to the magnetic attraction, which exerts an estimable solid tumor location effect. Magnetic heating refers to applying alternating magnetic field (AMF) on magnetic nanoparticles, subsequently generating heat [63, 64]. Many magnetic localization nanoparticles have performed promising effects in vitro, however, yet a few of them have performed in vitro corresponding effects in vivo [1, 64]. Moreover, magnetic-based nanoparticles show the talent in the nucleic acids delivery, such as siRNA and DNA [64]. Such responsive release strategies are based on the ferrous nanostructures such as core-shell nanoparticles [65] (Fe_3O_4 as the core), magnet liposomes [66], and mesoporous metallic nanocapsules [67].

Compared to magnetic localization applications, researchers are taking more interests in magnetic heating. The mechanism that AMF could stimulate magnetic nanoparticles is, while applying the AMF after administrating, due to the hysteresis loss and Néel relaxation effect, that the nanoparticles transduce the energy to heat, which causes hyperthermia in tumor tissues [64]. The AMF is controllable and free from the limitation of penetration, unlike the PTT mentioned before, and its efficacy is decreasing, while penetration depth is increasing. Henceforth, a kind of magnetic thermosensitive nanoparticles that possess on-off switch ability could be fabricated via coating thermosensitive polymers and lipids [68, 69], which could be used in achieving on-demand drug release wirelessly. In clinical practice, most solid tumors are removed via surgery directly, yet some hemorrhagic and high injury risk tumors are unremovable, such as encephaloma. In these cases, magnetic nanoparticles could exert their unique superiority compared to other strategies [1, 64]. Chen et al. had designed a kind of wireless magnetothermal brain blood barrier (BBB) penetrator, which could achieve on-demand BBB penetration and drug delivery [70]. Xuan et al. had developed a red blood cell membrane cloaked magnetic nanoparticles, which could exert long circulation and enhanced accumulation in tumor tissues [71].

More recently, it was found that magnetic nanoparticles could be employed to design artificial cells to regulate cell functions such as immune responses. Majedi et al. had utilized magnetic nanoparticles to fabricate an artificial antigen presenting cells (aAPCs), augmenting the activation of T-cell via magnetic oscillatory forces [72].

4.2.5 Ultrasound-Responsive Drug Release

Ultrasonic effect could facilitate to deliver drug, release drug, and promote drug penetration; the treatment schematic illustration is shown in Fig. 4.5 [73]. The mechanisms involve direct effect and secondary effect, which have been comprehensively researched. The cavitation phenomenon is a kind of secondary effect, which is thought to be the main effect of ultrasound rather than the direct effect.

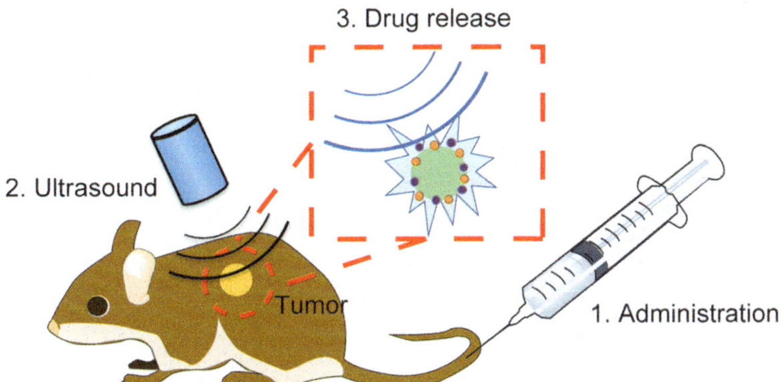

Fig. 4.5 Schematic of drug release through ultrasound: administration and latter apply ultrasound to promote drug release in tumor sites [73]

Besides, ultrasonic effect could inhibit proliferation of the tumor cells, which even induce the apoptosis [74].

Cavitation phenomenon is rapid generation and collapse of bubbles, or the high-frequency oscillation of the generated bubbles, both of which reveal enormous energy. Based on this theory, the bubbles occur cavitation phenomenon through four ways: acoustic streaming that generates shearing stress on tissues and cells, sudden collapse of bubbles that produces heat and free-radical, shock wave that disrupts the tumor tissues and enhances drug release, and liquid microjet that promotes drug penetration. These processes are also called sonodynamic effect that derives from the concept of photodynamic; hence the concepts of sonodynamic therapy (SDT) and sonosensitizers came for the same reason. Interestingly, many first- and second-generation photosensitizers could also serve as sonosensitizers [75]. And ultrasound could get through the joints between capillary endothelial cells temporarily and across BBB, which would applicate in encephaloma therapy [76]. Moreover, the cellular uptake could be enhanced through the ultrasound due to sonoporation phenomenon, that is, ultrasound lead to the temporary pore of membrane [73, 77].

The ultrasonic therapy could be adjusted by controlling external variable parameters, mainly including frequency and power of the ultrasound. Various functions could be realized through varying ultrasound frequencies, for example, frequencies of 25 kHz, 80–180 kHz, and 1–5 MHz are suitable for SDT, across BBB, and hyperthermia therapy, respectively [73].

Ultrasound-responsive drug release strategies have shown potential in the combination therapy; the sonoporation could exert synergistic effects while cooperating with functional nanomaterials. For instance, Paris et al. have fabricated an ultrasonic-responsive MSN which was coated by thermosensitive polymer, and the drug release could be controlled by hyperthermia which generated from ultrasonic stimulation [78]. Liu et al. have designed a cross-linked self-assembled nanoparticles that combined chemotherapy and sonodynamic therapy, which showed synergistic therapeutic effects and inhibited tumor metastasis by activating immune responses [79]. Song et al. have developed a kind of ultrasound responsive nanodroplets to solve the problem of tumoral hypoxia which limited the effect of PDT [80].

4.2.6 pH-Responsive Drug Release

pH-responsive drug release strategies are able to respond the pH changes; such pH-responsive nanoparticles are able to maintain stable at physiological pH (around 7.4) while release drug at acidic pH (around 4–5) [81, 82]. Many reports have demonstrated that tumor tissues could sustain lower extracellular pH_{ex} environment than normal tissues through pH microelectrodes method, while tumor intracellular pH_{in} environment is similar to normal cell [81, 83]. Besides, after the nanocarriers entering cell by endocytosis, the nanoparticles would undergo acidic endosomes that pH low to 4.5 [81]. Both of physiological acidic environments are exploited to design pH-responsive drug delivery system.

The mechanisms of pH-responsive materials mainly include protonation, acid labile chemical bonds cleavage, and charge switch.

Protonation refers to the process that molecules receive protons and hence causes property changing. Some polymers contain ionizable chemical groups, such as poly (acrylic acid) (PAA), poly(methacrylic acid) (PMAA), polydopamine (PDA), and poly(N,N-dimethylamino) ethyl methacrylate (PDMAEMA). These polymers can sustain deionized status at physiological pH whereas transfer to ionized status at acidic pH due to protonation. Therefore, the pH-responsive polymer-constructed systems can occur polymer ionizations in acidic environment, and latter the ionizations of the polymers lead to destabilization, aggregation or dissociation of the systems, which finally realize drug release [84]. The pH-sensitive liposomes contain pH-sensitive cholesteryl hemisuccinate (CHEMS); CHEMS is essential for sustaining stable liposome structure. Hence, when such CHEMS-constructed liposomes encounter the acidic environment, the carboxylic groups on CHEMS could be protonated, and subsequently the structure of liposome would be changed, which leads to drug release [81, 85]. In addition, pH-sensitive polysaccharides and peptides could also be utilized to design protonation mechanism based pH-responsive drug delivery systems. Besides, the materials mentioned above could form nano systems through coating on nanoparticles, or forming nanoparticles alone.

Acid labile chemical bonds cleavage refers to the pH-sensitive bond which breaks in acid environment; such bonds are often designed in pH-responsive drug release. The acid-catalyzed hydrolysis is the rate-determining step that is able to be adjusted by designing suitable chemical structures [81, 86]. The most popular pH-sensitive chemical bonds include hydrazone bond, amide bond, and imine bond; they could serve as linker between the drug molecule and carrier. For instance, the $-NH_2$ on the doxorubicin (Dox) could form imine bond that is sensitive to acidic pH.

Charge switch systems refers to charge switchable nanoparticles could change the negative charge to positive charge with the change of pH. The cells are more likely to endocytose the positive charged nanoparticles beacuse the cellular membrane is negative charged [87]. For instance, PCL copolymers and PMEMA-core micelles are commonly used for charge switchable materials [84].

These mechanisms could be adopted alone or combine to fabricate pH-responsive nanoparticles. Liu et al. have developed a phosphorylcholine-based pH-responsive nanoparticles, which could be utilized as protonation materials and acid labile chemical bonds simultaneously [88]. Zeng et al. have fabricated pH-sensitive PDA-coated MSNs for drug delivery; the coated PDA served as a gatekeeper to guarantee the release of antitumoral drug [89]. The problems are the drug might release in other acidic tissues such as muscle, or the drug releases in advance as a result of pH sensitivity is not accurate enough.

Other materials have also been employed to serve as gatekeeper, for instance, chitosan swelling in the acid environment. Deng et al. have constructed a silica core -based chitosan shell system to treat breast cancer, and the drugs are encapsulated in the silica core that are coated by chitosan shell [90]. The tight shell of chitosan could prevent drug from releasing in blood stream, but after swelling to loose shell in the tumoral acid environment, it allowed the drugs to release from the silica core.

4.2.7 Glutathione-Responsive Drug Release

Countless physiological reduction and oxidation processes are taking place in cell any time; there are variable differences of redox potential that as much as 1000-folds between different sites; hence redox potential responsive drug release strategies exhibit unique capacity in accuracy releasing [91]. The most commonly researched redox systems include glutathione system, oxygen/superoxide system, nicotinamide adenine dinucleotide phosphate (NADPH) system, and thioredoxin systems [92].

Glutathione is a kind of peptide that almost exists in every cell especially in tumor cells; glutathione has GSH form (reductive) and GSSG form (oxidative), both of which make up glutathione system, also called GSH/GSSG system, which is a typical reduction-responsive system. In recent years, the glutathione systems are developed due to the enormous concentration differences, that is, high intracellular GSH concentration (up to 10 mM), whereas low extracellular concentration (low to 1 μM), such concentration difference, provides better selectivity for drug releasing [93]. The mechanisms mainly include reacting with GSH through thiol-disulfide exchange, thiol-thioester exchange, and Michael addition [92]. The process of GSH-responsive is shown in Fig. 4.6: GSH-responsive nanoparticles can enter cell through endocytosis; after escape from lysosome, the nanoparticles can respond to the GSH in the cytosol.

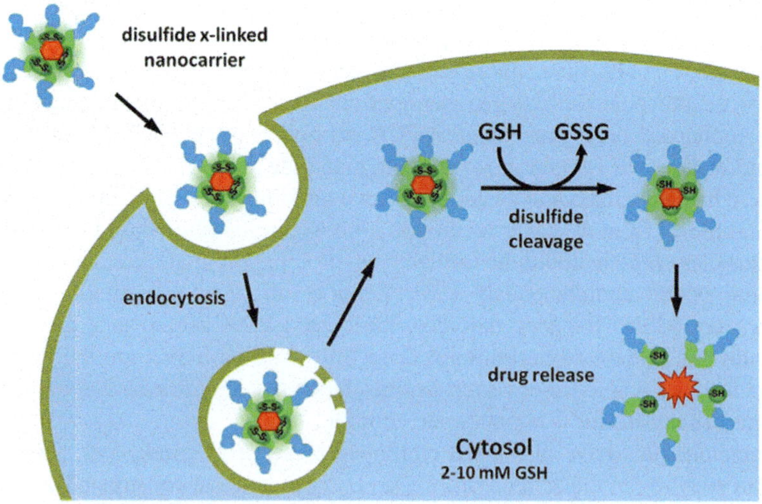

Fig. 4.6 Schematic illustration of GSH-triggered drug release process [94]

Based on these mechanism, many materials possessing glutathione-responsive ability have been designed, such as pyridyl disulfide acrylate (PDSA) [95], copolymer of N,N-dimethylacrylamide (DMA) [96], and copolymer of 2-hydroxyethyl methacrylate (HEMA) [97], respectively.

The glutathione-responsive materials are utilized to fabricate glutathione-responsive system by various methods. Polymers (such as PMAA) are employed to coat on the surface of nanoparticles to fabricate such glutathione-responsive systems through layer-by-layer method, and the coated polymers could dissolve to release drug in the high GSH concentration environment [98]. Besides, glutathione-responsive materials (like disulfide-linked cyclodextrin) could serve as a gatekeeper in porous inorganic materials (like MSNs), which could dissolve and then allow the drug release from the core [99]. Thiol-modified Pluronic could form micelles that encapsulated the drug; the micelles could break down in the presence of GSH [100]. Some drug molecule could even link with carrier by GSH-responsive bond, which also leads to reduction-responsive drug release.

The glutathione-responsive strategies not only exhibit potentials in chemotherapy drugs delivery but also show excellent effect in genetic drugs delivery [92]. Some genetic drugs require to protect the nucleic acid from destroying and releasing in cytosol; and glutathione-responsive system becomes candidates because it has suitable release behaviors that releasing in cytosol. Besides, polymers such as polyetherimide (PEI) are chose to realize genetic drug delivery due to positively charged surface. Kim et al. have fabricated a 2D PEI-PEG-based delivery system, which could deliver gene under the stimulation of GSH [101].

4.2.8 ROS-Responsive Drug Release

Reactive oxygen species (ROS) are extraordinary-activating oxygen ions, which possess oxidative ability mainly includes hydrogen peroxide (H_2O_2), singlet oxygen ($^1O_2$), hydroxyl radical (·OH), and superoxide (O_2^-). ROS are generated from endogenous and exogenous sources; the endogenous ROS includes incomplete reduction reaction of NADPH and oxygen in mitochondria plasma membrane [102], whereas exogenous ROS includes singlet oxygen from the PDT. In physiological environment, the relative low level of ROS modulates the ordinary cellular functions, such as cellular signal and proliferation; however, high ROS level is harmful [103]. Therefore, ROS is not only could be employed as oxidative-responsive strategies but also could attack cell and biomolecule even DNA directly.

Hydrogen peroxide (H_2O_2)-responsive strategy is most commonly employed to design such oxidation-responsive systems, and it is also a form of redox-responsive systems. Many different types of linkers could respond to H_2O_2, such as organic chalcogen linker, arylboronic ester linker, and sulfur-containing linker; hence H_2O_2-responsive systems could be constructed by these linkers [104].

Chalcogen elements are sensitive to oxidation potential, and organic chalcogen linker includes selenium (Se) and tellurium (Te); their mechanisms are shown in

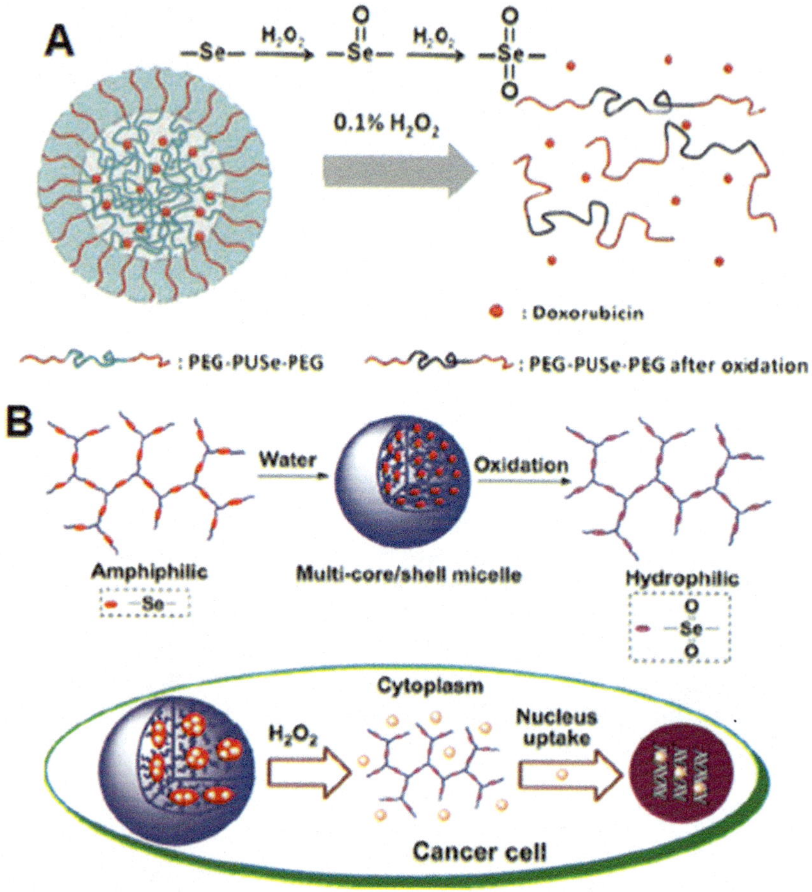

Fig. 4.7 Schematic mechanisms of 2 kinds of Se-containing H_2O_2-responsive drug release systems [104]

Fig. 4.7. The Se-Se bond energy (around 240 KJ/mol) is lower than C-Se (around 170 KJ/mol) bond energy, i.e., Se-Se bond in organic selenium linker is more likely to cleavage under the attacking of H_2O_2 (the -Se- is oxidized to -Se=O and finally oxidized to O=Se=O) [105]. Therefore, polymers that possess ~C-Se-Se-C~ structure are designed to respond high H_2O_2 environment. Organic tellurium (Te) linker could approach the same functions for the similar reason [106]. Cao et al. have synthesized a Te-linked polymer, the polymer self-assembled into micelles that could respond to H_2O_2 and degradation [106].

Arylboronic esters are easily oxidized by H_2O_2 and subsequently generate boronic acid and phenols; besides, the arylboronic esters exhibit excellent binding capacity with various functional groups. Based on these excellent characteristics, arylboronic esters become candidates for H_2O_2-responsive system. The oxidation of arylboronic ester might include two effects: one is the degradation of the skeleton, and the other is the changes of the physical chemistry properties [107]. Su et al. have designed an arylboronic ROS-triggered system which consisted of TPGS-linker-HA structure, and the system realized enhancing ROS level intracellularly, because the TPGS could activate the mitochondrial respiratory complex II that generating ROS [108].

Sulfur-containing linkers mainly include thioether and thioketal; the reaction mechanisms are similar to chalcogen linkers [109]. Thioether such as poly(propylene sulfide) (PPS) possesses reductivity and could transfer to hydrophilic from hydrophobic while encountering H_2O_2, which endows applications in H_2O_2-responsive systems [110]. Thioketal is stable in physiological environment but unstable in the presence of H_2O_2; hence it can sustain stable and preserve drug from degradation in circulation [111]. Liu et al. have developed a thioether-based ROS responsive micelle system for multi-level drug release, i.e., the micelle system firstly released polymer-conjugated drugs in mitochondria, and latter the polymer-conjugated drugs decomposed accurately in the presence of H_2O_2 [112]. Other linkers include aryloxalate and ferrocene, both of which exhibit considerable responsive ability.

Except these linker-based chemical reactions, the nanomaterials might exhibit other physical and chemical changes, including the hydrophobic/hydrophilic balance switch, solubility switch, and shape-changing, which are most commonly employed to prepare micelles and hydrogels. Porous inorganic materials such as MSNs are employed as carrier, and the drugs are linked to carrier via linkers, or employ linked polymers as gatekeepers on the surface of MSNs [104].

4.2.9 Enzyme-Responsive Drug Release

Enzyme is a protein or RNA that can catalyze bioreactions, which distributes almost everywhere in our body. Various enzymes are essential for cell normal functions, such as metabolizing, proliferating, biosynthesizing, and so on. Because of the rapid proliferation of tumor cells and DNA mutation, the enzyme types and levels might different from normal cells, which are utilized for enzyme-responsive drug release. Compared with other strategies (such as acidic pH and redox potential), enzyme-responsive strategy is much more efficacy because the enzyme can decrease activation energy to make the reaction easier to occur. Moreover, the superior selectivity and biospecificity of the enzyme provide considerable applications in accuracy drug delivery [113].

The mechanisms of enzyme-responsive releasing are mainly involved in different kinds, as shown in Fig. 4.8. Drugs could be linked to carriers through enzyme sensitive bond, encapsulated into the frameworks or formed special structure with carriers, which could release drugs directly by the nanostructure disintegrating under the catalyzation of specific enzyme. Besides, gatekeeper strategies could also stand in this stage (Fig. 4.8a, c). Carriers are designed to link with targeting ligand, so as to realize intracellular release (Fig. 4.8b). Moreover, pro-drug strategies are also

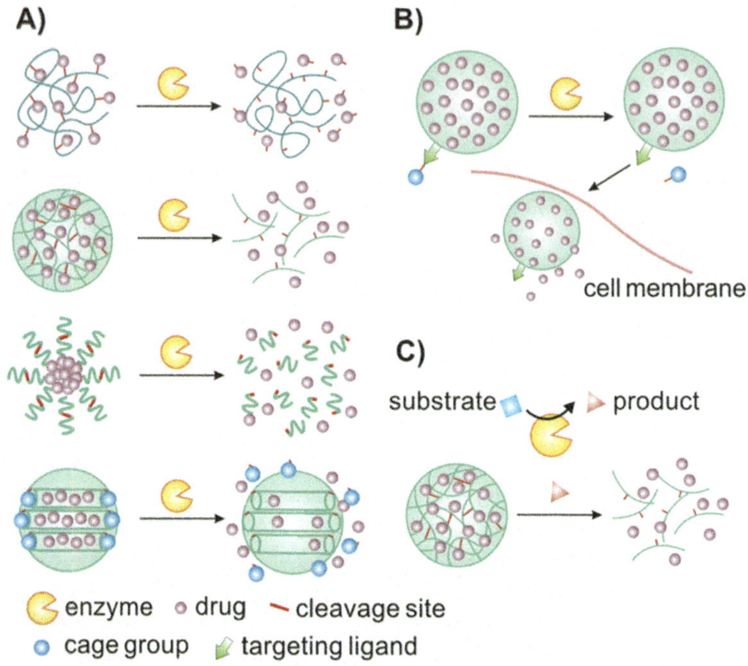

Fig. 4.8 The mechanism of enzyme-responsive drug release [114]

employed to deliver the nontoxic pro-drugs to the tumor cells and subsequently product drugs under the catalyzation of enzyme [114].

The cells handle countless proteins, lipids, and carbohydrates every day especially in tumor cells; therefore, the enzymes adopted as responder are classified to proteases, lipases, and oxidases.

Proteases are kinds of protein hydrolysis enzyme that can recognize specific peptide sequence then hydrolyze. Based on this mechanism, peptides and peptide complex materials that own specific sequence are allowed to fabricate enzyme-responsive system, such as peptide-polymer complex, peptide-drug complex, and many other versatile complexes. Cathepsin B and matrix metalloproteinases (MMPs) are widely researched in these years. Cathepsin B is located in intracellular lysosome and relates to protein degradation; hence enzyme-responsive nanoparticles based on Cathepsin B could be employed only when nanoparticles enter cells via endocytosis pathway [115]. MMP is a series of extracellular zinc-containing endopeptidases and relates to remodeling and degrading the extracellular matrix (ECM); the expression of MMPs is strictly regulated in healthy cells, whereas the regulation is imbalance in tumor cells [116, 117]. Tumor cells highly expressing MMPs are more likely to occur metastasis, because MMPs could hydrolyze ECM extracellularly, and such process provides chances for tumor metastasis. The MMPs-responsive system is one of the most widely researched enzyme-responsive systems, including peptide-linked systems and size-changeable systems. Peptide-linked systems are mainly used to deliver chemotherapy drugs, which release drugs through peptide linker cleavage or structure disassembly in the presence of MMPs. Morphology-changeable systems are able to change their morphology (like size shrinkage, size expansion and morphology transformation) in the presence of enzyme, and the morphology changes often lead to drug release or nanoparticle accumulate and penetrate in tumor [116]. Liu et al. have designed a peptide linker gatekeeper-based MSNs, which released the DOX while meeting with MMP-2 and caused cellular apoptosis [118]. Yao et al. have developed a peptide-PEG surface-linked polymeric micelle, which exhibited long circulation in systemic and robust releasing in MMP-2 highly expressed tumor tissue [119].

Lipases, such as phospholipase A_2 (PLA_2), are upregulated in the presence of danger signals which include inflammation, cardiovascular disease, and tumor [114]; especially in prostate tumor, group IIa $sPLA_2$ is found to 22-fold higher than normal tissues [120]. Liposomes and micelles are wildly utilized to fabricate such system, and the releasing patterns are similar to proteases.

Oxidases include glycosidase and oxidoreductases, carbohydrate nanogels and hydrogels are commonly used in glycosidase responsive system, and oxidoreductase-responsive system is included in redox-responsive system.

Other enzymes could be utilized via similar methodology; more recently, dual-responsive releasing strategies between enzyme-responsive strategies and other strategies are being popularly investigated, which exhibit enhanced therapy effects.

4.2.10 Membrane Receptor-Responsive Drug Release

The nanoparticles must undergo cellular crossing before reaching the intracellular action sites; however, passive strategies which relying solely on Brownian motion are inefficient, no doubt that active strategies which enhancing the combination between membrane and nanoparticles are more effective.

Receptors are variety kinds of cellular structure that located on or crossed in the surface of cellular membrane, which mediates signals between chemical matters and biological effects. In order to enhance the interaction between the cellular membrane and nanoparticles, molecules that can be recognized and linked through receptors are conjugated on the nanoparticles surface, which induces the receptor-mediated endocytosis [121]. Receptor-mediated endocytosis refers to the molecular-conjugated nanoparticles can bind with the specific receptor to induce the formation of pits (as shown in Fig. 4.9). The pits are finally pinched off to form vesicles, which subsequently fuse with endosomes to form early endo/lysomes [122].

Many tumor cells can overexpress different kinds of receptors according to types of mutations. The most commonly studied receptors include folate receptor, epidermal growth factor receptors (EGFR), transferrin receptors, asialoglycoprotein receptor (ASGPR), and scavenger receptor, which are employed to enhance the uptake of nanoparticles by receptor-mediated endocytosis.

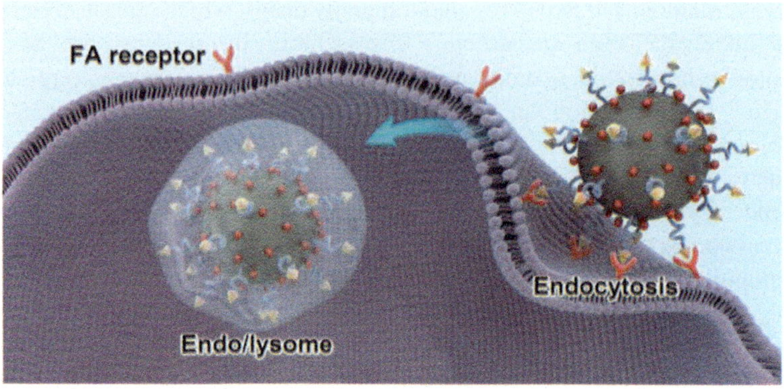

Fig. 4.9 Schematic illustration of the receptor-mediated endocytosis [122]

Folate receptor-based releasing is one of the most exhaustively investigated strategies; many tumor cells overexpress folate receptors due to the vigorous DNA synthesis [123]. Generally, the folate is linked via PEG to the surface of the nanoparticles. Other receptor-responsive strategies are developed in similar way.

More importantly, membrane receptor-responsive strategies possess excellent compatibility, which plays an important role in combination therapy.

4.2.11 *Other Novel Tumor-Responsive Drug Release*

The responsive strategies discussed above have been clarified by researchers for a long time; interestingly, many other novel-responsive drug release strategies are put out recently; these strategies have potential for further investigation and application.

4.2.11.1 Autonomous Movement Strategy

Autonomous movement strategy also called self-propelled strategy refers to the nanoparticles that possess the mobile ability through gas producing or other chemical reactions [124]. In order to catalyze the substrates, a catalyzation unit is decorated on nanoparticles to catalyze "fuels"; fuels such as H_2O_2 decomposition and urea decomposition are employed; the gas generated from fuels is commonly regarded as propelling force. Because of the propelling force, autonomous movement strategy has dramatically enhanced the motility of nanoparticles; however, this strategy exhibits some limitations. For instance, it is hard to control the movement direction because the fuels' concentration gradient is not as uniform as ideally imagined; although the autonomous movement strategies exhibit nice behaviors in vitro, they often fail to realize expected effects in vivo due to the low selectivity between normal cells and tumor cells.

4.2.11.2 Immune-Responsive Strategies

Immune-responsive strategies include in situ vaccine strategy and artificial functional cell strategy. In situ vaccine is based on immunotherapy and nano drug delivery systems, which refers to stimulate immune system to produce robust immune reaction in vivo and subsequently kill the tumor cells by means of activating immune cells and releasing immune factors. The commonly employed immune-responsive strategy is to design nanoparticles containing stimuli factors (like antigen and adjuvant). After being administrated, the nanoparticles could release immune stimuli factors to recruit dendritic cells to capture the specific antigen. The antigen-capturing dendritic cells can travel to lymph nodes and present the antigens to cytotoxic T cells to induce antitumor immune reactions [125]. Artificial functional cell refers to bionic nano systems; these systems possess functions that similar to specific cell to achieve antitumor effect. The most widely researched is artificial antigen present cells (aAPCs), which serve as antigen present cells to enhance antigen present progress in immune reaction. Ye et al. have developed a surgical tumor cell membrane derived photothermal vaccine to induce a robust immunoresponse. In such vaccine system, the tumor cell membrane provided specific antigens for APCs and the photothermal effect improved the immunoresponse. The activated immune systems could solve the problem that surgical unremoved residue tumor tissue might cause tumor recurrence.

4.2.11.3 Self-Assembly In Situ

In situ self-assembly is different from traditional self-assembly, it can responds stimulations and assembly in vivo, i. e., the molecules remain unassembled status before entering tumor tissues or cells, but the self-assembly occurs when the molecules are distributed to tumor sites. The in situ assembly strategy is thought promising to solve the problems that multi-drug resistance, side effect and low selectivity. Yao et al. have synthesized precursors that could be catalyzed by tumor phosphatase and assembled intracellularly, such in situ assembly could selectively activate the prodrug [126].

4.2.11.4 Hypoxia-Responsive Strategy

Tumor tissues are thought to be hypoxia due to the high oxygen consumption rate in tumor cells, which is an obstacle of PDT therapy. Fortunately, this obstacle is conquered through hypoxia-sensitive molecule. Azobenzene (AZO) is widely used in hypoxia-responsive drug release system; the N=N bond is broken into two -NH_2 via accepting four electrons [127]. Therefore, various hypoxia-sensitive polymers are synthesized to encapsulate drugs or carriers to realize hypoxia response. Yan et al. have designed a Poly(4,4′-azodianiline) (pDAB) to encapsulate MSNs for hypoxia-responsive release; the pDAB membrane degradation occurred, while the nanoparticles enter hypoxia area in the solid tumor, which led to model drug release [128].

4.2.11.5 Tumor Signal-Responsive Strategy

Tumor cells could secrete inflammatory mediators that indicate the dangerous signal. For this reason, mesenchymal stem cells (MSCs) are chose as natural cellular carrier due to their unique homing ability which responds the mediators and migrates to tumor cells [129]. Besides, recent researches found that MSCs could across BBB to enter brain. Due to this distinct function, MSCs are employed to uptake the nanoparticles and take the nanoparticles to brain tumor sites, which enhance the targeting efficacy via live cells in a novel way [130].

4.2.12 Combinations of Different Strategies Enhance the Antitumor Effect

With the extended investigations of tumor-responsive drug release strategies, more and more researchers have realized that single therapy not only cannot satisfy the clinical practice but also cannot beat the residue tumor cells. Combination between different types of strategies may obtain synergy effect, and the most employed combination is physical-/chemical-responsive strategy (such as PTT/pH and PTT/ROS) and chemical-/biological (such as pH/enzyme)-responsive strategy. However, it is notable that combination therapy is not simple permutation and combination; all the strategies are supposed to design through suitable mechanism and aim at unique tumor conditions.

4.3 Problems Commonly Encountered in Practice

All kinds of tumor-responsive drug release strategies involve three common progress: administrating into body, targeting to tumor tissues or cells (except intratumoral injection), and killing tumor. Most of tumor-responsive drugs may not exhibit satisfy therapeutic effect, therefore, it is urgent to design drug delivery systems with lower toxicity and better tumor killing effect.

There are some major problems in the tumor-responsive drug release strategies: too complex to industrialize, lack of stability, and unable to replicate in different patients.

Many drug delivery systems require too much prepare progresses; this could lead to nonuniform nanoparticle and finally result in obstacle of industrialization. This problem can be solved through simplifying preparation progresses and improving the yield by appropriate method.

Bio-degradable materials constructed drug delivery systems exhibit good bio-compatibility and decreased toxicity, but some of them are instable. The instability of them might impede their clinical application. Choosing property stable materials or adding stabilize agent in the systems can ameliorate the problem.

There are some reasons why the therapy effect cannot be easily replicate. The whole therapy is tightly relating among patients, clinical practice, and drugs. For patients, the individual differences between patients are confusing for disease diagnosis and treatment. For clinical practice, the administration and therapy operation are supposed to be easy, especially for physical stimulation-responsive strategies such as phototherapy and magnetic therapy. Selecting appropriate time interval between administration and applying physical stimulation (laser or magnetic field) are essential for therapy. Besides, the animal experiments cannot directly extrapolate to human because species differences and xenograft tumor models are different from the actual tumor.

Moreover, the toxicity of the nanomaterials is also an unignorable problem; acute and long-term toxicity experiments should be performed.

As a result, easy to practice, property stable, and effective drugs are urgently needed.

4.4 Conclusions and Perspectives

The past decades have witnessed the rapid development of the tumor-responsive strategies, which benefits from the advances in fundamental science and technology; these strategies show excellent therapy effect and diagnosis ability in the war of tumor. This chapter has discussed the mostly employed tumor-responsive strategies and gives examples of typical nanomaterial-based drug delivery systems. These versatile systems are designed to selectively respond to tumor stimulations and release drug, or combine with other external stimulations. Cases such as the pathological conditions, drug properties, therapy programs, and strategy feasibility should be considered while fabricating the precise drug delivery system; at the same time, we need to recognize the limitation of different strategies and explore novel ways to realize better effect.

Although many revolutionary steps have made in tumor-responsive drug release strategies, there are still some problems waiting to solve. The advanced platforms which emerging in recent years are providing opportunities for researchers to utilize novel strategies; the better strategies will definitely emerge in the future. As Newton said: *Standing on the shoulders of giants.*

References

1. Qiao Y, Wan J, Zhou L, Ma W, Yang Y, Luo W, Yu Z, Wang H (2019) Stimuli-responsive nanotherapeutics for precision drug delivery and cancer therapy. Wiley Interdiscip Rev Nanomed Nanobiotechnol 11(1):e1527
2. Hanahan D, Weinberg RA (2011) Hallmarks of cancer: the next generation. Cell 144 (5):646–674
3. Ojha T, Pathak V, Shi Y, Hennink WE, Moonen CTW, Storm G, Kiessling F, Lammers T (2017) Pharmacological and physical vessel modulation strategies to improve EPR-mediated drug targeting to tumors. Adv Drug Deliv Rev 119:44–60
4. Yin T, Wang P, Li J, Wang Y, Zheng B, Zheng R, Cheng D, Shuai X (2014) Tumor-penetrating codelivery of siRNA and paclitaxel with ultrasound-responsive nanobubbles hetero-assembled from polymeric micelles and liposomes. Biomaterials 35(22):5932–5943
5. Chen H, He W, Guo Z (2014) An H O-responsive nanocarrier for dual-release of platinum anticancer drugs and O: controlled release and enhanced cytotoxicity against cisplatin resistant cancer cells. Chem Commun 50(68):9714–9717
6. Shi C, Guo X, Qianqian Q, Tang Z, Wang Y, Zhou S (2014) Actively targeted delivery of anticancer drug to tumor cells by redox-responsive star-shaped micelles. Biomaterials 35 (30):8711–8722
7. Matsumura Y, Maeda H (1986) A new concept for macromolecular therapeutics in cancer chemotherapy: mechanism of tumoritropic accumulation of proteins and the antitumor agent smancs. Cancer Res 46(12 Part 1):6387–6392
8. Nichols JW, Bae YH (2014) EPR: evidence and fallacy. J Control Release 190:451–464
9. Danhier F (2016) To exploit the tumor microenvironment: since the EPR effect fails in the clinic, what is the future of nanomedicine? J Control Release 244(Pt a):108–121
10. McDonald DM, Baluk P (2002) Significance of blood vessel leakiness in cancer. Cancer Res 62(18):5381–5385
11. Bergers G, Benjamin LE (2003) Tumorigenesis and the angiogenic switch. Nat Rev Cancer 3 (6):401–410
12. Fang J, Nakamura H, Maeda H (2011) The EPR effect: unique features of tumor blood vessels for drug delivery, factors involved, and limitations and augmentation of the effect. Adv Drug Deliv Rev 63(3):136–151
13. Tee JK, Yip LX, Tan ES, Santitewagun S, Prasath A, Ke PC, Ho HK, Leong DT (2019) Nanoparticles' interactions with vasculature in diseases. Chem Soc Rev 48(21):5381–5407
14. Dreher MR, Liu W, Michelich CR, Dewhirst MW, Yuan F, Chilkoti A (2006) Tumor vascular permeability, accumulation, and penetration of macromolecular drug carriers. JNCI: J Natl Cancer Inst 98(5):335–344
15. Cabral H, Matsumoto Y, Mizuno K, Chen Q, Murakami M, Kimura M, Terada Y, Kano MR, Miyazono K, Uesaka M, Nishiyama N, Kataoka K (2011) Accumulation of sub-100 nm polymeric micelles in poorly permeable tumours depends on size. Nat Nanotechnol 6 (12):815–823
16. Perry JL, Reuter KG, Luft JC, Pecot CV, Zamboni W, DeSimone JM (2017) Mediating passive tumor accumulation through particle size, tumor type, and location. Nano Lett 17 (5):2879–2886
17. Li SD, Huang L (2008) Pharmacokinetics and biodistribution of nanoparticles. Mol Pharm 5 (4):496–504
18. Tan J, Li H, Hu X, Abdullah R, Xie S, Zhang L, Zhao M, Luo Q, Li Y, Sun Z, Yuan Q, Tan W (2019) Size-tunable assemblies based on ferrocene-containing DNA polymers for spatially uniform penetration. Chem 5(7):1775–1792
19. Peng F, Men Y, Tu Y, Chen Y, Wilson DA (2018) Nanomotor-based strategy for enhanced penetration across vasculature model. Adv Funct Mater 28(25)
20. Llopis-Lorente A, Garcia-Fernandez A, Murillo-Cremaes N, Hortelao AC, Patino T, Villalonga R, Sancenon F, Martinez-Manez R, Sanchez S (2019) Enzyme-powered gated

mesoporous silica nanomotors for on-command intracellular payload delivery. ACS Nano 13 (10):12171–12183
21. Maeda H (2012) Macromolecular therapeutics in cancer treatment: the EPR effect and beyond. J Control Release 164(2):138–144
22. Jain RK (2003) Molecular regulation of vessel maturation. Nat Med 9(6):685–693
23. Maeda H (2015) Toward a full understanding of the EPR effect in primary and metastatic tumors as well as issues related to its heterogeneity. Adv Drug Deliv Rev 91:3–6
24. Setyawati MI, Tay CY, Chia SL, Goh SL, Fang W, Neo MJ, Chong HC, Tan SM, Loo SC, Ng KW, Xie JP, Ong CN, Tan NS, Leong DT (2013) Titanium dioxide nanomaterials cause endothelial cell leakiness by disrupting the homophilic interaction of VE-cadherin. Nat Commun 4:1673
25. Setyawati MI, Tay CY, Docter D, Stauber RH, Leong DT (2015) Understanding and exploiting nanoparticles' intimacy with the blood vessel and blood. Chem Soc Rev 44 (22):8174–8199
26. Zhang Y, Yang WX (2016) Tight junction between endothelial cells: the interaction between nanoparticles and blood vessels. Beilstein J Nanotechnol 7:675–684
27. Setyawati MI, Tay CY, Bay BH, Leong DT (2017) Gold nanoparticles induced endothelial leakiness depends on particle size and endothelial cell origin. ACS Nano 11(5):5020–5030
28. Mehta D, Malik AB (2006) Signaling mechanisms regulating endothelial permeability. Physiol Rev 86(1):279–367
29. Leckband D, Prakasam A (2006) Mechanism and dynamics of cadherin adhesion. Annu Rev Biomed Eng 8:259–287
30. Setyawati MI, Mochalin VN, Leong DT (2016) Tuning endothelial permeability with functionalized nanodiamonds. ACS Nano 10(1):1170–1181
31. Tan J, Shah S, Thomas A, Ou-Yang HD, Liu Y (2013) The influence of size, shape and vessel geometry on nanoparticle distribution. Microfluid Nanofluidics 14(1–2):77–87
32. Wang J, Zhang L, Peng F, Shi X, Leong DT (2018) Targeting endothelial cell junctions with negatively charged gold nanoparticles. Chem Mater 30(11):3759–3767
33. Sun H, Su J, Meng Q, Yin Q, Chen L, Gu W, Zhang Z, Yu H, Zhang P, Wang S, Li Y (2017) Cancer cell membrane-coated gold nanocages with hyperthermia-triggered drug release and homotypic target inhibit growth and metastasis of breast cancer. Adv Funct Mater 27(3)
34. Ferjaoui Z, Jamal Al Dine E, Kulmukhamedova A, Bezdetnaya L, Soon Chang C, Schneider R, Mutelet F, Mertz D, Begin-Colin S, Quiles F, Gaffet E, Alem H (2019) Doxorubicin-loaded thermoresponsive superparamagnetic nanocarriers for controlled drug delivery and magnetic hyperthermia applications. ACS Appl Mater Interfaces 11 (34):30610–30620
35. Zhang Z, Wang J, Chen C (2013) Near-infrared light-mediated nanoplatforms for cancer thermo-chemotherapy and optical imaging. Adv Mater 25(28):3869–3880
36. Liu Y, Bhattarai P, Dai Z, Chen X (2019) Photothermal therapy and photoacoustic imaging via nanotheranostics in fighting cancer. Chem Soc Rev 48(7):2053–2108
37. Landsman ML, Kwant G, Mook GA, Zijlstra WG (1976) Light-absorbing properties, stability, and spectral stabilization of indocyanine green. J Appl Physiol 40(4):575–583
38. Zheng M, Zhao P, Luo Z, Gong P, Zheng C, Zhang P, Yue C, Gao D, Ma Y, Cai L (2014) Robust ICG theranostic nanoparticles for folate targeted cancer imaging and highly effective photothermal therapy. ACS Appl Mater Interfaces 6(9):6709–6716
39. Tsai MF, Chang SH, Cheng FY, Shanmugam V, Cheng YS, Su CH, Yeh CS (2013) Au nanorod design as light-absorber in the first and second biological near-infrared windows for in vivo photothermal therapy. ACS Nano 7(6):5330–5342
40. Novoselov KS, Geim AK, Morozov SV, Jiang D, Zhang Y, Dubonos SV, Grigorieva IV, Firsov AA (2004) Electric field effect in atomically thin carbon films. Science 306 (5696):666–669
41. Sun Z, Xie H, Tang S, Yu XF, Guo Z, Shao J, Zhang H, Huang H, Wang H, Chu PK (2015) Ultrasmall black phosphorus quantum dots: synthesis and use as photothermal agents. Angew Chem Int Ed Engl 54(39):11526–11530

42. Luo M, Fan T, Zhou Y, Zhang H, Mei L (2019) 2D black phosphorus–based biomedical applications. Adv Funct Mater 29(13)
43. Huang X, Tang S, Mu X, Dai Y, Chen G, Zhou Z, Ruan F, Yang Z, Zheng N (2011) Freestanding palladium nanosheets with plasmonic and catalytic properties. Nat Nanotechnol 6(1):28–32
44. Yang W, Guo W, Le W, Lv G, Zhang F, Shi L, Wang X, Wang J, Wang S, Chang J, Zhang B (2016) Albumin-bioinspired Gd:CuS Nanotheranostic agent for in vivo photoacoustic/magnetic resonance imaging-guided tumor-targeted photothermal therapy. ACS Nano 10 (11):10245–10257
45. Yang T, Tang Y, Liu L, Lv X, Wang Q, Ke H, Deng Y, Yang H, Yang X, Liu G, Zhao Y, Chen H (2017) Size-dependent Ag2S Nanodots for second near-infrared fluorescence/photoacoustics imaging and simultaneous photothermal therapy. ACS Nano 11(2):1848–1857
46. Song X, Gong H, Yin S, Cheng L, Wang C, Li Z, Li Y, Wang X, Liu G, Liu Z (2014) Ultra-small Iron oxide doped polypyrrole nanoparticles for in vivo multimodal imaging guided photothermal therapy. Adv Funct Mater 24(9):1194–1201
47. Zha Z, Yue X, Ren Q, Dai Z (2013) Uniform polypyrrole nanoparticles with high photothermal conversion efficiency for photothermal ablation of cancer cells. Adv Mater 25 (5):777–782
48. Zhu X, Feng W, Chang J, Tan YW, Li J, Chen M, Sun Y, Li F (2016) Temperature-feedback upconversion nanocomposite for accurate photothermal therapy at facile temperature. Nat Commun 7:10437
49. Kinnear C, Moore TL, Rodriguez-Lorenzo L, Rothen-Rutishauser B, Petri-Fink A (2017) Form follows function: nanoparticle shape and its implications for nanomedicine. Chem Rev 117(17):11476–11521
50. Lu X, Yuan P, Zhang W, Wu Q, Wang X, Zhao M, Sun P, Huang W, Fan Q (2018) A highly water-soluble triblock conjugated polymer for in vivo NIR-II imaging and photothermal therapy of cancer. Polym Chem 9(22):3118–3126
51. Albanese A, Tang PS, Chan WC (2012) The effect of nanoparticle size, shape, and surface chemistry on biological systems. Annu Rev Biomed Eng 14:1–16
52. Zhang X, Luo L, Li L, He Y, Cao W, Liu H, Niu K, Gao D (2019) Trimodal synergistic antitumor drug delivery system based on graphene oxide. Nanomedicine 15(1):142–152
53. Tao W, Zhu X, Yu X, Zeng X, Xiao Q, Zhang X, Ji X, Wang X, Shi J, Zhang H, Mei L (2017) Black phosphorus nanosheets as a robust delivery platform for cancer theranostics. Adv Mater 29(1)
54. Agostinis P, Berg K, Cengel KA, Foster TH, Girotti AW, Gollnick SO, Hahn SM, Hamblin MR, Juzeniene A, Kessel D, Korbelik M, Moan J, Mroz P, Nowis D, Piette J, Wilson BC, Golab J (2011) Photodynamic therapy of cancer: an update. CA Cancer J Clin 61(4):250–281
55. Dolmans DE, Fukumura D, Jain RK (2003) Photodynamic therapy for cancer. Nat Rev Cancer 3(5):380–387
56. Luo D, Carter KA, Miranda D, Lovell JF (2017) Chemophototherapy: an emerging treatment option for solid tumors. Adv Sci (Weinh) 4(1):1600106
57. Lucky SS, Soo KC, Zhang Y (2015) Nanoparticles in photodynamic therapy. Chem Rev 115 (4):1990–2042
58. Allison RR, Moghissi K (2013) Photodynamic therapy (PDT): PDT mechanisms. Clin Endosc 46(1):24–29
59. Wachowska M, Muchowicz A, Firczuk M, Gabrysiak M, Winiarska M, Wańczyk M, Bojarczuk K, Golab J (2011) Aminolevulinic acid (ALA) as a prodrug in photodynamic therapy of cancer. Molecules 16(5):4140–4164
60. Pan C, Ou M, Cheng Q, Zhou Y, Yu Y, Li Z, Zhang F, Xia D, Mei L, Ji X (2019) Z-scheme heterojunction functionalized pyrite nanosheets for modulating tumor microenvironment and strengthening photo/Chemodynamic therapeutic effects. Adv Funct Mater 30:3
61. Wang T, Zhu D, Liu G, Tao W, Cao W, Zhang L, Wang L, Chen H, Mei L, Huang L, Zeng X (2015) DTX-loaded star-shaped TAPP-PLA-b-TPGS nanoparticles for cancer chemical and photodynamic combination therapy. RSC Adv 5(62):50617–50627

62. Lv R, Yang D, Yang P, Xu J, He F, Gai S, Li C, Dai Y, Yang G, Lin J (2016) Integration of Upconversion nanoparticles and ultrathin black phosphorus for efficient photodynamic theranostics under 808 nm near-infrared light irradiation. Chem Mater 28(13):4724–4734
63. Liu T-Y, Hu S-H, Liu D-M, Chen S-Y, Chen IW (2009) Biomedical nanoparticle carriers with combined thermal and magnetic responses. Nano Today 4(1):52–65
64. Mura S, Nicolas J, Couvreur P (2013) Stimuli-responsive nanocarriers for drug delivery. Nat Mater 12(11):991–1003
65. Zhang L, Wang T, Yang L, Liu C, Wang C, Liu H, Wang YA, Su Z (2012) General route to multifunctional uniform yolk/mesoporous silica shell nanocapsules: a platform for simultaneous cancer-targeted imaging and magnetically guided drug delivery. Chemistry 18 (39):12512–12521
66. Plassat V, Wilhelm C, Marsaud V, Ménager C, Gazeau F, Renoir J-M, Lesieur S (2011) Anti-estrogen-loaded superparamagnetic liposomes for intracellular magnetic targeting and treatment of breast cancer tumors. Adv Funct Mater 21(1):83–92
67. Zhang F, Braun GB, Pallaoro A, Zhang Y, Shi Y, Cui D, Moskovits M, Zhao D, Stucky GD (2012) Mesoporous multifunctional upconversion luminescent and magnetic "nanorattle" materials for targeted chemotherapy. Nano Lett 12(1):61–67
68. Huang H-Y, Hu S-H, Chian C-S, Chen S-Y, Lai H-Y, Chen Y-Y (2012) Self-assembling PVA-F127 thermosensitive nanocarriers with highly sensitive magnetically-triggered drug release for epilepsy therapy in vivo. J Mater Chem 22(17):8566–8573
69. Katagiri K, Imai Y, Koumoto K, Kaiden T, Kono K, Aoshima S (2011) Magnetoresponsive on-demand release of hybrid liposomes formed from Fe3 O4 nanoparticles and thermosensitive block copolymers. Small 7(12):1683–1689
70. Chen R, Romero G, Christiansen MG, Mohr A, Anikeeva P (2015) Wireless magnetothermal deep brain stimulation. Science 347(6229):1477–1480
71. Xuan M, Shao J, Zhao J, Li Q, Dai L, Li J (2018) Magnetic mesoporous silica nanoparticles cloaked by red blood cell membranes: applications in cancer therapy. Angew Chem Int Ed Engl 57(21):6049–6053
72. Majedi FS, Hasani-Sadrabadi MM, Thauland TJ, Li S, Bouchard LS, Butte MJ (2019) Augmentation of T-cell activation by oscillatory forces and engineered antigen-presenting cells. Nano Lett 19(10):6945–6954
73. Mitragotri S (2005) Healing sound: the use of ultrasound in drug delivery and other therapeutic applications. Nat Rev Drug Discov 4(3):255–260
74. Bai W-K, Shen E, Hu B (2012) Induction of the apoptosis of cancer cell by sonodynamic therapy: a review. Chin J Cancer Res 24(4):368–373
75. Chen H, Zhou X, Gao Y, Zheng B, Tang F, Huang J (2014) Recent progress in development of new sonosensitizers for sonodynamic cancer therapy. Drug Discov Today 19(4):502–509
76. Aryal M, Arvanitis CD, Alexander PM, McDannold N (2014) Ultrasound-mediated blood-brain barrier disruption for targeted drug delivery in the central nervous system. Adv Drug Deliv Rev 72:94–109
77. Wood AK, Sehgal CM (2015) A review of low-intensity ultrasound for cancer therapy. Ultrasound Med Biol 41(4):905–928
78. Paris JL, Cabanas MV, Manzano M, Vallet-Regi M (2015) Polymer-grafted mesoporous silica nanoparticles as ultrasound-responsive drug carriers. ACS Nano 9(11):11023–11033
79. Liu M, Khan AR, Ji J, Lin G, Zhao X, Zhai G (2018) Crosslinked self-assembled nanoparticles for chemo-sonodynamic combination therapy favoring antitumor, antimetastasis management and immune responses. J Control Release 290:150–164
80. Song X, Feng L, Liang C, Yang K, Liu Z (2016) Ultrasound triggered tumor oxygenation with oxygen-shuttle nanoperfluorocarbon to overcome hypoxia-associated resistance in cancer therapies. Nano Lett 16(10):6145–6153
81. Kanamala M, Wilson WR, Yang M, Palmer BD, Wu Z (2016) Mechanisms and biomaterials in pH-responsive tumour targeted drug delivery: a review. Biomaterials 85:152–167

82. Shi Z, Li Q, Lin M (2020) pH-Sensitive nanoscale materials as robust drug delivery systems for cancer therapy. Chin Chem Lett
83. Wike-Hooley JL, Haveman J, Reinhold HS (1984) The relevance of tumour pH to the treatment of malignant disease. Radiother Oncol 2(4):343–366
84. Kocak G, Tuncer C, Bütün V (2017) pH-responsive polymers. Polym Chem 8(1):144–176
85. Jiang T, Zhang Z, Zhang Y, Lv H, Zhou J, Li C, Hou L, Zhang Q (2012) Dual-functional liposomes based on pH-responsive cell-penetrating peptide and hyaluronic acid for tumor-targeted anticancer drug delivery. Biomaterials 33(36):9246–9258
86. Tang R, Ji W, Panus D, Palumbo RN, Wang C (2011) Block copolymer micelles with acid-labile ortho ester side-chains: synthesis, characterization, and enhanced drug delivery to human glioma cells. J Control Release 151(1):18–27
87. Yuan YY, Mao CQ, Du XJ, Du JZ, Wang F, Wang J (2012) Surface charge switchable nanoparticles based on zwitterionic polymer for enhanced drug delivery to tumor. Adv Mater 24(40):5476–5480
88. Liu G, Tsai HI, Zeng X, Zuo Y, Tao W, Han J, Mei L (2017) Phosphorylcholine-based stealthy nanocapsules enabling tumor microenvironment-responsive doxorubicin release for tumor suppression. Theranostics 7(5):1192–1203
89. Cheng W, Nie J, Xu L, Liang C, Peng Y, Liu G, Wang T, Mei L, Huang L, Zeng X (2017) pH-sensitive delivery vehicle based on folic acid-conjugated polydopamine-modified mesoporous silica nanoparticles for targeted cancer therapy. ACS Appl Mater Interfaces 9 (22):18462–18473
90. Deng Z, Zhen Z, Hu X, Wu S, Xu Z, Chu PK (2011) Hollow chitosan-silica nanospheres as pH-sensitive targeted delivery carriers in breast cancer therapy. Biomaterials 32 (21):4976–4986
91. Huo M, Yuan J, Tao L, Wei Y (2014) Redox-responsive polymers for drug delivery: from molecular design to applications. Polym Chem 5(5):1519–1528
92. Quinn JF, Whittaker MR, Davis TP (2017) Glutathione responsive polymers and their application in drug delivery systems. Polym Chem 8(1):97–126
93. Smith CV, Jones DP, Guenthner TM, Lash LH, Lauterburg BH (1996) Compartmentation of glutathione: implications for the study of toxicity and disease. Toxicol Appl Pharmacol 140 (1):1–12
94. Fleige E, Quadir MA, Haag R (2012) Stimuli-responsive polymeric nanocarriers for the controlled transport of active compounds: concepts and applications. Adv Drug Deliv Rev 64(9):866–884
95. Bulmus V, Woodward M, Lin L, Murthy N, Stayton P, Hoffman A (2003) A new pH-responsive and glutathione-reactive, endosomal membrane-disruptive polymeric carrier for intracellular delivery of biomolecular drugs. J Control Release 93(2):105–120
96. Liu L, Wu L, Tan J, Wang L, Liu Q, Liu P, Liu L (2015) "Reduction" responsive thymine-conjugated biodynamers: synthesis and solution properties. Polym Chem 6(21):3934–3941
97. Wang X, Liu L, Luo Y, Shi H, Li J, Zhao H (2012) Comb-shaped glycopolymer/peptide bioconjugates by combination of RAFT polymerization and thiol-ene "click" chemistry. Macromol Biosci 12(11):1575–1582
98. Zelikin AN, Quinn JF, Caruso F (2006) Disulfide cross-linked polymer capsules: en route to biodeconstructible systems. Biomacromolecules 7(1):27–30
99. Kim H, Kim S, Park C, Lee H, Park HJ, Kim C (2010) Glutathione-induced intracellular release of guests from mesoporous silica nanocontainers with cyclodextrin gatekeepers. Adv Mater 22(38):4280–4283
100. Tao Y, Han J, Ye C, Thomas T, Dou H (2012) Reduction-responsive gold-nanoparticle-conjugated Pluronic micelles: an effective anti-cancer drug delivery system. J Mater Chem 22(36):18864–18871
101. Kim J, Kim H, Kim WJ (2016) Single-layered MoS2-PEI-PEG nanocomposite-mediated gene delivery controlled by photo and redox stimuli. Small 12(9):1184–1192

102. Xu Q, He C, Xiao C, Chen X (2016) Reactive oxygen species (ROS) responsive polymers for biomedical applications. Macromol Biosci 16(5):635–646
103. Brieger K, Schiavone S, Miller FJ Jr, Krause KH (2012) Reactive oxygen species: from health to disease. Swiss Med Wkly 142:w13659
104. Saravanakumar G, Kim J, Kim WJ (2017) Reactive-oxygen-species-responsive drug delivery systems: promises and challenges. Adv Sci (Weinh) 4(1):1600124
105. Ma N, Li Y, Xu H, Wang Z, Zhang X (2010) Dual redox responsive assemblies formed from diselenide block copolymers. J Am Chem Soc 132(2):442–443
106. Cao W, Gu Y, Li T, Xu H (2015) Ultra-sensitive ROS-responsive tellurium-containing polymers. Chem Commun (Camb) 51(32):7069–7071
107. Broaders KE, Grandhe S, Frechet JM (2011) A biocompatible oxidation-triggered carrier polymer with potential in therapeutics. J Am Chem Soc 133(4):756–758
108. Su Z, Chen M, Xiao Y, Sun M, Zong L, Asghar S, Dong M, Li H, Ping Q, Zhang C (2014) ROS-triggered and regenerating anticancer nanosystem: an effective strategy to subdue tumor's multidrug resistance. J Control Release 196:370–383
109. Ma N, Li Y, Ren H, Xu H, Li Z, Zhang X (2010) Selenium-containing block copolymers and their oxidation-responsive aggregates. Polym Chem 1(10):1609
110. Allen BL, Johnson JD, Walker JP (2011) Encapsulation and enzyme-mediated release of molecular cargo in polysulfide nanoparticles. ACS Nano 5(6):5263–5272
111. Kim JS, Jo SD, Seah GL, Kim I, Nam YS (2015) ROS-induced biodegradable polythioketal nanoparticles for intracellular delivery of anti-cancer therapeutics. J Ind Eng Chem 21:1137–1142
112. Liu B, Wang D, Liu Y, Zhang Q, Meng L, Chi H, Shi J, Li G, Li J, Zhu X (2015) Hydrogen peroxide-responsive anticancer hyperbranched polymer micelles for enhanced cell apoptosis. Polym Chem 6(18):3460–3471
113. Hu J, Zhang G, Liu S (2012) Enzyme-responsive polymeric assemblies, nanoparticles and hydrogels. Chem Soc Rev 41(18):5933–5949
114. Hu Q, Katti PS, Gu Z (2014) Enzyme-responsive nanomaterials for controlled drug delivery. Nanoscale 6(21):12273–12286
115. Xu Y, Geng J, An P, Xu Y, Huang J, Lu W, Liu S, Yu J (2015) Cathepsin B-sensitive cholesteryl hemisuccinate–gemcitabine prodrug nanoparticles: enhanced cellular uptake and intracellular drug controlled release. RSC Adv 5(9):6985–6992
116. Yao Q, Kou L, Tu Y, Zhu L (2018) MMP-responsive 'Smart' drug delivery and tumor targeting. Trends Pharmacol Sci 39(8):766–781
117. Nagase H, Visse R, Murphy G (2006) Structure and function of matrix metalloproteinases and TIMPs. Cardiovasc Res 69(3):562–573
118. Liu J, Zhang B, Luo Z, Ding X, Li J, Dai L, Zhou J, Zhao X, Ye J, Cai K (2015) Enzyme responsive mesoporous silica nanoparticles for targeted tumor therapy in vitro and in vivo. Nanoscale 7(8):3614–3626
119. Yao Q, Choi JH, Dai Z, Wang J, Kim D, Tang X, Zhu L (2017) Improving tumor specificity and anticancer activity of dasatinib by dual-targeted polymeric micelles. ACS Appl Mater Interfaces 9(42):36642–36654
120. Graff JR, Konicek BW, Deddens JA, Chedid M, Hurst BM, Colligan B, Neubauer BL, Carter HW, Carter JH (2001) Expression of group IIa secretory phospholipase A2 increases with prostate tumor grade. Clin Cancer Res 7(12):3857–3861
121. Sorkin A, Von Zastrow M (2002) Signal transduction and endocytosis: close encounters of many kinds. Nat Rev Mol Cell Biol 3(8):600–614
122. Luo M, Cheng W, Zeng X, Lin M, Liu G, Deng W (2019) Folic acid-functionalized black phosphorus quantum dots for targeted chemo-photothermal combination cancer therapy. Pharmaceutics 11(5):242
123. Scomparin A, Salmaso S, Eldar-Boock A, Ben-Shushan D, Ferber S, Tiram G, Shmeeda H, Landa-Rouben N, Leor J, Caliceti P, Gabizon A, Satchi-Fainaro R (2015) A comparative study

of folate receptor-targeted doxorubicin delivery systems: dosing regimens and therapeutic index. J Control Release 208:106–120
124. Ismagilov RF, Schwartz A, Bowden N, Whitesides GM (2002) Autonomous movement and self-assembly. Angew Chem Int Ed 41(4):652–654
125. Ye X, Liang X, Chen Q, Miao Q, Chen X, Zhang X, Mei L (2019) Surgical tumor-derived personalized photothermal vaccine formulation for cancer immunotherapy. ACS Nano 13 (3):2956–2968
126. Yao Q, Lin F, Fan X, Wang Y, Liu Y, Liu Z, Jiang X, Chen PR, Gao Y (2018) Synergistic enzymatic and bioorthogonal reactions for selective prodrug activation in living systems. Nat Commun 9(1):5032
127. Kiyose K, Hanaoka K, Oushiki D, Nakamura T, Kajimura M, Suematsu M, Nishimatsu H, Yamane T, Terai T, Hirata Y, Nagano T (2010) Hypoxia-sensitive fluorescent probes for in vivo real-time fluorescence imaging of acute ischemia. J Am Chem Soc 132 (45):15846–15848
128. Yan Q, Guo X, Huang X, Meng X, Liu F, Dai P, Wang Z, Zhao Y (2019) Gated mesoporous silica nanocarriers for hypoxia-responsive cargo release. ACS Appl Mater Interfaces 11 (27):24377–24385
129. Li M, Zhang F, Chen K, Wang C, Su Y, Liu Y, Zhou J, Wang W (2016) Nanoparticles and mesenchymal stem cells: a win-win alliance for anticancer drug delivery. RSC Adv 6 (43):36910–36922
130. Wang X, Chen H, Zeng X, Guo W, Jin Y, Wang S, Tian R, Han Y, Guo L, Han J, Wu Y, Mei L (2019) Efficient lung cancer-targeted drug delivery via a nanoparticle/MSC system. Acta Pharm Sin B 9(1):167–176

Chapter 5
Tumor Diagnosis Patterns

Xinwei Li and Cong Li

Abstract Cancer has been one of the human's nightmares for quite a long time which scientists and doctors are longing to defeat. Early diagnosis of tumor greatly helps patients with getting aware of their conditions as well as improving their chance of recovery. In this process, the participation of imaging techniques is urgently needed to realize tumor visualization. Various imaging modalities have been developed over the past decades, either serving the patients in clinical practice or offering references and inspirations with great value in preclinical studies. Here, we first introduce the history of the development of the imaging techniques. Then, each of their typical features and applications is compared and discussed. Recent advances involve the combination of both functional and anatomical information provided by several imaging modalities at the same time, namely, multimodal imaging, among which some of the working pairs such as PET/CT and PET/MRI are introduced with more details. In spite of the evolution of techniques and machines, the development of nanosystems also benefits the imaging results of tumor diagnosis along with more functions including image-guided tumor surgery, drug delivery, and therapeutic evaluation of tumor treatments, which will definitely keep promoting molecular imaging for tumor diagnosis.

Keywords Tumor diagnosis · Imaging patterns · Multimodal imaging · Imaging probes · Nanosystems

5.1 Introduction

Cancer is among the leading causes of death for human worldwide, and the situation is getting worse partially due to the aging population and the adoption of new unhealthy lifestyles. According to the newest reports, about 3,804,000 new cancer

X. Li · C. Li (✉)
Key Laboratory of Smart Drug Delivery, Ministry of Education, School of Pharmacy, Fudan University, Shanghai, China
e-mail: congli@fudan.edu.cn

R. Huang, Y. Wang (eds.), *New Nanomaterials and Techniques for Tumor-targeted Systems*, https://doi.org/10.1007/978-981-15-5159-8_5

cases were diagnosed in China in 2014, while the crude incidence rate was 278.07/ 100,000 [1]. And there were 9.6 million deaths from cancer around the world in 2018 recorded by the International Agency for Research on Cancer [2]. Horrible statistics are distinctly reminding people for preventing the tragedy as much as possible. During this process, early detection of tumor plays an important role as it is capable of saving countless lives by greatly increasing the chances of successful treatment. The results may go beyond our imagination. More than 90% of women with breast cancer can survive their disease for at least 5 years if they can be diagnosed at the earliest stage. Over 80% of lung cancer patients may survive for at least a year if the disease is confirmed in the earliest stage, while the number drops to only 15% when diagnosed in the most advanced stage. A primary tumor is the name of the cancerous mass at certain site in the body where a cancer starts, and the early stage it stays is essential for medical treatment as some cancers can sometimes proceed and spread to other parts of the body through blood circulation or lymphatic system which is called metastasis [3]. More complicated therapy and pain come to the patients when this happens, usually followed by poor prognosis and higher death rate. As a result, techniques for tumor diagnosis that are capable of helping people get aware of their conditions at an early stage were urgently needed. Among them, molecular imaging has risen to be an indispensable tool in clinical cancer detection and treatment in a rapid pace for its main advantage is the ability to characterize the tumor tissue noninvasively instead of traditional biopsies or surgical procedures.

Interestingly but not surprisingly, the history of the medical imaging development, especially molecular imaging, could almost be retold through the winning order of the famous Nobel Prize (Fig. 5.1). The first highlighted event which has to be mentioned in this path should be the discovery of X-radiation which was named by the German physician Wilhelm Röntgen in 1895. Since then, X-ray became a powerful tool for physical experiments and examinations of the body's interior. Wilhelm Conrad Röntgen thus won the Nobel Prize in Physics 1901 in recognition of his extraordinary services by the discovery of the remarkable rays subsequently named after him. This discovery not only stimulated a veritable sensation but also promoted the research of radiance which caused far-reaching influence to the medical imaging. A number of scientists continued working on X-rays and won the Nobel Prizes as well, including the Nobel Prize in Physics 1914 and 1915 [4]. The next award which was associated with imaging techniques was the Nobel Prize in Physics 1951. The pioneer work of the winners, John Cockcroft and Ernest T.S. Walton, on the transmutation of atomic nuclei by artificially accelerated atomic particles was a sound foundation for the development of positron emission tomography (PET). With the theoretical basis, innovative imaging apparatuses were developed. In the same year, Frank R. Wrenn at Duke University published the results of his study which used copper-64 placed within a brain preserved inside its skull, which was also the first study indicating the medical application possibilities of positron emitters [5]. And then, work by Dr. Brownell and Dr. Sweet at the Massachusetts General Hospital contributed significantly to the development of PET technology which included the first demonstration of annihilation radiation for medical imaging [6]. In 1961, James Robertson and his associates at Brookhaven

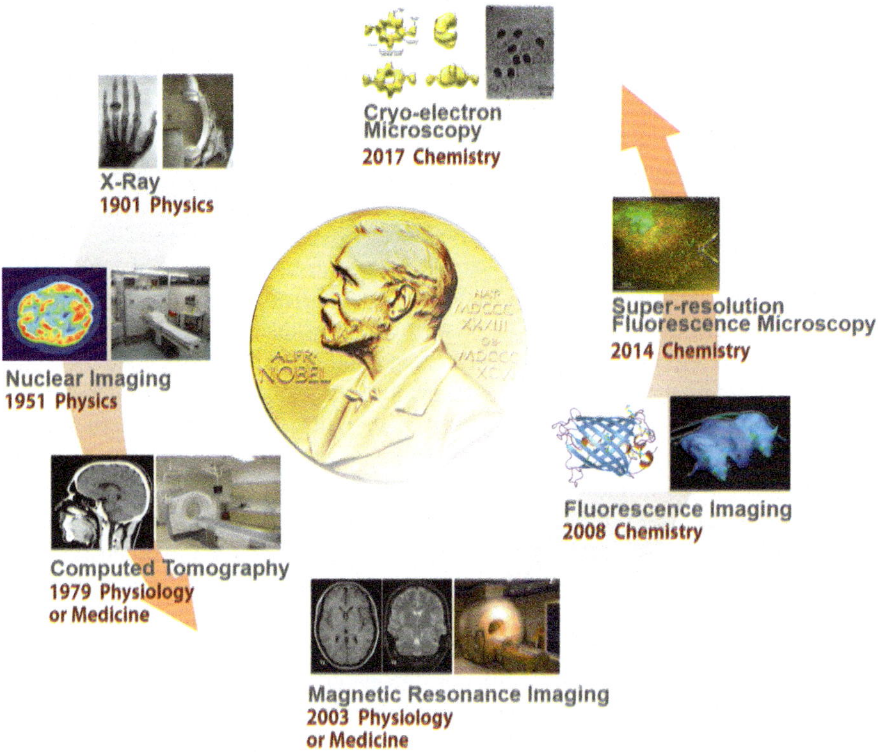

Fig. 5.1 The Nobel Prizes related molecular imaging technologies over the past 100 years

National Laboratory built the first single-plane PET instrument, which consisted of a positron camera named the "head-shrinker" and a direct forerunner of the present positron emission tomography [5]. Commercialization and large-scale application of PET were subsequently conducted later in the 1970s. The next technique emerged was computed tomography (CT) which made good use of X-rays by combining the power of X-rays with computer systems to show an axial view of the tested body. It was used for brain detection in the beginning by British engineer Godfrey Hounsfield and South Africa-born physicist Allan Cormack in 1972, and soon the range extended to be able to scan the whole body. As expected, Hounsfield and Cormack were later awarded the Nobel Prize in Physiology or Medicine for the development of computer-assisted tomography in 1979. A few years later, in 2003, Nobel laureates in Physiology or Medicine, Paul C. Lauterbur and Peter Mansfield, were awarded for making their original discoveries in the use of magnetic resonance to visualize different structures. Their work has led to the development of modern magnetic resonance imaging, MRI, which represents a breakthrough in medical diagnostics and researches [7]. Besides, the basis of these discoveries also won the Nobel Prize in Physics in 1952 for the laureates' development of new methods for nuclear magnetic precision measurements and discoveries in connection therewith.

In the next decades, MRI turned out to be a routine method in medical diagnostics with remarkable evolutions assisted by advanced computer systems and softwares. These imaging techniques, with their distinguished advantages and disadvantages, were soon combined to a variety of pairs in preclinical researches and clinical use in the 2000s to overcome certain limitations.

Except for the discoveries associated with atomic particles and energy emission generating black and white images, newborn optical imaging holds an important part of medical imaging with more colored pictures, including bioluminescence, fluorescence, Raman imaging, and many others. The most brilliant moment in this part of history may be the awarding of the Nobel Prize in Chemistry 2008. Three scientists were awarded for their discovery and development of the green fluorescent protein (GFP), which is a representative tool of in vivo fluorescent imaging. Since then, GFP has been developed into multitudinous applications in many areas of science and medicine, and one intriguing example was a study in which researchers used transgenic mice under the strategy that randomly mixed green, yellow, and cyan fluorescent proteins in individual neurons, thus creating a "brainbow" to map the neural circuits of the brain [8]. In fact, since George Gabriel Stokes studied fluorescence and proposed the use of it as an analytical tool, fluorescent imaging has benefitted biological researches as well as medical imaging. It first contributed to cell biology through the invention of fluorescence microscopy and then was experimented through animal models to finally come to human applications. Both molecular chemical probes like indocyanine green and biological reporters like encoded fluorescent proteins have been applied for in vivo cancer imaging research. Another related award was for the development of super-resolved fluorescence microscopy. The Nobel Prize in Chemistry 2014 has been given to three pioneers of biomedical imaging, Eric Betzig, Stefan W. Hell, and William E. Moerner, whose work has enabled nanoscale features within cells to be captured in exquisite detail [9]. Their work circumvented the problem of the "diffraction limit" which indicates the inability of light microscopy to distinguish between structures smaller than half the wavelength of visible light. This advance allowed nanoscale structures including individual molecules to be visualized within cells alive. Most recently, the Nobel Prize in Chemistry 2017 was awarded jointly to Jacques Dubochet, Joachim Frank, and Richard Henderson for developing cryo-electron microscopy for the high-resolution structure determination of biomolecules in solution [10, 11]. Though still in preclinical use, this technique holds the promise to further investigate our body and the diverse biological procedures in tumor progression and metastasis. And there are many other superior techniques that have been developed lately such as ultrasound imaging (USI), photoacoustic imaging, and Raman imaging making improvements in tumor diagnosis and treatment.

With all the efforts made by scientists and researchers, tumor early diagnosis is now far more convenient and popular in our life than they ever were. In this chapter, we would like to first introduce these representative imaging techniques commonly used in clinical tumor diagnosis and summarize their features and disadvantages together with their practical applications in detecting different tumors. Then, related contents of some emerging methods and multimodal imaging techniques are presented. Some typical and innovative imaging probes based on nanosystems will be discussed, and some examples of multitask interventions are introduced to give a more comprehensive view of the recent researches in tumor diagnosis. These well-designed systems are able to not only locate the tumors precisely but also assist drug delivery, surgical navigation, and therapeutic evaluation, bringing a brighter future for all patients.

5.2 In Vivo Imaging Techniques and Their Applications

Typically, the certain choice of imaging modality for certain patient depends on the purpose of the examination, the site of tumors, the biochemical processes of interest, and the concerns for the patients with different situations. Here, we are introducing some of the most commonly used techniques with their characteristics, developments, and some typical applications. Their distinctive features are summarized in Table 5.1 for a more direct and clearer view.

5.2.1 Computed Tomography (CT)

Shortly after its discovery by Wilhelm Röntgen, X-rays have been generally applied to medical use in the 1970s. X-ray absorption differs in compositions of human body like water, the bone, fat, and air, based on which scientists would be able to acquire structural information through the brightness of different regions of obtained images. And with the aid of computer calculation systems, scientists could integrate X-rays coming from different angles for 3D pictures, such as CT which provides the three-dimensional views of the organs or body regions of interest. Generally, a low-energy X-ray source and a detector are required for the detection which rotate within a circular opening around the patient body, producing a spiral or helical (also known as spiral) scan [12] (Fig. 5.2a). The detector receives the X-rays transmitted through the layer and converts it into visible light, which is transformed into an electrical signal by photoelectric conversion and then converted by an analog/digital converter. The resulting CT image is just a single slice of image among which a cross section is commonly used. Finally, in order to display the entire organ, multiple consecutive slices of images would be required.

The high-spatial resolution, fast acquisition time, and relatively lower cost have made CT a popular clinical testing technology which is user-friendly for both the operators and the patients. Generally, CT scans are mainly for the bone, vascular, and lung as CT has relatively low soft tissue contrast. Furthermore, it remains the current gold standard imaging modality to diagnose the presence of acute intracranial hemorrhage, calcifications, and osseous anatomy. In the field of tumor detection, traditional CT is mostly applied for imaging lung tumors and bone metastasis, while iodinated contrast agents are able to provide some additional help. It was reported

Table 5.1 Characteristics of imaging techniques most commonly used in clinical and preclinical diagnostics

Imaging technique	Basic principle	Technical characteristics	Shortcoming	Applications
Computed tomography	Computer-processed combinations of many X-ray measurements from different angles to produce tomographic images of specific areas of a scanned object	Quick testing	Unsuitable for soft tissue	Ideal for lung cancer diagnosis and bone metastasis
		Suitable for early detection	Radiance exposure	Mainly for bone, vascular, and lung imaging
		High-spatial resolution		
		High tissue penetration depth		
		Common contrast agents: Iodinated compounds		
Ultrasound imaging	Collecting the echo of the sound wave which encounters targeting tissues with different acoustical impedance	Cheap and fast	Blocked by air and bones	Showing tumors and guiding doctors for biopsies or treatments
		Consecutive and dynamic detection available	Contrast agents limited to vasculature	Diagnosis of liver nodules in hepatocellular carcinoma suspects
		Safe with little discomfort	Accuracy affected by operators	
		Enabled continuous and dynamic scans	Partial imaging only	
		Common contrast agent: Microbubbles		
Nuclear imaging (PET and SPECT)	Nuclear medicine functional imaging techniques detecting gamma rays emitted directly or indirectly by a radionuclide	High sensitivity and high specificity	Poor spatial resolution	Diagnosis, staging, and monitoring treatment of cancers through the labeling of different biomarkers
		Quantitative analysis available	Radiance exposure	Best tracing methods for detecting cancer metastatic spread
		Unlimited depth penetration		
		Radiative contrast agents required		
		Common contrast agents		
		PET – ^{11}C, ^{18}F		
		SPECT – ^{99m}Tc, ^{123}I		

(continued)

Table 5.1 (continued)

Imaging technique	Basic principle	Technical characteristics	Shortcoming	Applications
MRI	Manipulation of the inherent nuclear magnetic moment of endogenous nuclei and identification of differences in spin density among different tissues	Better for soft tissue imaging	Slow acquisition	Detections of central nervous system involving brain and spine studies
		Good penetrability with the possibility to image the whole body	Poor spatial resolution	^{1}H-MRS in clinical neurology
		Ability to image vascular structures	Relatively expensive	
		Free of radiation exposure	Specific patient restrictions	
		Common contrast agent: Gadolinium, iron oxide particles		
Fluorescence imaging	Fluorescent agents excited by external light of appropriate wavelength emitting colorful lights used for imaging	No radiation exposure	Limited penetration depth	Fluorescence-guided surgery for brain tumor
		Designed "switchable" ability	High background noise caused by autofluorescence	Retinal angiography
		Real time		Intraoperative tumor imaging
		Possible for functionalized		
		Common contrast agent: FDA-approved molecular dyes		
Raman imaging	Based on the inelastic collision between incident monochromatic radiation and molecules to provide information about the vibrational modes of the detected molecules	Structural analysis	Restrained penetration depth	Skin cancer detection via fully automated Raman system
		Both qualitative and quantitative	Low sensitivity later solved by SERS	Applications in ophthalmology
		Multiplexing capabilities due to the narrow peaks in spectra	Long acquisition time and slow data processing	Intraoperative imaging
				Real-time Raman endoscopic system

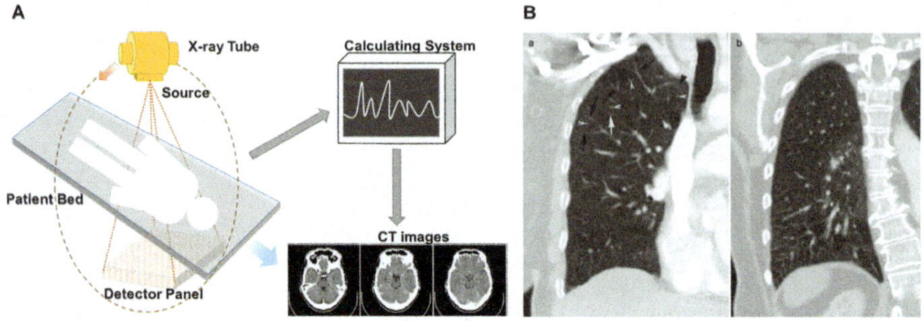

Fig. 5.2 The basic principle of CT scanning and its typical applications. (**a**) The simplified process of a CT scan. Electric signals acquired from the scanning machine which contains the structural information of the patient would be translated to several flat images by customized computer systems. (**b**) Different features can be observed in CT images of a normal lung (*a*) and the lung of a patient with lymphangitic spread of cancer and thickening of the interlobular septa (*b*) for cancer diagnosis

that low-dose CT screening could reduce mortality from lung cancer by the US National Lung Screening Trial Research Team [13]. There are several appearance patterns of lung disease which can be seen in CT images that are summarized in a review [14], and it would be easier to perform diagnostic tests when comparing the CT scans of healthy lungs and sick lungs together (Fig. 5.2b). For the benefits CT can bring, there has always been much enthusiasm in digging its potential because of the rapid development and emerging capabilities of the technology [15]. Improvements have been made in both spatial (approximately 0.4 mm in plane and approximately 0.5 mm in the z-axis direction) and temporal (>128 images per second) resolution, which made it possible to generate high-quality visualization of fine (submillimeter) anatomic structures and develop high-quality four-dimensional dynamic imaging, such as coronary angiography and cardiac imaging [16, 17].

One aspect of the progress is made in the improvement of X-ray source, and the successful outcome is a dual-source computed tomography (DSCT) system emerged in 2000s.The system consists of two X-ray tubes and two corresponding detectors, all settled onto the rotating gantry with a mechanical offset of 90° [18]. It provides temporal resolution equivalent to a quarter of the gantry rotation time and allows dual-energy acquisitions in subsecond scan times [19]. Another part of the progress is largely due to the innovative design of the scanner [20] (e.g., spiral and dual-source CT). Multi-detector row computed tomography (MDCT) and flat-panel volume CT are among these improved devices which are equipped with uniquely designed scanners and capable of high-spatial-resolution volumetric imaging and dynamic scanning. MDCT is comprehensible through its naming which indicates that multiple detectors have replaced the original one detector in the system. In many medical institutions, MDCT is routinely used as the most important preoperative examination in patients with suspected pancreatic cancer as it has good spatial and temporal resolution with wide anatomic coverage and thus permits both comprehensive local and distant disease assessment during a single session [21, 22]. As for flat-panel CT, the conventional detector is substituted by an area detector which is called the detector panel, consisting of rows of detector elements that can rotate to image the target from any angle. Its wide coverage is beneficial for imaging entire organs in one axial scan, and the extremely high-spatial resolution in all x, y, and z directions which is even superior to multi-detector CT enables more clinical applications like imaging calcified atherosclerotic plaque and brain aneurysms in addition to analyzing skeletal structures [23]. Volumetric scanning and unlimited imaging views make flat-panel volume CT promising for both diagnostic and interventional clinical purposes. One application has utilized a novel cone-beam volume CT breast imaging technique for breast tissue detection, and the proposed imaging technique provided significantly better low-contrast detectability of breast tumors and more accurate location of breast lesions [24].

Contrast agents for X-ray CT imaging are employed in numerous applications for enhancing image quality. Gas phase contrast agents, especially the inhaled Xe-contrast agents, are helpful in the noninvasive imaging of ventilation flow in the lung, which can be used as early symptoms of various lung diseases [25]. Besides, iodinated derivatives are also among the most popular contrast enhancers, which are soluble small molecules. Different forms of iodinated compounds, including the monomeric ionic agent diatrizoate and nonionic iopromide as well as the dimeric ones iotralan and ioxaglate, are already popularized in clinical CT imaging. However, their high osmolality causes side effects for patients, and disadvantages such as limited imaging time, the need of catheterization, and occasional renal toxicity have prevented further developments [26]. Nowadays, this dilemma has been overcome by the development of iso-osmolar media and iodinated contrast media encapsulation [27]. As an example, a liposomal nanoscale contrast agent encapsulating a high concentration of iodine (83–105 mg I/mL) was evaluated, demonstrating extra high attenuation within submillimeter vessels and long residence time in the blood with insignificant renal clearance [28]. Liposomal iodinated contrast agents have been used for detecting not only tumors but also inflammations and infections with improved results [29], and iodinated nanoparticles functioned as local drug delivery systems were used to monitor the diffusion of the loaded anticancer drug [30].

As CT requires a fairly high concentration of contrast agents to achieve satisfying images, the encapsulation of high dose of contrast agents in a relatively small carrier is beneficial for effective targeting and contrast enhancement in vivo. Under this thought, gold nanoparticles (AuNPs) have been chosen for the job as they have been employed for various medical purposes with guaranteed safety, and the high X-ray absorption of colloidal gold (Au) provide a great potential in X-ray imaging. Compared with conventional iodine-based contrast agents, the negligible osmolality of AuNPs and the higher absorption coefficient of Au than iodine make it a better choice for CT imaging. Besides, AuNPs experience less interference of bones and tissues, so they are able to exhibit improved contrast at lower doses. Together with the low viscosity and longer circulation time of AuNPs, close attention has been paid for the validation of their practical performance. Pharmacokinetics of AuNPs and iodine contrast agent in mice were studied after AuNPs of 1.9 nm in diameter were suspended in a phosphate-buffered saline (PBS) and injected via a tail vein into the mice bearing EMT-6 subcutaneous mammary tumors [31]. Also, with the advantages of nanosystems, there have already been commercial nanoparticulate contrast agents for preclinical CT imaging such as ExiTron nano 6000 for liver/spleen imaging and ExiTron nano 12,000 for angiography [32].

5.2.2 Ultrasound Imaging

After the development of X-radiography, ultrasound became the most common one of the medical imaging technologies as it is very good at getting pictures of some soft tissue diseases that don't often show up well on X-rays. Nowadays, ultrasonic imaging has already developed into a mature medical technology which is mostly used for fetus examinations (Fig. 5.3a). Also, ultrasound has been helpful in detecting tumors, visualizing the vasculature or microcirculation inside [33] (Fig. 5.3b), and guiding doctors to do biopsies or treatments [34]. It was reported in 2011 that nearly 22% of all the imaging procedures carried out in 2009–2010 in National Health Service hospitals in England (population 51 million) were ultrasonic investigations [35]. Modern medical ultrasound scanners are used for imaging nearly all soft tissue structures in the body. In traditional ultrasonic imaging, the signals which are then translated into the image are basically owing to the reflection and scattering of ultrasound. When a sound wave encounters a material (tissue) with different acoustical impedance, part of it would be reflected as an echo and travel back to the detector. The duration of this process is measured to calculate the depth of the tissue interface that causes the echo. The greater the difference between acoustic impedances, the larger the echo is. Then the computer calculation system would be able to construct an image based on these echo waveforms [36]. The anatomy can be studied from gray-scale B-mode images, where the reflectivity and scattering strength of the tissues are displayed. The imaging is in real time with a speed of 20–100 images/s.

Though conventional 2D ultrasound (US) has been widely used, the lack of the anatomy and orientation information limits its diagnostic accuracy. That's when 3D US was proposed for acquiring a better understanding of the spatial anatomic relationship. Data acquisition should be the initial challenge. First generations have been introduced with different types of mechanical scans to acquire the 3D volume which can be either a linear translation of the transducer, a rotation, or a tilting of the array [36, 37] (Fig. 5.3c), and the following developments include using a 2D ultrasound array for 3D image acquirements and mechanical 3D probes. Aside from quality and rate of data acquisition, quick and accurate volume reconstruction is in need of appropriate algorithms as well as volume rendering techniques which are well reviewed in Huang's work [38].

Ultrasonic imaging is highly acceptable to most patients as exposures used in current practice are considered to be safe and the cost is generally less than that of

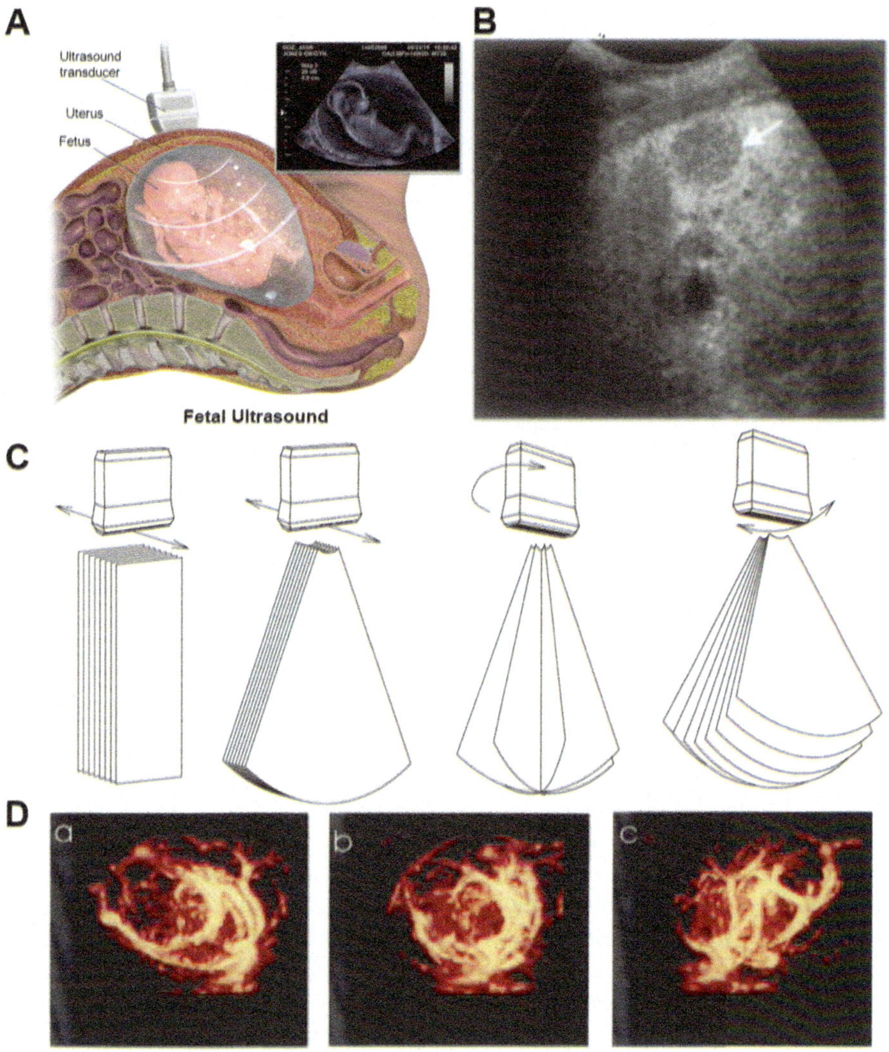

Fig. 5.3 Medical applications and advances of ultrasound imaging in oncology. (**a**) A scheme of the most popular application for medical ultrasound – fetal ultrasound. (**b**) Ultrasound image of the microvascular hemodynamics of a liver metastasis (arrow) using the method of interval delay imaging. (**c**) Different methods for mechanical 3D ultrasound scanning. (**d**) Example of three-dimensional reconstructions of a VX2 tumor growing in a rabbit thigh

other imaging technologies. However, the traditional US imaging is limited by penetration depth due to the attenuation of ultrasound in soft tissues as the transducer which sends out the sound waves and picks up the echoes is pushed against the skin surface, and normally the depth is only up to several centimeters.

Advanced improvements have been made by enhancing the acoustic signal to be detected using intravascular microbubble contrast agents which is called contrast-enhanced US (CE US) that are particularly useful in noninvasively characterizing liver lesions as well as monitoring angiogenesis and inflammatory responses [39] and sending the transducer inside a body, for example, through esophagus or rectum, which turns into an interventional imaging. CE US is a preferred modality for the difficult task of diagnosis of liver nodules detected on surveillance scans in those at risk for hepatocellular carcinoma [40]. Besides, advantages of contrast ultrasound for microcirculation detection include the increase of the power of the echo backscattered by the blood and the unique signature microbubbles offer to distinguish between true circulation and other tissues. Thus, visualization of the vasculature in tumor systems can be enhanced in CE US along with the help of 3D ultrasound technology (Fig. 5.3d). Moreover, when combined with some suitable ligands, sensitive detection of intravascular targets can be achieved, and this technique helps imaging angiogenesis with strong potential both in tumor biology studies and antiangiogenic therapy developments. Using RGD for targeting $\alpha_v\beta_3$ integrins is one of the typical methods to visualize tumor angiogenesis by contrast ultrasound imaging [41, 42], and enhanced contrast signals with a high tumor-to-background ratio were detected which facilitated the quantitative visualization of tumor angiogenesis [43].

5.2.3 Nuclear Imaging (PET/SPECT)

Here we summarize positron emission tomography (PET) and single-photon emission computed tomography (SPECT) as nuclear imaging as they share many common features [44]. Both techniques work on the same basic principle and require delivery of a gamma-emitting radioactive tracer into the patient, normally through injection into the bloodstream. These radioactive tracers are made up of carrier molecules that are bonded tightly to a radioactive atom, and these carrier molecules vary greatly depending on the purpose of the scan. Then, the gamma camera is rotated around the patient as the detection process begins, acquiring multiple 2-D images from multiple angles [45]. A computer is used to reconstruct these images which are also called projections to yield a 3-D data set. This data set can be manipulated to show slices along the body which shows a 2-D view of 3-D distribution of a radionuclide. As in this process, SPECT and PET are similar in the requirement of radioactive tracer material and detection of gamma rays. But, in contrast with PET, the tracers used in SPECT emit gamma radiation that is measured directly, whereas PET tracers emit positrons that annihilate with electrons which causes two gamma photons to be emitted in two opposite directions. Thus, a PET scanner detects these emissions "coincident" in time, which provides more radiation event localization information, resulting in higher spatial resolution images than SPECT [46]. And the sensitivity of PET at detecting molecular species is relatively high, in the range of 10^{-11} – 10^{-12} M [47]. However, SPECT is more widely available, because the radioisotope used is longer lasting and far less expensive and the gamma scanning equipment is less expensive as well. The most commonly used radioisotope in PET is likely to be fludeoxyglucose (FDG) which is the best available tracer for detecting cancer and its metastatic spread in the body [48], while in SPECT, it's some ^{99m}Tc-containing tracer. ^{99m}Tc is extracted from relatively simple technetium-99 m generators, which can be delivered to hospitals and scanning centers weekly to supply fresh radioisotope, whereas FDG is made in an expensive medical cyclotron or some "hot lab" (automated chemistry lab for radiopharmaceutical manufacture) and then delivered immediately to scanning sites because of the natural short 110-min half-life of fluorine-18, largely increasing the costing of PET. Anyhow, whether PET or SPECT is better suited for clinical applications still remains to be a hang jury as there are many factors to be considering and more advances continually made.

PET and SPECT scans provide a noninvasive way to evaluate the health of certain parts of the body, most commonly the heart, brain, and bones as they show the metabolic conditions rather than structure information [49]. There are several cardiac neuroreceptors used for developing radiotracers in PET and SPECT molecular imaging including ^{11}C-hydroxyephedrine (^{11}C-HED) and ^{123}I-meta-iodobenzylguanidine (^{123}I-MIBG), and labeled metabolic substrates such as ^{11}C -acetate, ^{11}C-glucose, or ^{18}FDG can be used to assess metabolism in particular organs or tumors [50, 51]. A typical PET scanning with the tracer fluorine-18 (F-18) fluorodeoxyglucose (FDG), called FDG-PET, is widely used in clinical oncology. It is a glucose analog taken up by glucose-using cells and phosphorylated by hexokinase. As the phosphate could not be removed in most tissues, FDG is usually trapped in any cell that takes it up until it decays, since phosphorylated sugars cannot exit from the cell due to their ionic charge. This results in intense radiolabeling of tissues with high glucose uptake, such as the normal brain, liver, kidneys, and most cancers. As a result, FDG-PET can be used for diagnosis, staging, and monitoring treatment of cancers [52]. Choline, a component of phospholipids making up the cell membrane, is an excellent biomarker to imaging proliferating cancer cells, and when labeled with ^{11}C, it can be used to image prostate cancer as it has also been reported to be more sensitive than ^{18}FDG PET in the detection of primary lesions, regional lymph node, and bone metastases [53, 54].

The long scan time is one of the drawbacks for PET or SPECT imaging. As projections are acquired at defined points during the rotation, typically every 3–6 degrees, and a full 360-degree rotation is often needed to obtain an optimal reconstruction, the total scan time is the product of each time it takes to obtain a single projection (typically 15–20 s) and the times the gamma camera stops to record the images. It varies from 15 to 20 min to several hours according to the different settings of the instruments such as the number of gamma cameras. In addition, attentions should be paid as both PET and SPECT produce radiation exposure to patients which may later result in harmful effects, but in spite of the disadvantages of PET and SPECT, they are indispensable imaging techniques with strong clinical potential, and more molecular radioactive tracers are in developing for tumor detection with higher specificity and sensitivity.

5.2.4 MRI

Magnetic resonance imaging (MRI) is a noninvasive, tomographic imaging modality that offers exquisite soft tissue contrast. Since its development in the 1970s and 1980s, MRI has proven to be a highly versatile imaging technique. In the late 1970s, physicists Peter Mansfield and Paul Lauterbur developed MRI-related techniques, like the echo-planar imaging (EPI) technique [55]. And as mentioned above, Mansfield and Lauterbur were awarded the 2003 Nobel Prize in Physiology or Medicine for their "discoveries concerning magnetic resonance imaging."

MRI is now widely used in hospitals and clinics for medical diagnosis, staging of disease, and follow-up without exposing the body to radiation. MRI is achieved by placing a subject in a strong magnetic field, typically 1.5 or 3 Tesla for human scanners, which aligns the hydrogen nuclei spins in a direction parallel to the field. It is based on the manipulation of the inherent nuclear magnetic moment of endogenous nuclei (most commonly ^1H in H_2O). By exposing nuclei to a static magnetic field and perturbing a steady-state equilibrium with time and space varying magnetic fields within, MRI images are obtained. All nuclei relax by two unique and codependent relaxation mechanisms after the perturbation: T1 (spin-lattice relaxation) and T2 (spin-spin relaxation), the results of which may translate to T1 and T2 images by the system [56]. And the difference in spin density among different tissues in a heterogeneous specimen enables the excellent tissue contrast of MRI [57]. High signals of tissues like fat and protein-rich fluid together with some paramagnetic substances such as gadolinium can be seen in T1-weighted images more clearly as bright regions. However, they look dark in T2-weighted images, while tissues with more water content as in edema, tumor, or inflammation are brighter. Thus, T1- and T2-weighted images of the same object look quite opposite to each other. For one example in practical application, the contrast provided between gray and white matter makes MRI the best choice for many conditions of the central nervous system, and as a matter of fact, brain and spine studies make up more than 50% of all studies [58].

A few shortcomings for MRI lie in its slow acquisition and poor spatial resolution. Expensive instruments also raise the cost of MRI testing for patients and medical institutions. And there are several occasions where patients should not accept MRI scans. Patients who have any metal parts inside their bodies are not suitable to have an MRI, which involve implants like cardiac pacemakers, aneurysm clips, and metal prosthesis as well as other things like bullets. Besides, those critical patients can't be scanned when they still need to be monitored by other medical devices as those devices are not allowed in the MRI operation room.

Recent advances in MRI facilities (higher field strengths, optimized pulse sequences, and better coil design especially for the breast) have made this modality a procedure of choice for imaging many cancers. The most widely studied and most commonly employed of these include MR spectroscopy, perfusion MRI, and diffusion MRI techniques, providing distinct and complementary diagnostic information that can help better characterizing brain tumors. Here we intend to introduce some of the emerging techniques in this field. MR spectroscopy (MRS) is based on the MR phenomena of spin-spin coupling effects and chemical shift to identify and quantify metabolites within a certain volume of interest, yielding a characteristic resonance frequency across the spectrum determined by the atomic nucleus of interest, which is usually the hydrogen proton. Traditional MRI detects the nuclear magnetic resonance spectra of water in bodies, while MRS detects chemical compounds other than water. Most developed use of MRS has been beneficial for brain cancer studies. ^{1}H-MRS has become the universal "gold standard" in clinical neurology due to the better volume resolution it shows [59]. Commonly measured metabolites include choline (Cho), lactate (Lac), lipids (Lip), N-acetyl aspartate (NAA), and creatine and creatine phosphate (Cr), although other minor metabolites may also be measured. And by comparing the characteristic changes in the metabolite profile in certain tumors to the normal CNS profile, MRS can potentially offer additional diagnostic biochemical information which is not shown through conventional MRI [60]. For example, gliomas will typically show increased Cho due to the accelerated cell membrane synthesis [61, 62], and decreased Cr as greater consumption of energy occurs in tumor metabolism. Meanwhile, increasing Cho/Cr and Cho/Cho (normal) ratios can be seen with increasing glioma grades [63]. Functional MRI (fMRI) is based on the principle that areas of neuronal activation within gray matter utilize oxygen supplied in the form of oxygenated blood to a greater degree than areas of gray matter at rest (or in a state of inhibition). It can localize regions of motor activation and language activation near or within a lesion, and this can be quite useful for pre-surgical evaluation with the goal of maximizing the extent of tumor resection and minimizing neurological deficits [64–66]. Dynamic contrast-enhanced MRI (DCE-MRI) is now widely used in the diagnosis of cancer and is becoming a promising tool for monitoring tumor response to treatment. DCE-MRI analyzes the temporal enhancement pattern of a tissue following the introduction of a paramagnetic contrast agent (CA) into the vascular system. By the comparison of acquired baseline images without contrast enhancement and a series of images acquired over time (usually over a few minutes) during and after the arrival of the CA in the tissue of interest, a time-intensity curve is generated. And with the analysis of the curve, physiological properties related to microvascular blood flow can be obtained. DCE-MRI has already been applied for various tumor imaging such as breast tumors, cerebral tumors, cervical cancer, and prostate cancer [67–70].

5.2.5 Optical Imaging

Typical optical imaging technologies include bioluminescence imaging, fluorescence imaging, and Raman imaging. Bioluminescence imaging detects light produced by the enzymatic reaction of a luciferase enzyme with its substrate, luciferin, which enters an animal through intraperitoneal injection. Thus, it is commonly applied for preclinical research in small animals for cellular and molecular imaging.

5.2.5.1 Fluorescence Imaging

Optical molecular imaging techniques have become essential tools for studying small-animal models, providing unique insights into disease pathogenesis, drug development, and effects of therapy [71]. Fluorescence imaging relies on the fluorescent agents that will emit light when excited with suitable wavelength, which can be used in both macroscopic and microscopic fields. An external light of appropriate wavelength is used to excite a fluorescent molecule, followed by the release of a longer-wavelength, lower-energy light for imaging. This fluorescent target may be endogenous molecules (such as collagen or hemoglobin), fluorescent proteins (such as genetically transfected luciferase or green fluorescent protein (GFP) and related molecules), or some exogenous fluorescent molecules which are injected into the test body. In 1942, red fluorescence by tumors was observed after intravenous administration of porphyrins [72], and the first use of fluorescein (which emits green light) to improve the detection of brain tumors was reported in 1948 [73]. The first clinical trials in the field of oncology using fluorescence imaging were published in 2008–2009 by Frangioni's and Sevick-Muraca's groups, employing this technique as an image-guiding intraoperative method for sentinel lymph node resection [74, 75].

The superiority of fluorescence imaging lies in several aspects. It has sensitivity comparable to those of PET or SPECT while safer for offering no radiation exposure. Real-time imaging can be realized due to the quick acquisition speed, and the optical instrumentation and computing needed are quite simple, so their handling requires moderate levels of training and protection. More importantly, as the fluorescent agents can also be "targeted" when conjugated to targeting ligands like the contrast agents for other imaging techniques, they have the unique "switchable" ability to be silent until certain environments turn them on, which enables higher target-to-background ratios especially for detecting tumors with distinct microenvironment [76, 77]. Thus, proper design and synthesis of fluorescent probes with low toxicity can be quite useful in oncology.

However, clinical applications of in vivo optical imaging techniques are currently confined to the surface of the body such as skin or ocular imaging due to the limited depth of tissue penetration (a few millimeters) for both excitation and emission lights in the visible range [78, 79]. In addition, autofluorescence from elastin [80], collagen [81], and other biological fluorophores has emission in the visible (<600 nm) and can cause high background signal which interferes strongly with probes incorporating short wavelength emitters. As a result, novel optical probes with emission and/or excitation windows in the near-infrared windows (NIR, 650–1400 nm) have been in rapid development in recent years, whose fluorescence is minimally absorbed by hemoglobin, muscle, and fat, minimizing autofluorescence of tissues at the same time [82, 83].

A few modern techniques have also been developed in preclinical small animal experiments. Tomographic fluorescence systems (fluorescence molecular tomography, FMT) allow relative quantification of imaging signals using transillumination to reconstruct 3D maps of fluorescent agents based on sophisticated reconstruction algorithms [84]. It enables whole-animal imaging with 1- to 3-mm spatial resolution at an imaging depth <10 cm [85]. Fluorescence reflectance imaging (FRI) systems consist of an excitation source, filters, and a charge-coupled device (CCD) camera to obtain planar images. They have limited depth resolution beyond 3–5 mm from the surface and thus are useful for imaging events in surface tumors (xenografts), surgically exposed organs, or for intraoperative imaging, but unlike FMT, are not quantitative.

Only numbered fluorescent dyes are approved for clinical use now, mostly for tumor imaging and fluorescence-guided surgery, which are well-summarized in the table of a review [86]. Indocyanine green (ICG, emission ≈800 nm), presently the typical near-infrared fluorescent dye with absorbing/emitting wavelengths >700 nm, has been approved by the FDA (US Food and Drug Administration) for human use [87]. It has been used for decades for retinal angiography from early the 1970s [88, 89], and other established medical applications include liver clearance testing and cardiac output monitoring. There have also been quite a few reviews [90] introducing ICG used for macular hole surgery [91], liver monitoring [92], burn wound assessment [93], and sentinel lymph node biopsy (SLNB) [94] together with cancer imaging and surgery [95–97]. Neurosurgery is ideal for fluorescence angiography as well because these operations are originally done with the assistance of a microscope (and camera), and the blood veins located on the brain surface are mostly exposed and thus can be seen more directly. It was reported early in 1993 that ICG was able to demarcate tumor margins in a rat glioma model [98]. In another clinical research, 23 patients with brain tumors were analyzed as ICG videoangiography was applied during the surgical resection process. It was demonstrated that ICG videoangiography could visualize tumoral and peri-tumoral circulation before the surgery as well as check the patency of remaining surrounding vessels when the resection was over [99]. Recently, in 2017, FDA approved 5-aminolevulinic acid (5-ALA; Gleolan®) as an oral intraoperative optical imaging agent for patients with gliomas which is also the first authorized imaging agent for brain tumor fluorescence-guided surgery in US [100].

5.2.5.2 Raman Imaging

Unlike fluorescence and bioluminescence, Raman imaging mainly relies on Raman spectroscopy in which we see lines and peaks instead of colorful lights and shades. Raman spectroscopy was first reported in 1928 and named in the honor of the scientist C.V. Raman who discovered Raman scattering [101], which is a relatively weak optical process that provides information about the unique vibrational modes of molecules. Raman spectra arise due to inelastic collision through the interaction between incident monochromatic radiation and molecules of sample. The sample is illuminated with a monochromatic laser beam which interacts with the molecules of sample and originates a scattered light [102]. The scattered light having a frequency different from that of incident light (inelastic scattering) is used to construct a Raman spectrum.

However, when a monochromatic radiation strikes at sample, it scatters in all directions, and only a small fraction of scattered radiation (1 in 100 million incident photons) has a frequency different from frequency of incident radiation and constitutes Raman scattering. Therefore, in general, Raman spectroscopy has been considered a technique for structural analysis, rather than a method for ultrasensitive trace detection [103]. This low sensitivity due to weak Raman scattering is the major problem to be solved.

A modified technique, surface-enhanced Raman spectroscopy (SERS), helps enhancing the sensitivity greatly. SERS was first reported in 1974 when the researchers discovered that Raman signals from pyridine were significantly enhanced when adsorbed onto a roughened Ag electrode [104]. In SERS, sample is absorbed on a colloidal metallic surface that can support localized surface plasmon resonances. It leads to strongly enhanced Raman scattering signals of molecules that are in several orders of magnitude [105–107] (10^4–10^6) higher than in conventional Raman scattering, providing the adequate sensitivity for bioanalytical and biomedical applications and even enabling optical detection of single molecules which is used for detection of a variety of medical samples in vitro [108].

A Raman spectrum is presented as an intensity-versus-wavelength shift which can be used for both qualitative and quantitative purposes. Raman peaks are typically spectrally narrow (a few wavenumbers) and in many cases can be associated with the vibration of a particular chemical bond (or normal mode dominated by the vibration of a single functional group) within a molecule [109]. Qualitative analysis can be performed by measuring the frequency of scattered radiations, while quantitative analysis can be performed by measuring the intensity of scattered radiations. Such quantitative or qualitative assessment in turn can be used to infer specific biochemical changes associated with tissue pathology or physiology for diagnosis or monitoring [110–112].

SERS fingerprints of individual molecules provide superior multiplexing capabilities due to the narrow width of Raman peaks in the spectra, and this benefits in vitro cancer diagnostics targeting cancer biomarkers. As there are diverse ingredients in the blood including proteins, lipids, nucleic acids, carbohydrates, and vitamins, Raman spectroscopic techniques are able to detect the changes in their compositions, conformations, and interactions from blood samples when diseases are ongoing. Immunoassays that rely on the recognition of biomarkers (cell surface markers, membrane receptors) with antibodies conjugated to SERS substrates are a common approach as well. Magnetic beads used as antibody-supporting materials make it possible for immunocomplexes to form in solution; thus this SERS immunoassays can be applied to complex samples from cancer patients (e.g., plasma [113], serum [114], and saliva [115]). For instance, MUC4 might be used as a serum marker for early detection of pancreatic cancer using a quantitative SERS-based platform [116]. And recently, SERS-based immunoassays for cancer detection have experienced several technical improvements including simplified structures, reduced sample volumes and assay times, as well as maximized sensitivity [117–

120]. Meanwhile, SERS-based assays have also been used for the detection of bloodstream circulating tumor cells (CTCs) in human whole blood which requires only a blood sample of 100 μl with a limit of detection of 50 cells/mL with 99.7% confidence.

As for in vivo imaging, like other optical imaging methods, Raman spectroscopy is also restrained to detections for surrounding tissues near the surface. A review has summarized the most recent clinical studies using Raman spectroscopy for cancer diagnosis [121]. Presently, real-time in vivo Raman spectroscopy has been reported by several research groups with diagnosis time ranging from minutes to less than 1 s [122–125]. A fully automated Raman system for rapid skin cancer detection (Verisante Aura™) has also received regulatory approval and is currently in clinical uses in Canada and Europe [126]. In skin cancer melanoma diagnosis, spectral features between 1055 cm^{-1} and 1800 cm^{-1} offered best discrimination between melanomas and nonmelanoma pigmented lesions, while the 500–1800 cm^{-1} region provided optimal separation between cancerous and precancerous lesions from benign skin lesions. Lieber et al. developed a portable confocal Raman system with a handheld probe for nonmelanoma skin cancer diagnosis [127]. Raman spectroscopy is also used in ophthalmology, concentrating on detections of tissues like cornea, lens, and retina. Bauer et al. first successfully obtained the Raman spectroscopy of cornea by noninvasive confocal Raman spectroscopic technique, which has the potential to assess the axial corneal water gradient with reasonable sensitivity and reproducibility [128].

Another application is Raman system for in vivo tumor diagnosis during surgery procedures. Haka et al. reported the use of in vivo real-time Raman system for intraoperative margin assessment during partial mastectomy surgery procedure [129]. With the aid of endoscopy, Raman system can be applied for in vivo organ examination such as an in vivo narrowband image-guided Raman endoscopy for gastric dysplasia detection [130] and real-time Raman endoscopic system for in vivo and online diagnosis of gastric cancer [131]. More of the SERS technologies have progressed toward application in microscopy and small-animal in vivo imaging with the help of various molecular probes. The potential noninvasive utility of SERS is highly valuable for live imaging, offering excellent resolution for monitoring of intracellular microenvironments and tracking of the cellular distribution of extrinsic molecules [132].

5.3 Multimodal Imaging for Cancer Imaging and Therapy

With diverse imaging techniques developing rapidly, scientists have been trying to combine their advantages and minimize the deficiency in practical use, which gives birth to a number of novel hybrid imaging patterns. And this combination is also called multimodality imaging, which has been attracted growing attention in the last 20 years. Multimodality imaging is widely considered to employ two or more imaging techniques in a single examination to allow acquisition of co-registered complementary data from tissue and overcome the limitations of the independent techniques.

Ideally, multimodality imaging would provide functional and molecular information (PET or SPECT) which are projected on data sets with anatomical information acquired using imaging techniques such as CT and MRI. In this process, two approaches were taken: fusing the images which are taken at different times through digital imaging manipulation techniques (asynchronous) or simultaneously acquiring images which can be merged automatically (synchronous) [133]. The former one suffers from more restraints, especially the deviation from different positioning of patients in the separated machines in the scans. Thus, synchronous acquisition of images with a hybrid device is largely preferred. The introducing of the hybrid devices began in 1990 when Hasegawa et al. built a device that simultaneously detected photons (SPECT) and X-rays [134], and the first proof-of-concept combined positron emission tomography and computed tomography (PET/CT) system started to operate in 1998 [135]. Since the commercial introduction of SPECT/CT and PET/CT, the application of multimodal imaging equipment has expanded. As PET and SPECT share some common features, SPECT/CT and PET/CT are largely alike, so here we focus more on the introduction of PET/CT.

The integration of PET and CT provides precise localization of the lesions on PET scans within the anatomic reference frame provided by CT, resulting in hardware-fused PET/CT images available at the end of the examination. All currently available data indicate that combined PET/CT is more sensitive and specific than either of its constituent imaging methods alone and probably more so than images obtained from separate PET and CT systems and viewed side by side. Aside from the optimized imaging results, it also accelerates the testing speed. With conventional PET scanners, attenuation correction is achieved by using data from a radioactive transmission source rotating around the patient which can also be derived from the fast CT data. This makes PET/CT 25–30% faster than PET alone with standard attenuation-correction methods, leading to higher patient throughput and a more comfortable examination, which typically last 30 min or less [136].

The most commonly used tracer at present is the^{18}F-labelled glucose analog FDG. FDG accumulation in tissue is proportional to the amount of glucose utilization. Increased consumption of glucose is characteristic of most cancers and is in part related to overexpression of the GLUT glucose transporters and increased hexokinase activity. FDG PET has been proven to be a sensitive imaging modality for detection, staging and restaging, and therapy response assessment in oncology [137–140]. FDG PET/CT provides essential information for radiation treatment planning, helping with critical decisions when delineating tumor volumes [141, 142], and identifies unexpected sites of metastases or recurrent tumors more accurately than conventional modalities [sensitivity, specificity, positive predictive value (PPV), and negative predictive value (NPV) with FDG-PET vs. CI: 97, 82, 87, 96 vs. 84, 60, 73, 75%] [143–146]. Additionally, FDG-PET/CT was found to be more sensitive than CI for detecting lymph node and bone metastasis. This hybrid technique is mature enough that the clinical operation procedures have been introduced as standard guidelines [147]. In lung cancer, whole-body FDG PET plays an important role in the evaluation of solitary lung nodules, in preoperative staging, in the diagnosis of recurrent disease, and in planning radiation treatment. In a clinical study, CT and PET alone, visually correlated PET and CT, and integrated PET/CT were performed separately in 50 patients with proven or suspected non-small-cell lung cancer, and their diagnostic accuracy were compared. It turned out that integrated PET/CT provided additional information in 20 of 49 patients, and tumor staging was significantly more accurate with integrated PET/CT than other methods [148].

However, the diagnosis accuracy can still differ when using separate probes even with the same imaging pattern for the same kind of disease. For prostate cancers, the cell surface peptidase prostate-specific membrane antigen (PSMA) shows intense overexpression in the majority of both primary and metastatic prostate cancers, and a positive correlation of PSMA with traditional adverse prognostic factors makes this structure a promising target for molecular imaging [149, 150]. A study demonstrated that^{68}Ga-PSMA-PET/CT detects lesions in 43.8% of biochemically recurring prostate cancer patients with PSA recurrence but negative 18F-choline scans [151]. And adding^{68}Ga-PSMA PET/CT in patients with negative choline scans improved the overall recurrence detection rate from 74.4% to 85.6% [151].

There are also a few studies comparing their diagnostic accuracy for various diseases, looking forward to find out the best solution [152–154]. But still, with no doubt, the popularization of PET/CT and SPECT/CT have benefited a large number of patients with their improved tumor identification and localization compared with conventional single-used techniques.

Radiation dose is certainly a concern in PET/CT imaging, especially in children that are more radiosensitive for carcinogenesis than adults due to their higher cell division rate. In contrast to CT, MR does not include any ionizing radiation, which makes PET/MR favorable in terms of radiation dose compared to PET/CT. Additionally, MR is known to exhibit better contrast over the capabilities of CT in determining different soft tissue structures. Thus, not only the field of oncology profits from the soft tissue contrast offered by MR, also neurology applications need the anatomical as well as functional imaging capabilities provided by PET/MR.

Although first patent applications for combined PET/MR have been filed in 1990, even before the first commercial PET/CT systems, it has only been applied for clinical use only in recent years. A main reason is that combining PET and MR is indeed a technical challenge. Different design approaches have been summarized in a review [155], and today, solutions based on semiconductor-based PET detectors are being made. The clinical systems have always been influenced by preclinical developments due to the similar fundamental instruments. Hence, as commercial sequential PET/MR devices for small animals are now available, preclinical researches are more likely to be translated into the clinic in the near future [156]. PET data and MRI data have been retrospectively combined for detection and staging of gliomas [157, 158] as well as for identification of areas with critical neurofunction in the vicinity of tumors, which is important for planning surgery [159]. Though developed later, PET/MR has shown its positive roles for clinical applications in several prospective studies recently [160]. As a result, researchers have put effort to compare the diagnostic accuracy from both PET/CT and PET/MR for a variety of diseases in order to provide a more proper choice for patients. However, the results can be much different and often depends on the circumstances. For example, one study was performed which intended to evaluate the feasibility of PET/MRI with the aid of the tracer ^{68}Ga-labelled HBED-CC-PSMA for imaging recurrent prostate cancer [161]. With investigations of 20 patients, the results indicated that recurrent prostate cancer was detected more easily and accurately with Ga-PSMA PET/MRI than with PET/CT, and improved contrast was usually observed in images of patients with lymph node metastases (Fig. 5.4). At the same time, other studies concluded that the integrated PET/MRI could not provide distinct advantages in comparison with MRI alone in breast cancer and that PET/CT and PET/MRI could perform equally well in all kinds of cancer diagnosis [162, 163]. Therefore, it may not be wise to give a conclusion based on current researches, and the value of the multimodal imaging techniques is still to be explored.

5.4 Nanosystems for Multitask Imaging and Treatment

Aside from the conventional imaging techniques and contrast agents, there have always been developments of novel nanosystems expecting for better imaging results and multitask accomplishments against cancer. Various probes at the nanometer scale have been developed for tumor diagnosis and even together with image-guided surgery and cancer therapy at the same time as the nanosystems exhibit a series of unique properties which are not possessed by traditional contrast agents. Though most of these designs remain in the preclinical studies, they are of certain value for future optimization and translation, providing reference and inspirations for researchers.

5.4.1 Nanoprobes for Single Imaging Modality

For tumor visualization by a single imaging modality, nanosystems have shown their superiority in plentiful ways. In the designs of nuclear imaging agent, a new dual-modality tumor-targeted agent NOTA-OA-IONP (Fig. 5.5a) was designed and radiolabeled with ^{68}Ga to realize dual-modality PET/MRI [164]. Iron oxide nanoparticles (IONP), which was Fe_3O_4 nanoparticles coated with PEG, were used as MRI contrast agents. The functional amine groups of the PEG phospholipid then provided reaction sites for both the radioactive chelating ligand NOTA for radioactive isotopes binding and the new tumor-targeted molecule oleanolic acid (3b-hydroxy-olea-12-en-28-oic acid, OA) which has been shown to induce apoptosis in cancer cells and inhibit their proliferation [165, 166]. With the aid of OA, the

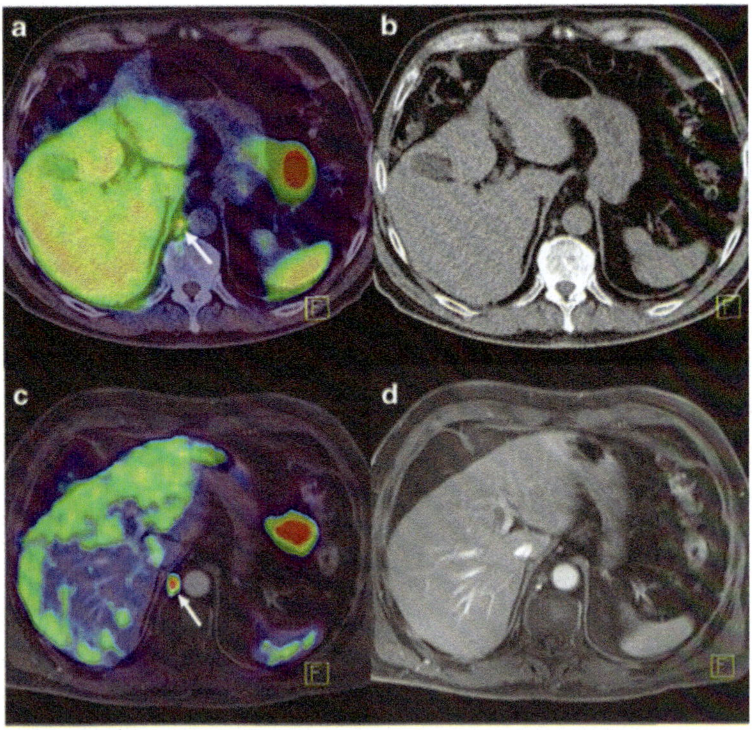

Fig. 5.4 Multimodal imaging in tumor diagnosis.^{68}Ga-PSMA PET/CT and PET/MRI images in a patient with lymph node metastases detected by PET/MRI due to a lower background signal and a time-dependent increase in PSMA tracer uptake in metastases. (**a**) PET/CT fusion image, (**b**) CT image, (**c**) PET/MRI fusion image, (**d**) MR image (T1 with contrast medium and fat saturation)

nanoparticle was able to achieve significant higher cellular uptake by HT-29 cancer cells and tumor-targeted ability. This combination of MRI contrast agents, tumor-targeted molecules, and radioactive labeling components gives it potential for PET/MRI-based diagnosis of tumors, and hybrid PET/MR imaging was performed in living mice with the new radiolabeled nanoparticle ^{68}Ga-NOTA-OA-IONP. Similarly, an RGD-conjugated IONP radiolabeled with ^{64}Cu was developed for dual PET/MRI images of tumor by integrin αvβ3 targeting, and both small-animal PET and T2-weighted MRI showed integrin-specific delivery of conjugated RGD-PASP-IO nanoparticles and prominent reticuloendothelial system uptake which allows early tumor detection with a high degree of accuracy [167]. Gong et al. also developed a targeted drug delivery and PET/MRI dual-imaging nanocarriers comprising radiolabeled IONP, doxorubicin (antitumor agent), and cRGD [168]. Magnetic resonance contrast agents aiming at enhancing signal intensity as well as offering specific targeting are also vigorously developing, and magnetic nanoparticles are favored for their engineered physicochemistry and tailored surface properties which can offer diverse clinical applications [169]. As Gd^{3+} is the most

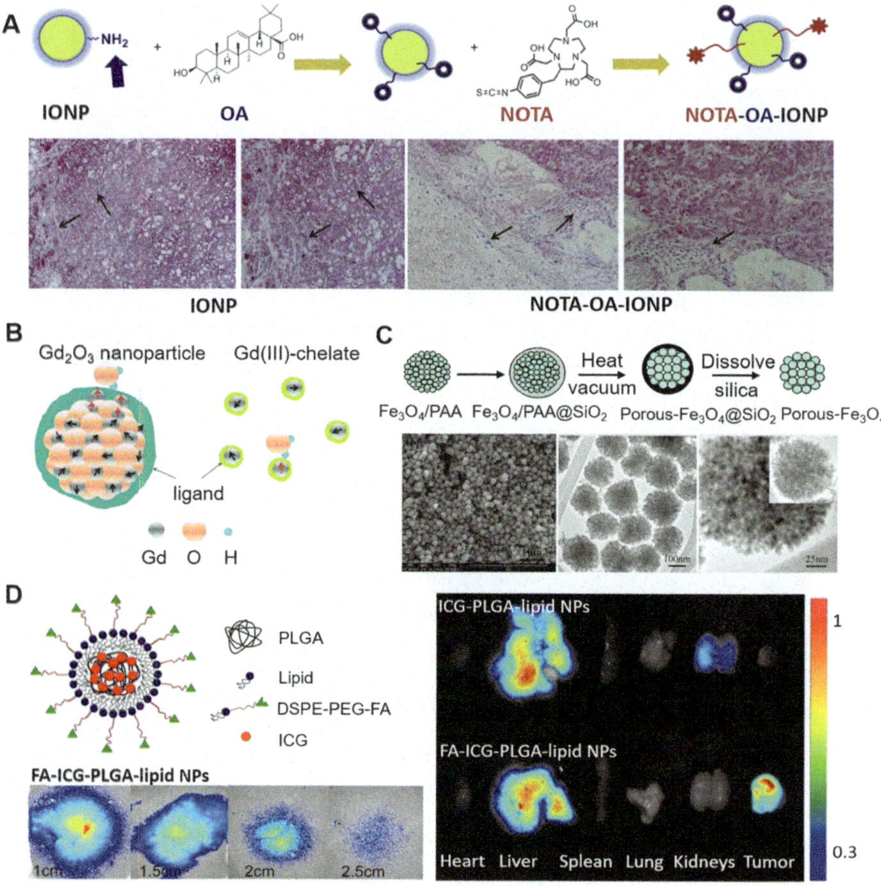

Fig. 5.5 Nanoprobes giving better performance than small-molecular probes in cancer diagnosis. (**a**) The hybrid PET/MR imaging agent NOTA-OA-IONP created with an OA-conjugated nanoparticle offers both tumor-targeting ability and aids in tumor clearance by inducing apoptosis of tumor cells and infiltration of immune cells in the tumor. (**b**) Schematic diagram showing the differences of longitudinal relaxation of the water proton in Gd-containing nanoparticle and individual Gd(III)–chelates. (**c**) The fabrication of mesoporous Fe_3O_4 nano/microspheres and the SEM & TEM images of the as-prepared mesoporous Fe_3O_4 microspheres. (**d**) The tumor-targeting NIR nanoprobe (FA-ICG-PLGA-lipid NPs) for NIR molecular imaging. The detectable penetration depths of FA-ICG-PLGA-lipid NPs were extended to 2.5 cm which was deeper than free ICG, and fluorescence images of organs and tumors showed highly selective tumor localization and prolonged circulation time in vivo

commonly used and preferred for T_1 contrast agents by forming Gd chelates, there are already nearly ten types of Gd chelates for clinical use now [170–172]. It was demonstrated that the inorganic nanoparticles containing Gd^{3+} could exhibit much higher longitudinal relaxivities r_1 per particle when compared with Gd complexes for the presence of a large amount of paramagnetic Gd^{3+} ions in each particle

[173]. A schematic diagram showing that four-surface Gd(III) ions as an example cooperatively induce the longitudinal relaxation of the water proton, whereas such an effect does not exist in individual Gd(III) – chelates (Fig. 5.5b) [174]. What's more, the longer blood circulation time of nanoparticles along with proper surface modifications gives more convenience for in vivo tumor imaging [175]. And other metal nanoparticles also function as magnetic resonance imaging agents showing excellent contrast and even drug loading efficiency. A Chinese team reported the fabrication of a biocompatible, mesoporous Fe_3O_4 nano/microspheres with the role of both drug carriers and MRI contrast agents [176]. The fabrication of this system began from the synthesis of poly(acrylic acid) (PAA)-entangled Fe_3O_4 nanospheres (Fe_3O_4/PAA). Then, the SiO_2 layer formed onto the surface of Fe_3O_4/PAA nanospheres which were core/shell-like nanostructures. After decomposition of the PAA polymer using a vacuum thermal treatment which was followed by etching of the SiO_2 layer, mesoporous Fe_3O_4 nanospheres were successfully synthesized (Fig. 5.5c). The average size of the microsphere is approximately 200 nm, and they are free of surface cracks and intersphere adherence. A high surface area value (163 m^2/g, Brunauer-Emmett-Teller) was calculated along with a high magnetic saturation value ($M_s = 48.6$ emu/g) which made the nanosphere suitable for MRI contrast agents ($r_2 = 36.3\ s^{-1}\ mM^{-1}$). This biocompatible, large-surface area, and soft magnetic Fe_3O_4 nano/microspheres with mesoporous nanostructure were then loaded with hydrophilic ibuprofen and hydrophobic ZnPC drug and have demonstrated good capacity for simultaneous MRI and drug delivery.

A few small-molecular fluorescent probes have already been introduced to clinical applications as discussed before. However, these fluorescent agents are still facing several challenges in wilder practical applications including the poor aqueous stability, rapid elimination, lack of affinity for a specific site, and respondence for certain chemical species. Thus, fluorescent nanosystems are emerging in preclinical studies which are promising for solving the problems. It was discussed in reviews [177] that several characteristics are required for fluorescent nanosystems development. First, NIR nanoparticles are preferred, and high extinction coefficient and high quantum yield are both essential for gaining stronger fluorescence signal and better sensitivity. Good photostability and solubility in aqueous environment of the nanoparticles are also necessary for noninvasive monitoring of cancer progression as well as the efficacy of anticancer drugs during the treatment. Most importantly, the nanosystem is able to provide a versatile synthetic platform for diverse implements of sensing and targeting schemes which are favorable for tumor detection. Cai and his team have developed biodegradable tumor-targeted nanoprobes (FA-ICG-PLGA-lipid NPs) for tumor diagnosis and targeted imaging which solved some difficulties in ICG applications including its poor aqueous stability, concentration-dependent aggregation, rapid clearance in vivo, and lack of target specificity [178]. This nanosystem has combined PLGA-lipid NPs encapsulated ICG with folic acid (FA) as the targeting ligand (Fig. 5.5d), which could be efficiently internalized into the cells through the receptor-mediated endocytosis. While hybrid NPs with lipid monolayer shell and biodegradable polymer core were previously reported to be able to provide great targeting versatility

[179, 180], the characterization of FA-ICG-PLGA-lipid NPs confirmed it. The fluorescence intensity of FA-ICG-PLGA-lipid NPs was more stable than that of the free ICG, which permitted prolonged periods of excitation without significant degradation in fluorescence intensity. And the in vivo distribution of three groups provided further evidence for the tumor targeting of FA-ICG-PLGA-lipid NPs which showed strong signals in tumor rather than the other organs. Besides, NIR quantum dots [181–183] and some dye-doped nanoparticles (NPs) such as dye-doped silica [184–186] and dye-doped polymer particles [187, 188] have also shown great promises for in vivo imaging and sensitive cancer detection.

5.4.2 *Multitask Nanoprobes for Oncology*

The ability to integrate more than one imaging modalities in a relatively simple system is another bonus for fluorescent imaging as the chemical structures are often definite and easier to be modified in fewer additional reactions. Thus, fluorescent imaging has been combined with most of the imaging techniques which require the assistance of contrast agents such as MRI, PET, and Raman imaging, especially SERS. In this process, not only locating the tumors but trying to eliminate them simultaneously are the goals. Zhang et al. made good use of the disordered pH in organelle lumen and developed a series of theranostic probes for image-guided tumor ablation by precisely positioning and destroying tumor foci intraoperatively [189]. The probes showed pH switchable near-infrared fluorescence in acidic lysosomal lumen, while photocytotoxicity was observed when they were delivered to mitochondria which would cause cancer cell death. Similarly, Meng et al. constructed another small-molecular probe RhoSSCy which shows excellent tumor-targeted and dual-modal imaging ability via both near-infrared fluorescence and photoacoustic (PA). And its innovation lies in the possibility to its dual sensing for intracellular pH and biothiols [190]. A multifunction small-molecular photosensitizer with NIR imaging property for synchronous cancer photodynamic therapy (PDT) and photothermal therapy (PTT) by targeting mitochondria is developed, which provides the potential for simultaneous cancer targeting, imaging, and therapy [191]. They all make remarkable examples of a single molecular fluorescent probe to be modified for multiple purposes.

Still, the strategy of using a single molecular probe is relatively challenging due to the limited number of attachment points and the potential spatial interference with its receptor binding affinity. From this perspective, nanoparticles are more suitable with larger superficial area and sufficient binding sites for multiple function moieties. Zhang and his fellows designed the self-assembled fluorescent dipeptide nanoparticles (DNPs) which composed of natural aromatic amino acids, and the Trp-Phe self-assembly with Zn(II) coordination was found to have a narrow emission bandwidth which would result in enhanced detection details as well as less background noise and impurity influences. For applications, they first chose to modify the DNPs with MUC1 aptamers, and the product DNP/aptamer exhibited good capability for cancer cell targeting and sensing. Then, DNPs were functionalized with DOX, the chemotherapy drug, of which the release behaviors could be monitored through the variation of fluorescence intensity [192]. Thus, it's evident that different purposes, including tracing tumors, monitoring drug delivery at the lesion, and providing treatment, can be realized by means of only one imaging techniques. However, complicated manipulations can be the limitation for practical use especially when the number of patients is quite enormous.

Fluorescence imaging also benefits intraoperative image-guided tumor surgery as the fluorescent imaging contrast agents can be activated by light and the images are then acquired from the software system in a very short time to help surgeons distinguish between tumor lesion and normal surrounding tissues. Therefore, it's beneficial for precise tumor detection in real time during the surgery and has already been applied for treatment of liver metastases [193], breast cancer [194, 195], ovarian cancer [196], melanoma [197, 198], vulvar cancer [199, 200], and cervical cancer [201]. As penetration depth is a challenge in optical imaging, intraoperative multimodal data fusion provides possible solutions, and several studies have been performed based on clinical data [202–204]. Nanoparticles combining fluorescence imaging and magnetic resonance imaging are receiving attention as they ally the high sensitivity of the fluorescence to the high-spatial resolution of MRI [205]. Chen and Shang have reported the synthesis of a novel dual-modality MRI and near-infrared fluorescence (NIRF) nanoprobe for liver cancer imaging [206], which is comprised of superparamagnetic iron oxide (SPIO) as the T2 contrast agent and indocyanine green (ICG) as NIR organic dye. Both of them have been approved by the US Food and Drug Administration (FDA) for human medical use [207], making the nanoprobe safer and more reliable for clinical translation. As the liver is the main organ for probe elimination, nonspecific background signal shows up in liver lesion imaging. For this problem-solving, Arg-Gly-Asp peptides (RGD) were employed due to its specific binding ability to the integrin $\alpha v\beta 3$ receptor which plays an important role in the pathological development of most solid tumors including liver cancer [208]. The higher tumor uptake and better optical contrast of this targeted nanoprobe were observed. Both MRI and fluorescent images showed clear tumor delineation, and the large tumor tissue mass was successfully resected under the fluorescence navigation system. Dual-modality imaging of the probes also helped in detecting small tumor lesions (0.9 ± 0.5 mm in MRI and 0.6 ± 0.3 in NIRF imaging) as well as primary liver tumors with intrahepatic metastasis.

Another novel drug delivery carrier based on a multimodal imaging agent, mFLAME, was developed for dual-modal imaging of fluorescence imaging and MRI together with the drug delivery capacity for cancer treatment [209]. It composed of a mesoporous silica shell with high stability and ease of functionalization and a liquid-phase perfluorocarbon (PFC) core serving as a highly sensitive ^{19}F MRI contrast agent owing to its multiple fluorine nuclei. Here, the researchers chose ^{19}F MRI rather than the most commonly used ^{1}H MRI owing to the difficulty for tracking the drug carrier in living bodies with the disturbance of the high background signals from water and lipids. The dual-modal detection of folate receptor-mediated cellular uptake via ^{19}F MRI and fluorescence microscopy was realized, and the carrier loaded with DOX was evaluated.

Gold nanoparticles have received significant interests recently for use in multiple imaging technologies [210]. In spite of the negligible toxicity, high-atomic number and high X-ray absorption coefficient, the synthesis is also technically easy and cost-effective [211]. For example, a unique nanoprobe, M-NPAPF-Au, was reported for dual-modal fluorescence/CT imaging. As gold nanoparticles (AuNPs) are an intensely explored CT contrast agent while it is also a strong quencher of fluorescence dyes [212], Liang and the team co-loaded the aggregation-induced emission (AIE) dye NPAPF and gold nanoparticles into 1,2-distearoyl-sn-glycero-3-phosphoethanolamine-N-[methoxy(polyethylene glycol)-2000] (DSPE-PEG2000) micelles as the AIE dye became highly emissive in aqueous solution and overcame the fluorescence quenching caused by AuNPs [213]. Both in vitro and in vivo studies demonstrated that M-NPAPF-Au has tumor-targeted properties and superior fluorescence and CT imaging effects. Besides CT, gold nanoparticles can also be applied for Raman detection. Some multimodal imaging nanosystems involving Raman AuNPs have recently been reported as the combination might overcome some restrictions of SERS (e.g., signal attenuation, low contrast, and temporal resolution) for in vivo deep tissue imaging. In a study intending to evaluate the in vivo biodistribution of SERS-active AuNPs in mice after different administration routes, gold surface-enhanced Raman scattering nanoparticles were radiolabeled with ^{64}Cu to form a bimodal PET-SERS probes [214], which provide potential for supporting the clinical translation of Raman spectroscopy as an endoscopic imaging tool. Also, pure optical in vivo imaging was achieved through a dual-modal SERS-fluorescent probe. Qian et al. functionalized gold nanorods (GNRs) with both NIRF and SERS, which showed enhanced imaging contrast and deeper tissue penetration and was further used for in vivo sentinel lymph node (SLN) mapping and tumor targeting of live mice [215].

Certain nanosystems are designed for not only tumor visualization but also surgery guidance. As tumor resection remains the very first and vital step in tumor treatment, it is extremely crucial for the doctors to determine the accurate location and boundary of tumors in order to remove tumor lesion without additional damage to the surrounding normal tissues, especially with the assistance of image-guided techniques. Based on this purpose, Gao et al. developed a pair of gold nanoprobes for brain tumor surgery guidance via the activation of both magnetic resonance (MR) and surface-enhanced resonance Raman spectroscopy (SERRS) signals as they achieve self-assembly in the acidic tumor environment [216] (Fig. 5.6a). The increase in SERRS signal is due to the formation of numerous interparticle "hotspots" [217] and enhanced longitudinal MR signals due to the lengthening of the rotational correlation time (τR) of the conjugated Gd^{3+} chelators [218]. Together, the nanoprobes were able to preoperatively define orthotopic glioblastoma xenografts by magnetic resonance imaging (MRI) with high sensitivity in vivo and also intraoperatively guide tumor excision by its Raman signal using a handheld scanner (Fig. 5.6b), which are promising for improving the outcome of brain tumor surgery. As for tumor resection guidance, fluorescence imaging nanosystems have not only been utilized in tumor imaging but also made much contribution for guiding surgery of various types of tumors. As an example, Zhang and the team have reported in vivo assembly of the second near-infrared window (NIR-II) emitting downconversion nanoparticles (DCNPs) modified with DNA and targeting peptides to improve the image-guided surgery for metastatic ovarian cancer [219] (Fig. 5.6c). A pair of complementary DNA (L_1 and L_2) was chosen for the assembly [220], and follicle-stimulating hormone (FSH_β) peptide specific to the epithelium ovarian cancer [221] was anchored on DCNPs. The assembled superstructures of the nanoparticles enabled tumor retention period as long as 6 h, and accumulation of agents in organs of the reticuloendothelial system was reduced, resulting in weak background signals. During the image-guided surgery, both all the large tumor boundary (Nos. 1–8) and invisible small metastatic lesion (Nos. 9–13) were identified by NIR-II fluorescence bioimaging in the optimal surgical time window, and the H&E staining results confirmed the precise delineation of the tumor margin, and that tumor lesions were thoroughly removed (Fig. 5.6d). In summary, potential benefit of intraoperative tumor-specific NIR-II fluorescence imaging in staging and debulking surgery for ovarian cancer was shown using the nanoprobes in vivo assembly strategy. And the combination of optical imaging technologies with tumor-targeted strategies offers a unique opportunity to intraoperatively detect and quantify tumor growth and intra-abdominal spread.

Furthermore, there are a few researches developing triple-modality imaging to gain better results. An outstanding example of this is a unique triple-modality MRI-Raman-photoacoustic imaging nanoparticle for noninvasive brain tumor detection and resection [222]. It incorporated the complementary strengths of the three imaging modality which includes the fine recognition of brain tumor border of MRI, three-dimensional capabilities of photoacoustic imaging, and the high sensitivity along with high specificity of Raman imaging. After intravenous injection in glioblastoma-bearing mice, the probes accumulated in tumor areas and enabled accurate delineation of brain tumor margins preoperatively by MRI and intraoperatively by the others. Cancerous foci were able to be detected without omission, and the images acquired by three different methods could corroborate each other for more convincing evaluation. Similarly, triple-modal imaging magnetic nanocapsules were capable of fluorescence/MR/nuclear imaging in vivo for visualizing solid tumors [223], and a theranostic nanoprobe for in vivo triple modal fluorescence, photoacoustic, and magnetic resonance imaging was developed and used for effective photothermal ablation of tumors in mice [224]. In general, multimodal imaging provides a path for scientists to unite the existing imaging techniques and make full use of their strengths at the same time, which may lead to surprisingly satisfactory results.

In spite of these imaging techniques applied to image tumor directly, nanoparticles are developed for cancer therapy and drug delivery on tumor sites while requiring the aid of imaging methods for tracing the probes in vivo and monitoring the drug efficacy. Gao et al. developed a turn-on theranostic fluorescent nanoprobe P-CDs/HA-Dox which was obtained by electrostatic assembly of polyethylenimine (PEI)-modified carbon dots (P-CDs) and hyaluronic acid (HA)-

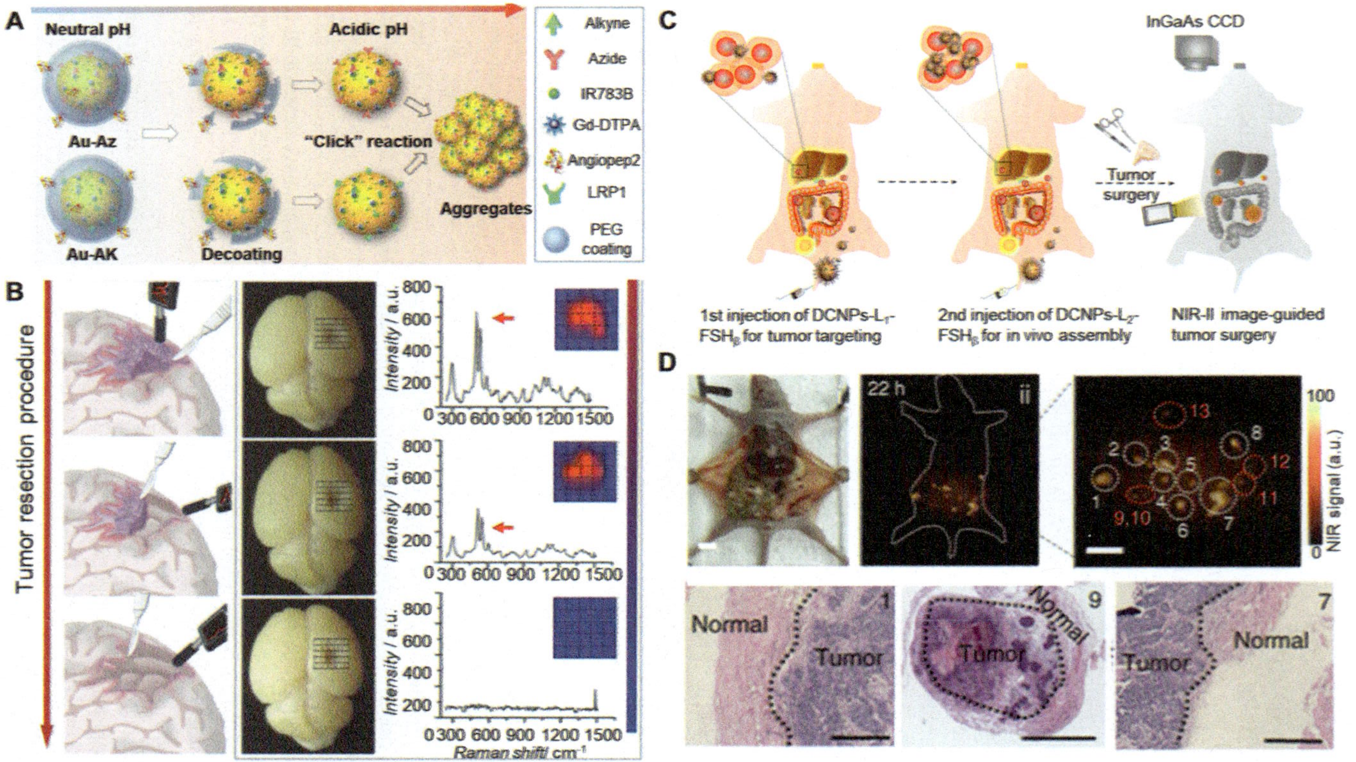

Fig. 5.6 Nanosystems for both tumor detection and resection guidance. (**a**) Guiding brain tumor surgery by acid-responsive gold nanoprobes. The cleavage of the PEG coating in physiological acidity triggers aggregation between Au–AZ and Au–AK via "click" cycloaddition reactions. (**b**) Au–AZ/Au–AK mixture intraoperatively guiding glioma resection using a handheld Raman detector. (**c**) Schematic illustration of NIR-II nanoprobes fabrication for ovarian metastasis surgery under NIR-II bioimaging guidance. (**d**) Tumors resection under NIR-II fluorescence bioimaging guidance and the H&E staining results of tumor margins after the operation. (Reproduced with permission)

conjugated doxorubicin (Dox) for hyaluronidase (HAase) detection, self-targeted imaging, and drug delivery [225]. As HA exhibits a high specific affinity for CD44 receptors overexpressed on the surfaces of many cancer cells and can enter cells through receptor-mediated endocytosis, it was designed as cell-targeting moiety for the nanoprobe [226]. Hyaluronidase (HAase) which is a family of endoglycosidases that is expressed increasingly in cancer cells rather than normal cells can degrade HA by cleaving its internal β-N-acetyl-D-glucosamine bonds [227]. Thus, the level of HAase in different cells allows it to self-target and distinguish cancer cells from normal cell. As shown in Fig. 5.7a, the theranostic fluorescent nanoprobe produces a negligible fluorescence signal due to the Förster resonance energy transfer from P-CDs to Dox but then could be activated by HAase after entering cancer cells, resulting in release of Dox for targeted drug delivery and subsequent recovery of fluorescence of PCDs. Upon P-CDs/HADox incubation, HeLa cells exhibited strong fluorescence signals in the blue and red channels (Fig. 5.7b), representing for fluorescence from P-CDs and Dox, respectively. The result indicated that P-CDs/HA-Dox had internalized into HeLa cells, where the HA-Dox chain was decomposed and separated from P-CDs, resulting in fluorescence recovery of P-CDs and the release of Dox. Besides, in the cytotoxicity assay, P-CDs/HA-Dox exhibits higher cytotoxicity than free Dox owing to its good ability of targeted drug delivery which showed effective treatment for HeLa cells at a relatively low concentration (0.5–2.5 μg/mL). As for normal cells, P-CDs/HA-Dox presents rather low cytotoxicity as the lack of CD44 receptors on NIH-3T3 cell surface can inhibit its cellular uptake. Generally, the combination of targeted bioimaging, hyaluronidase analysis, and drug delivery through electrostatic assembly P-CDs and HA-Dox offer a synergistic improvement in both cancer diagnosis and therapeutic efficiency, which makes the fluorescent nanoprobe a quite promising theranostic strategy for noninvasive self-targeted and image-guided chemotherapy. Similarly, two-dimensional nano-graphene oxide which offers unique electronic and thermal properties along with improved solubility and compatibility was loaded with doxorubicin via π-stacking which was able to selectively kill cancer cells in vitro [228].

As for cancer therapy monitoring, multiple methods involving therapeutic evaluation of cancer through imaging results after certain treatments have been developed. Several teams have focused on studying clinical response of breast cancer after primary systemic therapy [229–231] as it has become a standard treatment for patients with locally advanced or inflammatory breast cancer [232] in order to reduce further tumor progression and to facilitate operability [233, 234]. As an example, a research was conducted to evaluate a PERCIST-based, but simplified binominal PET score (PET/CT Method 1) and a novel, combined PET/CT score (PET/CT Method 2) supplemented with the morphological results of the RECIST system when defining clinical complete remission (CR) [235], and the efficiency of both methods in predicting pathological complete remission (pCR) was tested. This kind of evaluation could last for a couple of years to follow up, and in view of the earlier findings of Groheux et al., the presence of different biological subtypes of breast tumors could also limit the value of response evaluation with FDG-PET/CT [236, 237]. Moreover, multimodal imaging has been more beneficial for the evaluation of some novel

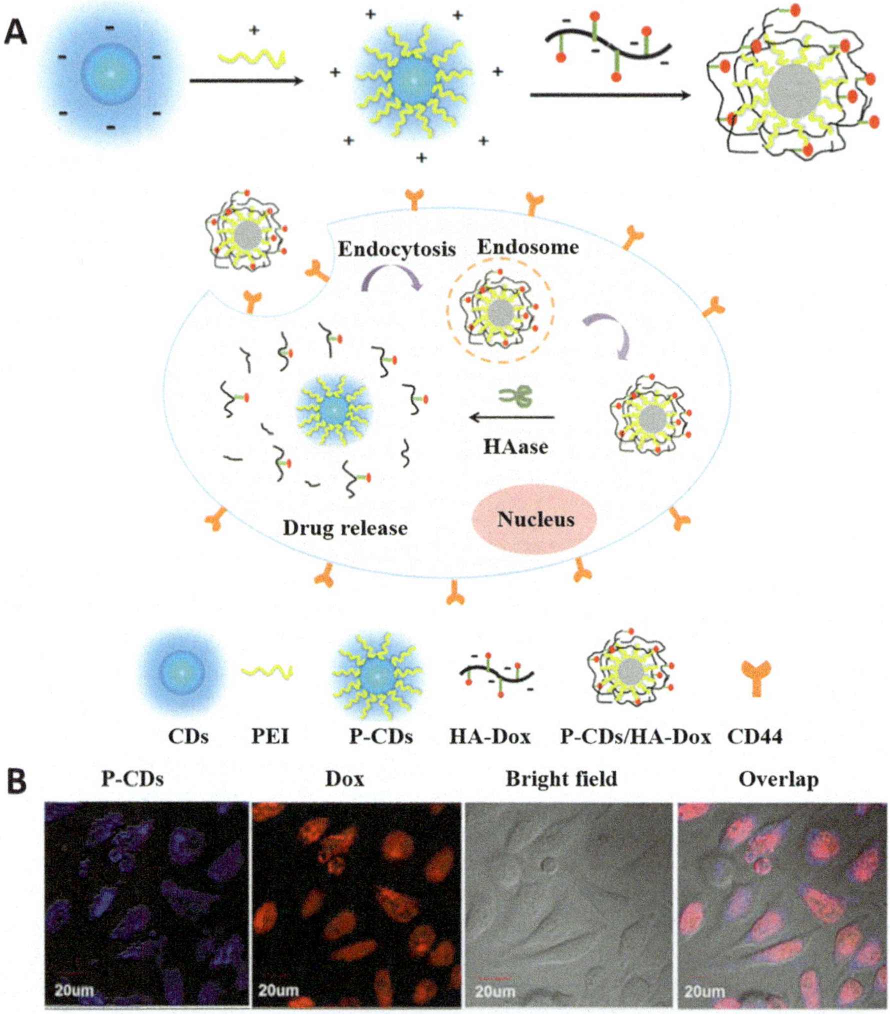

Fig. 5.7 The activatable fluorescent nanoprobe used for targeted cancer cell imaging and drug delivery. (**a**) Schematic illustration of the formation of PEI-CDs/HA-Dox. (**b**) Confocal fluorescence microscopy images of HeLa cells treated with P-CDs/HA-Dox for 2 h, indicating successful entry into HeLa cells where the HA-Dox chain was decomposed, resulting in fluorescence recovery of P-CDs and the release of Dox. (Adapted with permission)

therapeutic strategies. Radiolabeled annexin V may provide an early indication of the prognosis of anticancer therapy noninvasively on a patient-by-patient basis as an in vivo marker of tumor cell killing [238–240], so simultaneous in vivo measurements of tumor burden and uptake of radiolabeled annexin V in the syngeneic orthotopic murine BCL1 lymphoma model were conducted using in vivo bioluminescence imaging (BLI) and small animal single-photon emission computed tomography (SPECT). This multimodality imaging method helped revealing the temporal patterns of tumor cell loss and annexin V uptake, providing a better understanding of the timing of radiolabeled annexin V uptake for its development as a marker of therapeutic efficacy. Also, I-124-labelled 2′-fluoro-2′-deoxy-1β-D-arabino-furanosyl-5-iodo-uracil ([^{124}I]-FIAU), a specific marker substrate for gene expression of HSV-1-*tk*, was used for identifications of vector-mediated HSV-1-*tk* gene expression in five patients who has undergone gene therapy for recurrent glioblastoma in a clinical trial [241]. PET with fluorine-18-labelled 2-fluoro-2-deoxy-D-glucose (FDG), carbon-11-labelled methionine (MET-PET), and [^{124}I]-FIAU were shown in Fig. 5.8a, which demonstrated that after ganciclovir treatment, signs of necrosis could be observed by FDG-PET and MET-PET within the volume of specific FIAU trapping indicating the HSV-1-tk-mediated therapeutic response. These results indicated that FIAU-PET imaging of HSV-1-tk expression in patients is feasible and that vector-mediated gene expression may predict the therapeutic effect, which can be a representative for the application of molecular imaging for treatment evaluation of cancer. What's more, some well-designed nanosystems may realize therapeutic effect and imaging for assessment and the responses at the same time. Among the treatment modality, photothermal therapy (PTT) has emerged as an important one with the advantages of minimal invasiveness and excellent selectivity. It uses light-absorbing materials, usually in forms of nanoparticles, as transducers which will be delivered to tumors prior to the treatment and cause thermal ablation of cancer cells under photoirradiation [242, 243]. Zhang et al. developed GO-IONP-PEG as the PTT transducers and then explored the feasibility of using diffusion-weighted magnetic resonance imaging (DW-MRI) as a predicative tool for monitoring tumor response after PTT [244] (Fig. 5.8b). They injected GO or IONP-laden GO into 4T1-bearing mice and then photoirradiated the tumors at different drug-light intervals to initiate PTT. Both T_2-weighted MRI and DW-MRI were applied for monitoring the therapy response. T_2-weighted MR images were taken at different time points after PTT, and the most effective treatment was observed with the GO-IONP-PEG+MF group, while the best drug-light interval was 24 h which led to efficient tumor shrinkage with the primary tumors almost disappeared after 2 weeks. As for DW-MR imaging, the acquired images were displayed together with H&E and CD31 staining with tumors after PTT treatments. Meanwhile, the mean ADC values in tumors were also measured from DW-MRI results. And distinct ADC increase was observed in the GO-IONP-PEG+MF groups, which was mainly caused by PTT-induced necrosis [245–247] resulting in damaged cancer cell structures and thus increased water proton mobility. Generally, it was suggested that tumor ADC increase can be treated as an accurate early prognosis marker for efficient PTT and that DW-MRI would be able to assist future PTT studies as a therapy response and treatment prediction tool.

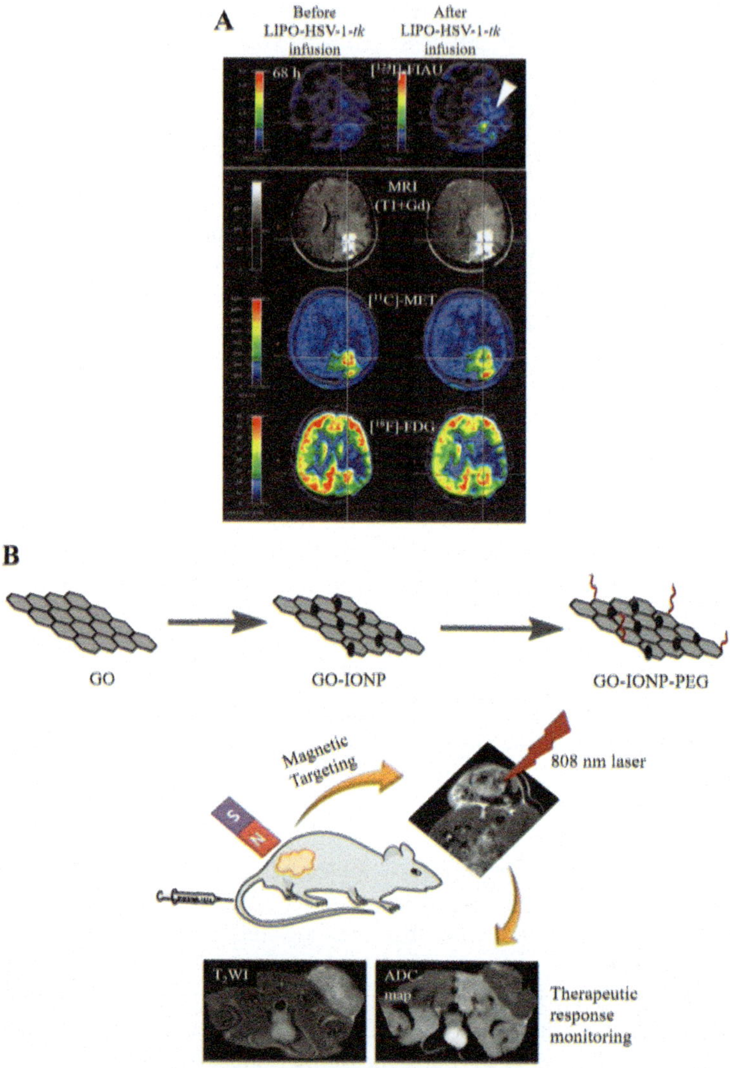

Fig. 5.8 Multimodal imaging offers assistance for tumor monitoring and therapeutic evaluation. (**a**) Coregistration of FIAU-PET, MET-PET, FDG-PET, and MRI before (left column) and after (right column) vector application, indicating therapeutic response of the glioma after gene therapy. (**b**) Schematic illustration of using DW-MRI to monitor tumor response after GO-mediated PTT. GO-IONP-PEG was prepared and administered into tumor-bearing animals. The tumor targeting was facilitated by an external magnetic field applied to the tumor area. Under photo-irradiation, the GO-IONP-PEG-mediated PTT was inflicted on cancer cells, and the therapy response was monitored by both T2-weighted MRI and DW-MRI

5.5 Conclusions and Future Prospects

Molecular imaging techniques have benefited many related scientific fields, especially oncology. Both the developed techniques, like CT, MRI, and nuclear imaging, and the emerging ones, including fluorescence imaging and Raman imaging, are all right hands of effective cancer imaging. It has become more and more convenient and painless for people to get cancer screening for early diagnosis and better chance to defeat this disease. For more precise imaging results and better sensitivity along with specificity, single kind of imaging agent or single modality has seemed to be insufficient when multimodal imaging showed up. Certain modifications are needed for enhanced signals and tumor-targeting ability, for which diverse nanosystems have shown their advantages. Moreover, the combination of multiple imaging modalities in the same nanostructure has largely improved the accuracy and efficiency of cancer imaging. With the technology advancing, more revolutions are being seen in the development of imaging instruments, contrast agents, and novel techniques, which all promote the practical applications of molecular imaging in cancer detection. The responsiveness of probes has been a hotspot for molecular imaging in recent years, and with the continuous exploration of tumor microenvironment coming up, developing environment-responsive probes will become one of the main focuses in the future. As nanostructures have a relatively large surface area with various reactive sites for functional binding, innovative nanoprobes with the ability of responding to multiple parameters or biological processes in a short period of time will be preferred. Anyhow, the rapid development of nanosystems will definitely keep promoting practical applications of molecular imaging not only in tumor diagnosis but also in tumor surgical navigation, anticancer drug delivery, and cancer therapeutic evaluation.

References

1. Chen W, Sun K, Zheng R, Zeng H, Zhang S, Xia C et al (2018) Cancer incidence and mortality in China, 2014. Chin J Cancer Res 30(1):1
2. Bray F, Ferlay J, Soerjomataram I, Siegel RL, Torre LA, Jemal A (2018) Global cancer statistics 2018: GLOBOCAN estimates of incidence and mortality worldwide for 36 cancers in 185 countries. CA Cancer J Clin 68(6):394–424
3. Valastyan S, Weinberg RA (2011) Tumor metastasis: molecular insights and evolving paradigms. Cell 147(2):275–292
4. Galli S (2014) X-ray crystallography: one century of Nobel Prizes. J Chem Educ 91 (12):2009–2012
5. Vaughan D (1997) A vital legacy: biological and environmental research in the atomic age. Lawrence Berkeley National Laboratory, Berkeley
6. Sweet WG (1953) Localization of brain tumors with positron emitters. Nucleonics 11:40–45
7. Slavkovsky P, Uhliar R (2004) The Nobel Prize in physiology or medicine in 2003 to Paul C. Lauterbur, Peter Mansfield for magnetic resonance imaging. Bratisl Lek Listy 105 (7–8):245–249

8. Zimmer M (2009) GFP: from jellyfish to the Nobel Prize and beyond. Chem Soc Rev 38 (10):2823–2832
9. Möckl L, Lamb DC, Bräuchle C (2014) Super-resolved fluorescence microscopy: Nobel Prize in chemistry 2014 for Eric Betzig, Stefan Hell, and William E. Moerner. Angew Chem Int Ed 53(51):13972–13977
10. Henderson R, Unwin PNT (1975) Three-dimensional model of purple membrane obtained by electron microscopy. Nature 257(5521):28
11. Adrian M, Dubochet J, Lepault J, McDowall AW (1984) Cryo-electron microscopy of viruses. Nature 308(5954):32
12. Brenner DJ, Hall EJ (2007) Computed tomography – an increasing source of radiation exposure. N Engl J Med 357(22):2277–2284. https://doi.org/10.1056/NEJMra072149
13. Team NLST (2011) Reduced lung-cancer mortality with low-dose computed tomographic screening. N Engl J Med 365(5):395–409
14. Verschakelen JA, De Wever W (2007) Computed tomography of the lung: a pattern approach. Springer, Berlin/Heidelberg
15. Kalender WA (2003) The use of flat-panel detectors for CT imaging. Radiologe 43 (5):379–387
16. Gupta R, Grasruck M, Suess C, Bartling SH, Schmidt B, Stierstorfer K et al (2006) Ultra-high resolution flat-panel volume CT: fundamental principles, design architecture, and system characterization. Eur Radiol 16(6):1191–1205
17. Dewey M, Laule M, Krug L, Schnapauff D, Rogalla P, Rutsch W et al (2004) Multisegment and halfscan reconstruction of 16-slice computed tomography for detection of coronary artery stenoses. Invest Radiol 39(4):223–229
18. Flohr TG, McCollough CH, Bruder H, Petersilka M, Gruber K, Süβ C et al (2006) First performance evaluation of a dual-source CT (DSCT) system. Eur Radiol 16(2):256–268
19. Petersilka M, Bruder H, Krauss B, Stierstorfer K, Flohr TG (2008) Technical principles of dual source CT. Eur J Radiol 68(3):362–368
20. Kalender WA (2006) X-ray computed tomography. Phys Med Biol 51(13):R29
21. Lee ES, Lee JM (2014) Imaging diagnosis of pancreatic cancer: a state-of-the-art review. World J Gastroenterol: WJG 20(24):7864
22. Brennan DD, Zamboni GA, Raptopoulos VD, Kruskal JB (2007) Comprehensive preoperative assessment of pancreatic adenocarcinoma with 64-section volumetric CT. Radiographics 27 (6):1653–1666
23. Obert M, Ahlemeyer B, Baumgart-Vogt E, Traupe H (2005) Flat-panel volumetric computed tomography: a new method for visualizing fine bone detail in living mice. J Comput Assist Tomogr 29(4):560–565
24. Chen B, Ning R (2002) Cone-beam volume CT breast imaging: feasibility study. Med Phys 29 (5):755–770
25. Lam WW, Holdsworth DW, Du LY, Drangova M, McCormack DG, Santyr GE (2007) Micro-CT imaging of rat lung ventilation using continuous image acquisition during xenon gas contrast enhancement. J Appl Physiol 103(5):1848–1856
26. Ahn S, Jung S, Lee S (2013) Gold nanoparticle contrast agents in advanced X-ray imaging technologies. Molecules 18(5):5858–5890
27. Hallouard F, Anton N, Choquet P, Constantinesco A, Vandamme T (2010) Iodinated blood pool contrast media for preclinical X-ray imaging applications–a review. Biomaterials 31 (24):6249–6268
28. Mukundan S Jr, Ghaghada KB, Badea CT, Kao C, Hedlund LW, Provenzale JM et al (2006) A liposomal nanoscale contrast agent for preclinical CT in mice. Am J Roentgenol 186 (2):300–307
29. Torchilin VP (2005) Recent advances with liposomes as pharmaceutical carriers. Nat Rev Drug Discov 4(2):145

30. Mawad D, Mouaziz H, Penciu A, Méhier H, Fenet B, Fessi H et al (2009) Elaboration of radiopaque iodinated nanoparticles for in situ control of local drug delivery. Biomaterials 30 (29):5667–5674
31. Hainfeld JF, Slatkin DN, Focella TM, Smilowitz HM (2006) Gold nanoparticles: a new X-ray contrast agent. Br J Radiol 79(939):248–253
32. Boll H, Nittka S, Doyon F, Neumaier M, Marx A, Kramer M et al (2011) Micro-CT based experimental liver imaging using a nanoparticulate contrast agent: a longitudinal study in mice. PLoS One 6(9):e25692
33. Foster FS, Burns PN, Simpson DH, Wilson SR, Christopher DA, Goertz DE (2000) Ultrasound for the visualization and quantification of tumor microcirculation. Cancer Metastasis Rev 19(1–2):131–138
34. Gress F, Gottlieb K, Sherman S, Lehman G (2001) Endoscopic ultrasonography–guided fine-needle aspiration biopsy of suspected pancreatic cancer. Ann Intern Med 134(6):459–464
35. Wells PNT, Liang HD (2011) Medical ultrasound: imaging of soft tissue strain and elasticity. J R Soc Interface 8(64):1521–1549. https://doi.org/10.1098/rsif.2011.0054
36. Jensen J (2007) Medical ultrasound imaging. Prog Biophys Mol Biol 93(1–3):153–165. https://doi.org/10.1016/j.pbiomolbio.2006.07.025
37. Nikolov SI (2001) Synthetic aperture tissue and flow ultrasound imaging. PhD thesis, Ørsted• DTU, Technical University of Denmark, 2800, Lyngby, Denmark
38. Huang Q, Zeng Z (2017) A review on real-time 3D ultrasound imaging technology. Biomed Res Int 2017:1–20
39. Kaufmann BA, Lindner JR (2007) Molecular imaging with targeted contrast ultrasound. Curr Opin Biotechnol 18(1):11–16
40. Wilson SR, Burns PN (2010) Microbubble-enhanced US in body imaging: what role? Radiology 257(1):24–39. https://doi.org/10.1148/radiol.10091210
41. Brooks PC, Montgomery AM, Rosenfeld M, Reisfeld RA, Hu T, Klier G et al (1994) Integrin αvβ3 antagonists promote tumor regression by inducing apoptosis of angiogenic blood vessels. Cell 79(7):1157–1164
42. Jun HY, Park SH, Kim HS, Yoon K (2010) Long residence time of ultrasound microbubbles targeted to integrin in murine tumor model. Acad Radiol 17(1):54–60
43. Anderson CR, Hu X, Zhang H, Tlaxca J, Declèves A, Houghtaling R et al (2011) Ultrasound molecular imaging of tumor angiogenesis with an integrin targeted microbubble contrast agent. Invest Radiol 46(4):215–224. https://doi.org/10.1097/RLI.0b013e3182034fed
44. Rahmim A, Zaidi H (2008) PET versus SPECT: strengths, limitations and challenges. Nucl Med Commun 29(3):193–207. https://doi.org/10.1097/MNM.0b013e3282f3a515
45. Rosenthal MS, Cullom J, Hawkins W, Moore SC, Tsui B, Yester M (1995) Quantitative SPECT imaging: a review and recommendations by the focus committee of the society of nuclear medicine computer and instrumentation council. J Nucl Med 36(8):1489–1513
46. Madsen MT (2007) Recent advances in SPECT imaging. J Nucl Med 48(4):661–673. https://doi.org/10.2967/jnumed.106.032680
47. Gambhir SS (2002) Molecular imaging of cancer with positron emission tomography. Nat Rev Cancer 2(9):683–693. https://doi.org/10.1038/nrc882
48. Gambhir SS, Czernin J, Schwimmer J, Silverman DH, Coleman RE, Phelps ME (2001) A tabulated summary of the FDG PET literature. J Nucl Med 42(5 suppl):1S–93S
49. Kherlopian AR, Song T, Duan Q, Neimark MA, Po MJ, Gohagan JK et al (2008) A review of imaging techniques for systems biology. BMC Syst Biol 2(1):74. https://doi.org/10.1186/1752-0509-2-74
50. Dobrucki LW, Sinusas AJ (2010) PET and SPECT in cardiovascular molecular imaging. Nat Rev Cardiol 7(1):38
51. Henneman MM, Bengel FM, van der Wall EE, Knuuti J, Bax JJ (2008) Cardiac neuronal imaging: application in the evaluation of cardiac disease. J Nucl Cardiol 15(3):442–455. https://doi.org/10.1016/j.nuclcard.2008.02.023

52. Bailey DL, Maisey MN, Townsend DW, Valk PE (2005) Positron emission tomography. Springer, Secaucus
53. Apolo AB, Pandit-Taskar N, Morris MJ (2008) Novel tracers and their development for the imaging of metastatic prostate cancer. J Nucl Med 49(12):2031–2041
54. Jadvar H (2011) Prostate cancer: PET with 18F-FDG, 18F-or 11C-acetate, and 18F-or 11C-choline. J Nucl Med 52(1):81–89
55. Mansfield P, Grannell PK (1975) "Diffraction" and microscopy in solids and liquids by NMR. Phys Rev B 12(9):3618
56. Condeelis J, Weissleder R (2010) In vivo imaging in cancer. Cold Spring Harb Perspect Biol 2 (12):a3848
57. Hornak JP (2005) The basics of MRI. Interactive Learning Software, New York. http://www.cis.rit.edu/htbooks/mri
58. Rinck P (ed) (2014) Magnetic resonance: a critical peer-reviewed introduction. Magnetic resonance in medicine. The basic textbook of the European magnetic resonance forum
59. Shah N, Sattar A, Benanti M, Hollander S, Cheuck L (2006) Magnetic resonance spectroscopy as an imaging tool for cancer: a review of the literature. J Am Osteopath Assoc 106(1):23–27
60. Kwock L, Smith JK, Castillo M, Ewend MG, Cush S, Hensing T et al (2002) Clinical applications of proton MR spectroscopy in oncology. Technol Cancer Res Treat 1(1):17–28
61. Castillo M, Kwock L, Scatliff J, Mukherji SK (1998) Proton MR spectroscopy in neoplastic and non-neoplastic brain disorders. Magn Reson Imaging Clin N Am 6(1):1–20
62. Tamiya T, Kinoshita K, Ono Y, Matsumoto K, Furuta T, Ohmoto T (2000) Proton magnetic resonance spectroscopy reflects cellular proliferative activity in astrocytomas. Neuroradiology 42(5):333–338
63. Young RJ, Knopp EA (2006) Brain MRI: tumor evaluation. J Magn Reson Imaging 24 (4):709–724
64. Bizzi A, Blasi V, Falini A, Ferroli P, Cadioli M, Danesi U et al (2008) Presurgical functional MR imaging of language and motor functions: validation with intraoperative electrocortical mapping. Radiology 248(2):579–589
65. Pillai JJ (2010) The evolution of clinical functional imaging during the past 2 decades and its current impact on neurosurgical planning. Am J Neuroradiol 31(2):219–225
66. Ulmer JL, Salvan CV, Mueller WM, Krouwer HG, Stroe GO, Aralasmak A et al (2004) The role of diffusion tensor imaging in establishing the proximity of tumor borders to functional brain systems: implications for preoperative risk assessments and postoperative outcomes. Technol Cancer Res Treat 3(6):567–576
67. Padhani AR, Gapinski CJ, MACVICAR DA, Parker GJ, Suckling J, Revell PB et al (2000) Dynamic contrast enhanced MRI of prostate cancer: correlation with morphology and tumour stage, histological grade and PSA. Clin Radiol 55(2):99–109
68. Gong QY, Brunt JN, Romaniuk CS, Oakley JP, Tan LT, Roberts N et al (1999) Contrast enhanced dynamic MRI of cervical carcinoma during radiotherapy: early prediction of tumour regression rate. Br J Radiol 72(864):1177–1184
69. Murray AD, Redpath TW, Needham G, Gilbert FJ, Brookes JA, Eremin O (1996) Dynamic magnetic resonance mammography of both breasts following local excision and radiotherapy for breast carcinoma. Br J Radiol 69(823):594–600
70. Cha S, Knopp EA, Johnson G, Wetzel SG, Litt AW, Zagzag D (2002) Intracranial mass lesions: dynamic contrast-enhanced susceptibility-weighted echo-planar perfusion MR imaging. Radiology 223(1):11–29
71. Luker GD, Luker KE (2008) Optical imaging: current applications and future directions. J Nucl Med 49(1):1–4. https://doi.org/10.2967/jnumed.107.045799
72. Auler H, Banzer G (1942) Untersuchungen über die Rolle der Porphyrine bei geschwulstkranken Menschen und Tieren. Z Krebsforsch 53(2):65–68
73. Moore GE, Peyton WT, French LA, Walker WW (1948) The clinical use of fluorescein in neurosurgery: the localization of brain tumors. J Neurosurg 5(4):392–398

74. Sevick-Muraca EM, Sharma R, Rasmussen JC, Marshall MV, Wendt JA, Pham HQ et al (2008) Imaging of lymph flow in breast cancer patients after microdose administration of a near-infrared fluorophore: feasibility study. Radiology 246(3):734–741
75. Troyan SL, Kianzad V, Gibbs-Strauss SL, Gioux S, Matsui A, Oketokoun R et al (2009) The FLARE™ intraoperative near-infrared fluorescence imaging system: a first-in-human clinical trial in breast cancer sentinel lymph node mapping. Ann Surg Oncol 16(10):2943–2952
76. Funovics M, Weissleder R, Tung C (2003) Protease sensors for bioimaging. Anal Bioanal Chem 377(6):956–963
77. Schaeffter T (2005) Imaging modalities: principles and information content. In: Imaging in drug discovery and early clinical trials. Springer, Basel, pp 15–81
78. Kim HL (ed) (2009) Optical imaging in oncology. In: Urologic oncology: seminars and original investigations. Elsevier
79. Zysk AM, Nguyen FT, Oldenburg AL, Marks DL, Boppart SA (2007) Optical coherence tomography: a review of clinical development from bench to bedside. J Biomed Opt 12 (5):51403
80. Boumaza S, Arribas SM, Osborne-Pellegrin M, McGrath JC, Laurent S, Lacolley P et al (2001) Fenestrations of the carotid internal elastic lamina and structural adaptation in stroke-prone spontaneously hypertensive rats. Hypertension 37(4):1101–1107
81. Lee WK, Bell J, Kilpatrick E, Hayes M, Lindop GB, Dominiczak MH (1993) Collagen-linked fluorescence in human atherosclerotic plaques. Atherosclerosis 98(2):219–227
82. Nolting DD, Gore JC, Pham W (2011) Near-infrared dyes: probe development and applications in optical molecular imaging. Curr Org Synth 8(4):521–534
83. Kim S, Lim YT, Soltesz EG, De Grand AM, Lee J, Nakayama A et al (2004) Near-infrared fluorescent type II quantum dots for sentinel lymph node mapping. Nat Biotechnol 22(1):93
84. Ntziachristos V, Ripoll J, Wang LV, Weissleder R (2005) Looking and listening to light: the evolution of whole-body photonic imaging. Nat Biotechnol 23(3):313
85. Graves EE, Ripoll J, Weissleder R, Ntziachristos V (2003) A submillimeter resolution fluorescence molecular imaging system for small animal imaging. Med Phys 30(5):901–911
86. Nagaya T, Nakamura YA, Choyke PL, Kobayashi H (2017) Fluorescence-guided surgery. Front Oncol 7. https://doi.org/10.3389/fonc.2017.00314
87. Alander JT, Kaartinen I, Laakso A, Pätilä T, Spillmann T, Tuchin VV et al (2012) A review of indocyanine green fluorescent imaging in surgery. Int J Biomed Imag 2012:1–26. https://doi.org/10.1155/2012/940585
88. Kogure K, Choromokos E (1969) Infrared absorption angiography. J Appl Physiol 26 (1):154–157
89. Engel E, Schraml R, Maisch T, Kobuch K, König B, Szeimies R et al (2008) Light-induced decomposition of indocyanine green. Investig Ophthalmol Vis Sci 49(5):1777–1783
90. Alander JT, Kaartinen I, Laakso A, Pätilä T, Spillmann T, Tuchin VV et al (2012) A review of indocyanine green fluorescent imaging in surgery. Int J Biomed Imaging 2012:1–26. https://doi.org/10.1155/2012/940585
91. Stanescu-Segall D, Jackson TL (2009) Vital staining with indocyanine green: a review of the clinical and experimental studies relating to safety. Eye 23(3):504
92. Paumgartner G (2010) Biliary physiology and disease: reflections of a physician-scientist. Hepatology 51(4):1095–1106
93. Kaiser M, Yafi A, Cinat M, Choi B, Durkin AJ (2011) Noninvasive assessment of burn wound severity using optical technology: a review of current and future modalities. Burns 37 (3):377–386
94. Polom K, Murawa D, Rho YS, Nowaczyk P, Hünerbein M, Murawa P (2011) Current trends and emerging future of indocyanine green usage in surgery and oncology: a literature review. Cancer 117(21):4812–4822
95. Schaafsma BE, Mieog JSD, Hutteman M, van der Vorst JR, Kuppen PJK, Löwik CWGM et al (2011) The clinical use of indocyanine green as a near-infrared fluorescent contrast agent for

image-guided oncologic surgery. J Surg Oncol 104(3):323–332. https://doi.org/10.1002/jso.21943

96. Te Velde EA, Veerman T, Subramaniam V, Ruers T (2010) The use of fluorescent dyes and probes in surgical oncology. Eur J Surg Oncol (EJSO) 36(1):6–15
97. Luo S, Zhang E, Su Y, Cheng T, Shi C (2011) A review of NIR dyes in cancer targeting and imaging. Biomaterials 32(29):7127–7138
98. Hansen DA, Spence AM, Carski T, Berger MS (1993) Indocyanine green (ICG) staining and demarcation of tumor margins in a rat glioma model. Surg Neurol 40(6):451–456
99. Kim EH, Cho JM, Chang JH, Kim SH, Lee KS (2011) Application of intraoperative indocyanine green videoangiography to brain tumor surgery. Acta Neurochir 153 (7):1487–1495. https://doi.org/10.1007/s00701-011-1046-x
100. Hadjipanayis CG, Stummer W (2019) 5-ALA and FDA approval for glioma surgery. J Neurooncol:1–8
101. Raman CV, Krishnan KS (1928) A new type of secondary radiation. Nature 121(3048):501
102. Bumbrah GS, Sharma RM (2016) Raman spectroscopy – basic principle, instrumentation and selected applications for the characterization of drugs of abuse. Egypt J Forensic Sci 6 (3):209–215. https://doi.org/10.1016/j.ejfs.2015.06.001
103. Kneipp J, Kneipp H, Kneipp K (2008) SERS—a single-molecule and nanoscale tool for bioanalytics. Chem Soc Rev 37(5):1052. https://doi.org/10.1039/b708459p
104. Fleischmann M, Hendra PJ, McQuillan AJ (1974) Raman spectra of pyridine adsorbed at a silver electrode. Chem Phys Lett 26(2):163–166
105. Stiles PL, Dieringer JA, Shah NC, Van Duyne RP (2008) Surface-enhanced Raman spectroscopy. Annu Rev Anal Chem 1:601–626
106. Sharma B, Frontiera RR, Henry A, Ringe E, Van Duyne RP (2012) SERS: materials, applications, and the future. Mater Today 15(1–2):16–25
107. McAughtrie S, Faulds K, Graham D (2014) Surface enhanced Raman spectroscopy (SERS): potential applications for disease detection and treatment. J Photochem Photobiol C Photchem Rev 21:40–53
108. Schlücker S (2009) SERS microscopy: nanoparticle probes and biomedical applications. ChemPhysChem 10(9–10):1344–1354
109. Ferraro JR, Nakamoto K (1994) Introductory Raman spectroscopy. Academic/Harcourt Brace & Company, Publishers, San Francisco
110. Pence I, Mahadevan-Jansen A (2016) Clinical instrumentation and applications of Raman spectroscopy. Chem Soc Rev 45(7):1958–1979
111. Willard HH, Merritt LL Jr, Dean JA, Settle FA Jr (1988) Instrumental methods of analysis. Wadsworth, Belmont
112. Skoog DA, Holler FJ, Crouch SR (2017) Principles of instrumental analysis: cengage learning. Cengage Learning
113. Feng S, Chen R, Lin J, Pan J, Wu Y, Li Y et al (2011) Gastric cancer detection based on blood plasma surface-enhanced Raman spectroscopy excited by polarized laser light. Biosens Bioelectron 26(7):3167–3174
114. Lin D, Feng S, Pan J, Chen Y, Lin J, Chen G et al (2011) Colorectal cancer detection by gold nanoparticle based surface-enhanced Raman spectroscopy of blood serum and statistical analysis. Opt Express 19(14):13565–13577
115. Kah JCY, Kho KW, Lee CGL, Richard CJ (2007) Early diagnosis of oral cancer based on the surface plasmon resonance of gold nanoparticles. Int J Nanomedicine 2(4):785
116. Wang G, Lipert RJ, Jain M, Kaur S, Chakraboty S, Torres MP et al (2011) Detection of the potential pancreatic cancer marker MUC4 in serum using surface-enhanced Raman scattering. Anal Chem 83(7):2554–2561
117. Tabakman SM, Chen Z, Casalongue HS, Wang H, Dai H (2011) A new approach to solution-phase gold seeding for SERS substrates. Small 7(4):499–505
118. Yang J, Wang Z, Zong S, Song C, Zhang R, Cui Y (2012) Distinguishing breast cancer cells using surface-enhanced Raman scattering. Anal Bioanal Chem 402(3):1093–1100

119. Lee M, Lee S, Lee J, Lim H, Seong GH, Lee EK et al (2011) Highly reproducible immunoassay of cancer markers on a gold-patterned microarray chip using surface-enhanced Raman scattering imaging. Biosens Bioelectron 26(5):2135–2141
120. Dinish US, Fu CY, Soh KS, Ramaswamy B, Kumar A, Olivo M (2012) Highly sensitive SERS detection of cancer proteins in low sample volume using hollow core photonic crystal fiber. Biosens Bioelectron 33(1):293–298
121. Wang W, Zhao J, Short M, Zeng H (2015) Real-time in vivo cancer diagnosis using Raman spectroscopy. J Biophotonics 8(7):527–545
122. Bergholt MS, Lin K, Zheng W, Huang Z, Lau D (2012) In vivo, real-time, transnasal, image-guided Raman endoscopy: defining spectral properties in the nasopharynx and larynx. J Biomed Opt 17(7):77002
123. Kaminaka S, Ito T, Yamazaki H, Kohda E, Hamaguchi HO (2002) Near-infrared multichannel Raman spectroscopy toward real-time in vivo cancer diagnosis. J Raman Spectros 33 (7):498–502
124. Huang Z, Zeng H, Hamzavi I, McLean DI, Lui H (2001) Rapid near-infrared Raman spectroscopy system for real-time in vivo skin measurements. Opt Lett 26(22):1782–1784
125. Motz JT, Gandhi SJ, Scepanovic OR, Haka AS, Kramer JR, Dasari RR et al (2007) Real-time Raman system for in vivo disease diagnosis. J Biomed Opt 10(3):31113
126. Lui H, Zhao J, McLean D, Zeng H (2012) Real-time Raman spectroscopy for in vivo skin cancer diagnosis. Cancer Res 72(10):2491–2500
127. Lieber CA, Majumder SK, Ellis DL, Billheimer DD, Mahadevan JA (2008) In vivo nonmelanoma skin cancer diagnosis using Raman microspectroscopy. Lasers Surg Med 40 (7):461–467
128. Bauer NJ, Wicksted JP, Jongsma FH, March WF, Hendrikse F, Motamedi M (1998) Noninvasive assessment of the hydration gradient across the cornea using confocal Raman spectroscopy. Invest Ophthalmol Vis Sci 39(5):831–835
129. Haka AS, Volynskaya Z, Gardecki JA, Nazemi J, Lyons J, Hicks D et al (2006) In vivo margin assessment during partial mastectomy breast surgery using Raman spectroscopy. Cancer Res 66(6):3317–3322
130. Huang Z, Bergholt MS, Zheng W, Lin K, Ho K, Yeoh K (2010) In vivo early diagnosis of gastric dysplasia using narrow-band image-guided Raman endoscopy. J Biomed Opt 15 (3):37017
131. Duraipandian S, Bergholt MS, Zheng W, Ho KY, Teh M, Yeoh KG et al (2012) Real-time Raman spectroscopy for in vivo, online gastric cancer diagnosis during clinical endoscopic examination. J Biomed Opt 17(8):81418
132. Vendrell M, Maiti KK, Dhaliwal K, Chang Y (2013) Surface-enhanced Raman scattering in cancer detection and imaging. Trends Biotechnol 31(4):249–257. https://doi.org/10.1016/j.tibtech.2013.01.013
133. Townsend DW (2008) Dual-modality imaging: combining anatomy and function. J Nucl Med 49(6):938–955
134. Hasegawa BH, Gingold EL, Reilly SM, Liew S, Cann CE (ed) (1990) Description of a simultaneous emission-transmission CT system. In: Medical imaging IV: image formation. International Society for Optics and Photonics
135. Kinahan PE, Townsend DW, Beyer T, Sashin D (1998) Attenuation correction for a combined 3D PET/CT scanner. Med Phys 25(10):2046–2053
136. von Schulthess GK, Steinert HC, Hany TF (2006) Integrated PET/CT: current applications and future directions. Radiology 238(2):405–422
137. Larson SM, Schwartz LH (2006) 18F-FDG PET as a candidate for "qualified biomarker": functional assessment of treatment response in oncology. J Nucl Med 47(6):901–903
138. Fletcher JW, Djulbegovic B, Soares HP, Siegel BA, Lowe VJ, Lyman GH et al (2008) Recommendations on the use of 18F-FDG PET in oncology. J Nucl Med 49(3):480–508

139. Borst GR, Belderbos JS, Boellaard R, Comans EF, De Jaeger K, Lammertsma AA et al (2005) Standardised FDG uptake: a prognostic factor for inoperable non-small cell lung cancer. Eur J Cancer 41(11):1533–1541
140. Bastiaannet E, Groen H, Jager PL, Cobben D, Van Der Graaf W, Vaalburg W et al (2004) The value of FDG-PET in the detection, grading and response to therapy of soft tissue and bone sarcomas; a systematic review and meta-analysis. Cancer Treat Rev 30(1):83–101
141. Thorwarth D, Beyer T, Boellaard R, De Ruysscher D, Grgic A, Lee JA et al (2012) Integration of FDG-PET/CT into external beam radiation therapy planning. Nuklearmedizin 51 (04):140–153
142. Gregoire V, Chiti A (2010) PET in radiotherapy planning: particularly exquisite test or pending and experimental tool? Radiother Oncol 96(3):275
143. Gallowitsch H, Kresnik E, Gasser J, Kumnig G, Igerc I, Mikosch P et al (2003) F-18 fluorodeoxyglucose positron-emission tomography in the diagnosis of tumor recurrence and metastases in the follow-up of patients with breast carcinoma: a comparison to conventional imaging. Invest Radiol 38(5):250–256
144. Litmanovich D, Gourevich K, Israel O, Gallimidi Z (2009) Unexpected foci of 18 F-FDG uptake in the breast detected by PET/CT: incidence and clinical significance. Eur J Nucl Med Mol Imaging 36(10):1558–1564
145. Vranjesevic D, Filmont JE, Meta J, Silverman DH, Phelps ME, Rao J et al (2002) Whole-body 18F-FDG PET and conventional imaging for predicting outcome in previously treated breast cancer patients. J Nucl Med 43(3):325–329
146. Jung NY, Yoo IR, Kang BJ, Kim SH, Chae BJ, Seo YY (2016) Clinical significance of FDG-PET/CT at the postoperative surveillance in the breast cancer patients. Breast Cancer 23 (1):141–148
147. Boellaard R, Delgado-Bolton R, Oyen WJ, Giammarile F, Tatsch K, Eschner W et al (2015) FDG PET/CT: EANM procedure guidelines for tumour imaging: version 2.0. Eur J Nucl Med Mol Imaging 42(2):328–354
148. Lardinois D, Weder W, Hany TF, Kamel EM, Korom S, Seifert B et al (2003) Staging of non–small-cell lung cancer with integrated positron-emission tomography and computed tomography. N Engl J Med 348(25):2500–2507
149. Sterzing F, Kratochwil C, Fiedler H, Katayama S, Habl G, Kopka K et al (2016) 68 Ga-PSMA-11 PET/CT: a new technique with high potential for the radiotherapeutic management of prostate cancer patients. Eur J Nucl Med Mol Imaging 43(1):34–41
150. Sooriakumaran P, Nyberg T, Akre O, Haendler L, Heus I, Olsson M et al (2014) Comparative effectiveness of radical prostatectomy and radiotherapy in prostate cancer: observational study of mortality outcomes. BMJ 348:g1502
151. Bluemel C, Krebs M, Polat B, Linke F, Eiber M, Samnick S et al (2016) 68Ga-PSMA-PET/CT in patients with biochemical prostate cancer recurrence and negative 18F-choline-PET/CT. Clin Nucl Med 41(7):515
152. de Camargo Etchebehere ECS, de Oliveira SA, Gumz B, Vicente A, Hoff PG, Corradi G et al (2014) 68Ga-DOTATATE PET/CT, 99mTc-HYNIC-octreotide SPECT/CT, and whole-body MR imaging in detection of neuroendocrine tumors: a prospective trial. J Nucl Med 55 (10):1598–1604
153. Jambor I, Kuisma A, Ramadan S, Huovinen R, Sandell M, Kajander S et al (2016) Prospective evaluation of planar bone scintigraphy, SPECT, SPECT/CT, 18F-NaF PET/CT and whole body 1.5 T MRI, including DWI, for the detection of bone metastases in high risk breast and prostate cancer patients: SKELETA clinical trial. Acta Oncol 55(1):59–67
154. Franzius C, Hermann K, Weckesser M, Kopka K, Juergens KU, Vormoor J et al (2006) Whole-body PET/CT with 11C-meta-hydroxyephedrine in tumors of the sympathetic nervous system: feasibility study and comparison with 123I-MIBG SPECT/CT. J Nucl Med 47 (10):1635–1642

155. Wehrl HF, Sauter AW, Judenhofer MS, Pichler BJ (2010) Combined PET/MR imaging — technology and applications. Technol Cancer Res Treat 9(1):5–20. https://doi.org/10.1177/153303461000900102
156. Wehrl HF, Sauter AW, Divine MR, Pichler BJ (2015) Combined PET/MR: a technology becomes mature. J Nucl Med 56(2):165–168. https://doi.org/10.2967/jnumed.114.150318
157. Chen W (2007) Clinical applications of PET in brain tumors. J Nucl Med 48(9):1468–1481
158. Jacobs AH, Kracht LW, Gossmann A, Rüger MA, Thomas AV, Thiel A et al (2005) Imaging in neurooncology. NeuroRx 2(2):333–347
159. Thiel A, Habedank B, Herholz K, Kessler J, Winhuisen L, Haupt WF et al (2006) From the left to the right: how the brain compensates progressive loss of language function. Brain Lang 98 (1):57–65
160. Park SY, Zacharias C, Harrison C, Fan RE, Kunder C, Hatami N et al (2018) Gallium 68 PSMA-11 PET/MR imaging in patients with intermediate-or high-risk prostate cancer. Radiology 288(2):495–505
161. Afshar-Oromieh A, Haberkorn U, Schlemmer HP, Fenchel M, Eder M, Eisenhut M et al (2014) Comparison of PET/CT and PET/MRI hybrid systems using a 68 Ga-labelled PSMA ligand for the diagnosis of recurrent prostate cancer: initial experience. Eur J Nucl Med Mol Imaging 41(5):887–897
162. Spick C, Herrmann K, Czernin J (2016) 18F-FDG PET/CT and PET/MRI perform equally well in cancer: evidence from studies on more than 2,300 patients. J Nucl Med 57(3):420–430
163. Grueneisen J, Nagarajah J, Buchbender C, Hoffmann O, Schaarschmidt BM, Poeppel T et al (2015) Positron emission tomography/magnetic resonance imaging for local tumor staging in patients with primary breast cancer: a comparison with positron emission tomography/computed tomography and magnetic resonance imaging. Invest Radiol 50(8):505–513
164. Kim S, Chae MK, Yim MS, Jeong IH, Cho J, Lee C et al (2013) Hybrid PET/MR imaging of tumors using an oleanolic acid-conjugated nanoparticle. Biomaterials 34(33):8114–8121. https://doi.org/10.1016/j.biomaterials.2013.07.078
165. Yan S, Huang C, Wu S, Yin M (2010) Oleanolic acid and ursolic acid induce apoptosis in four human liver cancer cell lines. Toxicol In Vitro 24(3):842–848
166. Juan ME, Wenzel U, Ruiz-Gutierrez V, Daniel H, Planas JM (2006) Olive fruit extracts inhibit proliferation and induce apoptosis in HT-29 human colon cancer cells. J Nutr 136 (10):2553–2557
167. Lee H, Li Z, Chen K, Hsu AR, Xu C, Xie J et al (2008) PET/MRI dual-modality tumor imaging using arginine-glycine-aspartic (RGD)–conjugated radiolabeled iron oxide nanoparticles. J Nucl Med 49(8):1371–1379
168. Yang X, Hong H, Grailer JJ, Rowland IJ, Javadi A, Hurley SA et al (2011) cRGD-functionalized, DOX-conjugated, and 64Cu-labeled superparamagnetic iron oxide nanoparticles for targeted anticancer drug delivery and PET/MR imaging. Biomaterials 32 (17):4151–4160
169. Gao Z, Ma T, Zhao E, Docter D, Yang W, Stauber RH et al (2016) Small is smarter: nano MRI contrast agents – advantages and recent achievements. Small 12(5):556–576. https://doi.org/10.1002/smll.201502309
170. Lewis M, Yanny S, Malcolm PN (2012) Advantages of blood pool contrast agents in MR angiography: a pictorial review. J Med Imaging Radiat Oncol 56(2):187–191
171. Caravan P, Ellison JJ, McMurry TJ, Lauffer RB (1999) Gadolinium (III) chelates as MRI contrast agents: structure, dynamics, and applications. Chem Rev 99(9):2293–2352
172. Chanyaputhipong J, Low SA, Chow PK (2011) Gadoxetate acid-enhanced MR imaging for HCC: a review for clinicians. Int J Hepatol 2011
173. Dong C, Korinek A, Blasiak B, Tomanek B, van Veggel FC (2012) Cation exchange: a facile method to make NaYF4: Yb, Tm-NaGdF4 core–shell nanoparticles with a thin, tunable, and uniform shell. Chem Mater 24(7):1297–1305
174. Park JY, Baek MJ, Choi ES, Woo S, Kim JH, Kim TJ et al (2009) Paramagnetic ultrasmall gadolinium oxide nanoparticles as advanced T1 MRI contrast agent: account for large

longitudinal relaxivity, optimal particle diameter, and in vivo T1 MR images. ACS Nano 3 (11):3663–3669. https://doi.org/10.1021/nn900761s
175. Sandiford L, Phinikaridou A, Protti A, Meszaros LK, Cui X, Yan Y et al (2012) Bisphosphonate-anchored PEGylation and radiolabeling of superparamagnetic iron oxide: long-circulating nanoparticles for in vivo multimodal (T1 MRI-SPECT) imaging. ACS Nano 7(1):500–512
176. Xuan S, Wang F, Lai JMY, Sham KWY, Wang YJ, Lee S et al (2011) Synthesis of biocompatible, mesoporous Fe3 O4 nano/microspheres with large surface area for magnetic resonance imaging and therapeutic applications. ACS Appl Mater Interfaces 3(2):237–244. https://doi.org/10.1021/am1012358
177. Wolfbeis OS (2015) An overview of nanoparticles commonly used in fluorescent bioimaging. Chem Soc Rev 44(14):4743–4768
178. Zheng C, Zheng M, Gong P, Jia D, Zhang P, Shi B et al (2012) Indocyanine green-loaded biodegradable tumor targeting nanoprobes for in vitro and in vivo imaging. Biomaterials 33 (22):5603–5609. https://doi.org/10.1016/j.biomaterials.2012.04.044
179. Liu Y, Li K, Pan J, Liu B, Feng S (2010) Folic acid conjugated nanoparticles of mixed lipid monolayer shell and biodegradable polymer core for targeted delivery of Docetaxel. Biomaterials 31(2):330–338
180. Chan JM, Zhang L, Yuet KP, Liao G, Rhee J, Langer R et al (2009) PLGA–lecithin–PEG core–shell nanoparticles for controlled drug delivery. Biomaterials 30(8):1627–1634
181. Soltesz EG, Kim S, Laurence RG, DeGrand AM, Parungo CP, Dor DM et al (2005) Intraoperative sentinel lymph node mapping of the lung using near-infrared fluorescent quantum dots. Ann Thorac Surg 79(1):269–277
182. Morgan NY, English S, Chen W, Chernomordik V, Russo A, Smith PD et al (2005) Real time in vivo non-invasive optical imaging using near-infrared fluorescent quantum dots1. Acad Radiol 12(3):313–323
183. Parungo CP, Colson YL, Kim S, Kim S, Cohn LH, Bawendi MG et al (2005) Sentinel lymph node mapping of the pleural space. Chest 127(5):1799–1804
184. He X, Duan J, Wang K, Tan W, Lin X, He C (2004) A novel fluorescent label based on organic dye-doped silica nanoparticles for HepG liver cancer cell recognition. J Nanosci Nanotechnol 4(6):585–589
185. Santra S, Zhang P, Wang K, Tapec R, Tan W (2001) Conjugation of biomolecules with luminophore-doped silica nanoparticles for photostable biomarkers. Anal Chem 73 (20):4988–4993
186. Santra S, Liesenfeld B, Dutta D, Chatel D, Batich CD, Tan W et al (2005) Folate conjugated fluorescent silica nanoparticles for labeling neoplastic cells. J Nanosci Nanotechnol 5 (6):899–904
187. Ueno H, Hihara J, Shimizu K, Osaki A, Yamashita Y, Yoshida K et al (2005) Experimental study on fluorescent microspheres as a tracer for sentinel node detection. Anticancer Res 25 (2A):821–825
188. Stsiapura V, Sukhanova A, Artemyev M, Pluot M, Cohen JH, Baranov AV et al (2004) Functionalized nanocrystal-tagged fluorescent polymer beads: synthesis, physicochemical characterization, and immunolabeling application. Anal Biochem 334(2):257–265
189. Zhang J, Liu Z, Lian P, Qian J, Li X, Wang L et al (2016) Selective imaging and cancer cell death via pH switchable near-infrared fluorescence and photothermal effects. Chem Sci 7 (9):5995–6005. https://doi.org/10.1039/C6SC00221H
190. Meng X, Yang Y, Zhou L (2017) Dual-responsive molecular probe for tumor targeted imaging and photodynamic therapy. Theranostics 7(7):1781
191. Luo S, Tan X, Fang S, Wang Y, Liu T, Wang X et al (2016) Mitochondria-targeted small-molecule fluorophores for dual modal cancer phototherapy. Adv Funct Mater 26 (17):2826–2835

192. Fan Z, Sun L, Huang Y, Wang Y, Zhang M (2016) Bioinspired fluorescent dipeptide nanoparticles for targeted cancer cell imaging and real-time monitoring of drug release. Nat Nanotechnol 11(4):388
193. van der Vorst JR, Schaafsma BE, Hutteman M, Verbeek FP, Liefers GJ, Hartgrink HH et al (2013) Near-infrared fluorescence-guided resection of colorectal liver metastases. Cancer 119 (18):3411–3418
194. Sugie T, Sawada T, Tagaya N, Kinoshita T, Yamagami K, Suwa H et al (2013) Comparison of the indocyanine green fluorescence and blue dye methods in detection of sentinel lymph nodes in early-stage breast cancer. Ann Surg Oncol 20(7):2213–2218
195. Mieog JSD, Troyan SL, Hutteman M, Donohoe KJ, Van Der Vorst JR, Stockdale A et al (2011) Toward optimization of imaging system and lymphatic tracer for near-infrared fluorescent sentinel lymph node mapping in breast cancer. Ann Surg Oncol 18(9):2483–2491
196. Van Dam GM, Themelis G, Crane LM, Harlaar NJ, Pleijhuis RG, Kelder W et al (2011) Intraoperative tumor-specific fluorescence imaging in ovarian cancer by folate receptor-α targeting: first in-human results. Nat Med 17(10):1315
197. Van der Vorst JR, Schaafsma BE, Verbeek F, Swijnenburg RJ, Hutteman M, Liefers GJ et al (2013) Dose optimization for near-infrared fluorescence sentinel lymph node mapping in patients with melanoma. Br J Dermatol 168(1):93–98
198. Fujiwara M, Mizukami T, Suzuki A, Fukamizu H (2009) Sentinel lymph node detection in skin cancer patients using real-time fluorescence navigation with indocyanine green: preliminary experience. J Plast Reconstr Aesthet Surg 62(10):e373–e378
199. Crane L, Themelis G, Arts H, Buddingh KT, Brouwers AH, Ntziachristos V et al (2011) Intraoperative near-infrared fluorescence imaging for sentinel lymph node detection in vulvar cancer: first clinical results. Gynecol Oncol 120(2):291–295
200. Hutteman M, Van Der Vorst JR, Gaarenstroom KN, Peters AA, Mieog JSD, Schaafsma BE et al (2012) Optimization of near-infrared fluorescent sentinel lymph node mapping for vulvar cancer. Am J Obstet Gynecol 206(1):81–89
201. Crane LM, Themelis G, Pleijhuis RG, Harlaar NJ, Sarantopoulos A, Arts HJ et al (2011) Intraoperative multispectral fluorescence imaging for the detection of the sentinel lymph node in cervical cancer: a novel concept. Mol Imaging Biol 13(5):1043–1049
202. Peloso A, Franchi E, Canepa MC, Barbieri L, Briani L, Ferrario J et al (2013) Combined use of intraoperative ultrasound and indocyanine green fluorescence imaging to detect liver metastases from colorectal cancer. HPB 15(12):928–934
203. Luo T, Huang P, Gao G, Shen G, Fu S, Cui D et al (2011) Mesoporous silica-coated gold nanorods with embedded indocyanine green for dual mode X-ray CT and NIR fluorescence imaging. Opt Express 19(18):17030–17039
204. Yuasa Y, Seike J, Yoshida T, Takechi H, Yamai H, Yamamoto Y et al (2012) Sentinel lymph node biopsy using intraoperative indocyanine green fluorescence imaging navigated with preoperative CT lymphography for superficial esophageal cancer. Ann Surg Oncol 19 (2):486–493
205. Jennings LE, Long NJ (2009) 'Two is better than one'—probes for dual-modality molecular imaging. Chem Commun 24:3511. https://doi.org/10.1039/b821903f
206. Chen Q, Shang W, Zeng C, Wang K, Liang X, Chi C et al (2017) Theranostic imaging of liver cancer using targeted optical/MRI dual-modal probes. Oncotarget 8(20):32741
207. Na HB, Song IC, Hyeon T (2009) Inorganic nanoparticles for MRI contrast agents. Adv Mater 21(21):2133–2148
208. Semela D, Dufour J (2004) Angiogenesis and hepatocellular carcinoma. J Hepatol 41 (5):864–880
209. Nakamura T, Sugihara F, Matsushita H, Yoshioka Y, Mizukami S, Kikuchi K (2015) Mesoporous silica nanoparticles for19 F magnetic resonance imaging, fluorescence imaging, and drug delivery. Chem Sci 6(3):1986–1990. https://doi.org/10.1039/C4SC03549F
210. Thakor AS, Jokerst J, Zavaleta C, Massoud TF, Gambhir SS (2011) Gold nanoparticles: a revival in precious metal administration to patients. Nano Lett 11(10):4029–4036

211. Saha K, Agasti SS, Kim C, Li X, Rotello VM (2012) Gold nanoparticles in chemical and biological sensing. Chem Rev 112(5):2739–2779
212. Li F, Pei H, Wang L, Lu J, Gao J, Jiang B et al (2013) Nanomaterial-based fluorescent DNA analysis: a comparative study of the quenching effects of graphene oxide, carbon nanotubes, and gold nanoparticles. Adv Funct Mater 23(33):4140–4148
213. Zhang J, Li C, Zhang X, Huo S, Jin S, An F et al (2015) In vivo tumor-targeted dual-modal fluorescence/CT imaging using a nanoprobe co-loaded with an aggregation-induced emission dye and gold nanoparticles. Biomaterials 42:103–111
214. Zavaleta CL, Hartman KB, Miao Z, James ML, Kempen P, Thakor AS et al (2011) Preclinical evaluation of Raman nanoparticle biodistribution for their potential use in clinical endoscopy imaging. Small 7(15):2232–2240
215. Qian J, Jiang L, Cai F, Wang D, He S (2011) Fluorescence-surface enhanced Raman scattering co-functionalized gold nanorods as near-infrared probes for purely optical in vivo imaging. Biomaterials 32(6):1601–1610
216. Gao X, Yue Q, Liu Z, Ke M, Zhou X, Li S et al (2017) Guiding Brain-Tumor Surgery via Blood-Brain-Barrier-Permeable Gold Nanoprobes with Acid-Triggered MRI/SERRS Signals. Adv Mater 29(21):1603917. https://doi.org/10.1002/adma.201603917
217. Braun G, Pavel I, Morrill AR, Seferos DS, Bazan GC, Reich NO et al (2007) Chemically patterned microspheres for controlled nanoparticle assembly in the construction of SERS hot spots. J Am Chem Soc 129(25):7760–7761
218. Holbrook RJ, Rammohan N, Rotz MW, MacRenaris KW, Preslar AT, Meade TJ (2016) Gd (III)-Dithiolane Gold Nanoparticles for T 1-Weighted Magnetic Resonance Imaging of the Pancreas. Nano Lett 16(5):3202–3209
219. Wang P, Fan Y, Lu L, Liu L, Fan L, Zhao M et al (2018) NIR-II nanoprobes in-vivo assembly to improve image-guided surgery for metastatic ovarian cancer. Nat Commun 9(1):2898
220. Li L, Wu P, Hwang K, Lu Y (2013) An exceptionally simple strategy for DNA-functionalized up-conversion nanoparticles as biocompatible agents for nanoassembly, DNA delivery, and imaging. J Am Chem Soc 135(7):2411–2414
221. Zhang X, Chen J, Zheng Y, Gao X, Kang Y, Liu J et al (2009) Follicle-stimulating hormone peptide can facilitate paclitaxel nanoparticles to tàrget ovarian carcinoma in vivo. Cancer Res 69(16):6506–6514
222. Kircher MF, de la Zerda A, Jokerst JV, Zavaleta CL, Kempen PJ, Mittra E et al (2012) A brain tumor molecular imaging strategy using a new triple-modality MRI-photoacoustic-Raman nanoparticle. Nat Med 18(5):829–834. https://doi.org/10.1038/nm.2721
223. Bai J, Wang JT, Rubio N, Protti A, Heidari H, Elgogary R et al (2016) Triple-modal imaging of magnetically-targeted nanocapsules in solid tumours in vivo. Theranostics 6(3):342
224. Yang K, Hu L, Ma X, Ye S, Cheng L, Shi X et al (2012) Multimodal imaging guided photothermal therapy using functionalized graphene nanosheets anchored with magnetic nanoparticles. Adv Mater 24(14):1868–1872
225. Gao N, Yang W, Nie H, Gong Y, Jing J, Gao L et al (2017) Turn-on theranostic fluorescent nanoprobe by electrostatic self-assembly of carbon dots with doxorubicin for targeted cancer cell imaging, in vivo hyaluronidase analysis, and targeted drug delivery. Biosens Bioelectron 96:300–307
226. Kim KS, Hur W, Park S, Hong SW, Choi JE, Goh EJ et al (2010) Bioimaging for targeted delivery of hyaluronic acid derivatives to the livers in cirrhotic mice using quantum dots. ACS Nano 4(6):3005–3014
227. Lokeshwar VB, Selzer MG (2009) Hyaluronidase: both a tumor promoter and suppressor. In: Hyaluronan in cancer biology. Elsevier, pp 189–206
228. Sun X, Liu Z, Welsher K, Robinson JT, Goodwin A, Zaric S et al (2008) Nano-graphene oxide for cellular imaging and drug delivery. Nano Res 1(3):203–212
229. Wang Y, Zhang C, Liu J, Huang G (2012) Is 18F-FDG PET accurate to predict neoadjuvant therapy response in breast cancer? A meta-analysis. Breast Cancer Res Treat 131(2):357–369

230. Mghanga FP, Lan X, Bakari KH, Li C, Zhang Y (2013) Fluorine-18 fluorodeoxyglucose positron emission tomography–computed tomography in monitoring the response of breast Cancer to neoadjuvant chemotherapy: a meta–analysis. Clin Breast Cancer 13(4):271–279
231. Martincich L, Montemurro F, De Rosa G, Marra V, Ponzone R, Cirillo S et al (2004) Monitoring response to primary chemotherapy in breast cancer using dynamic contrast-enhanced magnetic resonance imaging. Breast Cancer Res Treat 83(1):67–76
232. Giordano SH (2003) Update on locally advanced breast cancer. Oncologist 8(6):521–530
233. Kaufmann M, Morrow M, von Minckwitz G, Harris JR (2010) Locoregional treatment of primary breast cancer: consensus recommendations from an International Expert Panel. Cancer 116(5):1184–1191
234. Kinoshita T (2011) Preoperative therapy: recent findings. Breast Cancer 18(2):80–84
235. Tőkés T, Kajáry K, Szentmártoni G, Lengyel Z, Györke T, Torgyík L et al (2017) Predictive and prognostic value of FDG-PET/CT imaging and different response evaluation criteria after primary systemic therapy of breast cancer. Breast Cancer 24(1):137–146
236. Groheux D, Giacchetti S, Hatt M, Marty M, Vercellino L, De Roquancourt A et al (2013) HER2-overexpressing breast cancer: FDG uptake after two cycles of chemotherapy predicts the outcome of neoadjuvant treatment. Br J Cancer 109(5):1157
237. Groheux D, Hindié E, Giacchetti S, Delord M, Hamy A, de Roquancourt A et al (2012) Triple-negative breast cancer: early assessment with 18F-FDG PET/CT during neoadjuvant chemotherapy identifies patients who are unlikely to achieve a pathologic complete response and are at a high risk of early relapse. J Nucl Med 53(2):249–254
238. Schellenberger EA, Bogdanov A Jr, Petrovsky A, Ntziachristos V, Weissleder R, Josephson L (2003) Optical imaging of apoptosis as a biomarker of tumor response to chemotherapy. Neoplasia 5(3):187–192
239. van Heerde WL, Robert-Offerman S, Dumont E, Hofstra L, Doevendans PA, Smits JF et al (2000) Markers of apoptosis in cardiovascular tissues: focus on Annexin V. Cardiovasc Res 45 (3):549–559
240. Blankenberg FG, Katsikis PD, Tait JF, Davis RE, Naumovski L, Ohtsuki K et al (1999) Imaging of apoptosis (Programmed cell death) with 99mTcAnnexin V.
241. Jacobs A, Voges J, Reszka R, Lercher M, Gossmann A, Kracht L et al (2001) Positron-emission tomography of vector-mediated gene expression in gene therapy for gliomas. Lancet 358(9283):727–729
242. Huang X, El-Sayed IH, Qian W, El-Sayed MA (2006) Cancer cell imaging and photothermal therapy in the near-infrared region by using gold nanorods. J Am Chem Soc 128 (6):2115–2120
243. O'Neal DP, Hirsch LR, Halas NJ, Payne JD, West JL (2004) Photo-thermal tumor ablation in mice using near infrared-absorbing nanoparticles. Cancer Lett 209(2):171–176
244. Fu G, Zhu L, Yang K, Zhuang R, Xie J, Zhang F (2016) Diffusion-weighted magnetic resonance imaging for therapy response monitoring and early treatment prediction of photothermal therapy. ACS Appl Mater Interfaces 8(8):5137–5147. https://doi.org/10.1021/acsami.5b11936
245. Chen C, Kuo L, Chang C, Hwu Y, Huang C, Lee S et al (2010) In situ real-time investigation of cancer cell photothermolysis mediated by excited gold nanorod surface plasmons. Biomaterials 31(14):4104–4112
246. Pattani VP, Shah J, Atalis A, Sharma A, Tunnell JW (2015) Role of apoptosis and necrosis in cell death induced by nanoparticle-mediated photothermal therapy. J Nanopart Res 17(1):20
247. Tong L, Zhao Y, Huff TB, Hansen MN, Wei A, Cheng JX (2007) Gold nanorods mediate tumor cell death by compromising membrane integrity. Adv Mater 19(20):3136–3141

Chapter 6
Tumor Therapeutic Modes

Yu Zhong Peng, Li Jun Yang, Hang Hong Lo, Betty Yuen Kwan Law, and Vincent Kam Wai Wong

Abstract Cancer, also known as a malignant tumor, has become one of the top burdens of disease for few decades, and it's also the second lethal cause globally. With extensive efforts of oncology researches and clinical trials in the modern era, scientists have developed numbers of therapeutic approaches in the treatment of cancer. In this chapter, we introduce the existing conventional cancer therapeutic methods in two categories, the traditional cancer treatments and the novel tumor therapeutic modes. The former includes surgery, radiotherapy, hormone therapy, chemotherapy, and stem cell transplant, and the latter involves immunotherapy, targeted therapy, and gene therapy. We first give a brief introduction to each therapy from their history to the definition and list several examples for illustration. Then we discuss the combination of each treatment with other cancer therapies and risks of certain remedies. Particularly, we briefly demonstrate the nanomaterials in the application of targeted therapy. Besides, we also state some current obstructions of targeted therapy that needed to be overcome. Finally, there is the introduction of complementary and alternative medicine for the use of antineoplastic therapy, which we provide some information including the overview of some traditional medicine or whole medical systems.

6.1 Introduction

The word cancer, which derived from a Greek word, is used to describe carcinoma by a physician Hippocrates (460–370 B.C). And some of the earliest evidence of human bone cancer indicates that the time of human finding cancer could go back to 1500 B.C. Besides, the world's oldest records of cancer hailed from ancient Egypt in 1600 BC, the hieroglyphics recorded a probable case of breast cancer: "a bulging tumor. . .like touching a ball of wrappings," which also described cancer as an incurable disease [1, 2].

Y. Z. Peng · L. J. Yang · H. H. Lo · B. Y. K. Law · V. K. W. Wong (✉)
State Key Laboratory of Quality Research in Chinese Medicine, Macau University of Science and Technology, Macau, China

R. Huang, Y. Wang (eds.), *New Nanomaterials and Techniques for Tumor-targeted Systems*, https://doi.org/10.1007/978-981-15-5159-8_6

Cancer has become a non-negligible problem both in developed and developing countries, and it is the second cause of death after cardiovascular diseases mostly associated with genetic mutations and unhealthy lifestyles or other environmental factors [3, 4]. Cancer is the second leading cause of death, and according to the statistics in United States, deaths from cancers decreased by 27% in recent 25 years [5, 6]. It is found that early diagnosis, probing assisted precise operation, and improvement of the technique in these therapies have a significant consequence in cancer patients' survival [3].

Cancer (cancerous) is mostly described as a tumor (neoplasm), but it doesn't mean that cancer is the same as the tumor or cancers and tumors are synonyms [7]. The tumor is a type of excessive and abnormal cell growth which can form a mass. And if these cells spread or potentially expand and invade other parts of the body, it could be cancers also known as malignant neoplasms [8, 9]. Considering the benign neoplasms, a mass of cells that lack the ability to metastasize or invade adjacent tissues, which has the characteristics, are contrary to cancer; thus not all tumors can be life-threatening. Of note, unlike most benign tumors elsewhere in the body, benign brain tumors could be lethal [10]. Just like the evidence of that not all the tumors are cancerous, cancer could happen without tumor growth. For example, blood cancer, also known as leukemia, is not involved with tumor [7].

In the ancient era, cancers were treated with surgery, herbal therapies, or heat. In 1600 BC, Egyptian physician had identified three cases as cancer according to *the Ebers medical papyrus* [11]. Medical practitioners in ancient India used regional or whole-body hyperthermia as treatments to heal medical maladies [12]. From an old Egyptian textbook on trauma surgery, it demonstrated eight cases of breast tumors which were removed by cauterization with a tool called fire drill [13]. Early treatments for cancer were either impractical or too awful to contemplate healer even performed the operation like resection without anesthetic and in unhygienic conditions. Accordingly, patients usually died from infection, but not this terrible disease. With the progression of human modern medicine and oncology, such as the invention of anesthetic and the discovery of oncogenes, people bravely stepped into the gate of modern cancer therapy and strugglingly moved forward to the campaign of defeating cancer [2, 13].

Cancer patients may be treated with one of the treatments in some case. Mostly, people will have a combination of treatment. This chapter will list comprehensive and detailed modern tumor therapeutic methods based on professional research including some traditional treatments, which are involved with surgery like radiation therapy, hormone therapy, stem cell transplants, chemotherapy, and also some advanced or novel therapeutic tools such as immunotherapy, targeted therapy, gene therapy, and some information of complementary and alternative medicine (Fig. 6.1) [14].

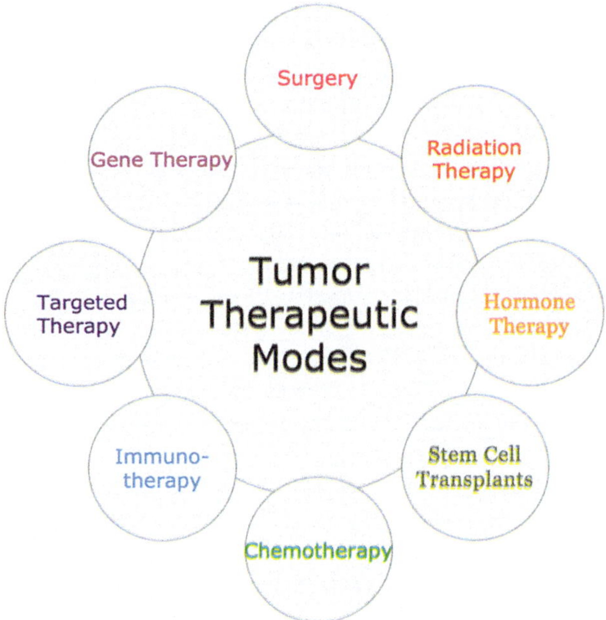

Fig. 6.1 Chapter catalogs tree

6.2 Traditional Cancer Treatment

6.2.1 Surgery

6.2.1.1 Introduction

Surgery is a medical specialty which surgeon uses operative implement and instrumental techniques on a patient to remove tumor tissue from affected organs of cancer patients.

In 1600 BC, *Edwin Smith Papyrus* from Egypt described the earliest human known surgery procedure to treat breast cancer. In 400 BC, Hippocrates, the man who is acknowledged as the "Founder of Western Medicine," insisted on asking the surgeon to clean himself before the operation and using scientific methods in medicine. He also became the first person to distinguish benign and malignant breast tumors, advocating some types of tumors are not lethal [15, 16]. In 50 AD, surgeon Aulus Cornelius Celsus finished his work for compiling a treatise named *De Medicina* including a description of the anatomy of breast cancer "dilated tortuous veins surround the breast" [17]. Meanwhile, Leonides conducted the first excision of a breast tumor with cautery to stop bleeding [18]. Surgery treatments have remained immense pain for patients and carried extreme risk throughout the subsequent 1800 years. Even there were many new achievements in this field, but surgery was still a rare and unconsidered treatment for fighting with cancer. Until the middle of the nineteenth century, Scottish surgeon John Hunter (scientific surgery), American dentist William T. G. Morton (anesthesia), and British surgeon Joseph Lister (aseptic) successfully created the breakthrough in the evolution of surgery, thereby making surgery become an effective and widely accepted intervention thereafter. However, long-term survival rates remain stagnant because of recurrence and metastasis of the tumor [19–21]. Of note, there was a huge improvement of surgical techniques and the combination of other cancer therapies with surgery, successfully reduced the surgical morbidity and mortality. In the future, there will be more precise therapies and tumor target strategies for individual medicine [22].

The types of surgery can be divided into two categories: open surgery or minimally invasive surgery. Open surgery is the most traditional and well-established surgery, which makes a large cut in patients' body and then directly removes the tumor. Minimal invasive surgery uses an approach that does not cut a large incision. Usually, the surgeons take the endoscope to look at the disease area inside the patient's body and then use specific fine instruments to remove these tumor tissues [23, 24].

In this part, 12 types of surgery including open surgery, endoscopic surgery, cryosurgery, laser surgery, hyperthermia surgery, photodynamic therapy, robotic surgery, electrosurgery, high-intensity focused ultrasound, Mohs surgery, stereotactic surgery, and percutaneous ethanol injection will be described.

6.2.1.2 Types of Surgery

6.2.1.2.1 Open Surgery

Open surgery is the most traditional type of surgery which requires surgical scalpels and other sharp tools to cut patient body. Meanwhile, these cuts can make a wide incision which often requires cuts through the skin, muscles, and even bone. The width of the incision may achieve 3–4 inches or even larger, depending on the surgical needs [25].

During the surgery, surgeons normally use anesthesia to keep patients painless. There are three types of anesthesia. Local anesthesia suppresses the sensory feeling of the patient in a particular small area, and regional anesthesia suppresses the sensory feeling in a part of the body. General anesthesia provides a complete loss of awareness of the whole body [23].

The operation of open surgery is declining because of the newly developed new techniques, which cause less invasive to the human body. Current operations reveal that minimally invasive surgery like laser surgery or photodynamic therapy becomes more popular. Nevertheless, the traditional strategy using surgical scalpel is still preferable in the following situations:

- Repairs simply cannot be made effectively using minimally invasive techniques.
- Large visual area is inevitably required to completely remove tissues or accurately diagnose a disease condition.
- Require the insertion of large surgical materials [25].

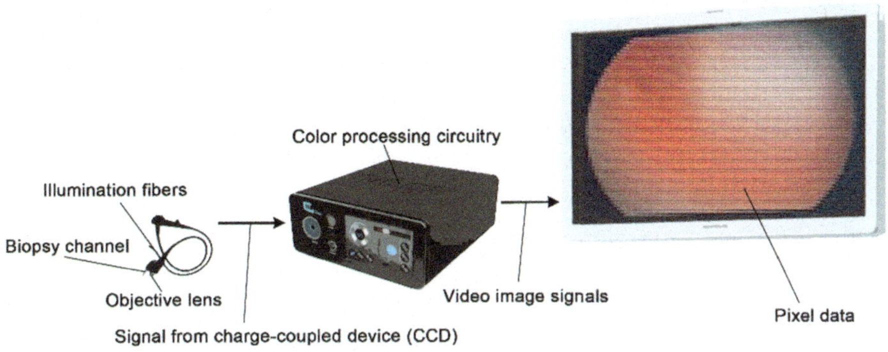

Fig. 6.2 System design of modern video endoscopy

6.2.1.2.2 Endoscopic Surgery

Endoscopic surgery is one of the most popular minimally invasive surgery that is performed using a small scope and goes through into patient's body with a flexible tube connecting with a camera and light source at the end. The whole setup allows the surgeon to see clearly inside and perform the surgical operation without making an unnecessary large incision (Fig. 6.2) [26].

In comparison with open surgery, the main benefit of endoscopic surgery is that it only requires a small incision, usually three small incisions (each ¼ to ½ inch), allowing the patients to experience less pain and short recovery time. In addition, there are many other advantages such as the lower risk of bleeding, the decrease of painful, significant smaller scar, less possibility of pathogens contaminants, and reduced risk of postoperative infection [27]. Besides, there is a type of surgical procedure which is very similar to the endoscopy, called laparoscopy. Laparoscopy is generalized used to examine the organs inside the abdomen, and it is a specific term of endoscopy which the endoscopy also involve thoracoscopy and arthroscopy, etc. [28].

6.2.1.2.3 Cryosurgery

Cryosurgery, also called cryotherapy, is a commonly used surgical procedure that employs freeze condition to destroy tissues like malignant lesions or tumors. Cryosurgery is widely used to treat external tumors, such as skin cancer. In the operation, liquid nitrogen is applied directly to the tumor lesions with a cotton swab or spraying device. Recently, this technique is further improved by incorporating new imaging technologies [29, 30].

The mechanism of this technique can be concluded in three phases.

1. Heat transfer: The heat transfer means that rapidly transfer the heat of patients' lesion area to produce extreme cold which can quickly cool down the patient's tumor to destroy the targets cells by liquid nitrogen or argon which has a boiling point around −200 °C.
2. Cell injury by thawing: The cell damage occurs during the thawing of frozen cells or tissues. It's the most rapid procedure for optimum destruction.
3. Inflammation: This process is the response to the consequence of cell death and cell destruction, which usually observed as erythema and edema [29].

Cryosurgery is used to treat several types of cancer and some precancerous or noncancerous diseases such as retinoblastoma, early-stage skin cancers, precancerous skin growths, cervical intraepithelial neoplasia, AIDS-related Kaposi sarcoma. Besides, researchers are estimating the effectiveness of the use of cryosurgery in the treatment of breast, colon and kidney cancer [31].

6.2.1.2.4 Laser Surgery

Laser surgery is a type of surgery that uses high-intensity light beams to burn cancer cells and tissues. LASER stands for "light amplification by the stimulated emission of radiation" and was first developed in 1960. Unlike other light sources, the laser is powerful to focus on a small spot area and destroy the tumor tissues by intense energy. In surgery, it allows surgeons to work at high precision [32, 33].

Laser therapy is usually adopted through an endoscope which is fitted with an optical fiber or used alone to treat superficial cancers. Lasers could be given to treat some skin cancer and the early stages of cancers such as cervical, penile, vaginal, vulvar, and non-small cell lung cancer [34].

Of note, there are several types of lasers for cancer treatment, such as carbon dioxide lasers, neodymium:yttrium-aluminum-garnet (Nd:YAG) lasers, laser-induced interstitial thermotherapy, and argon lasers [33].

The advantages of laser surgery are that lasers give more precise and less damage to adjacent normal tissues compared with the standard surgical tools. However, surgeons must be specifically trained and qualified to use laser device and strictly follow safety instruction for laser usage. In the future, many types of surgeries can be treated by laser therapy such as prostate and brain cancer which have already been tested in clinical trials [34].

6.2.1.2.5 Hyperthermia Surgery

Hyperthermia is a type of cancer treatment in which body tissue is exposed to high temperatures (up to 113 °F or 45 °C). Hyperthermia is one of the oldest cancer therapeutic methods, and the history of this treatment can trace back to ancient India [35].

It has been shown that high temperature can damage the abnormal cancer cells or tissues but minimally injury to healthy tissues, and hyperthermia could promote the shrinkage of tumors. Recent studies revealed the relationship between temperature and immune system, e.g., the properly high temperature may boost up the immune system of the human body [35–37].

There are several types of hyperthermia including local, regional, and whole-body hyperthermia [37]. However, owing to regional differences in tissue characteristics, high temperature may occur hurt in the body. And the balance of a proper temperature still needs to make sure in case any complications or side effects [35].

Nowadays, hyperthermia is still undergoing to evaluate its effectiveness and is not widely used. In the future, there are numbers of challenges need to be overcome before hyperthermia can be considered a standard treatment for cancer [36].

6.2.1.2.6 Photodynamic Therapy

Photodynamic therapy (PDT) is a type of treatment that uses a light-sensitive medication, called a photosensitizer or photosensitizing agent, and/or a particular light source. When photosensitizers are exposed to a specific wavelength of light, there will be forming abundant of reactive oxygen species (ROS) that can kill nearby cells. PDT appears to shrink or destroy tumor cells in two other ways. Because the photosensitizers have the ability to damage blood vessels in the tumor, it can prevent cancer from receiving necessary nutrients and oxygen. Besides, some evidence showed that PDT may activate the immune system to assault the tumor cells [38].

The first procedure of PDT is to inject a photosensitizing agent into the bloodstream, and the agent will stay inside the body and be absorbed by the cells. After 24–72 h, the photosensitizer in the tumor tissues absorbs light excitation and produces ROS for destroying adjacent cancer cells. Besides, the patients who are treated by PDT will be advised to avoid direct sunlight for 6 weeks because of the photosensitizers [38–40].

PDT can be applied to treat many types of cancers including esophageal cancer, mouth cancer, and lung cancer in early stage [39]. PDT is often performed as an outpatient therapy which also combined with other treatments, such as surgery and chemotherapy [38, 41, 42].

Similar to minimally invasive surgeries, PDT normally causes less damage and more accurate operation in the surgery. However, the photosensitizers can't access more than one-third of an inch of tissue. As such, researchers continue to investigate and improve the effectiveness of PDT and further expand its usage in other types of cancers [38].

6.2.1.2.7 Robotic Surgery

Robotic surgery, also called robot-assisted surgery, is a type of new and emerging technology which allows doctors to perform many complex and difficult surgeries with more accuracy, precision, and flexibility (Fig. 6.3) [43–45].

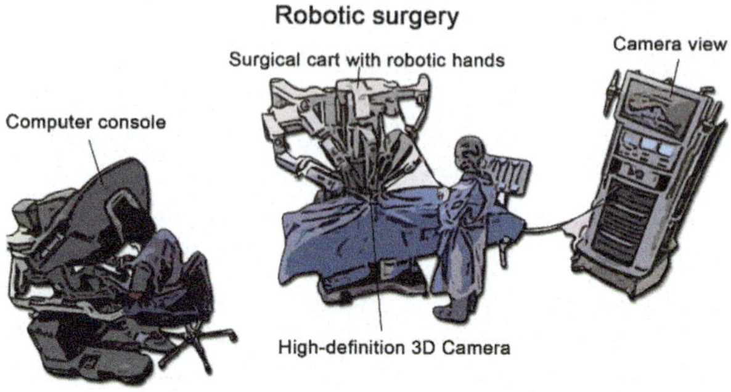

Fig. 6.3 Simple drawing of robotic surgery

The most widely used clinical robotic surgical system includes a camera arm and mechanical arms connecting with surgical instruments, and it allows the surgeon seats in front of the computer console near the operating table. The surgeon can operate many arms at the same time through a stereoscopic high-definition monitor that literally places him inside the patient body, which can give the surgeon a better, magnified, 3D view of the surgical site [43, 44]. Robotic surgery generally causes fewer complications, less pain and lower blood loss, and smaller scars and therefore offers a better recovery [43, 44]. Of noted, robotic surgery has been criticized for its expense, and this type of surgery has not been approved by the FDA for cancer treatment in 2019. But then, people are applying robotic surgery in the treatment of breast cancer including mastectomy and other cancer-related surgeries [46].

6.2.1.2.8 Electrosurgery

Electrosurgery is a procedure that uses an electric current to cut and destroy abnormal tissues or cancer cells. After William T. Bovie developed this technique in the early 1920s, electrosurgery has become one of the most popular procedures for cancer treatment [47, 48]. Electrosurgery can be used to treat basal skin cancer by using electric current and make it pass through the body tissue so as to create the desired clinical effect. Electrosurgery is a cheaper cancer treatment in comparison with other types of surgery like laser surgery and usually causes less bleeding throughout the operation. However, utilization of electric currents may have electric shocks that this method is not completely safe and the risk should also be considered [48].

6.2.1.2.9 High-Intensity Focused Ultrasound (HIFU)

A high-intensity focused ultrasound (HIFU) is a new technique that utilizes focused ultrasound to destroy prostate tissue, which commonly used to treat prostate cancer. In particular, medical doctor adopts the HIFU sound waves to point through the wall of the patients' rectum, providing a non-invasive, radiation-free alternative approach to kill cancer cells [49, 50].

There are many advantages of using HIFU:

- Focused ultrasound is a non-invasive, less pain treatment which the patients can recover quickly.
- Precise targeting minimizes damage to adjacent healthy tissue.
- It has lower possibility to lead complication.
- The procedure of HIFU operation does not involve ionizing radiation, so it is possible to offer it to the patients who fail in radiation treatment.
- Treatment can be a complement to drug therapy, enabling the enhanced delivery of chemotherapy or immunotherapy to tumor sites.
- Focused ultrasound may potentially induce an antitumor immune response. Reports indicated that focuses ultrasound can cause an immune response for antitumor function [51].

Until 2019, two focused ultrasound systems for destroying abnormal prostate tissue are approved by the FDA in the United States. In the future, the application of HIFU will be further developed to treat many other types of cancers [50].

6.2.1.2.10 Mohs Surgery

Mohs surgery is a precise surgery for treatment of skin cancer, and this method is mainly used to remove the cancer cells while saving the adjacent healthy tissue [52, 53]. Frederick Mohs developed this treatment in the 1930s, and Mohs surgery is considered as the most effective treatment for basal cell carcinoma (BCC) and squamous cell carcinoma (SCC) and other skin cancer [52].

The surgeon will remove the visible part of the patient's cancer after the anesthetic has taken effect and continue to remove a thin layer of tissue under the tumor and put on a temporary bandage. The surgeon will take this tissue for sectioning and perform the pathological analysis under the microscope. If the biopsy is confirmed as a tumor, more skin layers will be removed until the last tissue sample removed shows there are no cancer characteristics [52, 53].

The advantages of Mohs surgery are significantly clear that it can directly monitor where the cancer cells are removed. Mohs surgery has a high cure rate and allows patients to keep much more healthy skin because of this type of removing strategy [54].

Although Mohs surgery is generally considered as the safest method, there are still some risks, such as bleeding from the site of surgery, pain feeling in the surgical area, and potential infection [52].

6.2.1.2.11 Stereotactic Surgery

Stereotaxic surgery, also called stereotactic surgery or stereotaxy, is a three-dimensional surgical technique that enables the surgeon to clearly control the focus or lesions in the deep tissue of the patient. And the first stereotaxic apparatus was invented by Canadian neurologist, Aubrey Mussen, in 1918 [55].

Stereotaxic surgery is often used to locate lesions in the brain and to combine with radiation therapy. The specialized equipment focuses on a big amount of small beams of radiation on a tumor by using the 3D image to locate the target. With the improvement of the utility of stereotaxy, it will be more precise and advanced in the treatment of brain cancer [56].

6.2.1.2.12 Percutaneous Ethanol Injection (PEI)

Percutaneous ethanol injection (PEI) is the first ablative technique used for hepatocellular carcinoma (HCC), which was described for the first time in 1983 [57, 58]. PEI is performed with the local anesthesia using a special needle and coordinated with ultrasound or computed tomography. Ethanol diffuses into the tumor cells and eventually causes tumor cell damage and protein denaturation and then leads to coagulative necrosis [59].

Ethanol injection is also an inexpensive and well-tolerated procedure in HCC treatment. And there some criteria have been made to judge if it's proper to use PEI.

- It's not appropriate for resection or transplantation, or the patients had refused surgery.
- The examination results of patients show that no extrahepatic metastasis or vascular invasion [60].

So far, PEI has not been compared with surgery in a randomized method. Nevertheless, the 3-year survival rates with PEI and surgery were 71% and 79% [58]. It's remarkable that there can be some new ablation therapies will achieve similar or even better long-term results compared with PEI in HCC [60].

6.2.1.3 Surgery Works with Other Cancer Treatment

For a better outcome to eliminate cancer cells, surgery is usually used with many other types of cancer treatments, such as radiation therapy and chemotherapy. For example, radiotherapy, chemotherapy, and immunotherapy are widely given after surgery for breast cancer [61].

Usually, the goals for performing such a combination of surgery and other treatments are decreasing the risk of cancer recurrence and the possibility of cancer metastasis. However, it may induce a worse consequence, such as nausea, infertility, and cardiovascular diseases [62].

The following cases indicated some specific cancers that were treated by combination therapy with both surgery and other types of cancer treatment:

- Malignant melanoma is often treated by using chemotherapy as an adjacent treatment. For example, most patients in the United States will be treated with interferon alpha 2b in some cases [63].
- There is a drug named fluorouracil which is effective chemotherapy for preventing outcomes of micrometastatic disease from colorectal cancer or patients with microsatellite instability [64, 65].
- Recent experiments have revealed that immune checkpoint inhibitors like programmed death 1 (PD-1) can be used for treating exocrine pancreatic cancer as a way to increase the patients' survival rate [66].
- A comprehensive analysis indicated that the non-small cell lung cancer (NSCLC) patients treated with additional chemotherapy or radiotherapy have longer life span than those with single surgery [67].
- It reports that bladder cancer patients who were given platinum-based chemotherapy before surgery, this treatment also called neoadjuvant therapy, exhibited a higher overall survival rate. But this type of cancer strategy still exists controversy [67].
- Scientists suggested that extra platinum-based chemotherapy can improve survival in cervical cancer [68].
- Adjacent chemotherapy demonstrated a stable position in the treatment of breast cancer for its contribution to increasing patients' overall survival rate with low relapse possibility in more than 30 years [69]. Recent research shows that combination adjuvant chemotherapy for breast cancer can be a better choice, such as some agents, doxorubicin, and cyclophosphamide followed by cyclophosphamide, methotrexate, and fluorouracil [70].

6.2.1.4 Risk of Surgery

Surgeons are professional and well-trained, but there are still some risks because of the surgery itself, drugs used, and the health status of patients [71].

The most common surgical risks include:

- If a patient has an overreaction to the prescribed drugs like anesthetics or other medicines used during the operation, there will be terrible complications. For example, some patients can have an overreaction to anesthesia that the temperature can rise rapidly. It's a type of malignant hyperthermia which can lead to life-threatening conditions.
- Bleeding usually happens either internally or externally in the procedure of surgery. But if the bleeding amount is too large, it can cause severe crisis, and surgeons will make a necessary transfusion.
- Blood clots can form in the beginning area of surgery or because of inactivity. If the clots break loose and travel through the bloodstream to another part of bodies,

it will be a serious problem which the clots may deposit in the lung causing a pulmonary embolus or brain causing a stroke.

- Pains and infections widely appeared in the past, because of the insanitary medicine condition and poor surgical techniques. However, even nowadays the surgeon operates in a relevant clear environment and usually coordinates with the pain-relieving drug, it still has a chance to have pains and infections. A surgical incision can lead to possible infection, and the surgeon uses antibiotics to deal with this situation. Besides some pains happen in surgery, it exists risks of injury that some parts of the body will be damaged in the procedure and even lead to death.
- In some cases, patients will take more time for recovery. For example, diabetic patients must turn the situation into minds that the purposes and risks of having surgery because the surgery can influence blood sugar levels which will increase the healing time.
- There may be some long-term side effects of cancer surgery because of the surgery item. For example, severe scarring can be caused by a large and wide incision for needs in surgery. And as a poor result, men who are taking radical prostatectomy to remove his prostate will have risks of incontinence and impotence [71, 72].

As the risks of surgery in cancer mentioned above, it's wise for patients to discuss the details and problems of the health conditions with the medical team to reduce the operational risks as much as possible [71]. In the treatment of cancer, there have been more and more multidisciplinary teams which are cooperating with each other to perform risk prediction and find the relationship between it and patient management [73].

6.2.2 Radiation Therapy

6.2.2.1 Introduction of Radiation Therapy

Radiation therapy is a type of cancer treatment that uses ionizing radiation to shrink tumors and eliminate carcinoma; ionizing radiation can cause DNA breakage or damage, thereby inhibiting cancer cells growth [74, 75].

Radiation therapy for cancer treatment has been developed 100 years ago, which is originally invented by Wilhelm Röntgen in 1885. In 1896, a French physician, Victor Despeignes, was firstly used for the treatment of stomach cancer [76, 77]. In that early era of developing radiation, there are some controversies for using radiation item in cancer therapy. After all, the pioneers like Emil Grubbe and Thor Stenbeck had successfully demonstrated the use of X-rays for treatment of various types of cancer, such as epithelioma and breast cancer. Besides, some scientists have studied the application of radiation therapy in treating leukemia [78, 79]. Owing to the breakthrough of Marie Curie's, the discovery of two radioactive elements polonium and radium in 1898 further promoted the new start in medical treatment [80]. Many trials showed that radium therapy can cure some types of cancers in which X-ray didn't work. Plenty of methods of applying radium have been tested afterward, such as radium emanation or direct radium substance in the salt form [81]. And, some emerging commercial radiation products also involved some quackeries, which were invented for squeezing money from the public. For example, a product called Revigator, a radioactive water crock, was filled with water which is dissolved with radioactive elements such as radon [82]. Nowadays people can be astonished by people's insane practice at that time, but in fact, people really have a delusion like as radiation is a ready-made panacea because of the lack of scientific cognition. There was a professional and logical process for standardizing the radiation therapy operation until 1934; a talented French scientist Henri Coutard developed a requirement of fractionation and dosage for radiation [83]. Then in 1935, surgeons started following his treatment plan which was still generally followed in today's cancer treatment [84]. From the 1940s, medical linear particle accelerators had been developed and firstly used at the Hammersmith Hospital in 1953 which became fashionable and gradually replaced X-ray and other old therapies since the 1980s [85]. In 1971, computed tomography (CT) was developed by Godfrey Hounsfield, which further brought a new round of innovation of radiation therapy, and allowed the physicians and oncologists to have more accurately determined the radiation dosage and precisely target tumors [86].

There are two main ways of radiation delivery: external beam radiation therapy and internal radiation therapy. In external beam radiation therapy, the machine releases radiation at the tumor site in a patient's body. When the radiant substance needs to go inside of patient's body, a sealed form will be used and be allowed to treat with higher doses of localized radiation, this method also known as internal radiation therapy [75, 87].

6.2.2.2 Types of Radiation Therapy

6.2.2.2.1 External Beam Radiation Therapy

External beam radiation therapy is the most common type of radiotherapy in cancer treatment (Fig. 6.4) [87]. There are two similar ways to manipulate radiation in the treatment of cancer, such as the conventional external beam radiation therapy (2DXRT) and three-dimensional conformal radiation therapy (3D-CRT). The medical linear accelerator firstly generates a high-energy X-ray, and then the beam of radiation will be delivered to several directions for locally targeting tumors. 3D-CRT is like an upgraded version in comparison with 2DXRT, which allows the surgeon to capture images from CT, magnetic resonance imaging (MRI), and positron emission tomography (PET) scans and to precisely design the radiation beams for matching the shape of tumors [85, 88]. In recent medical applications, 2DXRT has been widely replaced by 3D-CRT which the latter is becoming the standard methods for radiotherapy [89, 90].

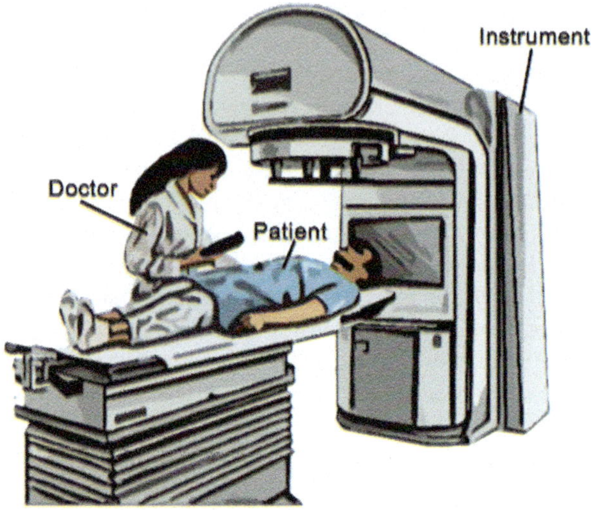

Fig. 6.4 Having external beam radiotherapy

Intensity-modulated radiation therapy (IMRT) is an advanced form of external beam radiation therapy, which creates more complicated body sites than 3D-CRT, especially for the central nervous system, prostate, and lung. Accordingly, the physician will be easier to confirm the different doses matched with those irregular tumors. Nevertheless, this process can allow higher doses of radiation to apply accurately on the targeted malignant tissues, thereby reducing side effects [91]. Tomotherapy, also named as helical tomotherapy, is the special equipment of IMRT for taking more precise images of the tumor. The patients will be asked to lie on a position through a donut, and a machine rotates during the treatment that gives radiation like some spirals layer by layer in order to offer precise treatment. However, it still waits for clinical trials before formal applicants [90, 92].

Stereotactic radiosurgery is used to treat small cancers in the brain or CNS, which are usually working with well-defined edges. A lot of small beams will focus on the tumors from several different directions that can have little effects on adjacent normal tissues. Gamma knife is a non-invasive instrument belonging to stereotactic radiosurgery that uses no less than 192 precise beams of radiation with the accuracy of less than one-tenth of a millimeter instead of traditional scalpels at all. And it can be given to treat a lot of brain cancers, such as meningiomas, pituitary tumors, and acoustic neuromas [93]. Stereotactic body radiation therapy (SBRT) can be a special type of SRS, which mainly targets small tumors outside of the brain. For example, its successful targeting rate could be up to 98% with SBRT better than traditional therapy for non-small cell lung cancer in some trials. SBRT may become a popular approach in radiation treatment in the future [94].

Image-guided radiation therapy (IGRT) can be a form of the pre-surgery check before each treatment that is always used with intensity-modulated radiation therapy (IMRT). CT scans and X-rays are the usual methods to help physicians precisely target cancers. In normal practice, patients will be asked to take images in the same position every day; as a result, the surgeon can see small changes in the shape or size of the tumors. For example, in lung cancer treatment, the radiation therapist can take images during the procedure of surgery. By analyzing these images, the oncologist can ensure the actual information of patients' cancers and made necessary adjustments to radiation delivery [95].

Some heavy particles are also commonly used in radiation treatment, such as electron, proton, and neutron; it's called particle therapy. Synchrocyclotrons and synchrotrons, two types of accelerated machine, generate such energetic ionizing particles to accurately target the malignant tumors [89, 96]. Proton therapy is the most common form of particle radiation, and protons have a unique pattern to deposit their energy at a distinct depth, known as Bragg's peak. It indicates that the proton beam can cause only a little damage to healthy tissues and allows bigger doses to be delivered in the tumors [96].

Volumetric modulated arc therapy (VMAT) is a new radiation approach based on rotational therapy, which method is firstly introduced in 2007, which allows modifying three parameters to control the machine and achieve high dose distribution in a wide coverage [97]. The principle is that the radiation delivers in continuous rotation, and the patient can to be treated by a 360° beam angle. The patients can get benefits from both on reduced radiation delivery time and less damage by radiation exposure [98, 99].

Auger therapy is a radiation treatment based on the Auger effect that destructs the cancer tissues by utilizing a large number of low-energy electrons [100, 101]. The low-energy feature allows the damage of these electrons controlled in less than the size of a single cell, even up to nanometer unit. The electronics will enter into the cell and emit radiation that induces damage of DNA, which permits an extremely high level of target therapy [102, 103].

6.2.2.2.2 Internal Radiation Therapy

Brachytherapy is a form of radiation therapy, which means short-range radiotherapy, usually a sealed source, is put inside the patient and then starts working [104, 105]. The radioactive implement is directly placed into or near the tumors. It is often used as an effective way to treat cancers of the head and neck, breast, cervix, prostate, breast, skin, and eye [106]. The brachytherapy includes three main types: low dose rate (LDR) brachytherapy, high dose rate (HDR) brachytherapy, and permanent brachytherapy [107].

In low dose rate (LDR) brachytherapy, the implant gives radiation only a very lower level, refer to *Brachytherapy Physics* [108]; it involves releasing radiation at a rate of up to 2 $Gy{\cdot}h^{-1}$. And the radiation sources will stay inside the same position for 1–7 days [107]. Patients will be asked to stay in the hospital until the procedure is finished [104]. Besides, when the rate of dose ranging between 2 $Gy{\cdot}h^{-1}$ and 12 $Gy{\cdot}h^{-1}$, it's characterized as medium-dose rate (MDR) brachytherapy [108]. If the rate of radiation dose exceeds 12 $Gy{\cdot}h^{-1}$, it will be the high dose rate (HDR) brachytherapy [108]. This is a type of brachytherapy using intense energy for generating a high dose of radiation to the target tissues over just 10–20 min once [105, 107]. The catheter, a delivering implant to emit the radiation sources, needs to be placed in the targeted area before each therapy and the whole treatment usually last for 2–5 weeks [107].

The permanent brachytherapy refers that the implant will remain in the patient's body all the time. Although it will emit radiation every time, the radiation can be weaker and weaker as time goes on. The patients need to take attention to the condition at the implant put inside firstly, in case of causing some bad effects on people or children around them [107].

Besides, there is a type of brachytherapy called contact X-ray brachytherapy (CXB) which uses kilovoltage X-rays to treat rectal cancer; usually the patients are in an early stage of rectal cancer and not suitable for surgery [109]. The X-ray tube will take through the anus into the rectum, then the radiation emitted into the tumor at two weekly intervals, so that it can damage cancer [109–111].

6.2.2.2.3 Other Types of Radiation Therapy

There are some other types of radiation therapy, such as systemic radioisotope therapy, intraoperative radiotherapy (IORT), and deep inspiration breath-hold (DIBH).

Intraoperative radiotherapy (IORT) stands for the process of using therapeutic doses of radiation to the tumor during the surgery so that the radiation can precisely deliver to the targeted area with less damage to surrounding tissues [112]. The research demonstrated that IORT can inhibit the effect of invasion in the surgical wound and impair stimulation of tumor cells proliferation in breast cancer [113].

Deep inspiration breath-hold (DIBH) is an approach to minimize radiation exposure to the other organs like lungs and hearts which is used for the treatment of breast cancer, lung cancer, and etc. [114, 115]. The surgeon will deliver the radiation when the patients hold their breath; this allows limiting radiation affects other healthy areas [116]. There are two primary methods of performing DIBH. Free-breathing breath-hold, the real-time position management (RPM) DIBH, puts an infrared camera to track the breath movement of patients. The other is spirometry-monitored breath-hold, also known as active breathing coordinator (ABC) DIBH systems; it controls the flow of air which can give more reproductive consequences [117].

Systemic radioisotope therapy is a form of radiotherapy that uses unsealed radiation source like isotopes to achieve target therapy. This is due to the chemical properties of these isotopes. Oncologists attach a molecular or antibody, which is used for treating targets, to the chosen isotopes, so they can be absorbed specifically. For example, iodine-131 is a radioisotope of iodine and can be absorbed by the thyroid gland better than other organs used for elimination of remnant thyroid tissue to treat thyroid cancer [118, 119].

6.2.2.3 Radiation Works with Other Cancer Treatment

Radiotherapy is usually combined with other types of treatment in cancer therapy, such as surgery, chemotherapy, and immunotherapy that can be given before, during, or after the other tumor therapies [75].

The radiation therapy may be the only treatment in the early stage of cancers, but it mostly works in combination with other types of treatments for cancer. There are many modalities involving with radiotherapy which are listed here:

- It's not an essential treatment in breast cancer, but the surgeon will suggest using an item in some cases. Genarally if cancer has spread to other parts of the body (cancer metastasis), or the breast tumor is too large (more than 5 cm) to remove (after mastectomy), radiation therapy can be a choice Besides, it will also be used to decrease the chance of cancer recurrence [120].
- The treatment plan for rectal and anal cancer depends on the stage of cancer, and surgery is the major approach when cancers have not yet spread to other tissues. If cancer has spread to nearby lymph nodes even parts of the body, patients will be treated with chemotherapy, radiotherapy, and surgery [121].
- Cervix carcinomas are treated with five standard treatments: surgery, radiation surgery, chemotherapy, targeted surgery, and immunology. Some new types of treatment are still been tested in clinical trials [122].
- Radiotherapy is an optional treatment in bladder carcinomas when transurethral resection (TURBT) is not the candidate because of patients' health problems [123].

6.2.2.4 Risk of Radiation Therapy

Radiotherapy can also affect healthy cells in the treatment. Because of the impacts of radiation, the side effects can happen anytime in the procedure of therapy including fatigue, skin problems, alopecia, loss of appetite and nausea, and low blood cell counts.

- Fatigue usually happens after the radiotherapy, which is correlated with nausea, vomit, and loss of appetite. Problems of such eating and sleeping can make fatigue worse.
- Some forms of radiotherapy need to use radiation to go through the skin and reach the targeted area; it can make damages as well as some permanent harms, such as scarring or skin darkening.
- Alopecia is the term of hair loss, and the severity of hair loss varies from person to person. The hairs can regrow in 3–6 months usually after radiotherapy is finished.

Low blood cell counts are common because the radiation affects the bone marrow. But this side effect generally doesn't cause big problems that blood cell counts can recover after several days after radiotherapy. Some late side effects also need to be noticed, such as brain, heart, and lung problems [124].

6.2.3 Hormone Therapy

6.2.3.1 Introduction of Hormone Therapy

Hormone therapy is a type of treatment that slows down or stops the growth of cancer cells by regulating the hormone [125, 126].

Hormone therapy is discovered by Thomas Beatson, a pioneer in the field of cancer treatment and has been acknowledged as "the father of endocrine ablation in cancer management" [127]. In 1878, Thomas discovered that if he moved the ovaries of some rabbits, they will not produce milk anymore. In his opinion, it can be evidence that there is a relationship between ovaries and breast [128]. Then he published a research article to illustrate his findings, titled *Treatment of Inoperable Cases of Carcinoma of the Mamma: Suggestions for a New Method of Treatment, with Illustrative Cases*. This article elaborated his daring attempt in the treatment of breast cancer that he used bilateral oophorectomy (removal of the ovaries) to treat three patients with advanced breast cancer [129]. Although he didn't repeat the same work for more results, the bilateral oophorectomy had become the standard operation for treating advanced breast cancer in the following years. Because of his findings that the secretion from ovaries can stimulate breast cancer in his research, even the female hormone (estrogen) has not been discovered at that time, he is considered as the father of hormone therapy in breast cancer who laid the groundwork for the modern method to treat such carcinomas with hormone treatment. Nowadays, scientists develop many drugs based on hormone therapy to treat or prevent breast cancer, such as tamoxifen and aromatase inhibitors [128].

Since the metastatic prostate cancer was remarkably regressed after the removal of the testicle, it can cause a consequence of androgen decrease. Charles Huggins is thought to be the first person to use a systemic way to treat prostate cancer, and his works provide a strong foundation for the modern approach to design some drugs that blocked male hormone [130]. From the perspective of the future, it will be more types of new hormone drugs to reduce the risk of breast cancer and prostate cancer or other potential cancer candidates that their growth related to the hormone.

The hormone treatments can be divided into two main types including inhibiting of the body's ability to produce hormone and interfere hormone's normal behavior in the body [126]. In this part, the author is going to introduce this form of cancer therapy based on this classification.

6.2.3.2 Types of Hormone Therapy

6.2.3.2.1 Prevent Hormone from Producing

One strategy for reducing the amount of hormone is the prevention of hormones secretion (Fig. 6.5).

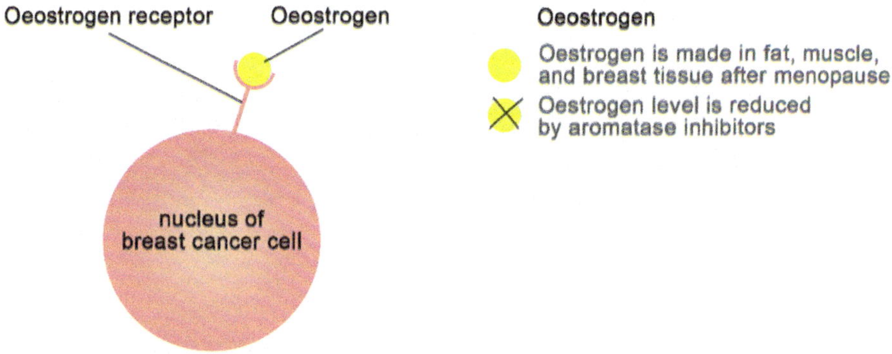

Fig. 6.5 The example strategy of inhibiting the production of the hormone

Luteinizing hormone-releasing hormone agonist (LHRH Agonist), as known as gonadotropin-releasing hormone blockers (GnRH blockers), is a type of hormone that can stop the process of producing the luteinizing hormone, which can reduce the expression of testosterone in males [131, 132]. The LHRH agonist is able to suppress the growth of cancer cells, and it reported that lower the levels of sex hormone up to 95% both in males and females [133]. There are some typical cases of medications such as goserelin, degarelix, and leuprolide [134].

Aromatase inhibitors are a class of compounds that can block the conversion of androgens to estrogens, leading to fewer estrogens production in the body. Estrogen can promote the growth of breast cancer, so the use of aromatase inhibitors had been proved to be an effective way to treat hormone-sensitive breast cancer in postmenopausal women [135]. Anastrozole, letrozole, and exemestane are some common aromatase inhibitor drugs [134].

17α-hydroxy/17,20-lyase (CYP17) is a critical enzyme that participates in the biosynthesis of androgens, which helps cells to produce androgens contributing to the growth of prostate cancer. CYP17 inhibitors block this enzyme and stop the process of androgen secretion such as ketoconazole and abiraterone [136].

Orchiectomy (surgical castration) directly removes the testicles to control the production of androgens that leads to a positive consequence of treating prostate cancers [137].

In the treatment of many types of cancer like breast, lung, colon, and prostate cancer, somatostatin analogs can suppress the cancer cells to release hormone such as adrenocorticotropic hormone and growth hormone-releasing hormone in lung cancer and to reduce symptoms of carcinoid syndrome without significant side effects [125, 138]. For example, octreotide is mimic natural somatostatin that is more effective than some natural hormones as an inhibitor [139].

6.2.3.2.2 Interfere with Hormone Behavior

Another approach is to interfere with hormone activity in our body (Fig. 6.6).

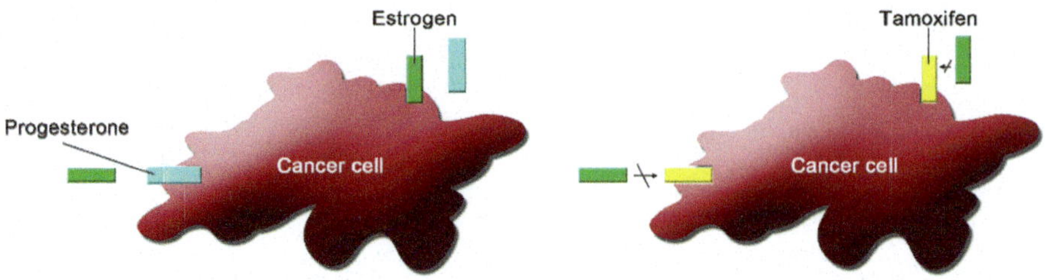

Fig. 6.6 The effect of tamoxifen and the progesterone

Progestin is a type of female hormone and is combined with estrogen as a means to treat breast cancer. Medroxyprogesterone acetate (MPA) and megestrol acetate (MA) are two common progestin drugs that are commonly used in the treatment of breast cancer [140, 141].

Nearly 80% of breast cancers are related to estrogen, which these types are called hormone-receptor-positive breast cancers, and they can be treated with endocrine therapy [142, 143]. A recent study showed that estrogen has been implicated as a potential mediator in the advancement of prostate cancer (150). Anti-estrogens including selective estrogen receptor modulators (SERMs) and estrogen blockers can bind and target the estrogen receptors so that estrogens fail to work normally and eventually affect the growth of cancer cells [125, 134]. Some agents like fulvestrant, tamoxifen, and toremifene are widely used in hormone therapy [134].

The mechanism of anti-androgen drugs is similar to anti-estrogen, which anti-androgen works by blocking of testosterone receptors, so as to stop the secretion of testosterone, the primary androgen. Some medicines like bicalutamide, nilutamide, and flutamide are commonly used in androgen deprivation therapy (ADT) for prostate cancer [144].

Some other drugs like fluoxymesterone, a type of androgen, are used to treat breast cancer. Diethylstilbestrol, a type of estrogen, is widely used in the treatment of advanced prostate cancer [125, 137].

6.2.3.3 Hormone Therapy Works with Other Cancer Treatment

Hormone therapy is commonly combined with many other anti-cancer treatments, such as chemotherapy and radiation therapy. There is a novel strategy for the cancer treatment called neoadjuvant therapy which is quite different from conventional adjuvant treatment. Hormone therapy can be applied before the main treatment as a neoadjuvant therapy [145]. For example, hormone therapy can be firstly utilized before the primary means, radical radiotherapy, prostatectomy, or chemotherapy for prostate adenocarcinoma [146].

Doctors also use hormone treatment as adjuvant therapy. In the treatment of hormone receptor-positive breast cancer, some patients will be treated with hormone medicines as hormone therapy [147]. Based on the research studies, the effects of the intervention of adjuvant hormonal therapy in postmenopausal women with breast cancer indicated 40% reduction in the risk of recurrence and 30% reduction in the risk of death with the application of hormone treatment as adjuvant therapy [148].

6.2.3.4 Risk of Hormone Therapy

The side effects of hormone therapy depend on the types of hormone treatment in cancer patients. Most patients will experience a lot of common side effects, such as fatigue, nausea and vomiting, weight gain, loss of interests in sex, hot flashes, muscle loss, diarrhea, loss of fertility, insomnia, etc. [149].

Besides, some side effects of hormone therapy may happen particularly for some types of cancers. Prostate cancer patients may have erectile dysfunction (impotence) and breast tenderness or swelling because of the properties of the hormone. And the women who receive hormone therapy for breast cancer may have some side effects including mood changes, change of menstruation (period), and vaginal dryness. Under certain circumstance, some dangerous conditions may happen without cautions of taking hormone therapy. For example, studies suggested that there is a risk of breast cancer in the treatment combined with estrogen and progestogen that per 1000 women take that combination of hormones will be five extra cases of breast cancer [150]. Other serious or potential risks include ovarian cancer, womb cancer, blood clots, and strokes [151].

6.2.4 Stem Cell Transplants

6.2.4.1 Introduction of Stem Cell Transplants

Stem cell transplants, also known as bone marrow transplantation, are the procedure of transplanting the multipotent hematopoietic stem cells derived from healthy individuals to the patients in order to restore their damaged bone marrow; this therapy is usually operated for treatment of leukemia, lymphoma, neuroblastoma, and multiple myeloma [152–154].

In 1868, Neumann from the University at Königsberg in Russia and Bizzozero from University of Pavia in Italy, respectively, reported their discoveries that bone marrow is the primary site of the new blood cell production in mammals [155]. In 1939, Gorer demonstrated the existence of antigen of tumor transplantation in his experiments, which contributed to the theoretical research of stem cell transplants [156]. After 10 years, Jacobson demonstrated that the mice can survive under lethal irradiation when their spleens were protected [157]. In the following years, lots of the investigators including Lorenz, Loutit, and Barnes began to discover the importance of the spleen or bone marrow, and many experimental findings indicated that the irradiation damaged mice can still survive after the bone marrow infusion [158, 159]. Barnes and his colleagues made a successful attempt to treat high-dose radiation-induced murine leukemia by marrow transplant in 1956 [160]. Georges Mathé, a French oncologist and immunologist, employed a marrow graft to treat some leukemia patients. Unfortunately, all of these transplants were rejected and eventually failed to survive due to serious immune reactions against the hosts, which the procedure is called graft-versus-host disease (GVHD) [161]. Notably, the understanding of complications of GVHD and the human leukocyte antigen (HLA) led a new period in the field of hematopoietic stem cell transplant (HSCT). Later on, three infants with primary immunodeficiency disorders underwent bone marrow transplants from HLA-matched relatives in 1969 and survived without any complications of GVHD [162]. In 1975, John Kersey performed the first successful bone marrow transplant to cure a 16-year-old lymphoma patient [163]. In 1979, E. Donnall Thomas who won his Nobel Prize award in 1990 and his colleagues reported their attempts to take hematopoietic cell transplants (HCT) as an effective strategy to treat acute non-lymphoblastic leukemia patients in their first remission [164, 165]. Afterward, researchers started a new trial to combine chemotherapy with HCT to prevent GVHD, such as the combination of cyclosporine with methotrexate [166, 167]. Since the National Marrow Donor Program was founded in 1987, the organization had successfully recruited more than 20 million HLA unrelated donors all around the world for the next 20 years [168, 169].

In the last decade of the twentieth century, patients who were lack of HLA-matched siblings can survive using unrelated donors' marrow transplants with the help of more and more volunteers contributing their bone marrow [170]. Some clinical transplant protocols have been modified to reduce the intensity of therapy, including combination low-dose radiotherapy with chemotherapy to eliminate the problems of GVHD [171]. In the future, the focus on adoptive immunotherapy and transplantation of elder or infirm patients will be taken more and more attention.

The types of transplant depend on where the stem cells come from or whether the patients have matched with the donors, which include autologous, allogeneic, and syngeneic transplants. However, umbilical cord blood transplants and haploidentical stem cell transplants are without suitably matched bone marrow from donators [152–154].

6.2.4.2 Types of Stem Cell Transplants

6.2.4.2.1 Autologous Transplant

Autologous transplant is the transplantation of stem cells from the patients themselves, and there are several procedures of autologous HSCT. Firstly, the doctors require isolation and purification of stem cells from patients' blood and then store them in a freezer. Sometimes, patients will be asked to take medication of granulocyte colony-stimulating factor to increase the number of stem cells before the harvest of stem cells [172, 173]. After that, the patients will receive high dose of chemotherapy with or without radiotherapy to eradicate the malignant cells, but the bone marrow and immune system will be damaged at the same time by the side effects [154]. When the treatment is finished, the doctors will thaw the patients' own stem cells and then injected them into their bloodstream [174]. After a few days, the patients can reconstruct their bone marrow with their newly injected stem cells so that they gradually produce new blood cells.

The recovery of autologous transplant needs to take a long time with lots of care supports in case of infection by returned stem cells, complications by chemotherapy or radiotherapy, and so on [173]. Because the donor and recipient are the same one, the risk of autologous transplant is quite low. However, the mortality of autologous treatment is high due to the probable relapse in acute myeloid leukemia. Under such circumstances, allogeneic treatment could be an optimized choice comparing with autologous therapy [175].

6.2.4.2.2 Allogeneic Transplant

The allogeneic transplant is different from autologous transplant; the recipients must have HLA-matched donors to prevent serious rejection [176, 177].

The first step is to find the proper donors who are the syngeneic (monozygotic) twins or the patients' siblings who inherited the same HLA genes. But these conditions are only less than 25% of people to match with, so that most of the patients have to find unrelated donors. There are some nonprofit organizations like the National Marrow Donor Program where the patient can search for a potential HLA-matched marrow [176]. Because the mismatch of HLA will lead to some serious consequences, the HLA-testing is necessary before each transplant. However, some infirm and elder patients can be given a mini-transplant, which they will receive preparative treatment of chemotherapy or radiotherapy with less toxicity or intensity to ensure their life safety [177]. There are two types of stem cell donation including the peripheral blood stem cell (PBSC) and bone marrow. Interestingly, a recent report indicated that race is concerned to be an important element in donors' recruitment because they are more possible to have matched HLA genes in the same ethnic groups [178, 179].

There are some novel attempts in allogeneic transplants.

A savior sibling is a child who is born to provide the bone marrow or organ for saving a sibling life, and the savior sibling is created by in vitro fertilization with the preimplantation genetic diagnosis. But this strategy is considered to be controversial even some countries have laws to rule [180, 181]. There are several commercialized stem cell drugs up to now, such as cartistem, pneumostem, and neurostem, some of which are already approved to use clinically by many countries [182, 183]. However, there are no FDA-approved stem cell drugs used for cancer treatment until now because of the safety and low drug efficacy [184].

6.2.4.2.3 Syngeneic Transplant

The stem cells are harvested from recipients identical (monozygotic) twins, which this procedure is called syngeneic transplant [185]. The advantage of syngeneic transplant is that the treatment will not cause GVHD, because they have identical tissue types. But the probability is lower to have an identical twin. And the disadvantage of syngeneic transplant is the lack of the graft versus leukemia (GVL) effect. This effect is caused by the transplanted stem cells including donors' T cells, which can recognize and attack the carcinoma [185, 186].

6.2.4.2.4 Umbilical Cord Blood Transplant

Umbilical cord blood is an essential source of stem cell transplant for children with hematologic diseases, besides bone marrow and peripheral blood stem cells [187, 188]. The blood is collected from newborns' umbilical after birth immediately [188]. Because the cells from newborns are less immunologically mature, there is less risk of GVHD. Hence, the umbilical cord blood transplant (UCBT) is not limited to the test of human leukocyte antigen (HLA) [189].

The UCBT is used more often in children because the number of stem cells is not enough in a single umbilical cord to treat adults. And in the research of using UCBT to treat children with acute leukemia, the results showed that the patients with UCBT have higher survival by comparing with allogeneic transplants [190]. However, the outcomes in adults were poor, which 40% of patients died after treated by UCBT for 100 days. Further studies of UCBT need to solve the problems of infection, particularly the viral infection in cancer treatment [191].

6.2.4.2.5 Haploidentical Stem Cell Transplant

The haploidentical stem cell transplant is a type of allogeneic transplant that the donor is half-matched with the patient. And the outcomes of the HLA test from the Johns Hopkins Kimmel Cancer Center showed that the haploidentical transplant can be performed as well as allogeneic transplants. Nearly 50% of siblings of the patients can be matched for the treatment, which the possibility to find a reliable bone marrow is extremely increased compared with HLA-matched transplants [185, 192, 193].

Nowadays, even though the advancement of medical development, the risk of disease complications and graft rejection between haploidentical transplants and traditional HLA-matched transplants are not improved. And the haploidentical stem cell transplant is a novel approach that is still in the clinical trial stage; more experiments and tests need to be carried out in the future [193].

6.2.4.3 Stem Cell Transplants Work with Other Cancer Treatments

Before the transplant, there is a conventional treatment of high-intensity chemotherapy or radiotherapy for several days to eradicate carcinoma. After transplantations, the patients will take some medications all the time until the new bone marrow completely starts working. For example, some antibiotics will be used to prevent infection, and some patients may get transfusions of irradiated platelet or blood. Besides, other special treatment ways may be employed to treat complications such as GVHD [194].

6.2.4.4 Risk of Stem Cell Transplants

The adverse effects of transplants are mainly caused by high-dose treatment and complications.

Because the patients will first undergo the chemotherapy or radiotherapy to clear up the malignant cells, it will damage the immune system and induce a series of side effects, such as infection, bleeding, anemia, and so forth.

Some complications are based on the types of transplants. For instance, the allogeneic stem cell transplant can cause GVHD and acute rejection. The former is that donors' healthy cells attack the cells of patients and the latter happens when recipients' bodies reject the transplanted cells, both of which will be lethal to patients.

Additionally, the recipients would have risks of graft failures, which the transplanted bone marrow doesn't work to make new blood cells, and they need to conduct another transplantation [195].

6.2.5 Chemotherapy

6.2.5.1 Introduction of Chemotherapy

Chemotherapy is a type of cancer treatment that uses drugs to eliminate malignant cells or suppress the growth of the tumor [196, 197]. The history of chemotherapy can trace back to 1600 BC, which *the Edwin Smith Papyrus* and *the Ebers Papyrus* are the ancient Egyptian medical papyrus involving descriptions of herbal medicines [198–200]. In 1025, the outstanding Persian physician Avicenna (Ibn Sīnā) wrote the book named *The Canon of Medicine (Al-Qanun fi al-Tibb)*, which is a medical encyclopedia including some descriptions of malignancies and the combinations of drugs [201, 202].

In the nineteenth century, surgery and radiotherapy are dominant in cancer treatment [203, 204], whereas chemotherapy evolved rapidly with the advance of technology and a new understanding of cancer in the following years [204]. In 1789, an English physician Thomas Fowler used a mixture of arsenic trioxide and potassium bicarbonate to treat a number of diseases, which was famous as "Fowler's Solution." In 1865, Heinrich Lissauer, a German scientist, used Fowler's approach in the treatment of leukemia and lymphoma [205]. The German chemist and physician Paul Ehrlich was the first person to use the term "chemotherapy" and identified many chemical compounds for testing their effectiveness against diseases by animal experiments. He was described as the "father of chemotherapy" for his contribution to cancer therapy using chemical drugs [206]. Based on Ehrlich's great work in the 1900s, George Clowes made the first transplantable tumors system in rodents in the early 1910s, which was considered a breakthrough in animal model development for drug testing purpose [207]. After that, there were a lot of the new model systems developed, such as Sarcoma 37 (S37), Sarcoma 180 (S180), and Walker 256, which all tumors were induced in mice. In 1935, Murray Shear sets up the most organized program for cancer drugs screening among the thousands of compounds from natural products and others [208, 209]. With reference to the effect of mustard gas that was previously used in the battlefield of World War I as chemical weapons, two pharmacologists from Yale University, Alfred Gilman and Louis Goodman, discovered that nitrogen mustard has potential antitumor activity against murine lymphoma and non-Hodgkin's lymphoma, and it was finally applied in patients with non-Hodgkin's lymphoma in 1943 [210, 211]. In 1948, Sidney Farber tested a drug now known as methotrexate to treat children with leukemia, and the result showed dramatic remissions [212]. The American Congress created the Cancer Chemotherapy National Service Center (CCNSC) in 1955 which mainly focused on discovering drugs with developed tools and models for cancer treatment [206]. In the 1960s, a Chinese-American oncologist and his colleague used methotrexate to treat choriocarcinoma, a rare germ cell malignancy, which cancer can be cured by chemotherapy alone [213]. In the meanwhile, the concept of combination chemotherapy had emerged and brought new hope. Clinical treatment began using combination chemotherapy to treat acute lymphocytic leukemia in children, which the agents named as VAMP (vincristine, methotrexate, 6-mercaptopurine, and prednisone) incredibly increased cure rate from 25% to 50% [206]. And MOMP (melphalan, methotrexate, vincristine, and prednisone) instead of methotrexate treated non-Hodgkin's lymphoma with a satisfactory improvement of remission from nearly zero to 80% [206]. In 1975, the Developmental Therapeutics Program which was established for cancer drug development made a new series of screening drugs models. And the Shear's old screening methods were substituted by the novel systems [203]. Norton and Simon manufactured a mathematical model of tumor growth in 1976 and proposed that the small tumors were easier to be eliminated than the large one. This point of view led to an idea that high-dose therapy could increase the survivals [214]. By the end of the twentieth century, the age of "target therapy" emerged. The first tyrosine kinase inhibitor imatinib mesylate was developed in 1996, which can specifically bind and suppress the tyrosine kinase to treat myelogenous leukemia [215]. Today, most of the human testicular cancers and children leukemia are curable with potent chemotherapy including more than 100 different chemical compounds [206, 216]. And the attempt of neoadjuvant chemotherapy had

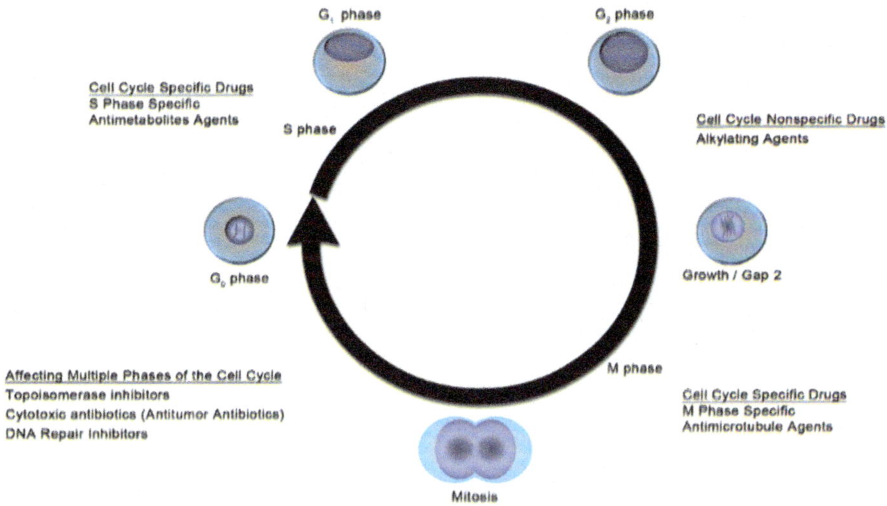

Fig. 6.7 Various types of chemotherapy agents in different cell cycles and cell cycle-independent representative drugs

made great success in lots of cancers, such as breast, bladder, gastroesophageal, rectal head and neck cancers [203, 217].

Currently, the strategies of chemical therapy are complicated. Based on the condition of cancer, doctors can use chemotherapy alone or combine with other cancer treatment [218]. Or the chemotherapy can be given after or before the main treatment as adjuvant or neoadjuvant therapy [204, 219].

According to how the agents work in patients' bodies, the chemotherapy can be categorized into several cases (Fig. 6.7). This part will demonstrate the types of chemotherapy as shown below.

6.2.5.2 Types of Chemotherapy

6.2.5.2.1 Alkylating Agents

Alkylating agents are the oldest type of chemotherapy that attaches an alkyl group (C_nH_{2n+1}) to DNA to stop cells from dividing [220]. Firstly, the alkylating agent can damage DNA by creating a covalent bond in cancer cells, and this modification of DNA will lead to apoptosis because the DNA strand would breakdown in the process of DNA replication during the cell division process [221].

These agents can work at any time during the cell cycle, which is known as cell cycle-independent drugs. They work in the process of cell division, cancer cells are more sensitive to the DNA-alkylate-induced damage, which is also the main mechanism of alkylating agents used for the cancer treatment [222]. But, they are toxic to healthy cells as well, especially the proliferative cells in the gastrointestinal tract, bone marrow and etc. And most of alkylating agents are carcinogenic and exhibit certain severe side effects [223].

The alkylating agents can be divided into several types, including the agents with alkyl groups (classical alkylating agents), such as cyclophosphamide, chlormethine, and uramustine. And platinum-based chemotherapeutic drugs act in a similar approach like alkylating agents, such as cisplatin, platinum, and carboplatin. On the other hand, there are some nonclassical alkylating agents comprising procarbazine, altretamine, and hexamethylmelamine [223, 224].

The alkylating agents are found to be limited by the DNA repair enzyme O-6-methylguanine-DNA methyltransferase (MGMT), which can inhibit the cross-link between agents and DNA strand during the DNA repair process [225].

6.2.5.2.2 Anti-microtubule Agents

Microtubule is an important skeleton structure of a cell that permanently moves in a dynamic state to facilitate cell division process. Anti-microtubule agents can inhibit cell division by blocking microtubule dynamic function [226, 227].

Vinca alkaloids and taxanes are the two main groups of anti-microtubule agents. Vinca alkaloids and podophyllotoxin prevent the microtubule formation, whereas taxanes stabilize microtubule structure by stopping their dynamic state. Of note, both types of these agents could effectively suppress the growth of the blood vessel, thereby decreasing the tumor progression [226].

Vinca alkaloids are extracted from Madagascar periwinkle, a species of flowering plant, and the original types of the agents are natural products including vincristine and vinblastine. The vinca alkaloids can bind to the tubulin to prevent it from forming of microtubule in S-phase, leading to the mitotic abnormality in M-phase. Other semisynthetic vinca alkaloids are well developed, such as vinorelbine, vindesine, and vinflunine for anti-cancer treatment [228].

Taxanes are semisynthetic chemical derived from the bark of yew tree including *Taxus brevifolia* and *Taxus baccata*. These types of agents, on the other hand, promote the stability of microtubules and prevent their disassembly for microtubule dynamic movement; the most classical types of taxanes include paclitaxel and docetaxel [226].

Podophyllotoxin, a lignan derived from the American Mayapple and Himalayan Mayapple, can inhibit mitosis by blocking microtubule formation that is similar to vinca alkaloid agents [229].

The scarcity of the sources can be a primary problem for the development of these drugs. For instance, the effective anti-microtubule chemical, taxanes, is rare in nature [230]. Nevertheless, the problems were solved by the total synthesis of the drugs [231].

6.2.5.2.3 Topoisomerase Inhibitors

Topoisomerase inhibitors inhibit the action of topoisomerases, which are a type of enzymes including topoisomerase I and topoisomerase II and can change the structure of DNA by their catalytic mechanism during the DNA replication process [232, 233].

The topoisomerase is important for normal cell cycle, since the double-stranded DNA is in helix form and is necessary to be untwisted in DNA replication or transcription process. And the topoisomerase can reduce the tension and catalyze the unwound of DNA into a single strand by breaking the phosphodiester bond of the DNA strand [234].

There are two types of topoisomerases inhibitors. And the type II topoisomerase inhibitors are split into two groups, topoisomerase II poisons and topoisomerase II inhibitors, both of which can block the effect of topoisomerase and cause cell death during replication or transcription [232].

Camptothecin, an active component isolated from Chinese ornamental tree *Camptotheca acuminata*, and two semisynthetical derivations of camptothecin, irinotecan and topotecan, are considered to be an effective topoisomerase I inhibitors. And they were approved by the US Food and Drug Administration (FDA) for the treatment of several cancers including colon and lung cancers [222, 235].

The topoisomerase II poisons target the complex of topoisomerase II and DNA and lead to DNA strand breaking and apoptosis. These drugs, such as amsacrine, etoposide, etoposide phosphate, teniposide, and doxorubicin, were employed to treat various types of cancers including testicular cancer, lung cancer, lymphoma, leukemia, neuroblastoma, and ovarian cancer [236, 237].

The topoisomerase II inhibitors can block the activity of the enzyme and inhibit DNA synthesis [236]. They include merbarone, novobiocin, and anthracycline aclarubicin [238, 239]. However, these drugs have other targets besides topoisomerase II, and the effectiveness of these targets in anti-cancer treatment is still not determined [236].

Notably, using topoisomerase inhibitors for cancer treatment might have the risk of causing secondary cancer such as leukemias [240]. Nevertheless, because of the significant difference in topoisomerase levels presented in cancer cells, the changes may indicate that the topoisomerase inhibitor can be a promising anti-cancer drug if used in a proper dosage [236].

6.2.5.2.4 Cytotoxic Antibiotics

Cytotoxic antibiotics prevent cell division and cause cell death, and these antibiotics are different from those used in anti-infection [241].

These types of antibiotics manifest their cytotoxic actions through interaction with DNA by insertion of their drug molecules into each planar base pair of DNA, resulting in DNA dysfunction and lesions. Furthermore, they also bind to the phosphate-sugar backbone of DNA and block cell replication or transcription and suppress topoisomerase-mediated unwind of DNA supercoiled strand. Besides, these chemicals can generate reactive oxygen free radicals which further contribute to the cytotoxicity to cancer cells [241].

Doxorubicin and daunorubicin are two common anthracycline antibiotics obtained from the bacterium *Streptomyces peucetius* [242]. Other analogs such as pirarubicin, aclarubicin, mitoxantrone, epirubicin, and idarubicin are provided to treat kinds of malignant neoplasms like breast cancer, lung carcinomas, Wilms' tumor, gestational choriocarcinoma, acute myeloblastic, and acute lymphoblastic leukemia [241, 243]. Actinomycin, bleomycin, and mitomycin are the other types of cytotoxic antibiotics agents [244–246], which exhibit cardiotoxic effects, leading to hypertensive, ischemic, and arrhythmic complications [247].

6.2.5.2.5 Antimetabolites

Antimetabolites interfere with the normal metabolism in cells by blocking DNA or RNA synthesis, and many of them usually have similar structures to nucleobases [248].

The regents manifest their antitumor effects by impeding the enzymes for DNA synthesis or incorporation of chemically altered nucleotides. Inhibition of the DNA synthetic enzymes can prevent DNA duplication. Also, the misincorporation of antimetabolite molecules into DNA can also cause cell damage and induce apoptosis [249]. As cancer cells are normally fast-dividing, antimetabolites are therefore more susceptible in cancer cells than in normal cells.

Antimetabolites are different from alkylating agents; they are cell cycle-dependent and only take an effect on the S-phase (the DNA synthesis phase).

Subtypes of these agents include four groups which are folate antagonists, purine analogs, pyrimidines analogs, and hydroxyurea [250].

Folic acid, also known as vitamin B9, is an essential vitamin in the human body as a cofactor to participate in the synthesis of thymidylate and purine that are critical for DNA synthesis. The anti-folates, such as methotrexate and pemetrexed, can impair folates biosynthesis machinery. To be specific, methotrexate blocks the function of dihydrofolate reductase (DHFR), an enzyme related to the regeneration of tetrahydrofolate, so that the cellular levels of the folates are decreased. Because of the lack of folate coenzymes, the DNA synthesis will be inhibited [251].

The purines and pyrimidines analogs include fluorouracil and thioguanine, both of which can masquerade as corresponding purines and pyrimidines, and prevent original substances from becoming the building blocks of DNA [248]. There are another deoxynucleoside analogs including cytarabine, gemcitabine, decitabine, azacitidine, and fludarabine. And capecitabine is a prodrug of fluorouracil that becomes active in cells [248, 252].

Hydroxyurea, also known as hydroxycarbamide, can diminish the production of deoxyribonucleotides via inhibition of ribonucleotide reductase, which is an essential enzyme for reduction of nucleoside diphosphates (NDPs) by cleaning the tyrosyl free radicals [253]. As such, it was used as an inhibitor of DNA replication. Other studies also indicated that hydroxyurea could block the process of DNA repair which provided a potential strategy in the treatment of cancer with a combination of alkylating agents [254]. However, antimetabolites exhibit toxicity in the rapidly dividing cells. And they are considered to have severe damage to mucous membranes of the mouth and gastrointestinal tract. Some other adverse effects include hair loss and skin disorders and the decrease of white blood cells [255].

6.2.5.2.6 DNA Repair Inhibitors

DNA repair inhibitors inhibit the DNA repair system through alteration of some DNA repair-related substance to prevent cell restoration after DNA damaging [256].

Currently available DNA repair inhibitors focus on five DNA repair pathways, which are direct repair (DR), nucleotide excision repair (NER), base excision repair (BER), mismatch repair (MMR), and homologous repair and nonhomologous end joining (HR/NHEJ) [257].

There are also five targets in each pathway.

In DR pathway, O6-methylguanine-DNA-methyltransferase (MGMT), a crucial "enzyme" used for genome stability, removes the alkyl group from the guanine base at the N7 position and is indicated as a significant protein in DNA alkylation repair [258]. Evidence showed that the increase in DNA repair capacity can enhance cancer cell viability [259]. Accordingly, several medications have been developed that are designed to target MGMT including Lomeguatrib, O(6)-(4-bromothenyl)-guanine, and 8-aza-O(6)-benzylguanine [257, 260].

Excision repair cross-complementation group 1 (ERCC1) is considered a key chemoresistance factor in NER. This protein generally forms a complex with DNA repair endonuclease XPF (ERCC4 or XPF) which allow the cleavage of DNA strand in damage condition or in many other DNA repair pathways [256]. Other researchers indicated that ERCC1 is overexpressed in cancer patients treated with platinum-based therapy [261]. Current ERCC1 inhibitors include tafluposide, UCN-01, and trabectedin [256].

DNA polymerase beta (DNA pol β and POLB) plays an important role in participating in base excision repair (BER) [256]. And the overexpression of POLB gene is thought to be correlated with a lot of cancers because POLB is demonstrated to be the most error-prone from all known eukaryotic DNA polymerases [262]. There are many inhibitors of POLB comparing with ERCC1, such as myristinin A, edgeworin, and stigmasterol [263, 264].

MMR is the machinery for recognizing and repairing the abnormality of DNA synthetic activity, which the errors include deletion, substitution, insertion, and point mutation [265]. Four of these genes including MSH2, MLH1, PMS2, and MSH6 are the major genes affected by mutations and predispose to a number of inherited cancers including hereditary nonpolyposis colorectal cancer (HPNCC or Lynch syndrome) in the MMR pathway [266, 267]. Because the mutations can be occurred by BER repression, it was indicated that Pol β (POLB) inhibitors are the promising medications by repairing DNA mismatches for cancer treatment, especially for

MMR-deficient tumors like HPNCC [265]. Besides, poly (ADP-ribose) polymerase (PARP) inhibitors are essential for the treatment of MSH3-deficient colorectal cancers [268].

Homology-directed repair (HDR) and nonhomologous end joining (NHEJ) are two major pathways in eukaryotic cells for the repair of double-stranded DNA lesions [269]. Homologous recombination is the most common form of HDR, whereas NHEJ repairs double-strand break (DSB) of DNA without the requirement of sequence homology [270, 271]. Synthetic lethality is used to describe a situation that the existence of two or more mutations, which either one of them can't induce a lethal phenotype, but the combination of deficiencies with expression will lead to death [272]. In this case, scientists developed the SL approach in the cancer treatment by combining two or more mutations before the regular treatment can form an additive effect for killing more tumor cells. Cells in the G1 phase rely on NHEJ, while HDR occurs in late S-phase and G2 phase. Because NHEJ appears to be more sensitive to mutants than HDR, thc researchers tend to focus on the inhibitors of NHEJ in enhancing radiosensitization [273]. For example, wortmannin, NU7026, and LY294002 targeted DNA-PKcs proteins, and SCR7 and L189 aimed at Ligase IV, which DNA-PKcs and Ligase IV are the two important repair proteins in NHEJ [274–277]. The HDR inhibitors including streptonigrin, UCN-01, and mirin can sensitize tumor but would not change the radiosensitivity of the normal tissues, whereas the NHEJ lacked the selectivity of malignant and normal cells [273, 278–281].

6.2.5.3 Chemotherapy Works with Other Cancer Treatment

Chemotherapy can be given alone or implemented along with many other types of treatment in cancer therapy, such as surgery, radiotherapy, stem cell transplant, and targeted therapy.

Medical doctors sometimes use chemotherapy as an auxiliary approach before convention treatment like surgery and radiation therapy to make tumors smaller. Or they take chemotherapy after regular treatment to completely destroy the remained cancer cells, which is called adjuvant chemotherapy.

Besides, chemotherapy helps with other treatments to prevent the spread of cancer cells because it works throughout the whole body. Furthermore, it's an important procedure in stem cell transplant, and the patients usually have high-dose chemotherapeutic treatment.

Sometimes, physicians will take chemotherapy as a palliative treatment for terminal cancer patients to prolong their lives or ease the pains but cannot cure the malady [282, 283].

6.2.5.4 Risk of Chemotherapy

Typically, the side effects of chemotherapy are due to the influence in normal cells. And usually the fast-growing cells from the bone marrow, mouth, gastrointestinal (GI) tract, and hair are damaged during the chemotherapy, thereby causing the side effects, such as low blood cell counts, mouth sores, nausea, and hair loss.

Moreover, most of the anti-cancer compounds show organ-specific toxicity toward the heart, kidney, lung, liver, lung, stomach, etc. [62, 282, 284].

6.3 Novel Tumor Therapeutic Modes

6.3.1 Immunotherapy to Treat Cancer

6.3.1.1 Introduction of Immunotherapy

Immunotherapy, also known as biologic therapy and immuno-oncology, is a novel form of cancer treatment that helps the immune system fight against cancer and has been investigated as the fourth "pillar" with the other three, chemotherapy, surgery, and radiation therapy by oncologists [285–289].

Since 1891, William B. Coley, an outstanding surgeon at Memorial Hospital, firstly developed immune-modulating therapy to harness the immune system for anti-cancer treatment. He bravely attempted to inoculate cancer patients with a filtered mixture of inactivated bacteria and bacterial lysates such as *Streptococcus pyogenes* and *Serratia marcescens*, named as "Coley's Toxins" [289–291]. In 1909, Paul Ehrlich, a Nobel prize-winning German physician, proposed that the immune system can inhibit the tumor growth, known as "immunosurveillance" hypothesis [292]. However, people in that period didn't pay much attention to the potential connection between Ehrlich's hypothesis and Coley's work [293].

There is little progress in immunotherapy after the great discovery in the early twentieth century. Until 1957, Lewis Thomas and Frank Macfarlane Burnet proposed a theory of cancer immunosurveillance that some cells, like lymphocytes, act as sentinels in surveillance against malignant cells [294]. There are many striking achievements of this field in 1957, when Charlotte Friend, a famous American virologist, isolated a virus from murine named the Friend leukemia virus (FLV), which cemented the virus theory of cancer [295]. In addition, Richmond T. Prehn and Joan Main demonstrated the specific antigens of tumors from genetically identical mice had methylcholanthrene-induced sarcomas [296]. And these outcomes from mice suggested the relation of the immune surveillance and tumors antigens [297]. Besides, Alick Isaacs and Jean Lindenmann characterized interferon, and it was demonstrated to be effective in the treatment of cancer, which suggested that the cytokines can be the substances to boost the immune system for treatment of cancer [298].

In 1959, Lloyd Old and Baruj Benacerraf used the Calmette-Guérin (BCG) to suppress tumor growth in mice, and it is still used to treat bladder cancer nowadays [299]. John Graham and Ruth Graham published their vaccine study on cancer patients in the same year [300]. Since 1967, the mysteries of the immune response were gradually resolved by extensive research. These included the existence of T cells reported by Jacques Miller [301], discovery of the usage of dendritic cells by Ralph Steinman [302], characterization of major histocompatibility complex (MHC) by Peter Doherty and Rolf Zinkernagel [303], and recordation of natural killer (NK) cells by Eva Klein [304]. After the long-term practice of interferon, Jordan U. Gutterman and his colleagues performed a clinical trial of interferon alpha in several cancer patients in 1978 [305].

In the 1990s, checkpoint inhibitor therapies continued to progress, and cytotoxic T-lymphocyte antigen 4 (CTLA-4) was then identified as a critical immune checkpoint by James P. Allison in 1995 [306].

Since the hybridomas cells were cultured for the production of monoclonal antibodies, research on antibody-based therapies finally developed some specific biological agents such as rituximab approved by US Food and Drug Administration (FDA) in 1997 [307]. In addition, the T-cell growth factor, interleukin 2 (IL-2), was approved by FDA to apply in metastatic melanoma in 1998 [290]. After 2000, there are more and more cancer vaccines and monoclonal antibodies which have been developing for cancer treatment with the rapid development of technology and medicine. Owing to the fewer side effects than existing drugs, immunotherapy is much more popular in cancer treatment in recent years and brings a hope to cure the carcinoma [308, 309]. Of note, the 2018 Nobel Prize in Medicine or Physiology was awarded to Tasuku Honjo of Kyoto University and James P. Allison of the University of Texas, for their works in checkpoint therapy by inhibition of negative immune regulation [310, 311]. Although there are still hundreds of hardship need to be overcome, the work on cancer immunotherapy greatly leaps forward in human oncology and causes a veritable revolution in tumor therapy. In the future, scientists and researchers need to completely find out what and how immunotherapies work and improve the current therapeutic approach. For instance, we should understand how the tumor microenvironment or the individual's microbiome influences the immune system or the immunotherapy [288]. There are five common types of immunotherapy in cancer treatment including monoclonal antibodies, adoptive cell transfer, nonspecific immunotherapy (cytokines), treatment vaccines, and immunomodulating drugs.

6.3.1.2 Types of Immunotherapy

6.3.1.2.1 Monoclonal Antibodies

Monoclonal antibodies (MABs) are produced by the immune cells that are the clones of a single ancestral cell [286, 312]. Monoclonal antibodies work by finding and binding of the antigens in order to discriminate the specific epitopes, and the immune system can attack and destroy the particular cells. MABs have the ability to recognize the specific receptor, which allows the antibodies to seek for the cancer cells but not harming the normal ones. For example, rituximab, a type of MABs, can target a particular protein called CD20 on the surface of the lymphoma and leukemia cells, so that it is commonly prescribed to treat the non-Hodgkin's lymphoma and chronic lymphocytic leukemia [312, 313].

The MABs mainly consist of three parts based on their function, including targeted antineoplastic agent, immune checkpoint inhibitors, and conjugated monoclonal antibodies.

- Targeted antineoplastic agents

There are at least 12 monoclonal antibodies approved by the FDA [314]. For example, trastuzumab is used to treat breast cancer by binding to human epidermal growth factor receptor 2 (HER2) [315]. And pertuzumab is usually utilized with trastuzumab in targeting HER2 [316]. Because vascular endothelial growth factor (VEGF) can increase the formation of blood vessels, it becomes a crucial mediator of angiogenesis in cancer [317]. Bevacizumab, a recombinant humanized monoclonal antibody, was approved in clinical to treat varieties of cancers including colorectal cancer, lung cancer, and glioblastoma by targeting VEGF [318–320]. And ranibizumab is another effective antibody for blocking of VEGF-induced angiogenesis in cancer treatment [321]. Besides, epidermal growth factor receptor (EGFR) is a tumor-specific protein which is the anti-cancer target as well, and several targeted EGFR monoclonal antibodies include cetuximab and necitumumab [322, 323].

- Immune checkpoint inhibitors

The normal cells can avoid the attacks from the immune system by some special proteins called immune checkpoints [286]. However, some cancer cells can masquerade by turning on their immune checkpoints, so that the immune system will be confused to recognize them. Checkpoint inhibitors are a type of drugs (monoclonal antibodies) that work by blocking the checkpoint proteins of the cancer cells which like a release of the "brakes," so that the function of the immune system can be restored and respond against cancer [324, 325].

Currently, there are two main immune checkpoint pathways including cytotoxic T-lymphocyte-associated protein 4 (CTLA4) and programmed cell death protein 1 (PD-1) and programmed death-ligand 1 (PD-L1) pathway [286]. CTLA4 is a protein receptor on T cells that functions as an immune checkpoint, and it is highly expressed in cancer patients [308]. Ipilimumab is the first CTLA-4 blockade monoclonal antibodies which were approved by the FDA in 2011 for treatment of melanoma [326]. In addition, another CTLA-4 inhibitor tremelimumab is designated as an orphan drug to treat mesothelioma, but it still undergoes the human trials and has not been approved by FDA for the treatment of other tumors [327].

PD-1 is another immune checkpoint, and it was reported that high expression of PD-1 is associated with the increase of tumor aggressiveness [328]. The cancer cells could evade the surveillance of the host immune system by attaching to the PD-1 of immune cells, and PD-1 ligand, PD-L1, highly expressed in cancer cells [329]. Many inhibitors for PD-L1 or PD-1 were extensively developed in the last several years. Nivolumab is a human anti-PD-1 monoclonal antibody that was approved to treat numbers of cancer, such as renal cell carcinoma, classical Hodgkin's lymphoma, and melanoma. And in 2018, Opdivo (nivolumab) was further approved by the FDA for the treatment of small cell lung cancer [330]. On June 15 of the same year, China's Drug Administration approved nivolumab as well, and it is the first immuno-oncology product that targeting PD-1 in China [331]. Other PD-1 inhibitors include pembrolizumab and spartalizumab [332, 333], whereas PD-L1 inhibitors contain atezolizumab, avelumab, and durvalumab [334].

- Conjugated monoclonal antibodies

Conjugated monoclonal antibodies are normally attached to radiation or chemicals, and it could be delivered to cancer cells directly [286]. The antibodies specifically bind to the tumor cells and then release the radiative or antitumor drug particles to destroy cancer. Normally, the conjugation of drugs and antibodies should be inactive before the conjugate linked to the tumor cells [335].

Up to now, there are several types of chemical drugs conjugated to monoclonal antibodies that are undergoing in clinical trials including enediyne antibiotics, maytansinoids, and anthracyclines [336]. And some of them have been approved by the FDA. For instance, brentuximab vedotin targets the protein CD30 on the surface of cancer cells for the treatment of Hodgkin's lymphoma [337]. And gemtuzumab ozogamicin is applied in patients with CD33-positive acute myeloid leukemia [338]. And trastuzumab emtansine is proved to be effective in the treatment of advanced HER2-positive breast cancer [339].

The monoclonal antibodies can also carry radioactive substance as a tracer to targeted tumors, and this approach is known as radioimmunotherapy (RIT) [286]. Recent studies in RIT have indicated that some methods should be employed to estimate the appropriate radiation dose for individualizing patient treatment, which can maximumly reduce the toxic injuries from radiation [340]. And ibritumomab is just a type of monoclonal antibody used in radioimmunotherapy. Generally, ibritumomab is combined with radioactive isotope including yttrium-90 and iodine-131. And the conjugation is usually delivered to treat non-Hodgkin's lymphoma [286]. Normally, the monoclonal antibodies are produced by hybridomas cells. The hybridomas are the fusion of myeloma cells and antibody-producing splenocytes, and the splenocytes are typically collected from mice. In that case, the generated cells are able to produce antibodies and proliferate unlimitedly. Then the clones will be cultured in HAT medium (selective medium) to screen and select the positive cells (fused cells), and they are collected to dilute and separate on microtiter wells to produce monoclonal antibodies [312].

6.3.1.2.2 Adoptive Cell Transfer

Adoptive cell transfer (ACT) is the transfer of T cells into the patients in order to fight cancer, and the cells are originally isolated from the recipients or other donors

[285, 341]. Since the first ACT-based immunotherapy was applied in 1988, it has emerged as one of the most effective treatments for metastatic melanoma [342, 343]. It usually takes 2–8 weeks for the culture of T cells in vitro with interleukin-2, and the T cells will be injected back to the patients [285].

There are two types of ACT. One is unmodified T cells isolated from resected tumors (tumor-infiltrating lymphocytes), and the other is genetically engineered T cells recognizing tumor antigens [344].

Tumor-infiltrating lymphocytes (TIL) are a type of immune cell population discovered in tumor lesions, and the fraction of the T-cell receptors (TCRs) can help them specifically direct the antigens expressed in the tumors cells and fight against tumor. This approach is used as an effective way to treat melanoma, and some types of neoplasms demonstrated the potential response to TIL therapy such as Hodgkin and non-Hodgkin's lymphoma, and gastric carcinoma. The major limitations of TIL therapy are the T cells specific to the antigens so that they can't recognize and exert function in other types of cancers. The genetically engineered T-cell therapy exhibits more widespread application in cancer treatment [344].

Genetically engineered T cells are firstly isolated from peripheral blood T cells, and the investigators develop some approaches to modify these cells that they will be allowed to specifically direct the tumor antigens and activate the adaptive immune to attack the tumors. Generally, there are two approaches to "redesign" the T cells, modification of genes with TCRs specifically directed against the tumor-associated antigens and introduction of chimeric antigen receptors (CARs) [344].

The TCR T-cell therapy modifies some specific genes in the case of the types of cancer. The TCR T-cell therapy was used for the treatment of metastatic melanoma modified with MART1, NY-ESO-1, and MAGE-A3 in melanoma and synovial sarcoma [345–347].

The limitation of TCR T-cell therapy is that the therapy is based on the recognition of the specific antigens in the malignant cells, and sometimes tumors decrease their expression of antigens to escape the T cells. Accordingly, CAR therapy is developed to overcome such limitation. The T cells will be changed to receive an antibody-derived single-chain variable fragment so that the T cells can recognize the antigens without MHC restriction [344]. In 2017, the FDA had approved two CAR T-cell therapies to treat acute lymphoblastic leukemia and advanced lymphomas [285].

Some clinical trials attempt to combine checkpoint inhibitors with adoptive T-cell therapy. The checkpoint inhibitors block the T-cell restriction from tumors serving the environment for ACT to exert the antitumor effects [344].

6.3.1.2.3 Nonspecific Immunotherapy (Cytokines)

Nonspecific immunotherapies do not target cancer cells specifically, which they mainly use cytokines to boost the immune system to attack cancer, and these cytokines are generally produced in the human body. In addition, it often works by the administration of interferons, interleukins, granulocyte colony-stimulating factor (G-CSF), and granulocyte-macrophage colony stimulating factor (GM-CSF) [286].

Notably, interleukin-2 (IL-2) is a T-cell growth factor; it can activate some tumor-reactive cells to fight several solid tumors. Besides, it stimulates and helps the body to make more types of immune cells and antibodies against cancer. Aldesleukin, the synthetic interleukin-2, is used for the treatment of kidney cancer and melanoma [348]. Interferons can also trigger a stronger immune response to resist cancers. It can also suppress the growth of cancer cells and angiogenesis. Among the three interferons, interferon-α is the only one used for cancer treatment, such as melanoma, non-Hodgkin's lymphoma, and chronic myelogenous leukemia [348].

G-CSF and GM-CSF can activate the immune system and stimulate the bone marrow to produce more granulocytes and macrophages. Some studies begin to apply G-CSF or GM-CSF to treat breast cancer as an adjuvant treatment [349]. And the administration of G-CSF or GM-CSF represents a new strategy to treat cancer that is related to restore the efficacy of granulocytes [350].

6.3.1.2.4 Treatment Vaccines

Vaccines are very common in people's daily life, and they are normally used to prevent diseases and infections. However, some cancers are induced by several viruses; therefore different types of vaccines have been developed to against these cancers. And vaccines that can be utilized for cancer treatment are also known as cancer vaccines [351].

Some cancer vaccines have been developed to direct against the specific viruses that can cause certain cancers. For instance, human papilloma virus (HPV) is associated with several cancers including cervical, anal, and throat cancer, and the hepatitis B virus (HBV) is proved to cause liver cancer. And it is estimated that the HPV vaccines can prevent approximately 70% of cervical cancers, 80% of anal cancers, 60% of vaginal cancers, 40% of vulvar cancer, and possibly some mouth cancer [352–354].

The other type of vaccines is therapeutic cancer vaccines, which can activate the immune system to attack the cancer cells. These vaccines are composed of cancer cells, antigens, or the immune cells taken from the patients. They will be prepared in the laboratory, and usually the vaccines are combined with some other agents or cells as the adjuvants to improve the drug efficacy [286].

Up to now, several therapeutic cancer vaccines have been designed and proved to be effective in cancer treatment. BiovaxID was reported to prolong remission of follicular lymphoma, and sipuleucel-T was demonstrated to improve the survival rate of prostate cancer [355, 356]. And Bacillus Calmette-Guérin (BCG) is developed to treat the early stage of bladder cancer by using weakened bacteria to stimulate the immune system [357]. Besides, some oncolytic viruses are also developed to treat cancer. For instance, talimogene laherparepvec, genetically engineered herpes virus, is used to treat melanoma that cannot normally receive operation [358].

The process of presenting the antigens to T cells is a critical step in vaccination, and dendritic cells are the most significant antigen-presenting cells. So, some clinical trials are focusing on the diversity of dendritic cells to discover the potential strategy to increase the efficiency of therapeutic vaccines [359, 360].

6.3.1.2.5 Immunomodulatory Drugs

Immunomodulatory drugs can improve the immune system, and generally, the immunomodulatory imide drugs (IMiDs) or thalidomide analogs are a class of chemicals that have been extensively studied [286].

Thalidomide is derived from glutamic acid; it is the first synthesized antiemetic agent in pregnancy but leading to the defects of births [361]. However, the investigators found that the ability of thalidomide is very powerful in some treatments of inflammation by inhibiting the production of TNF-α [362]. Subsequently, thalidomide was found to have immunomodulatory effects including stimulation of NK cells and activation of T cells and suppression of angiogenic property [363, 364]. This has drawn a surge of interest in thalidomide as a promising anti-cancer agent, and two IMiDs, lenalidomide and pomalidomide, have been developed later with increased anti-cancer potency and less toxicity [361].

Because IMiDs have powerful properties of immunomodulatory including the suppression of T regular cells, T-cell co-stimulation, and activation of NK cells and enhance the procedure of antibody-dependent cellular cytotoxicity (ADCC), these drugs are applied in the treatment of myeloma, and their development also facilitates relevant research to improve the understanding of myeloma biology [361].

With the increase of the public awareness of the health and immune, more and more agents that have the ability to modulate the immune system have received the attention. For example, probiotics and its combination with prebiotics have been confirmed to confer health benefits on the host through the immune-modulatory ability [365]. And some polysaccharides are found to have the anti-cancer properties through activating the immune system [366]. Besides, some traditional Chinese medicine is also demonstrated to be effective in immunological regulation, but the potential effects and complicated mechanisms still need more researches and evidence to test and achieve scientific validity [367].

6.3.1.3 Immunotherapy Works with Other Cancer Treatment

Immunotherapy has not yet as widely used when compared with some traditional tumor therapies because of too many uncertain problems, such as the immunotherapy doesn't work, and some terrible side effects [368]. In some types of immunotherapy, the patients will receive high-dose radiation therapy or chemotherapy before the main treatment. Sometimes, traditional cancer treatments don't work well against metastatic tumors, and then physicians will use immunotherapy to help them [369].

6.3.1.4 Risk of Immunotherapy

Most side effects of immunotherapies are mild, and sometimes it is caused by the immune system which may attack healthy cells.

The most common side effects include skin reactions and flu-like symptoms. Besides, some other side effects are common in cancer treatment, such as diarrhea, fever, fatigue, and many others. But the dangerous side effect can happen sometimes, which patients who received immunotherapy may have some serious disease such as pneumonitis, colitis, hepatitis, and complicated skin reactions [370].

6.3.2 *Targeted Therapy*

6.3.2.1 Introduction of Targeted Therapy

Targeted therapy is one of the novel modalities of cancer treatment that attacks specific molecules in order to block the growth of cancer cells [371–374]. Because of the great effectiveness, targeted cancer therapy has attracted much public attention and considered as a "magic bullet" to replace traditional chemotherapy [375].

Targeted therapy was firstly proposed in the 1940s. Physicians used radioactivity of iodine to treat thyroid cancer because thyroid cancer cells could uptake more iodine than the normal cells [376]. A more typical model of targeted therapy is tamoxifen, which had been approved to treat cancer since the 1970s. Tamoxifen is an estrogen antagonist that prevents estrogen from binding to the estrogen receptor, thereby suppressing the activity of estrogen receptors. Given that some breast cancer cells demand estrogen for cell proliferation, thus tamoxifen is commonly used to treat estrogen receptor-positive breast cancer [377]. Imatinib, also known as Gleevec, was one of the breakthroughs of targeted therapy developed in the 1990s. It's a type of tyrosine kinase inhibitor, and it was thought to revolutionize the treatment of chronic myeloid leukemia (CML). Before the discovery of imatinib, there was no drug that could manage such a disease like CML [378]. In the late two decades, the tremendous progress in the medical biology such as the identification of oncogenes and tumor suppression genes, has extremely fueled some advances in the understanding of the molecular mechanisms and pathogenesis of cancer. Accordingly, a significant number of new drugs have been developed to treat various types of tumors, while some of the new-targeted cancer therapies are still in trials to investigate the efficacy and safety [379].

Most targeted therapies utilise small-molecule drugs or monoclonal antibodies. Small-molecule drugs are low molecular weight organic compounds which can enter cells and interfere with certain enzymes and proteins to inhibit cancer cell growth. In contrast, monoclonal antibodies inhibit cancer cell growth by blocking targets outside the cells or on the cells surface [371].

In addition, nanotechnology is a relatively new type of subject that mainly modulates the structure, property, and behavior of nanometer materials [380]. In the past few years, the nanotechnological intervention has attracted significant attention in the therapy of cancer because its elaborated design greatly revolutionized the existing chemotherapeutic agents [381]. And the size of nanometer ideally serves a novel platform to surmount the current problems including low drug loading, undesirable drug distribution, and serious adverse effects [382].

In the US, the National Cancer Institute has established a comprehensive Molecular Targets Development Program (MTDP) to concentrate on identification and evaluation of molecular targets that may be potential candidates for drug development [371].

6.3.2.2 Types of Targeted Therapy

6.3.2.2.1 Small-Molecule Drugs

The small-molecule drugs are designed to enter the cells and act on intracellular targets and interfere with some signaling pathways. The targets could be some specific enzymes or growth factor receptors (GFRs), such as tyrosine kinases, mammalian target of rapamycin (mTOR), and angiogenesis, and there are also some apoptosis-inducing drugs that activate the cell apoptosis (Table 6.1).

Table 6.1 Small-molecule drugs targeted cancer therapies approved by FDA in US

Brand name	Compound name	Drug target	FDA-approved indications
Gleevec	Imatinib	BCR-ABL	Chronic myeloid leukemia
Iressa	Gefitinib	Epidermal growth factor receptor (EGFR)	Non-small cell lung cancer
Tarceva	Erlotinib	EGFR	Non-small cell lung cancer, pancreatic cancer
Nolvadex	Tamoxifen	Estrogen receptor (ER)	Breast cancer
Istodax	Romidepsin	Histone deacetylase (HDAC)	Cutaneous T-cell lymphoma
Avastin	Bevacizumab	Vascular endothelial growth factor (VEGF)	Colorectal cancer, non-small cell lung Cancer
Sprycel	Dasatinib	BCR-ABL, Src, c-Kit	Chronic myeloid leukemia, acute lymphocytic leukemia
Sutent	Sunitinib	c-Kit, PDGFR, VEGFR	Renal cell cancer, gastrointestinal stromal tumor
Velcade	Bortezomib	The 26S proteasome	Multiple myeloma, mantle cell lymphoma
Tykerb	Lapatinib	HER2	Breast cancer
Torisel	Temsirolimus	mTOR	Advanced renal cell carcinoma
Tasigna	Nilotinib	BCR-ABL	Philadelphia chromosome positive chronic myelogenous leukemia
Zolinza	Vorinostat	HDAC	Cutaneous T-cell lymphoma
Nexavar	Sorafenib	VEGFR, PDGFR	Renal cell carcinoma
Xalkori	Crizotinib	ALK	Non-small cell lung cancer
Venclexta	Venetoclax	Bcl-2	Chronic lymphocytic leukemia

- Tyrosine kinase inhibitors

Uncontrolled kinase activity is characterized as a major factor in which normal cells become cancerous cells, and the signaling pathways regulated by these kinases are the common targets with somatic mutations. The existing oncogenes are more than a hundred, and many of them encode protein tyrosine kinases [379]. Tyrosine kinase is a type of protein that acts as a switch of the death and growth of cells. Several receptor proteins related to tyrosine kinase are found to be mutated or overexpressed in human cancer. And these proteins involve the epidermal growth factor receptor (EGFR), human epidermal growth factor receptor 2 (HER2), insulin and insulin-like growth factor (IGF) receptors, and among others. By blocking the activity of the enzyme using specific kinase inhibitors, the aberrant activation of signaling pathways can be prevented. Imatinib targets BCR-ABL fusion protein by blocking the tyrosine kinase to inhibit chronic myeloid leukemia. Gefitinib specifically targets the cancer cells overexpressing the EGFR protein and inhibits the tyrosine kinase activity, and therefore it is clinically approved for the treatment of non-small cell lung cancer. Examples of other tyrosine kinase inhibitors include erlotinib, sunitinib, lapatinib, and ibrutinib [272, 374].

- mTOR inhibitors

mTOR participates in the regulation of cells growth and division, and it also plays an important role in the process of catabolism. Accordingly, mTOR is a key protein kinase controlling the cell cycle and protein synthesis. In some types of cancer, the mTOR pathway is highly upregulated, leading to the uncontrolled cancer cell growth and cell division. Thus, mTOR inhibitors are developed to target mTOR protein to stop cancer cells growth. Temsirolimus is a small-molecule drug targeting mTOR protein, which is used to treat advanced renal cell carcinoma. Everolimus is another mTOR inhibitor, which is approved for the treatment of kidney cancer [374, 383, 384].

- Angiogenesis inhibitors

Angiogenesis is the cellular process of growth of new blood vessels among the tissue and is confirmed as the fundamental step for the transition of tumors from a benign one to malignant neoplasms [385]. As such, the anti-angiogenic agents are applied to interrupt the tumor's blood supply by stopping the development of new blood vessels. Sunitinib, a type of tyrosine kinase inhibitor, blocks the vascular endothelial growth factor (VEGF) signaling and impedes blood vessel growth. It is approved by the FDA for the treatment of renal cell carcinoma (RCC) and imatinib-resistant gastrointestinal stromal tumor (GIST) [386]. Besides, thalidomide could also interfere with the activation signals for the proliferation and growth of blood vessels and is therefore utilized in the treatment of multiple myeloma [371].

- Apoptosis-inducing drugs

Apoptosis is a form of programmed cell death to control cell proliferation and survival. Typically, cancer cells can escape immune surveillance and avoid cell death. Apoptosis-inducing drugs can make cancer cells easier to be destroyed by restoring the signals of cell death. Oblimersen, a BCL2 inhibitor, can promote cancer cell survival by inducing drug resistance. Olaparib, a poly ADP-ribose polymerase (PARP) inhibitor, can inhibit PARP expression so that cancer cells can't repair DNA damage. Bortezomib, a proteasome inhibitor, is used for treating relapsed multiple myeloma in some regions [374, 387].

6.3.2.2.2 Monoclonal Antibodies

Monoclonal antibodies are the synthetic antibodies that can block a target outside of a certain cancer cell. Scientists have developed many monoclonal antibodies that target a wide range of proteins in different cancers. These antibodies take effect by recognizing and locking onto the protein to work properly. Besides, some antibodies act as transmitters to help some chemicals or radioactive particles getting into the cancer cells and direct to the targets, which is known as conjugated monoclonal antibodies. Currently, there are numerous monoclonal antibodies created by researchers and approved by the FDA for the treatment of various cancers (Table 6.2).

- Vascular endothelial growth factor (VEGF)

Clinical studies have proposed that monoclonal antibodies against VEGF are more effective compared with traditional therapy for cancer treatment [388]. VEGF is a secreted glycoprotein that contributes to the process of angiogenesis occurring in tumors and is usually overexpressed within the tumor environment. The high level of VEGF has been demonstrated to be correlated with poor survival in patients [389]. Multiple signaling pathways are also regulated by VEGF, and these pathways are involved in many biological processes of cells, such as cell apoptosis, differentiation, and proliferation [390]. Hence, some VEGF blockers can be used to suppress the abnormal expression of VEGF, which resulted in the elimination of cancers. Bevacizumab is a VEGF targeting antibody used for colorectal cancer and non-small cell lung cancer. And ramucirumab, a fully human monoclonal antibody (IgG), is synthesized for the treatment of solid tumors by targeting VEGFR2 and is approved to treat gastric cancer in the clinic [391, 392].

- Epidermal growth factor receptor (EGFR) and human epidermal growth factor receptor (HER2)

EGFR and HER2 are receptor tyrosine kinases that have been shown to be crucial regulators for the development and progression of various types of cancer. Overexpression of EGFR is associated with the development of a wide variety of tumors such as adenocarcinoma of the lung, anal cancers, and epithelial tumors of the head and neck [393]. And the mutations that lead to HER2 amplification are associated with the progression of certain aggressive types of breast cancer. It has become a critical biomarker and target for approximately 30% of breast cancer patients [394]. Several monoclonal antibodies have been developed to target

Table 6.2 Monoclonal antibodies targeted cancer therapies approved by the FDA in the US

Brand name	Compound name	Drug target	FDA-approved indications
Arzerra	Ofatumumab	CD-20	Chronic lymphocytic leukemia
Bexxar	Tositumomab	CD-20	Non-Hodgkin's lymphoma
Erbitux	Cetuximab	Epidermal growth factor receptor (EGFR)	Colorectal cancer, head and neck cancers
Rituxan	Rituximab	CD-20	Non-Hodgkin's lymphoma
Herceptin	Trastuzumab	Human epidermal growth factor receptor 2 (HER2)	Breast cancer
Avastin	Bevacizumab	Vascular endothelial growth factor (VEGF)	Colorectal cancer, non-small cell lung cancer
Mylotarg	Gemtuzumab	CD-33	Acute myeloid leukemia
Campath	Alemtuzumab	CD-52	B-cell chronic lymphocytic leukemia
Zevalin	Ibritumomab tiuxetan	CD-20	Non-Hodgkin's lymphoma
Vectibix	Panitumumab	EGFR	Colorectal cancer
Yervoy	Ipilimumab	Cytotoxic T-lymphocyte-associated protein 4 (CTLA-4)	Melanoma
Opdivo	Nivolumab	Programmed cell death protein 1 (PD-1)	Small cell lung cancer
Keytruda	Pembrolizumab	PD-1	Melanoma
Tecentriq	Atezolizumab	Programmed death-ligand 1 (PD-L1)	Advanced triple-negative breast cancer
Bavencio	Avelumab	PD-L1	Merkel cell carcinoma
Imfinzi	Durvalumab	PD-L1	Non-small cell lung cancer
Blincyto	Blinatumomab	CD-19	Precursor B-cell acute lymphoblastic leukemia
Adcetris	Brentuximab vedotin	CD-30	Hodgkin's lymphoma anaplastic large cell lymphoma
Darzalex	Daratumumab	CD-38	Multiple myeloma
Unituxin	Dinutuximab	GD2	Children with high-risk neuroblastoma
Empliciti	Elotuzumab	SLAMF7 (CD-319)	Multiple myeloma
Portrazza	Necitumumab	EGFR	Metastatic squamous non-small cell lung carcinoma
Mylotarg	Gemtuzumab ozogamicin	CD-33	Acute myeloid leukemia
Cyramza	Ramucirumab	VEGFR2	Gastric cancer

EGFR and HER2. Trastuzumab is developed for the treatment of breast cancer in HER2 receptor positive patients. Necitumumab, panitumumab, and cetuximab are the EGFR inhibitors used to treat several types of cancer [322, 323, 395].

- Programmed cell death protein 1 (PD-1) and programmed death-ligand 1 (PD-L1)

PD-1 and PD-L1 can combine with each other specifically. PD-1 protein is on the surface of T cells that act like a "checkpoint" to keep the immune system from killing other cells in the body (autoimmune diseases) [328]. PD-L1 can attach to PD-1 and is found on some cancer cells which help them hide from immune attack [329]. Some inhibitors targeting either PD-1 or PD-L1 can prevent the interaction between PD-1 and PD-L1, so that to enhance the T-cell responses for cancer therapy. In 2019, there are several PD-1 or PD-L1 inhibitors approved by FDA including atezolizumab, durvalumab, avelumab, pembrolizumab, and nivolumab [334, 396].

- Cytotoxic T-lymphocyte-associated protein 4 (CTLA-4)

CTLA-4 is another protein receptor which functions as a "checkpoint" like PD-1 and is also expressed on activated T cells. It acts as a "turnoff" switch that downregulates immune response, and the levels of CTLA-4 are particularly high in cancers [308]. Ipilimumab is an antagonistic antibody against CTLA and is approved for treating melanoma by inhibiting immune system tolerance to tumors [326].

- CD-19

CD-19 is a transmembrane protein expressed in all B cells, and it is considered as a biomarker for B lymphocyte development. Besides, CD-19 plays an active role in driving the growth of certain cancers such as acute lymphoblastic leukemia and chronic lymphocytic leukemia [397]. Blinatumomab is a biopharmaceutical drug, which is the monoclonal antibody and exerts action selectively and specifically targets on CD-19 antigen presented on B cells, and to activate the T cells to exert cytotoxic activity on the malignant B cells [398]. And it was granted approval by the FDA for the treatment of relapsed or refractory B-cell precursor acute lymphoblastic leukemia in adults and children in 2017 [399].

- CD-20

B-lymphocyte antigen CD-20 is a glycosylated phosphoprotein expressed on the surface of B cells, whereas the abnormal expression of CD-20 is associated with B-cell chronic lymphocytic leukemia [400]. Based on the structure and function of the CD-20 antigen, the preclinical work has investigated the effect of CD-20 which was found to engage in the progression of cancer. Thus, several monoclonal antibodies have been designed to target the CD-20. Tositumomab, ofatumumab, and rituximab are the classical anti-CD20 monoclonal antibodies used in the treatment of non-Hodgkin's B-cell lymphoma, B-cell chronic lymphocytic leukemia, and few B-cell-mediated cancers [401].

- CD-30

CD-30, a cell membrane protein, is a member of the tumor necrosis factor receptor family, and it is a tumor marker characteristically expressed in several hematopoietic tumors including Hodgkin's lymphoma and anaplastic large cell lymphoma. Brentuximab vedotin is an antibody-drug conjugate approved for the treatment of

Hodgkin's lymphoma and anaplastic large cell lymphoma. Of note, clinical trials also showed that the antibody is well tolerated in other CD30-related lymphomas [402].

- CD-33

CD-33, considered as a myeloid-specific receptor protein, is a member of the immunoglobulin supergene family expressed in myeloid cells. It is identified as an excellent myeloid marker utilized for the diagnosis of acute myeloid leukemia [403, 404]. Gemtuzumab is usually linked with the cytotoxic agents to manifest anti-cancer effects. For example, gemtuzumab ozogamicin is a drug-linked monoclonal antibody used in the treatment of acute myeloid leukemia [404, 405].

- CD-38

CD-38, known as cyclic ADP ribose hydrolase, is a transmembrane glycoprotein on the surface of immune cells and acts as a multifunctional ectoenzyme. Since the upregulation of CD-38 would unsavory the prognosis of chronic lymphocytic leukemia (CLL), this protein is also used as a prognostic marker for CLL [406, 407]. Besides, daratumumab, an anti-cancer monoclonal antibody targeting CD-38, has been approved in treating multiple myeloma. It can selectively bind and inhibit the enzymatic activity of CD-38 and induce apoptosis in malignant cells [408].

- CD-52

CD-52 is a glycoprotein expressed on the surface of the lymphoid cells, and the expression of CD-52 is associated with various types of lymphoma. Alemtuzumab is a type of monoclonal antibody, which is used for the treatment of B-cell chronic lymphocytic leukemia [409].

- SLAMF7

SLAM family member 7, an NK cell receptor, expresses on the immune cells, and it can regulate immune function. Because of the expression of the high level of SLAM7 in multiple myeloma, the protein has become a diagnostic marker and an effective target for immunotherapy in this type of tumor [410]. Elotuzumab is a promising SLAMF7-directed humanized monoclonal antibody used in relapsed multiple myeloma, and it was granted "Breakthrough Therapy" designation by the FDA in 2015 for the treatment of multiple myeloma [411].

- GD2

GD2 is a disialoganglioside expressed on tumors of neuroectodermal origin including human neuroblastoma and melanoma. GD2 is a robust target for monoclonal antibodies therapy, due to its relatively high expression in tumors [412]. For instance, dinutuximab is used as an effective drug for children with high-risk neuroblastoma [413].

6.3.2.3 Nanomaterials for Targeted Therapy

Owing to the unique characteristics of tumor-targeted of nanomaterials, these small particles with dimensions below several hundred nanometers have raised exciting and remarkable opportunities for cancer therapeutic and diagnostic applications. Compared with tissues or cells in microscale, the nanoparticles are typically in smaller and tinier size (Fig. 6.8). A wide range of nanomaterials have been recently emerging and reported as the potential and promising approaches including liposomes, dendrimers, polymeric micelles, and polymeric nanoparticles [381]. These artificially designed nanomaterials could significantly improve the application of nanotechnology in cancer treatment through developing multifunctional targeted treatment for both cancer diagnosis and therapy and greatly enhance the therapeutic efficacy of the medications, whereas the cytotoxicity will be reduced because of tumor-specific property [414]. Some nanoparticles have been developed with multiple functions, which have the ability to specifically deliver the drugs and direct them to the site of the tumor, thereby allowing the tumor microenvironment to be visualized for monitoring the therapeutic response [415].

The nanomaterials have lots of unique and significant advantages compared with conventional chemotherapy. First of all, these nanomaterials generally have a similar size to the biological macromolecules including peptides and nucleic acids and usually ranged around 10 nanometers in diameter which is 100–1000 times smaller than a cancer cell [381]. As the micro-sized particles are easily taken up by the body, these microparticles are rapidly cleaned so that many of them can't reach the target. However, the tiny nanomaterials offer great possibilities to circumvent the vascular barriers and immune systems so that the well-designed nanoparticles can possibly reach the targets or deliver drugs to the site of the tumors [416].

Secondly, nanomaterials are able to improve the therapeutic efficacy of the conventional anti-cancer agents and overcome many limitations of these existing drugs. For example, paclitaxel is widely used in the treatment of breast, lung, and other cancers, but it has poor solubility in water which leads to difficulties in the preparation of these drugs for antitumor therapy. Of note, nanomaterials have the ability to enhance the water solubility of paclitaxel by stably incorporating them in the hydrophobic microenvironment [417]. Additionally, these nanomaterials can reduce the toxicity of drugs in damaging healthy tissues and prevent drugs from metabolic degradation. Besides, the encapsulation of nanomaterials also improves the drugs therapeutic efficacy and bioavailability in the bloodstream [381].

Moreover, nanomaterials can load a huge amount of therapeutic drugs or imaging agents through their big surface area compared to their volume [381]. For instance, a polymeric nanoparticle with a diameter of 70 nm can contain about 2000 molecules of drugs [418]. Therefore, nanomaterials have a great advantageous and effective for the enhancement of therapeutic efficacy.

Recently, researchers have further developed many novel nanomaterials for cancer therapy, and several of polymeric nanomedicines have been approved by the FDA. Methoxy-polyethylene glycol (PEG)-poly-(d,l-lactide)-paclitaxel micelle,

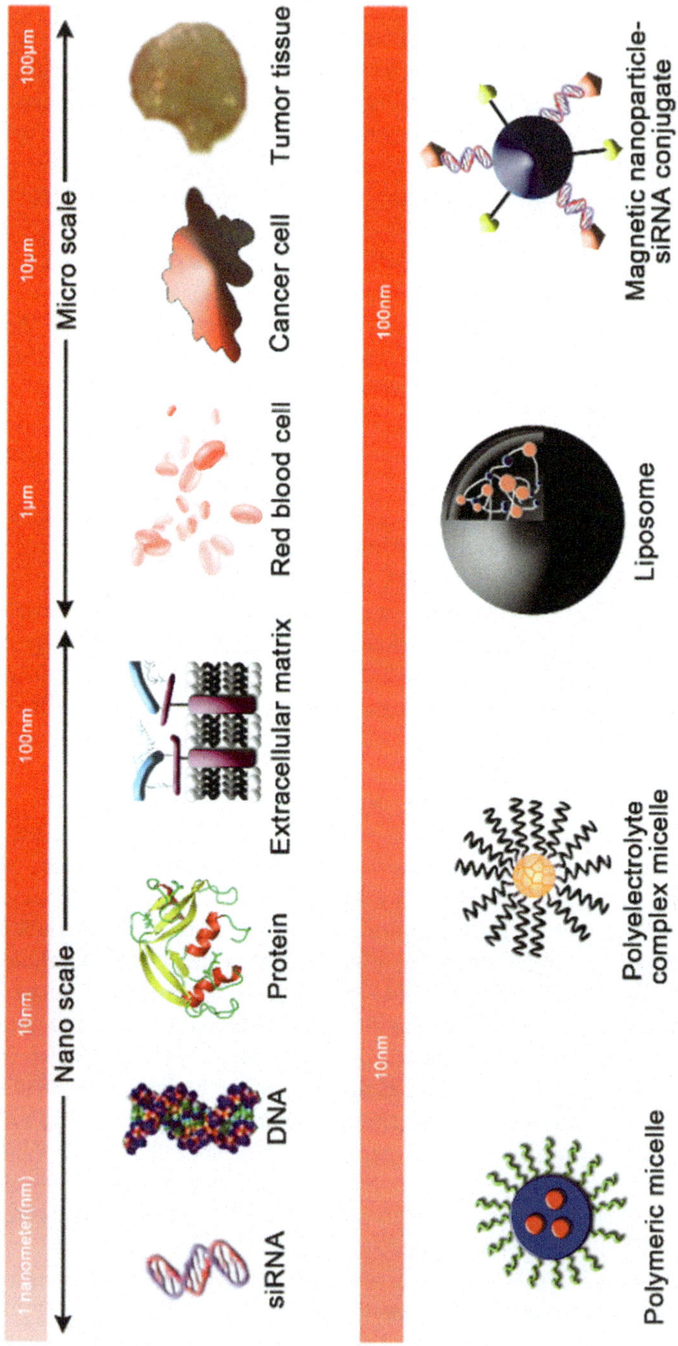

Fig. 6.8 A schematic diagram displaying several important biological systems and artificially engineered nanomaterials in a wide range of sizes

a polymeric micelle, is used for metastatic breast cancer. And PEG-l-asparaginase, a type of polymer-protein conjugate, is applied for the treatment of acute lymphoblastic leukemia [382]. In addition, multifunctional nanomaterials have been proposed to simultaneously play a therapeutic role and visualization of tumors [381]. For example, zinc oxide nanoparticles are a contrast agent material that can act as a carrier for some therapeutic agents including siRNA drugs and small molecules. And it also has some unique properties such as enhanced cytotoxicity, biocompatibility, and high selectivity which would facilitate the efficacy of anti-cancer agent [419]. Some magnetic nanoparticles including iron oxide and manganese oxide have been widely adopted in magnetic resonance imaging (MRI) for cancer imaging and simultaneously carrying the anti-cancer drugs [420, 421].

6.3.2.4 Future Prospective and Challenge to Overcome

In spite of the remarkable advancement, the targeted therapy still remains an unmet medical need, and several obstacles are necessary to be overcome.

- Tumor heterogeneity causing limited single-drug strategy

Tumor heterogeneity is defined as different tumor cells can show distinct morphological and phenotypic profiles [422]. And the tumors not only have differences among various patients but also have genetic and molecular diversities within a single patient. These phenomena occur very frequently, which the former is known as intertumoral heterogeneity, while the latter is referred to intratumoral heterogeneity. Recent studies showed that there is a wide range of intratumoral heterogeneity [423]. Because most tumors harbor numbers of mutation events that may generate additional subclones from an original tumor cell, which subclones have different molecular and genetic characteristics, these changes can lead to activation of many signaling pathways, thereby stimulating the progression and development of tumors [379, 423]. The presence of tumor heterogeneity has been reported in many cancers including breast and pancreatic cancer and bone and soft tissue sarcomas [424–426]. Another explanation for tumor heterogeneity is the existence of cancer stem cells which are resistant to traditional chemotherapies [375]. When the main population of the tumor is destroyed by conventional cancer treatment, these cancer stem cells may survive, leading to cancer remission. Hence, the work to identify this tumor population and find the potential biomarkers will be the most important step to overcome such a challenge in developing effective targeted therapies. And the understanding of cancer resistance by clarifying the tumor heterogeneity will ultimately contribute to personalized medicine with the guidance of the genomic context of each tumor [423].

- Cancer stem cells and epithelial-mesenchymal transition

As mentioned above, cancer stem cells (CSCs) grow slowly and are undifferentiated, which have the potential of differentiation into tumorigenic progenies, which will become resistant to existing chemotherapies [375]. Clinical studies revealed some markers for CSCs including CD44+/CD24(−/low) for breast cancer and CD133+ for endometrial cancer [427, 428]. Of note, the ATP binding cassette (ABC) and multidrug resistance (MDR) transporter are the leading cause for drug resistance toward these cytotoxic compounds [429]. Several strategies have been proposed to target CSCs including the inhibition of multiple survival pathways and the alternation of CSCs microenvironment [430].

Besides, epithelial-mesenchymal transition (EMT) is a process that the epithelial cells lose their ability of adhesion and cell polarity, so that the cells gain metastatic potential and invasive properties to become mesenchymal stem cells [431]. When cancer cells develop into a solid tumor, they can maintain an epithelial phenotype which has the firm junction to prevent metastasis. And new studies have indicated that chemotherapy may induce epithelial cells to occur EMT to acquire migratory capability [432]. Therefore, it provides a potential target to block EMT process to prevent tumor metastasis and recurrence [375].

- Inefficient development in clinical trials and insufficient preclinical models to test targeted medication efficacy

Nearly 20% of researches listed on ClinicalTrials.gov are related to cancer, but research analysis has estimated that only 10% of cancer drug candidates could finally receive FDA approval [433]. The failure and inefficiency of cancer drug development have greatly increased heavy burdens on patients [379]. However, some studies have proposed that drug developers could marshal and analyze preclinical and early stage trials to improve the efficiency of drug discovery [434].

Additionally, the complicated tumor context can't be completely replicated through existing transgenic or xenograft models. And the shortage of reliable preclinical models strongly limits the sensitivity of drugs in cancer therapy. Even though the engineering mouse tumor models provide a potential and promising approach to evaluate anti-cancer drugs efficacy, the validation of such models has not been confirmed in the current stage of clinical research [379].

6.3.2.5 Risk of Targeted Therapy

As well as immunotherapy, the risk of targeted therapy is relatively low. These possible side effects vary greatly depending on the drug efficacy and patient's body responses. For instance, the epidermal growth factor receptor (EGFR)-targeted drugs would induce some skin problems such as hives and intense itching [435].

Besides, other common side effects include tiredness, fever, joint aches, nausea, headaches, diarrhea, high blood pressure, and headaches. Nevertheless, some targeted drugs will cause serious side effects including heart or other organ damage and autoimmune reaction [371]. Notably, the tumors or lesions were resected and tested to find targets in treatment, which maybe give rise to the risks to disclose the privacy of the personal information such as genetic or other information [373].

6.3.3 *Gene Therapy*

6.3.3.1 Introduction of Gene Therapy

The concepts of gene therapy are arisen during the 1960s after some successful trails to introduce foreign DNA into the cells of interest. Roger S. and his colleagues are one of the first to demonstrate that the foreign DNA can be introduced to the cells of interest, and the viruses could be used to carry these foreign genetic materials [436]. In 1972, Friedmann T. and Roblin R. purposed that gene therapy could be leveraged to ameliorate some human genetic diseases, whereas the viruses could be modified to carry the designed genetic information for complementation of gene defects or correction of mutated genes [437]. Roger S. was the first person to conduct gene experiments on human as a therapy. However, the erroneous understanding of Shope papilloma virus, he failed to treat two girls with arginase deficiency (hyperargininemias). Indeed, he mistaken hypothesized that the Shope papilloma virus was naturally expressing the relevant genes encoded for arginase activity [438, 439]. In 1984 a murine retrovirus shuttle vector system was developed by Cepko CL et al.; this system could demonstrate a more efficient method in the application of genetic transformation [440]. In 1989, the US Food and Drug Administration (FDA) approved the first gene therapy protocol [441]. Blaese M. and his colleagues were approved to perform their first gene therapy trials in a patient with severe combined immunodeficiency disorder in 1990 [442]. After several successful clinical trials in cancer gene therapy, more commercially approved medications for gene therapy were released including Onyx-15, a recombinant oncolytic adenovirus originally developed by Onyx Pharmaceuticals (Emeryville, CA, USA) for treating head and neck cancer (2005), and Provenge, consisted of modified dendritic cells for metastatic prostate cancer (2010) [443, 444]. Up to 2019, more than 2600 gene therapy clinical trials have been performed worldwide, and the number of new trials is steady in a general uptrend [445, 446].

6.3.3.2 Gene Transfer Approaches and Vectors Used for Gene Therapy

Getting genes into target cells is one of the most difficult aspects, and several approaches have been developed to facilitate entry of genetic materials into target cells or tissues. They can be divided into two groups: (1) viral vectors and (2) non-viral methods.

6.3.3.2.1 Viral Vectors

Viral particles consist of either ribonucleic acid (RNA) or deoxyribonucleic acid (DNA) as their genetic material. The viral structure comprises the nucleic acid and a protein coat (capsid) which the outer shell of protein can help the virus bind to the receptor of host cells and protect itself from the cell nuclease enzymes digestion [444, 447]. The virus is unable to replicate by itself, it has to insert its hereditary information into the host cells, and therefore it can be utilized to carry the genetic material of interest for cells delivery.

There are many types of viral vectors used for gene transfer, such as adenoviruses, herpes simplex virus, lentiviruses, retroviruses, baculoviruses, adeno-associated viruses, and vaccinia viruses. These vectors are different from each other based on their cell tropisms, transgene capacities, duration of transgene expression, and expression profiles [441]. Viral vectors are regarded as the most effective gene delivery methods with minimal side effects, long-term gene expression, and high purification titers [444, 448].

6.3.3.2.2 Non-viral Methods

Since the use of viral vectors may have immunogenic and inflammatory potential, it is therefore required the development of new non-viral methods. The form of non-viral methods can be divided into several groups.

The physical methods are including electroporation, ultrasound, and gene gun deliveries. The DNA material is normally coated with a synthetic carrier like liposome or nanoparticles, and the high energy can shoot the genetic material into the target cells [441].

The chemical methods utilize synthetic or natural polymers as carriers to deliver genetic material into the target cells. Cationic liposomes are one of the most effective chemical methods to mediate gene transfer into the cells by endocytosis, which method also enables the carrying of most biology materials including drugs, proteins, and plasmids [449]. Interestingly, cationic liposomes can also lead to the inhibition of tumor proliferation when combined with small interfering RNA [450].

Some bacteria are able to specifically target tumor cells, such as *Clostridium*, *Listeria*, and *Escherichia coli*. These bacteria can be genetically modified to act as vectors for the application of genetic engineer. Researchers also considered bacteria as relatively safer, cheaper, and more practical tools for gene therapy in comparison with viral vectors [451].

Nevertheless, the efficacy of all mentioned gene delivery systems can be affected by the various extra- and intracellular barriers including cellular or nuclear uptake, endosomal escape, and some endothelial barriers. For this reason, it's still a major problem waiting for overcoming. With recent development in the area of non-viral gene delivery systems involving the inception of nanotechnology, it has made a new hope for conquering these technical barriers [452].

6.3.3.3 Types of Gene Therapy

Gene therapies can be classified into three categories: (1) boosting the immune response, (2) prodrug gene therapy, and (3) stemming the expression of oncogenes or targeting tumor suppressor gene [444].

6.3.3.3.1 Gene Therapeutic Approaches to Boost the Immune Response

Generally, the immunotherapy is implemented to enhance the immunogenicity of the tumor, leading the immune cell to recognize and eliminate them. And one of the approaches is to improve the presentation of tumor-associated antigens (TAAs), which are expressed on the surface of tumor cells and recognized by the immune system [444, 453]. However, the TAA's natural tolerance and the overwhelmingly immunosuppressive tumor microenvironment strongly challenge the conventional immunotherapies [441]. One of the strategies to overcome such a problem is the genetic engineering of T lymphocytes/cells [454]. For instance, a T-cell receptor transgene which is against a target of TAA could be introduced into T cells of the cancer patient, so that it is able to facilitate the T cell to recognize and eliminate the tumor cells [441]. In a clinical study of melanoma, researchers transformed the peripheral blood lymphocytes from cancer patient by using retrovirus encoding an anti-MART1 TCR transgene that was cloned from tumor-infiltrating lymphocytes of the patient. They demonstrated durable engraftment of the T cells at levels exceeding 10% of peripheral blood lymphocytes for at least 2 months after the infusion in 15 patients and observed high sustained levels of circulating that two patients demonstrated objective regression of metastatic melanoma lesions in 1 year after the infusion [455]. Other clinical studies revealed that the patients with metastatic synovial cell sarcoma and melanoma were treated by the T cells with TCR against the antigen NY-ESO-1. These findings demonstrated that the TCR-transduced T cells can be used successfully to cure cancer beyond melanoma tumors [456]. Similarly to the transduction of TCR, a chimeric antigen receptor (CAR) can be introduced into T cells. Different from TCR therapies, T cells are genetically modified to express specific T-cell receptor, whereas CAR therapies utilize the fusion receptor through the combination of both an antigen-binding region with transmembrane domain to anchor antigens to the T cells [457]. In clinic trials, CD19-specific CAR-redirected T lymphocytes have been leveraged as an approach against hematological malignancies and also assessed the potential and safety of CAR therapies [458, 459]. Besides, dendritic cells are also powerful antigen-presenting cells, which could be further modified in the genetic level and used for anti-cancer treatment [460]. For instance, the previous study showed that adenovirus MART-1-engineered autologous dendritic cells were used for treating metastatic melanoma [461]. Other clinical trials in gene therapy implemented for boosting the immune response include injecting recombinant tumor cells with fowlpox virus encoding human granulocyte-macrophage colony-stimulating factor (GM-CSF) into patients with carcinoembryonic antigen-expressing carcinomas and directly vaccinating the patients with metastatic breast carcinoma with a plasmid DNA encoding HER-2/neu together with low doses of GM-CSF and IL-2 to enhance the immunogenicity [462, 463]. Additionally, one example of gene therapy of renal cancer adopted recombinant adeno-associated virus encoding human endostatin (an anti-angiogenic gene), providing a new strategy for the tumor therapy by targeting on tumor microenvironment [464].

Up to March 2019, two CAR T-cell drugs have been approved by the FDA which are YESCARTA (axicabtagene ciloleucel) and KYMRIAH (tisagenlecleucel) for the treatment of lymphoma [465, 466]. And IMLYGIC (talimogene laherparepvec), a drug composed of modified herpes simplex virus type 1 which are engineered to express GM-CSF, won FDA approval for the therapy of melanoma [467, 468].

6.3.3.3.2 Prodrug Activating Suicide Gene Therapy

One of the aims of prodrug activating suicide gene therapy is to maximize treatment efficacy while reducing side effects [444, 469]. In general, the tumor cells will be transduced with genes that express a special enzyme and combined with a subsequent administration of a prodrug that can be converted into an active form by the mentioned enzyme, resulting in the death of the enzyme-producing cancer cells [441, 444, 469]. In addition, the death of these enzyme-producing cancer cells can also lead the death of neighboring non-transduced cells which the phenomenon is known as bystander effect [470]. One explanation for this concept in suicide gene theory is that the cell-to-cell diffusion of toxic metabolites can probably damage surrounding tumor cells [444]. Examples include the cytosine deaminase gene (CD) of *Escherichia coli* and the herpes simplex virus thymidine kinase gene (HSV-TK). After activation by CD, the prodrug 5-fluorocytosine (5-FC) can be converted to 5-fluorouracil (5-FU) which leads to cell cycle arrest and apoptosis of cancer cells. And the HSV-TK can convert ganciclovir (GCV) to ganciclovir monophosphate which further converted to the form of triphosphate by the endogenous kinases within the cancer cells. In such a form, the GCV-triphosphate can induce cell apoptosis by causing chain termination and DNA single-strand broke [469–472]. CD/5-FC system has been used for the treatment of different types of cancer including glioma, pancreatic carcinoma, and colon cancer [473–475]. Meanwhile, the HSV-TK system has been used to treat several types of cancer such as glioma, pancreatic adenocarcinoma, and hepatocellular carcinoma [476–478]. Unfortunately, the drawbacks in the gut and other problems limit the clinical usefulness of CD/5-FC system [479]. Unlike CD/5-FC system, the clinical research about HSV-TK system has made important advances. Sitimagene ceradenovec, a novel gene-based product, consisting of the HSV-TK enveloped in an Ad5 adenovirus vector, is designated but not yet approved by FDA as an orphan drug for the therapy of malignant glioma [480]. Other prodrug systems are including nitroreductase/CB1954 system, carboxypeptidase G2/nitrogen mustard system, cytochrome P450/

oxazaphosphorine system, and purine nucleoside phosphorylase/6-methylpurine deoxyriboside system [479].

Besides, it's worth to mention that the tumor-specific expression of the enzyme can be regulated by tumor-specific promoters. By applying these kinds of promoters, it enables the enzyme expression only to the tumor cells by restricting the adenoviral replication to cancer cells. An example is the promoter of human telomerase reverse transcriptase which has successfully entered into clinical trials [481].

6.3.3.3.3 Inhibiting the Expression of Oncogenes or Targeting Tumor Suppressor Genes

Oncogenes are a type of genes mutated from the wild-type proto-oncogenes. These proto-oncogenes promote cell growth and regulate cell division. When a mutation or mutations occur in oncogenes, it can become activated and excessively produce modified "products" which may lead to tumor growth [482]. Opposite to oncogenes, tumor suppressor genes (TSG) also known as anti-oncogenes negatively regulate cell growth and proliferation. Mutation in TSG may induce loss of function promoting uncontrolled cell division and tumor growth [483]. Below are lists of some known oncogenes and TSGs with each main function and predominantly associated cancer types (Tables 6.3 and 6.4) [484]. Besides, search results from the gene database of GeneCards® show that there are more than 10,000 genes related to oncogenes or TSGs [485, 486].

Table 6.3 Oncogenes table

Gene name	Function/production	Associated cancer	References
TRK	Receptor tyrosine kinase	Colon and thyroid carcinomas, glioblastoma, lung adenocarcinoma	[487]
TIAM1	Guanine nucleotide exchange factor	T-cell lymphoma	[488] [489]
SRC	Tyrosine kinase	Leukemia	[490]
SET/CAN	The fusion protein product of a chromosomal rearrangement	Acute myeloid leukemia	[491]
ROS	Tyrosine kinase	Lung cancer	[492]
SKI	Transcription factor	Pancreatic, melanoma, and esophageal cancer	[493]
RAR/PML	Fusion protein formed by the translocation	Acute promyelocytic leukemia	[494]
FOS	Transcription factor	Osteosarcoma, head and neck squamous cell carcinoma	[495] [496]
ABL	Tyrosine kinase	Leukemia	[497]
ERBB	Cell surface receptor that triggers cell growth through tyrosine kinase activity	Lung, breast, stomach, colorectal, head and neck, and pancreatic carcinomas and glioblastoma	[498]

Table 6.4 Tumor suppressor genes table

Gene name	Function\production	Associated cancer	References
P53	Nuclear transcription factor involved in the induction of cell cycle arrest and/or apoptosis	Over 50% of human cancers carry mutated p53 gene	[499]
PTEN	Lipid phosphatase can regulate cell survival	Breast, thyroid, kidney, and colorectal cancer	[500]
TSC2	Promoter of cell growth and cell cycle progression	Breast cancer	[501]
DPC4	Signal transduction protein	Pancreatic, colorectal, and prostate cancers	[502]
VHL	Regulation of other genes and control of cell division	Hemangioblastoma and renal cell carcinoma	[503]
NF1	Bevacizumab	Neurofibromatosis, melanoma	[504] [505]

Scientists have developed a series of strategies to control the activity of oncogenes at a DNA (transcription) or RNA (translation) level. There are three major gene silencing methods used for cancer treatment, antisense oligodeoxynucleotides (AONs), ribozymes, and RNA interference.

The conception of antisense technology is basing on the use of a sequence that is complementary by virtue of Watson-Crick base pair hybridization and utilizes this sequence to bind a specific mRNA to inhibit its expression [506]. A method based on targeting the c-myc gene has been designed for the treatment of human melanoma and leukemia. Researchers used a phosphorothioate c-myc AONs complexed with zinc injected into human melanoma and leukemia xenografts in immunocompromised mice and achieved the consequence of retardation in cell growth rate [507]. Besides, there are increasing numbers of clinical trials with AONs which are ongoing, such as Custirsen (OGX-011) for prostate and non-small cell lung cancer, EGFR antisense DNA for advanced head and neck squamous cell carcinoma, and Apatorsen (OGX-427) for prostate, pancreatic, and non-squamous non-small cell lung cancer [508].

Ribozymes are RNA enzymes that possess specific catalytic activities which are similar to that of protein enzymes [509]. Ribozymes are produced naturally, but they can also be artificially synthesized and engineered to target specific sequences as a power weapon [510]. For example, multiple anti-survivin hammerhead ribozymes are designed and successfully inhibit tumor growth in a hepatocellular carcinoma xenograft mouse model [511]. Additionally, ribozymes are also used for cleaving the p53 pre-messenger RNA resulting in suppression of lung cancer cell growth [512]. Other types of ribozymes are used in cancer treatment as well, such as group I intron ribozyme and hairpin ribozyme [513, 514].

Another big group of RNA is called non-coding RNA (ncRNA), which is used for gene disruption in cancer treatment and this process is also a normal mechanism as RNA interferences (RNAi) for gene silencing in mammals. RNAi-induced-specific gene silencing is a process of transcriptional regulation in the eukaryotic cell [515]. Nc RNA can be divided into two types, long non-coding RNA (lncRNA, >200 nucleotides long) and small interference RNA (sncRNA, 21–25 nucleotides long) [516]. In this section, we will majorly discuss the function of sncRNA and its subtypes. Some sncRNAs have been introduced for cancer treatment, such as small hairpin RNA (shRNA), bifunctional short hairpin RNA (bi-shRNA), and small interfering RNA (siRNA). The processing and maturation of shRNA are similar to microRNA (miRNA) which is synthesized in the nucleus and transported into the cytoplasm for final processing [517]. The formation and activation of the final shRNA are associated with the interaction of the RNA-interfering silencing complex (RISC) and RNA precursors. Generally, the shRNA-mediated RNA interference pathway starts from the delivery of the shRNA expression vector into the cytoplasm, and then the vector will enter into the nucleus to produce the primary transcripts (pre-shRNA). Then the pre-shRNA can be formed as pre shRNAs by the Drosha/DGCR8 complex. After that, it will be loaded onto the Dicer/TRBP/PACT complex to further process the mature shRNA that provides RNA interference function through mRNA cleavage and degradation or translational suppression by the interaction with Argonaute protein-containing RISC [518]. For instance, researchers successfully observed the suppression of tumor progression by MGAT1 shRNA knockdown in the PC-3-Yellow orthotopic prostate cancer xenograft model [519]. The bi-shRNA consists of two shRNAs for each targeted mRNA, one with a perfect match, the other with mismatches at the central location. Therefore, they can be loaded onto multiple types of RISCs, both the cleavage-dependent and the cleavage-independent RISCs, and are thus able to simultaneously induce degradation and translation repression of target mRNA [518]. In some clinical trials, bi-shRNA vectors have been used for targeting furin which is a proconvertase of transforming growth factors beta (TGF-β) 1 and 2 in the advanced liver and ovarian cancer treatment. Researchers carried out a novel treatment (vigil immunotherapy) by incorporating dual GM-CSF expressive and bi-shRNA interference vector. As a consequence, the safety justification and improvement of patients' recurrence-free survival supported further assessment of this therapy by bi-shRNA [520, 521]. Although both siRNA and shRNA can be applied to achieve similar functional effects, they are intrinsically different molecules [518]. The siRNA is naturally exciting in plants and mammals which the double-strand (ds) RNA can be cut to form siRNA by dicer [522]. More than 20 different siRNA therapeutics have reached clinical trials, but few of them are designed for the purpose of cancer treatment [523]. For example, ALN-VSP, an LNP formulation of siRNAs targeting VEGF and kinesin spindle protein, has been created for the treatment of liver cancer [524]. CALAA-01, a targeted nanoparticle-containing siRNA used for inhibiting RRM2 gene is applied to inhibit tumor and has been tested in several cancer types including melanoma, gastrointestinal, and prostate cancer [525]. Furthermore, Atu-027, a siRNA-lipoplex directed against protein kinase N3 (PKN3), is developed for inhibiting cancer progression [526]. The advantages of RNAi treatment compared with other cancer therapies are (1) can specifically target lots of undruggable gene products, (2) can be relatively safe therapeutics, (3) can poses high gene silencing activity with potent and efficient ability, and (4) can be easily designed for the cancer type of interest [527].

Apart from the way to silence the expression of genes, it's also an effective strategy to target the suppressor gene for cancer therapy. During the oncogenic process, TSGs could be rendered dysfunctional through mutation, deletion, genetic rearrangement, and epigenetic silencing of transcription [528, 529]. The strategies for targeting TSGs can be concluded into three types: (1) insertion of a wild-type TSG, (2) small molecules or compounds restore the activity of TSGs, and (3) targeting epigenetics for TSGs [530, 531].

One of the most direct and easiest way to restore the function of TSG is to reintroduce a wil-type TSG into cancer cells for expression [530]. For instance, a medication named ONYX-15 which is developed with the function of the E1B gene knocked out has been released as a gene therapy treatment for refractory head and neck cancer. It's an adenovirus modified and selectively replicate in and kill the cells which harbor p53 mutations, so the normal cells won't be affected and survive [531]. Besides, some types of recombinant adenovirus engineered to express wild-type-p53 have been approved for cancer treatment in China. As an example, Gendicine, a recombinant human p53 adenovirus, developed by Shenzhen SiBiono GeneTech Co. Ltd., is the world's first Adp53-based gene therapy product approved by the China Food and Drug Administration (CFDA) in 2003 for treating head and neck cancer [532].

One of the major developments in the field of developing small molecules or compounds for restoring the activity of TSGs is the finding of Nutlin, which is a small molecule with the function of disrupting MDM2-p53 interaction resembling an MDM2-binding peptide that competitively interacts with MDM2 [530, 533]. MDM2 is a major p53 inhibitor which plays a critical role in the degradation of p53 [534]. When the Nutlin successfully binds to the MDM2, it can prevent the association between MDM2 and p53 and lead to p53 stabilization and activation. And several MDM2 inhibitors are in clinical development for the treatment of various tumor types such as soft tissue sarcoma, advanced prostate cancer, and liposarcoma. Besides Nutlin, some other compounds that inhibit Mdm2 are MI-773, MK-8242, and RO5503781 [530, 535, 536]. Besides, some clinical studies are seeking for a method to convert existing mutant p53 proteins into wild types which harbor normal p53 functions. For example, PRIMA-1 has been shown to reactivate several missense mutants of p53 to regain the wild-type p53 to induce apoptosis and inhibit tumor growth [537, 538].

With the rapid increase of epigenetic researches for cancer therapy, some methods have been developed for targeting epigenetic silencing of TSGs. Recent pieces of evidence suggested that TSG silencing is an initiating and early stage in the oncogenic process [529]. For example, it's indicated that the silencing of CDKN2A correlates with an increased risk of many types of cancer, such as melanomas and gastric and breast cancer [538, 539]. Additionally, silencing of TSGs is also found to be correlated with oncogenic progression. For instance, a panel of TSGs (WIF1, TFPI2, RASSF1A, RARβ2, SOCS1, and GATA4) are found to be commonly silenced through DNA methylation in melanoma, which can be used for predicting the advanced clinical staging and metastasis [540]. Generally, TSGs are undergoing epigenetic silencing for all cancers; it can be a promising and potential therapeutic target to treat a spectrum of malignancies. In a review of epigenetics, Kazanets A. and his colleagues demonstrate five models to account for the mechanism of TSGs silencing: (1) ablation of transcription factor binding, (2) overexpression of DNA methyltransferases, (3) disruption of CTCF binding, (4) elevation the activity of polycomb protein Ezh2, and (5) aberrant expression of ncRNAs. They present a challenging and an attractive goal for targeting reactivation of epigenetically silenced TSGs because of the complexity of events initiating and maintaining the silenced state and the reversible nature of epigenetic modifications [529]. In current studies, a spectrum of epigenetic medications is still developing including DNA methyltransferase inhibitors (Zebularine), histone deacetylase inhibitors (Belinostat), histone demethylase inhibitors (KDOAM-20), and bromodomain inhibitors (OTX015) [541–544].

6.3.3.4 Future Prospective and Challenge to Overcome

It's hard to face the tragic case of Jesse Gelsinger, who was the first person publicly identified as having died in a clinical trial for gene therapy in 1999; he died of severe immune reaction to the adenoviral vectors [545]. As Alan Milstein (the Gelsinger family's lawyer) says, "We are at the crossroads." Twenty years later, the strategies not only used for the gene therapy but also for all the cancer treatments have made an earth-shaking difference. And the truth from all the clinical data may prove that we are actually making the right decision in that crossroads [446].

One of the most cautions that need to be considered is about the safety of virus vectors. For example, it may lead to an unwanted immune response by the viral vectors because they are typically human pathogens. And generally, there is still not much more long-term safety data of viral vectors used in humans [441]. Nevertheless, several meta-analyses already exist to provide some reliable shreds of evidence for demonstrating an adequate safety profile of viral vectors in humans [546].

In order to improve the safety, specificity, as well as transduction efficiency of viral vectors, there are existing some practical strategies including the development of targeting methods, modification or remove-replacement the surface proteins, and the use of tissue-specific or conditional promoters [441]. By the way, the studies of non-viral methods also provide new boundaries for the development of gene therapy.

Besides, the challenges and problems of cancer gene therapy are not only technical but also related to ethical and policy. As researchers, we should keep a clear head and be responsible for what we are doing even the insertion of a single gene.

Moreover, the combination of gene therapy with chemotherapy, radiotherapy, and immunotherapy can be considered as a promising approach for cancer treatment. Additionally, the wide use of patient and tumor genomic analysis and the assessment of host immunity of humoral and cellular will facilitate a better choice of the most appropriate gene therapy per patient for the purpose of precise treatment [444].

Finally, we should thank Dr. Albrecht Kossel, we knew the chemical composition of DNA and RNA; thanks to Dr. James Watson and Dr. Francis Crick, we concerned the molecular structure of nucleic acids; thanks to Dr. Paul Burg, Dr. Walter Gilbert, and Dr. Frederick Sanger, we can have the method to determine the sequence of DNA; thanks to Dr. Kary B. Mullis, we are able to study DNA from a small sample by amplifying them a technique known as PCR; thanks to Dr. Dana Carroll, we have the ability to carry out genome editing experiments; thanks to Dr. Emmanuelle Charpentier, Dr. Jennifer Doudna, and Dr. Feng Zhang, we can be easier and more specific and efficient to implement gene editing. With more and more discoveries of the secrets of cancer not only in genetics but also in epigenetics, we are optimistic that current impediments to effective systemic treatment of gene therapy, such as trapping in organs, nonspecific targeting, and neutralization by the immune system, will be finally overcome leading to further success in cancer treatment. The curtain has been opened, and the show must go on.

6.4 Complementary and Alternative Medicine (CAM)

6.4.1 Introduction of CAM

Complementary and alternative medicine (CAM) is a type of drugs which of them are not commonly used as the standard medical treatment [547, 548]. Usually, these medications are lack of practices to test the biological plausibility and fall outside of mainstream healthcare [547]. However, the definition of CAM is controversial through the opinions from prominent scientists and biomedical researchers, because the boundaries between CAM and biomedicine are overlapped and vague [549, 550]. The National Center for Complementary and Integrative Health (NCCIH) is a federal government's lead agency of United States for the goals of determining the problems in CAM that was established in 1998. And it has become one of the most major institutes in the exploration of CAM that its stated mission is: "To define, through rigorous scientific investigation, the usefulness and safety of complementary and alternative medicine interventions and their roles in improving health and health care." [551]. It has provided the theories to distinct the "complementary" and "alternative" and created a classification system that divides these CAMs into five groups. If the non-mainstream practices are single used instead of conventional and standard medicine, it will be defined as "alternative." When these drugs are used as the supplements together with conventional medicine, it will be considered "complementary" [547]. Aforesaid, NCCIH has made a classification system for branches of CAMs that classified into five general types including whole medical systems, mind-body medicine, "biology"-based practices, manipulative and body-based practices, and energy medicine [552].

In China, CAM is extremely dominated by traditional Chinese medicine (TCM) that lots of the universities provide TCM courses and some medical colleges are specially established for the education of TCM or the relevant and confer the degrees which are approved in most countries or regions of the world [553–555].

Some traditional ethnic medical system of CAM can have thousands of history, such as Ayurvedic medicine and traditional Chinese medicine (TCM). Recently, these traditional medicines have been investigated and gradually accepted by the mainstream scientific community through the effectiveness of application, but the belief and theory systems are not completely admitted. However, most of CAMs are labeled "irregular practices" because of the lack of formal regulation of the industry since the term of CAM is promoted at the beginning of the 1970s [556, 557].

Additionally, the national survey in the United States indicated that there is more than 30% of adults and about 12% of children using CAM in 2012 [558]. And studies of CAM with 835 Chinese respondents and 802 white Canadians respondents reported that about 60% of the Chinese and white respondents had consumed CAM within the past year, and more white people had used multiple CAM therapies. And herbal therapies are the predominantly treatments among the Chinese respondents [559].

6.4.2 Types of Complementary and Alternative Medicine

6.4.2.1 Whole Medical Systems

Whole medical systems are involving complete systems which are built upon abundant theories and practices [560]. These systems have evolved independently in different regions and cultures over time such as traditional Chinese medicine, Kampo, Ayurveda, Unani, homeopathy, naturopathy, and anthroposophic medicine. Usually, they are earlier than the conventional medical approach nowadays and are also apart from conventional medicine or Western medicine.

- Traditional Chinese medicine (TCM)

TCM is a style of CAM developed in China for more than 2500 years of numerous medical practices in connection with some philosophical traditions of Taoism (Lao Tzu, 605–531 BC) and Confucianism (Confucius: 551–479 BC). TCM is including various forms of massage (tuina), shiatsu, acupuncture, moxibustion, exercise (qigong and tai chi), herbal medicine, some medicinal animals, and dietary therapy [561]. Besides, some evidence shows that the TCM is recently influenced by modern conventional medicine and integrative medicine that combines traditional Chinese and Western medicine; it is therefore becoming a promising and emergent branch of learning [562].

Huangdi Neijing (*Yellow Emperor's Inner Canon*), an ancient Chinese medical book, has been treated as the most ancient work and also the fundamental doctrinal theory for TCM. And some basic tenets of TCM are also involving in such a monumental work, such as Qi, Yin, and Yang, and five elements (metal, wood, water, fire, and earth) [563]. The concepts of the body and disease and the body and spirit reflect its emphasis on dynamic processes and are similar to European humoral theory which the latter is almost 1000 years late than TCM [564].

Furthermore, TCM is a huge subject system with a long history in cancer treatment. And the ideas of cancer have appeared in several ancient and classical TCM works including Huangdi Neijing (Yellow Emperor's Inner Canon) and Shanghan Zabing Lun (Treatise on Cold Damage Diseases). The ideas of wholism of interconnection in the inner body and the concepts of syndrome differentiation have become the advantages of TCM for cancer treatment. And these heritages inheriting from ancient works are not only the foundation for the development of cancer treatment with TCM but also attract more and more attention all around the world [565].

- Kampo medicine

Kampo medicine is the Japanese unique system of diagnosis and therapy that was written at the beginning of the seventh century by studying traditional Chinese medicine. Some Japanese traditional medicine mostly focuses on learning several Chinese therapies including acupuncture and moxibustion, whereas Kampo medicine mainly researches on herbal medicine [566]. Currently, Kampo medicine is widely used in Japan and is integrated into the Japanese national healthcare system. Besides, there are 165 herbal ingredients used as the Kampo medicines listed in the 14th edition of the *Japanese Pharmacopoeia (JP, Nihon yakkyokuhō)* [567].

The approach of Kampo medicine used in cancer treatment is to prevent tumors by improving the body's defense mechanism and the power of natural healing, rather than directly eliminate the cancer cells [568]. Juzentaihoto is a typical Kampo medicine used for treating cancer, and the main efficacy of item is to promote the physical strength after conventional therapies and relieve the side effects in recovery such as fatigue, malaise, and anemia [569].

- Ayurveda

The ancient Indian system, also known as Ayurveda, which is referred to "the science of life," is basing on the ancient knowledge of natural healing and holistic approach to mental and physical health, and the origins of Ayurveda can be traced back to 6000 BC [560, 570, 571]. Some of the primary methods in Ayurveda include exercise (yoga), diet, herb, massage, and meditation [560].

Although there is no scientific evidence or studies to prove the absolute effectiveness of Ayurveda, few well-designed clinical trials or experiments and comprehensive research reviews indicated that Ayurvedic approaches may be effective [570]. For instance, researchers carried out the experiments to compare two Ayurvedic drugs from the plant with the natural product glucosamine sulfate and the agent celecoxib that used to treat knee osteoarthritis, and the results showed similar effects of pain reduction and disease improvement [570]. Besides, some previous studies showed that there were about 250,000 registered medical practitioners of the Ayurvedic system in India and more than 2000 Ayurveda plants are currently used for curing different disease conditions [572]. And there are more than 100 Ayurvedic medicine training centers or colleges in India which are qualified to confer bachelor degrees [573].

Cancer is described as the inflammatory or non-inflammatory swelling, and Granthi is referred to minor neoplasm, and Arbuda is referred to major neoplasm according to Sushruta Samhita (an ancient Sanskrit text on medicine and surgery). And the therapeutic approach of Ayurveda mainly uses herbal decoctions to heal and reduce the side effects and/or lower some cancer-associated complications. Additionally, many herbs are under clinical trials to investigate their potential abilities in anti-cancer including Annona atemoya, *Phyllanthus niruri*, *Piper longum*, *Podophyllum hexandrum*, and etc. [574].

- Unani medicine

Unani is a systematic Perso-Arabic traditional medicine which is popular in Mughal India and in Muslim culture in South Asia, basing on the Greek system of medicine, and is influenced by Indian and Chinese traditional systems [575, 576]. The basic tenets of Unani are like European humoral theory basing on the theory of the presence of some elements in the human body. And some therapies of Unani include aromatherapy, bathing, bloodletting, cupping, and some herbs which are totally 36 regimes mentioned in a famous Unani work [577].

Cancer is defined as the disease of the black bile humor in Unani medicine, and there are numbers of Unani herbs which have been used for the prevention and treatment of cancer including asgand, balela, deodār, halela, and etc. [578].

Leech therapy, a famous type of Unani medicine, was introduced by Hippocrates who was also a very famous Unani physician known as the father of medicine. And the modern leech therapy was developed by the surgeons M Derganc and F Zdravic, and the therapy was performed with the help of medicinal leeches as a blood-sucking process for the treatment of various diseases such as hypertension, thrombophlebitis, and musculoskeletal diseases [577]. In vitro studies have suggested that the leech saliva extracts can prevent coagulation and inhibit metastasis and tumor progression [579].

- Homeopathy

Homeopathy is a complete system of alternative medicine developed in 1796 by Samuel Hahnemann originated in Europe [560]. Homeopathy is a plausible system of treatment that it takes a different approach in diagnosing, classifying, and treating disease, and it also emphasizes the relationship between the emotional and mental states with person's health status and includes a wide range of research across physics, chemistry, biology, and psychology [560, 580]. The principles of homeopathy involve three key concepts: (i) homeopathy is administered in a minute or potentially nonexistent drug dosages because the healthy people will suffer from similar symptoms of illness when they undergoing larger doses; (ii) the primary aim of homeopathy is to stimulate the body's defense mechanisms for treatment of diseases; (iii) homeopathy is designed individually for each patient including emotional and mental states, the condition of health, and lifestyles [580].

Some radical opinions of such a therapy described homeopathy as a pseudoscience because of the lack of scientific validations [581]. On the contrary, homeopathy is prevalent and widely used in several countries including Scotland, France, Luxembourg, and India [582].

Mostly, the current studies of the homeopathy's effectiveness focus on comparisons of homeopathic methods and placebos, studies of the biological effects of ultrahigh dilutions of homeopathy, and tests of the effectiveness of homeopathy in some particular clinical conditions [560].

There are few studies to randomized clinical trials (RCT) to confirm the effectiveness of homeopathy in the area of cancer. From a study of the reviews about RCT, it conservatively demonstrated that there was no evidence that the homeopathic remedies exhibit anti-cancer effect [583].

- Naturopathy

Naturopathy, also known as naturopathic medicine, is a form of alternative medicine that focuses on health restoration and disease treatment. The principles of naturopathy are based on a belief in people's inner body to heal themselves by a magic power or a special vital energy. And the practice of naturopathy is based on several key concepts including the healing power of nature, treat the whole person, treat the cause of disease, the physician is a teacher, prevention is the best cure, and do no harm [560, 561, 580, 584].

Naturopaths usually suggest people to exposure to naturally occurring substances and recommend them to undergo several naturopathic treatments such as hydrotherapy, rolfing, iridology, colonic enemas, ozone therapy, and etc. [584]. And most of them are thought to be pseudosciences or quackeries, because they are considered to be dangerous to health with no verified benefits [585]. Besides, the naturopathy practitioners are opposed to vaccination and reject to use it because of the belief of nature appreciation [586].

Taking some herbs such as curcumin and nutrients such as vitamin D is thought to be effective in cancer treatment through the theory of naturopathy [587]. Other naturopathic cancer therapies include injections with high-dose vitamin C or the herb substance mistletoe and using ozone gas to treat the blood [588].

- Anthroposophic medicine

Anthroposophic medicine is an integrative multimodal treatment system based on occult notions and draws on spiritual philosophy by Rudolf Steiner and Ita Wegman in 1925 [561, 589]. The practitioners of anthroposophic medicine used a number of treatment techniques according to anthroposophic precepts, such as exercise, counseling, massage, and the pharmaceuticals from the guidelines of anthroposophy [590]. Besides, there are also some special therapies of anthroposophic medicine, such as art therapy and eurythmy therapy [589].

The key concepts of anthroposophic medicine are based upon four levels of formative forces including physical, life, soul, spirit, and threefold model of the human constitution including two being polar to each other (motor-metabolic and nerve-sense systems) and one being intermediate (rhythmic system) [589].

Anthroposophic medicine believes that mistletoe, a semiparasitic plant that grows on trees, can cure cancer, and the potential of the medical efficacy is influenced by the position of the sun, moon, and planets which it is significant to harvest the plant at the right time [590, 591].

Up to now, the varies of mistletoe-based drugs have not been approved for any purpose including cancer treatments by the US FDA, and as of 2015, there are no licensed mistletoe-based drugs available in the UK by the National Health Service [592, 593].

6.4.2.2 Mind-Body Medicine

Mind-body medicine is a type of treatment that focuses on the interactions among the brain, mind, body, spirit, and behavior, and it involves various powerful approaches in which the mind's capacity influenced by mental, social, and spiritual factors is able to directly affect body functions and symptoms [594].

Some traditional types of mind-body interventions include tai chi, yoga, and meditation. Since the 1960s, mind-body interactions have been widely studied [595]. The current list of some mind-body practices provided by the NCCIH is more than ten types, such as art therapy, tai chi and qigong, yoga, hypnotherapy chiropractic and osteopathic manipulation, and meditation [558].

Although mind-body interventions are defined as a type of complementary and alternative medicine or not conventional medicine, there are several scientific research documented potential benefits and advantages of mind-body interventions through randomized controlled trials [595]. It indicated that the central nervous system is able to influence immune, endocrine, and autonomic functioning that has an impact on health. Besides, mind-body interventions can contribute to ameliorate the symptoms induced by conventional cancer treatments and improve recovery time. Additionally, medicine is reported to improve the overall quality of life and can be effective adjuncts in the management of a variety of chronic conditions and some coronary artery disease [595–598].

From the mind-body research, it finds that most of these mind and body practices have a positive effect on most systems in the body to improve aspects of quality of life in cancer patients. It also indicates that mind-body practices are some healthy behaviors to manage stress and achieve a better balance in daily life and are some good ways to help patients manage the distress from cancer [599].

6.4.3 Problems and Future of CAM

Because of the negative outcomes of some CAM therapies and many of them are less-known, the CAMs are treatments that are outside of the mainstream healthcare

[547, 548]. Some proposed risks or problems include lack of adequacy of regulation and evaluation of the safety, the problems of interactions with conventional treatments, and rejection of science that can induce treatment delay or more serious results [600–602].

In addition, the growth of the pharmaceutical industry has not diminished the demands of natural products or other traditional medicine in many countries. By contrast, the demand for medicinal plants has been greatly expanded with the growth of the population and the increasing interests of these CAMs products. This need has been building up worldwide, and this is what pushed people into developing novel comprehensive regulation and varies clinical trials to discriminate the really helpful CAMs and the approaches of pseudoscience [603].

There five core objectives identified through NCCIH's strategic direction in complementary and integrative health research in 2016. This strategic plan involves the advancement of fundamental science and the development of new research methods, the improvement of care for hard-to-manage symptoms, the fosterage of disease prevention and health promotion, the enhancement of the population of professional complementary and integrative health researchers, and the spread of the objective evidence-based information to the public [604].

6.5 Conclusions and Future Prospects

With the improvement of the public awareness of the risk factors of cancer such as smoking and drinking and the detection techniques of cancer, the numbers of cancer diagnosis and deaths are decreased year by year. Notably, the outcomes of cancer statics also must owe to the development of numerous cancer treatments. These various treatments may develop at different times, based on distinctive mechanisms, but all of these therapies have a common goal, which is the elimination of cancer.

Since the first cancer treatment developed by the human, more and more cancer therapies have been created through history. This chapter attempted to summarize the history or evolution of these critical tumor therapeutic modes and introduce some basic knowledge of these treatments. Some common types of cancer treatments have been applied and developed for a long time including surgery, radiotherapy, hormone therapy, and stem cell transplant. Several novel therapies of cancer are mostly still on the early stage or under investigation such as immunotherapy, targeted therapy, and gene therapy. Besides, the complementary and alternative medicine (CAM) becomes more and more popular but without comprehensive research. Some types of CAMs such as Chinese traditional medicine and some natural products reveal the potential effectiveness for the use of cancer treatment, and it's urgent to conduct more and more evidence-based clinical trials to test these plausible treatments. Additionally, nanotechnology has performed the great effectiveness that emerged as a kind of targeted therapy. For example, some nanoparticles have shown great effects on treating tumors used singly or combined with other biomolecules or chemicals.

The era of the next time of fighting cancer is bright and hopeful, and the contemporaries are very possible to witness the radical cure for cancer. However, it is not an easy task, and nobody is able to reach the sky in a single bound that anti-cancer therapy is more difficult than other tough matters for hundreds of times from the lessons of history. Besides, the pathogenesis of cancer is still not completely investigated up to now, and the existing theories that used as the guideline to treat cancer are gradually challenged by more and more new discoveries from clinical trials or laboratory experiments. The conventional therapies have exhibited to be unmanageable to many types of cancers through thousands of novel research including the drug resistance and cancer metastasis or recurrence. And the novel treatments are not mature enough for widely used to combat various malignant tumors.

There still a long way to go for people to eliminate or eradicate cancer which is the emperor of all maladies. And it's our modern researchers' mission to develop the effective treatments for numerous patients suffered from cancer. I wish that we will see people can get triumphs over cancer in our lifetimes.

References

1. Sudhakar A (2009) History of Cancer, ancient and modern treatment methods. J Cancer Sci Ther 1(2):1–4
2. Mukherjee S (2010) The emperor of all maladies: a biography of cancer. Scribner, New York
3. Urruticoechea A, Alemany R, Balart J et al (2010) Recent advances in cancer therapy: an overview. Curr Pharm Des 16(1):3–10. https://doi.org/10.2174/138161210789941847
4. Anand P, Kunnumakara AB, Sundaram C et al (2008) Cancer is a preventable disease that requires major lifestyle changes. Pharm Res 25(9):2097–2116
5. Siegel RL, Miller KD, Jemal A (2019) Cancer statistics, 2019. CA Cancer J Clin 69(1):7–34. https://doi.org/10.3322/caac.21551
6. Stacy Simon (2019) American Cancer Society. Facts & figures 2019: US cancer death rate has dropped 27% in 25 years. https://www.cancer.org/latest-news/facts-and-figures-2019.html. Accessed 15 Apr 2019
7. Diffen. Cancer vs. Tumor. https://www.diffen.com/difference/Cancer_vs_Tumor. Accessed 15 Apr 2019
8. World Health Organization (2018) Cancer. https://www.who.int/cancer/en/. Accessed 15 Apr 2019
9. Nall R (2018) Medical news today. What to know about cancer. https://www.medicalnewstoday.com/articles/323648.php. Accessed 15 Apr 2019
10. National Cancer Institute (2015) What is cancer. https://www.cancer.gov/about-cancer/understanding/what-is-cancer. Accessed 15 Apr 2019
11. David AR, Zimmerman MR (2010) Cancer: an old disease, a new disease or something in between? Nat Rev Cancer 10(10):728–733. https://doi.org/10.1038/nrc2914
12. Wust P, Hildebrandt B, Sreenivasa G et al (2002) Hyperthermia in combined treatment of cancer. Lancet Oncol 3(8):487–497
13. Hawkes N (2015) History of cancer treatment. https://www.raconteur.net/healthcare/history-of-cancer-treatment. Accessed 15 Apr 2019
14. National Cancer Institute. Types of Cancer Treatment. https://www.cancer.gov/about-cancer/treatment/types. Accessed 15 Apr 2019

15. Garrison FH (1966) Contributions to the history of medicine. W.B. Saunders Company, Philadelphia
16. Retief FP, Cilliers L (2001) Tumours and cancers in Graeco-Roman times. S Afr Med J 91 (4):344–348
17. Introduction, Celsus, On Medicine. http://penelope.uchicago.edu/Thayer/E/Roman/Texts/Celsus/Introduction*.html. Accessed 15 Apr 2019
18. Retief FP, Cilliers L (2011) Breast cancer in antiquity. S Afr Med J 101(8):513–515
19. Evans CH (2007) John hunter and the origins of modern orthopaedic research. J Orthop Res 25 (4):556–560
20. Robinson DH, Toledo AH (2012) Historical development of modern anesthesia. J Investig Surg 25(3):141–149
21. Fox NJ (1988) Scientific theory choice and social structure: the case of Joseph Lister's antisepsis, humoral theory and asepsis. Hist Sci 26(4):367–397
22. Wyld L, Audisio RA, Poston GJ (2015) The evolution of cancer surgery and future perspectives. Nat Rev Clin Oncol 12(2):115–124. https://doi.org/10.1038/nrclinonc.2014.191
23. National Cancer Institute (2015) Surgery to treat cancer. https://www.cancer.gov/about-cancer/treatment/types/surgery#TS. Accessed 15 Apr 2019
24. Canadian Cancer Society. Types of surgery. http://www.cancer.ca/en/cancer-information/diagnosis-and-treatment/surgery/types-of-surgery/?region=on. Accessed 15 Apr 2019
25. Whitlock J (2018) Verywell health. What is open surgery? Is it right for you? https://www.verywellhealth.com/open-surgery-3157124. Accessed 15 Apr 2019
26. Medstarhealth. Endoscopic Surgery. https://www.medstarhealth.org/mhs/our-services/colon-and-rectal-surgery/treatments/endoscopic-surgery/. Accessed 15 Apr 2019
27. Mehra H (2017) Quora. Which is better, open surgery or laparoscopic surgery? https://www.quora.com/Which-is-better-open-surgery-or-laparoscopic-surgery. Accessed 15 Apr 2019
28. National Health Service. Overview – Laparoscopy (keyhole surgery). https://www.nhs.uk/conditions/laparoscopy/. Accessed 15 Apr 2019
29. Unger JG (2017) Medscape. Cryotherapy. https://emedicine.medscape.com/article/1125851-overview#a1. Accessed 15 Apr 2019
30. Rubinsky B (2000) Cryosurgery. Annu Rev Biomed Eng 2(1):157–187
31. National Cancer Institute (2003) National Cancer Institute Cryosurgery in Cancer Treatment. https://www.cancer.gov/about-cancer/treatment/types/surgery/cryosurgery-fact-sheet. Accessed 15 Apr 2019
32. Chi E (2016) Health line. Laser therapy. https://www.healthline.com/health/laser-therapy. Accessed 15 Apr 2019
33. Standford Health care. What Is Laser Surgery? https://stanfordhealthcare.org/medical-treatments/l/laser/types/laser-surgery.html. Accessed 15 Apr 2019
34. National Cancer Institute (2011) National Cancer Institute Lasers in Cancer Treatment. https://www.cancer.gov/about-cancer/treatment/types/surgery/lasers-fact-sheet. Accessed 15 Apr 2019
35. Hulse RM, Kenneth HL (1980) Hyperthermia in cancer therapy. West J Med 132(3):179–185
36. Van der Zee J (2002) Heating the patient: a promising approach? Ann Oncol 13(8):1173–1184
37. Desaulniers V (2019) The truth about cancer. Hyperthermia therapy: using heat to help heal cancer. https://thetruthaboutcancer.com/hyperthermia-treatment/. Accessed 15 Apr 2019
38. National Cancer Institute (2011) National Cancer Institute Photodynamic Therapy for Cancer. https://www.cancer.gov/about-cancer/treatment/types/surgery/photodynamic-fact-sheet. Accessed 15 Apr 2019
39. National Health Service. Photodynamic therapy (PDT). https://www.nhs.uk/conditions/photodynamic-therapy/. Accessed 15 Apr 2019
40. Dolmans DE, Fukumura D, Jain RK (2003) Photodynamic therapy for cancer. Nat Rev Cancer 3(5):380–387
41. Capella MAM, Capella LS (2003) A light in multidrug resistance: photodynamic treatment of multidrug-resistant tumors. J Biomed Sci 10(4):361–366

42. Wilson BC (2002) Photodynamic therapy for cancer: principles. Can J Gastroenterol Hepatol 16(6):393–396
43. Mayo Foundation for Medical Education and Research (2019) Robotic surgery. https://www.mayoclinic.org/tests-procedures/robotic-surgery/about/pac-20394974. Accessed 15 Apr 2019
44. NYU Langone Health. What is Robotic Surgery? https://med.nyu.edu/robotic-surgery/physicians/what-robotic-surgery. Accessed 5 May 2019
45. Lanfranco AR, Castellanos AE, Desai JP, Meyers WC (2004) Robotic surgery a current perspective. Ann Surg 239(1):14–21
46. US. Food and Drug Administration (2019) Caution when using robotically-assisted surgical devices in women's health including mastectomy and other cancer-related surgeries: FDA Safety. Communication. https://www.fda.gov/medical-devices/safety-communications/caution-when-using-robotically-assisted-surgical-devices-womens-health-including-mastectomy-and. Accessed 5 May 2019
47. Wallace D, Cure medical. What is electrosurgery and how does it help cure cancer. http://www.curemedicalglobal.com/electrosurgery-help-cure-cancer/. Accessed 5 May 2019
48. Massarweh NN, Cosgriff N, Slakey DP (2006) Electrosurgery: history, principles, and current and future uses. J Am Coll Surg 202(3):520–530
49. Ratini M (2019) WebMD. What is a HIFU procedure? https://www.webmd.com/prostate-cancer/prostate-cancer-hifu-surgery#1. Accessed 5 May 2019
50. National Health Service. Prostate Cancer. https://www.nhs.uk/conditions/prostate-cancer/. Accessed 5 May 2019
51. DocDoc. What is High Intensity Focused Ultrasound (HIFU): overview, benefits, and expected results. https://www.docdoc.com.sg/info/procedure/high-intensity-focused-ultrasound/. Accessed 5 May 2019
52. American Academy of Dermatology (2007) What is Mohs surgery? https://www.aad.org/public/diseases/skin-cancer/what-is-mohs-surgery. Accessed 5 May 2019
53. Mayo Foundation for Medical Education and Research (2017) Mohs surgery. https://www.mayoclinic.org/tests-procedures/mohs-surgery/about/pac-20385222. Accessed 5 May 2019
54. Swanson NA (1983) Mohs surgery: technique, indications, applications, and the future. Arch Dermatol 119(9):761–773
55. Encyclopaedia Britannica (2017) Stereotaxic surgery. https://www.britannica.com/science/stereotaxic-surgery. Accessed 5 May 2019
56. Mayo Foundation for Medical Education and Research (2019) Stereotactic radiosurgery. https://www.mayoclinic.org/tests-procedures/stereotactic-radiosurgery/about/pac-20384526. Accessed 5 May 2019
57. Ebara M, Okabe S, Kita K, Sugiura N, Fukuda H, Yoshikawa M et al (2005) Percutaneous ethanol injection for small hepatocellular carcinoma: therapeutic efficacy based on 20-year observation. J Hepatol 43(3):458–464
58. Cicalese L (2018) Medscape. What is the role of Percutaneous Ethanol Injection (PEI) in the treatment of Hepatocellular Carcinoma (HCC)? https://www.medscape.com/answers/197319-39253/what-is-the-role-of-percutaneous-ethanol-injection-pei-in-the-treatment-of-hepatocellular-carcinoma-hcc. Accessed 5 May 2019
59. Ansari D, Andersson R (2012) Radiofrequency ablation or percutaneous ethanol injection for the treatment of liver tumors. World J Gastroenterol 18(10):1003. https://doi.org/10.3748/wjg.v18.i10.1003
60. Shiina S et al (2012) Percutaneous ethanol injection for hepatocellular carcinoma: 20-year outcome and prognostic factors. Liver Int 32(9):1434–1442
61. Canadian Cancer Society. Surgery in cancer treatment. http://www.cancer.ca/en/cancer-information/diagnosis-and-treatment/surgery/?region=qc. Accessed 5 May 2019
62. American Cancer Society (2016) Chemotherapy side effects. https://www.cancer.org/treatment/treatments-and-side-effects/treatment-types/chemotherapy/chemotherapy-side-effects.html. Accessed 5 May 2019

63. Kirkwood JM, Strawderman MH, Ernstoff MS, Smith TJ, Borden EC, Blum RH (1996) Interferon alfa-2b adjuvant therapy of high-risk resected cutaneous melanoma: the Eastern Cooperative Oncology Group Trial EST 1684. J Clin Oncol 14(1):7–17
64. Ribic CM, Sargent DJ, Moore MJ, Thibodeau SN, French AJ, Goldberg RM et al (2003) Tumor microsatellite-instability status as a predictor of benefit from fluorouracil-based adjuvant chemotherapy for colon cancer. N Engl J Med 349(3):247–257
65. Boland CR, Goel A (2010) Microsatellite instability in colorectal cancer. Gastroenterology 138(6):2073–2087
66. Ryan DP, Hong TS, Bardeesy N (2014) Pancreatic adenocarcinoma. N Engl J Med 371 (11):1039–1049
67. Burdett S, Pignon JP, Tierney J, Tribodet H, Stewart L, Le Pechoux C et al (2015) Adjuvant chemotherapy for resected early-stage non-small cell lung cancer. Cochrane Database Syst Rev 2(3):CD011430. https://doi.org/10.1002/14651858.CD011430
68. Kumar L, Harish P, Malik PS, Khurana S (2018) Chemotherapy and targeted therapy in the management of cervical cancer. Curr Probl Cancer 42(2):120–128
69. Bonadonna G, Valagussa P (1981) Dose-response effect of adjuvant chemotherapy in breast cancer. N Engl J Med 304(1):10–15
70. Anampa J, Makower D, Sparano JA (2015) Progress in adjuvant chemotherapy for breast cancer: an overview. BMC Med 13(1):195
71. American Cancer Society (2018) Risks of cancer surgery. https://www.cancer.org/treatment/treatments-and-side-effects/treatment-types/surgery/risks-of-cancer-surgery.html. Accessed 5 May 2019
72. Whitlock J (2018) Verywell health. Understanding the risks involved when having surgery. https://www.verywellhealth.com/understanding-the-risks-involved-when-having-surgery-3156959. Accessed 5 May 2019
73. Chand M, Armstrong T, Britton G et al (2007) How and why do we measure surgical risk? J R Soc Med 100(11):508–512
74. Radiation Oncology Targeting Cancer. Radiation therapy. https://www.targetingcancer.com.au/radiation-therapy/brachytherapy/. Accessed 5 May 2019
75. National Cancer Institute (2019) Radiation therapy to treat cancer. https://www.cancer.gov/about-cancer/treatment/types/radiation-therapy. Accessed 5 May 2019
76. Foray N (2016) Victor Despeignes, the forgotten pioneer of radiation oncology. Int J Radiat Oncol Biol Phys 96(4):717–721
77. Thariat J et al (2013) Past, present, and future of radiotherapy for the benefit of patients. Nat Rev Clin Oncol 10(1):52
78. Leszczynski K, Boyko S (1997) On the controversies surrounding the origins of radiation therapy. Radiother Oncol 42(3):213–217
79. Del Regato JA (1991) Milestones in therapeutic radiology. In: The role of high energy electrons in the treatment of cancer. Karger Publishers, Basel, pp 4–9
80. Mould RF (1998) The discovery of radium in 1898 by Maria Sklodowska-Curie (1867-1934) and Pierre Curie (1859-1906) with commentary on their life and times. Br J Radiol. 71 (852):1229–1254
81. Metzenbaum M (1905) Radium: its value in the treatment of lupus, rodent ulcer, and epithelioma, with reports of cases. Int Clinics 14(4):21–31
82. Frame PW (1985) Radioactive curative devices and spas. Oak Ridger Newspaper
83. Holsti LR (1995) Development of clinical radiotherapy since 1896. Acta Oncol 34 (8):995–1003. https://doi.org/10.3109/02841869509127225
84. Mitchell G (2013) The rationale for fractionation in radiotherapy. Clin J Oncol Nurs 17(4). https://doi.org/10.1188/13.CJON.412-417
85. Thwaites DI, Tuohy JB (2006) Back to the future: the history and development of the clinical linear accelerator. Phys Med Biol 51(13):R343
86. Bhattacharyya KB (2016) Godfrey Newbold Hounsfield (1919–2004): the man who revolutionized neuroimaging. Ann Indian Acad Neurol 19(4):448

87. Canadian Cancer Society. Radiation therapy. http://www.cancer.ca/en/cancer-information/diagnosis-and-treatment/radiation-therapy/?region=qc. Accessed 5 May 2019
88. Hill R, Healy B, Holloway L, Kuncic Z, Thwaites D, Baldock C (2014) Advances in kilovoltage x-ray beam dosimetry. Phys Med Biol 59(6):R183
89. Baskar R, Lee KA, Yeo R, Yeoh K-W (2012) Cancer and radiation therapy: current advances and future directions. Int J Med Sci 9(3):193
90. Canadian Cancer Society. External beam radiation therapy for cancer. https://www.cancer.ca/en/cancer-information/diagnosis-and-treatment/radiation-therapy/external-radiation-therapy/?region=on. Accessed 5 May 2019
91. Radiation Oncology Targeting Cancer. Intensity-Modulated Radiation Therapy (IMRT). https://www.targetingcancer.com.au/radiation-therapy/ebrt/intensity-modulated-radiation-therapy-imrt/. Accessed 5 May 2019
92. National Cancer Institute (2018) External beam radiation therapy. https://www.cancer.gov/about-cancer/treatment/types/radiation-therapy/external-beam. Accessed 5 May 2019
93. Elekta (2019) Gamma knife treatment process. https://www.elekta.com/patients/gammaknife-treatment-process/. Accessed 5 May 2019
94. Radiation Oncology Targeting Cancer. Stereotactic Body Radiation Therapy (SBRT). https://www.targetingcancer.com.au/radiation-therapy/ebrt/stereotactic-body-radiation-therapy-sbrt/. Accessed 5 May 2019
95. Radiation Oncology Targeting Cancer. Image-Guided Radiation Therapy (IGRT). https://www.targetingcancer.com.au/radiation-therapy/ebrt/image-guided-radiation-therapy-igrt/. Accessed 5 May 2019
96. Kjellberg RN, Abe M (1988) Stereotactic Bragg peak proton beam therapy. In: Lunsford LD (ed) Modern stereotactic neurosurgery. Topics in neurological surgery. Springer, Boston, pp 463–470. https://doi.org/10.1007/978-1-4613-1081-5_36
97. Teoh M, Clark CH, Wood K, Whitaker S, Nisbet A (2011) Volumetric modulated arc therapy: a review of current literature and clinical use in practice. Br J Radiol 84(1007):967–970
98. Bertelsen A, Hansen CR, Johansen J, Brink C (2010) Single arc volumetric modulated arc therapy of head and neck cancer. Radiother Oncol 95(2):142–148
99. Van Gestel D, van Vliet-Vroegindeweij C, Van den Heuvel F, Crijns W, Coelmont A, De Ost B et al (2013) RapidArc, SmartArc and TomoHD compared with classical step and shoot and sliding window intensity modulated radiotherapy in an oropharyngeal cancer treatment plan comparison. Radiat Oncol 8(1):37
100. Unak P (2002) Targeted tumor radiotherapy. Braz Arch Biol Technol 45:97–110
101. Persson L (1994) The auger electron effect in radiation dosimetry. Health Phys 67(5):471–476
102. Kassis AI (2003) Cancer therapy with Auger electrons: are we almost there? J Nucl Med 44(9):1479–1481
103. Sastry KSR (1992) Biological effects of the auger emitter iodine-125: a review. Report no. 1 of AAPM nuclear medicine task group no. 6. Med Phys 19(6):1361–1370
104. American Cancer Society. Internal radiation therapy (Brachytherapy). https://www.cancer.org/treatment/treatments-and-side-effects/treatment-types/radiation/internal-radiation-therapy-brachytherapy.html. Accessed 5 May 2019
105. Limbergen EV, Skowronek J, Pötter R (2012) The GEC ESTRO handbook of brachytherapy. https://www.wco.pl/zb/files/publication/fd9f39.pdf. Accessed 5 May 2019
106. Mayo Foundation for Medical Education and Research (2018) Brachytherapy. https://www.mayoclinic.org/tests-procedures/brachytherapy/about/pac-20385159. Accessed 5 May 2019
107. National Cancer Institute (2019) Brachytherapy to treat cancer. https://www.cancer.gov/about-cancer/treatment/types/radiation-therapy/brachytherapy. Accessed 5 May 2019
108. Thomadsen BR, Williamson JF, Rivard MJ et al (2008) Anniversary paper: past and current issues, and trends in brachytherapy physics. Med Phys 35(10):4708–4723
109. Sun Myint A et al (2017) Dose escalation using contact X-ray brachytherapy (Papillon) for rectal cancer: does it improve the chance of organ preservation? Br J Radiol 90(1080):20170175

110. Myint AS (2014) Novel radiation techniques for rectal cancer. J Gastrointest Oncol 5 (3):212–217. https://doi.org/10.3978/j.issn.2078-6891.2014.031
111. Myint AS, Smith FML, Gollins S et al (2018) Dose escalation using contact X-ray brachytherapy after external beam radiotherapy as nonsurgical treatment option for rectal cancer: outcomes from a single-center experience. Int J Radiat Oncol Biol Phys 100(3):565–573. https://doi.org/10.1016/j.ijrobp.2017.10.022
112. Dutta SW, Showalter SL, Showalter TN, Libby B, Trifiletti DM (2017) Intraoperative radiation therapy for breast cancer patients: current perspectives. Breast Cancer Targets Ther 9:257
113. Belletti B, Vaidya JS, D'Andrea S, Entschladen F, Roncadin M, Lovat F et al (2018) Targeted intraoperative radiotherapy impairs the stimulation of breast cancer cell proliferation and invasion caused by surgical wounding. Clin Cancer Res 14(5):1325–1332
114. Bergom C et al (2018) Deep inspiration breath hold: techniques and advantages for cardiac sparing during breast cancer irradiation. Front Oncol 8:87
115. Rosenzweig KE et al (2000) The deep inspiration breath-hold technique in the treatment of inoperable non–small-cell lung cancer. Int J Radiat Oncol Biol Phys 48(1):81–87
116. GenesisCare. Deep inspiration breath hold. https://www.genesiscare.com/uk/treatment/cancer/radiotherapy/deep-inspiration-breath-hold/. Accessed 5 May 2019
117. Latty D et al (2015) Review of deep inspiration breath-hold techniques for the treatment of breast cancer. J Med Radiat Sci 62(1):74–81
118. Stokkel MPM, Junak DH, Lassmann M, Dietlein M, Luster M (2010) EANM procedure guidelines for therapy of benign thyroid disease. Eur J Nucl Med Mol Imaging 37 (11):2218–2228
119. Silberstein EB, Alavi A, Balon HR, Clarke SEM, Divgi C, Gelfand MJ et al (2012) The SNMMI practice guideline for therapy of thyroid disease with 131I 3.0. J Nucl Med 53 (10):1633–1651
120. American Cancer Society (2017) Radiation for breast cancer. https://www.cancer.org/cancer/breast-cancer/treatment/radiation-for-breast-cancer.html. Accessed 5 May 2019
121. American Cancer Society (2018) Treatment of rectal cancer, by Stage. https://www.cancer.org/cancer/colon-rectal-cancer/treating/by-stage-rectum.html. Accessed 5 May 2019
122. National Cancer Institute (2019) Cervical Cancer Treatment (PDQ®)–Patient Version. https://www.cancer.gov/types/cervical/patient/cervical-treatment-pdq#_180. Accessed 5 May 2019
123. American Cancer Society (2019) Treatment of bladder cancer, by stage. https://www.cancer.org/cancer/bladder-cancer/treating/by-stage.html. Accessed 5 May 2019
124. Canadian Cancer Society. Side effects of radiation therapy. http://www.cancer.ca/en/cancer-information/diagnosis-and-treatment/radiation-therapy/side-effects-of-radiation-therapy/?region=qc. Accessed 5 May 2019
125. Johns Hopkins Medicine. Hormonal therapy. https://www.hopkinsmedicine.org/breast_center/treatments_services/medical_oncology/adjuvant_hormonal_therapy.html. Accessed 5 May 2019
126. National Cancer Institute (2015) Hormone therapy to treat cancer. https://www.cancer.gov/about-cancer/treatment/types/hormone-therapy#THT. Accessed 5 May 2019
127. Beatson GT (1983) Classics in oncology: on the treatment of inoperable cases of carcinoma of the mamma: suggestions for a new method of treatment, with illustrative cases. CA Cancer J Clin 33(2):108–121
128. American Cancer Society (2014) Evolution of cancer treatments: hormone therapy. https://www.cancer.org/cancer/cancer-basics/history-of-cancer/cancer-treatment-hormone-therapy.html. Accessed 5 May 2019
129. Beatson GT (1896) On the treatment of inoperable cases of carcinoma of the mamma: suggestions for a new method of treatment, with illustrative cases. Trans Med Chir Soc Edinb 15:153
130. Huggins C, Hodges CV (1972) Studies on prostatic cancer. I. The effect of castration, of estrogen and androgen injection on serum phosphatases in metastatic carcinoma of the prostate. CA Cancer J Clin 22(4):232–240

131. Magon N (2011) Gonadotropin releasing hormone agonists: expanding vistas. Indian J Endocrinol Metab 15(4):261
132. Moreau J-P, Delavault P, Blumberg J (2006) Luteinizing hormone-releasing hormone agonists in the treatment of prostate cancer: a review of their discovery, development, and place in therapy. Clin Ther 28(10):1485–1508
133. Novara G, Galfano A, Secco S, Ficarra V, Artibani W (2009) Impact of surgical and medical castration on serum testosterone level in prostate cancer patients. Urol Int 82(3):249–255
134. Vachani C (2018) Oncolink. Hormone therapy: the basics. https://www.oncolink.org/cancer-treatment/hormone-therapy/hormone-therapy-the-basics. Accessed 5 May 2019
135. Howell A, Cuzick J, Baum M, Buzdar A, Dowset M (2005) Results of the ATAC (Arimidex, tamoxifen, alone or in combination) trial after completion of 5 years' adjuvant treatment for breast cancer. Lancet 365:60–62
136. Vasaitis TS, Bruno RD, Njar VCO (2011) CYP17 inhibitors for prostate cancer therapy. J Steroid Biochem Mol Biol 125(1–2):23–31
137. American Cancer Society (2016) Hormone therapy for prostate cancer. https://www.cancer.org/cancer/prostate-cancer/treating/hormone-therapy.html. Accessed 5 May 2019
138. Keskin O, Yalcin S (2013) A review of the use of somatostatin analogs in oncology. Onco Targets Ther 6:471
139. Maurer R, Gaehwiler BH, Buescher HH, Hill RC, Roemer D (1982) Opiate antagonistic properties of an octapeptide somatostatin analog. Proc Natl Acad Sci 79(15):4815–4817
140. Lange CA, Yee D (2008) Progesterone and breast cancer. Womens Health 4(2):151–162
141. Lundgren S (1992) Progestins in breast cancer treatment: a Review. Acta Oncol 31 (7):709–722
142. Brechon S (2012) The Maurer Foundation. Estrogen and breast cancer. https://www.maurerfoundation.org/estrogen-and-breast-cancer/. Accessed 5 May 2019
143. Lumachi F, Santeufemia DA, Basso SMM (2015) Current medical treatment of estrogen receptor-positive breast cancer. World J Biol Chem 6(3):231
144. OncoLink (2018) Taking Androgen Deprivation Therapy (ADT) for prostate cancer. https://www.oncolink.org/cancers/prostate/treatments/taking-androgen-deprivation-therapy-adt-for-prostate-cancer. Accessed 5 May 2019
145. Kent EC, Hussain MHA (2003) Neoadjuvant therapy for prostate cancer: an oncologist's perspective. Rev Urol 5(Suppl 3):S28
146. Trimble EL, Ungerleider RS, Abrams JA, Kaplan RS, Feigal EG, Smith MA et al (1993) Neoadjuvant therapy in cancer treatment. Cancer 72(S11):3515–3524
147. Johns Hopkins Medicine. Adjuvant hormonal therapy. https://www.hopkinsmedicine.org/breast_center/treatments_services/medical_oncology/adjuvant_hormonal_therapy.html. Accessed 5 May 2019
148. Shaikh AJ, Kumar S, Raza S, Mehboob M, Ishtiaq O (2013) Adjuvant hormonal therapy in postmenopausal women with breast cancer: physician's choices. Int J Breast Cancer. https://doi.org/10.1155/2012/849592
149. Canadian Cancer Society. Side effects of hormonal therapy. http://www.cancer.ca/en/cancer-information/diagnosis-and-treatment/chemotherapy-and-other-drug-therapies/hormonal-therapy/side-effects-of-hormonal-therapy/?region=on. Accessed 5 May 2019
150. Tavani A, La Vecchia C (1999) The adverse effects of hormone replacement therapy. Drugs Aging 114(5):347–357
151. National Health Service. Risks – Hormone replacement therapy (HRT). https://www.nhs.uk/conditions/hormone-replacement-therapy-hrt/risks/. Accessed 5 May 2019
152. National Cancer Institute. Stem cell transplants in cancer treatment. https://www.cancer.gov/about-cancer/treatment/types/stem-cell-transplant. Accessed 5 May 2019
153. Canadian Cancer Society. Stem cell transplant. http://www.cancer.ca/en/cancer-information/diagnosis-and-treatment/stem-cell-transplant/?region=on. Accessed 5 May 2019
154. American Society of Clinical Oncology (2018) What is a bone marrow transplant (Stem cell transplant)? https://www.cancer.net/navigating-cancer-care/how-cancer-treated/bone-

marrowstem-cell-transplantation/what-bone-marrow-transplant-stem-cell-transplant. Accessed 5 May 2019
155. Coller BS (2015) Blood at 70: its roots in the history of hematology and its birth. Blood 126 (24):2548–2560
156. Gorer PA (1938) The antigenic basis of tumour transplantation. J Pathol Bacteriol 47 (2):231–252
157. Jacobson LO, Marks EK, Robson MF et al (1949) Effect of spleen protection on mortality following X irradiation. J Lab Clin Med 34:58
158. Lorenz E, Uphoff D, Reid TR, Shelton E (1951) Modification of irradiation injury in mice and guinea pigs by bone marrow injections. J Natl Cancer Inst 12(1):197–201
159. Barnes DWH (1954) What is the recovery factor in spleen? Nucleonics 12:68–71
160. Barnes DW, Corp MJ, Loutit JF, Neal FE (1956) Treatment of murine leukaemia with X rays and homologous bone marrow; preliminary communication. Br Med J 2(4993):626–627
161. Mathe G, Amiel JL, Schwarzenberg L, Cattan A, Schneider M (1965) Adoptive immunotherapy of acute leukemia: experimental and clinical results. Cancer Res 1965(25):1525–1531
162. Bortin MM, Bach FH, Good RA (1994) 25th anniversary of the first successful allogeneic bone marrow transplants. Bone Marrow Transplant 14(2):211–212
163. Lorna Benson (2013) Minnesota public radio news. John Kersey, U of M cancer research pioneer, dies. https://www.mprnews.org/story/2013/03/15/health/university-of-minnesota-cancer-research-pioneer-dies. Accessed 05 May 2019
164. Thomas ED, Buckner CD, Clift RA, Fefer A, Johnson FL, Neiman PE et al (1979) Marrow transplantation for acute nonlymphoblastic leukemia in first remission. N Engl J Med 301 (11):597–599
165. Blume KG, Beutler E, Bross KJ, Chillar RK, Ellington OB, Fahey JL et al (1980) Bone-marrow ablation and allogeneic marrow transplantation in acute leukemia. N Engl J Med 302 (19):1041–1046
166. Prentice HG, Janossy G, Price-Jones L, Trejdosiewicz LK, Panjwani D, Graphakos S et al (1984) Depletion of T lymphocytes in donor marrow prevents significant graft-versus-host disease in matched allogeneic leukaemic marrow transplant recipients. Lancet 323 (8375):472–476
167. Storb R, Deeg HJ, Whitehead J, Appelbaum F, Beatty P, Bensinger W et al (1986) Methotrexate and cyclosporine compared with cyclosporine alone for prophylaxis of acute graft versus host disease after marrow transplantation for leukemia. N Engl J Med 314(12):729–735
168. Hansen JA, Clift RA, Thomas ED, Buckner CD, Storb R, Giblett ER (1980) Transplantation of marrow from an unrelated donor to a patient with acute leukemia. N Engl J Med 303 (10):565–567
169. Petersdorf EW (2010) The world marrow donor association: twenty years of international collaboration for the support of unrelated donor and cord blood hematopoietic cell transplantation. Bone Marrow Transplant 45(5):807–810. https://doi.org/10.1038/bmt.2010.10
170. Hansen JA, Gooley TA, Martin PJ, Appelbaum F, Chauncey TR, Clift RA et al (1998) Bone marrow transplants from unrelated donors for patients with chronic myeloid leukemia. N Engl J Med 338(14):962–968
171. Storb R, Yu C, Wagner JL, Deeg HJ, Nash RA, Kiem H-P et al (1997) Stable mixed hematopoietic chimerism in DLA-identical littermate dogs given sublethal total body irradiation before and pharmacological immunosuppression after marrow transplantation. Blood 89 (8):3048–3054
172. Cancer Research UK. Granulocyte colony stimulating factor (G-CSF). https://www.cancerresearchuk.org/about-cancer/cancer-in-general/treatment/cancer-drugs/drugs/g-csf. Accessed 5 May 2019
173. Memorial Sloan Kettering Cancer Center. Autologous transplantation. https://www.mskcc.org/cancer-care/diagnosis-treatment/cancer-treatments/blood-stem-cell-transplantation/autologous. Accessed 5 May 2019

174. Leukemia Foudation (2015) Autologous stem cell transplants. https://www.leukaemia.org.au/disease-information/transplants/autologous-transplants/. Accessed 5 May 2019
175. Bruno B, Rotta M, Patriarca F, Mordini N, Allione B, Carnevale-Schianca F et al (2007) A comparison of allografting with autografting for newly diagnosed myeloma. N Engl J Med 356 (11):1110–1120
176. Memorial Sloan Kettering Cancer Center. Allogeneic transplantation. https://www.mskcc.org/cancer-care/diagnosis-treatment/cancer-treatments/blood-stem-cell-transplantation/allogeneic. Accessed 5 May 2019
177. Leukemia Foundation (2018) What is an allogeneic stem cell transplant? https://www.leukaemia.org.au/disease-information/transplants/allogeneic-transplants/. Accessed 5 May 2019
178. National Marrow Donor Program. HLA basics. https://bethematch.org/transplant-basics/matching-patients-with-donors/how-donors-and-patients-are-matched/hla-basics/. Accessed 5 May 2019
179. Stavropoulos-Giokas C, Dinou A, Papassavas A (2012) The role of HLA in cord blood transplantation. Bone Marrow Res 2012:9
180. Bennett B (2015) Symbiotic relationships: saviour siblings, family rights and biomedicine. Aust J Fam Law 19(3):195–212
181. Spriggs M, Savulescu J (2002) Saviour siblings. J Med Ethics 28:289
182. Simaria AS, Farid S, Hassan S (2013) American pharmaceutical review. Cost-effectiveness of single-use technologies for commercial cell therapy manufacture. https://www.americanpharmaceuticalreview.com/Featured-Articles/134042-Cost-effectiveness-of-Single-Use-Technologies-for-Commercial-Cell-Therapy-Manufacture/. Accessed 5 May 2019
183. Park YB, Ha CW, Lee CH, Yoon YC, Park YG (2017) Cartilage regeneration in osteoarthritic patients by a composite of allogeneic umbilical cord blood-derived mesenchymal stem cells and hyaluronate hydrogel: results from a clinical trial for safety and proof-of-concept with 7 years of extended follow-up. Stem Cell Transl Med 6(2):613–621
184. Agrawal P (2015) Stem cell therapy in drug discovery and development. J Pharmacovigilance 3:e140
185. Johns Hopkins Medicine (2019) Types of bone marrow transplants. https://www.hopkinsmedicine.org/kimmel_cancer_center/centers/bone_marrow_transplant/types_transplants.html. Accessed 5 May 2019
186. Kolb HJ (2018) Graft-versus-leukemia effects of transplantation and donor lymphocytes. Blood 112(12):4371–4383
187. Ballen KK, Gluckman E, Broxmeyer HE (2013) Umbilical cord blood transplantation: the first 25 years and beyond. Blood 122(4):491–498
188. Cord blood and transplants. https://bethematch.org/transplant-basics/cord-blood-and-transplants/. Accessed 5 May 2019
189. Kurtzberg J (2009) Update on umbilical cord blood transplantation. Curr Opin Pediatr 21 (1):22–29
190. Eapen M, Rubinstein P, Zhang MJ, Stevens C, Kurtzberg J, Scaradavou A et al (2007) Outcomes of transplantation of unrelated donor umbilical cord blood and bone marrow in children with acute leukaemia: a comparison study. Lancet 369(9577):1947–1954
191. Ballen K, Ahn KW, Chen M, Abdel-Azim H, Ahmed I, Aljurf M et al (2016) Infection rates among acute leukemia patients receiving alternative donor hematopoietic cell transplantation. Biol Blood Marrow Transplant 22(9):1636–1645
192. Leukaemia Foundation. Haploidentical stem cell transplant. https://www.leukaemia.org.au/disease-information/transplants/haploidentical-transplant/. Accessed 5 May 2019
193. National Marrow Donor Program. Haploidentical transplant. https://bethematch.org/patients-and-families/about-transplant/what-is-a-bone-marrow-transplant/haploidentical-transplant/. Accessed 5 May 2019
194. Canadian Cancer Society. After the stem cell transplant. http://www.cancer.ca/en/cancer-information/diagnosis-and-treatment/stem-cell-transplant/after-stem-cell-transplant/?region=on. Accessed 5 May 2019

195. Canadian Cancer Society. Side effects of a stem cell transplant. http://www.cancer.ca/en/cancer-information/diagnosis-and-treatment/stem-cell-transplant/side-effects-of-stem-cell-transplant/?region=on. Accessed 5 May 2019
196. Canadian Cancer Society. Chemotherapy. http://www.cancer.ca/en/cancer-information/diagnosis-and-treatment/chemotherapy-and-other-drug-therapies/chemotherapy/?region=on. Accessed 5 May 2019
197. Memorial Sloan Kettering Cancer Center. Chemotherapy. https://www.mskcc.org/cancer-care/diagnosis-treatment/cancer-treatments/chemotherapy. Accessed 5 May 2019
198. Cunha FI (1949) The Ebers papyrus. Am J Surg 77(1):134–136
199. Atta HM (1999) Edwin Smith Surgical Papyrus: the oldest known surgical treatise. Am Surg 65(12):1190–1192
200. Papac RJ (2001) Origins of cancer therapy. Yale J Biol Med 74(6):391–398
201. Riddle JM (1985) Ancient and medieval chemotherapy for cancer. Isis 76(283):319–330
202. McGrew RE, McGrew MP (1985) Encyclopedia of medical history. McGraw-Hill, New York
203. DeVita VT, Chu E (2008) A history of cancer chemotherapy. Cancer Res 68(21):8643–8653
204. Morrison WB (2010) Cancer chemotherapy: an annotated history. J Vet Intern Med 24 (6):1249–1262
205. Antman KH (2001) Introduction: the history of arsenic trioxide in cancer therapy. Oncologist 6 (Supplement 2):1–2
206. DeVita VT, JaEC (2008) A history of cancer chemotherapy. Cancer Res 68(21):8643–8653. https://doi.org/10.1158/0008-5472.CAN-07-6611
207. Siegel JH, McDermott WV, Steele GD, Wilmore DW, Hirsch EF, Jenkins RL et al (1990) In memoriam: George HA Clowes, Jr, MD, 1915-1988. Arch Surg 125(4):491–492
208. Zubrod CG, Schepartz S, Leiter J, Endicott KM, Carrese LM, Baker CG (1996) The chemotherapy program of the National Cancer Institute: history, analysis and plans. Cancer Chemother Rep 50(7):349–540
209. Zubrod CG, Schepartz SA, Carter SK (1977) Historical background of the National Cancer Institute's drug development thrust. Natl Cancer Inst Monogr 45:7–11
210. Gilman A (1946) Therapeutic applications of chemical warfare agents. Fed Proc 5:285–292
211. Goodman LS, Wintrobe MM, Dameshek W, Goodman MJ, Gilman A, McLennan MT (1946) Nitrogen mustard therapy: use of methyl-bis (beta-chloroethyl) amine hydrochloride and tris (beta-chloroethyl) amine hydrochloride for Hodgkin's disease, lymphosarcoma, leukemia and certain allied and miscellaneous disorders. J Am Med Dir Assoc 132(3):126–132
212. Farber S, Diamond LK, Mercer RD, Sylvester RF Jr, Wolff JA (1948) Temporary remissions in acute leukemia in children produced by folic acid antagonist, 4-aminopteroyl-glutamic acid (aminopterin). N Engl J Med 238(23):787–793
213. Li MC, Hertz R, Bergenstal DM (1958) Therapy of choriocarcinoma and related trophoblastic tumors with folic acid and purine antagonists. N Engl J Med 259(2):66–74
214. Norton L, Simon R, Brereton HD, Bogden AE (1976) Predicting the course of Gompertzian growth. Nature 264(5586):542
215. DeVita VT Jr (1984) On special initiatives, critics, and the National Cancer Program. Cancer Treat Rep 68(1):1
216. Cancer Research UK (2017) What is chemotherapy? https://www.cancerresearchuk.org/about-cancer/cancer-in-general/treatment/chemotherapy/what-chemotherapy-is. Accessed 5 May 2019
217. Teshome M, Hunt KK (2014) Neoadjuvant therapy in the treatment of breast cancer. Surg Oncol Clin 23(3):505–523
218. Herskovic A, Martz K, Al-Sarraf M et al (1992) Combined chemotherapy and radiotherapy compared with radiotherapy alone in patients with cancer of the esophagus. N Engl J Med 326 (24):1593–1598. https://doi.org/10.1056/NEJM199206113262403
219. Goldie JH (1987) Scientific basis for adjuvant and primary (neoadjuvant) chemotherapy. Semin Oncol 14(1):1–7

220. Swift L, Golsteyn R (2014) Genotoxic anti-cancer agents and their relationship to DNA damage, mitosis, and checkpoint adaptation in proliferating cancer cells. Int J Mol Sci 15 (3):3403–3431
221. Woolley PV (1998) Mechanisms of resistance to alkylating agents. Cytotechnology 27 (1–3):165
222. Malhotra V, Perry MC (2003) Classical chemotherapy: mechanisms, toxicities and the therapeutic window. Cancer Biol Ther 2(4 Suppl 1):S2–S4
223. Corrie PG (2008) Cytotoxic chemotherapy: clinical aspects. Medicine 36(1):24–28
224. Takimoto CH, Calvo E (2008) In: Pazdur R, Wagman LD, Camphausen KA, Hoskins WJ (eds) Principles of oncologic pharmacotherapy: in cancer management, a multidisciplinary approach. PRR, Melville, New York
225. Shiraishi A, Sakumi K, Sekiguchi M (2000) Increased susceptibility to chemotherapeutic alkylating agents of mice deficient in DNA repair methyltransferase. Carcinogenesis 21 (10):1879–1883
226. Yue Q-X, Liu X, Guo D-A (2010) Microtubule-binding natural products for cancer therapy. Planta Med 76(11):1037–1043
227. Rowinsky EK, Donehower RC (1991) The clinical pharmacology and use of antimicrotubule agents in cancer chemotherapeutics. Pharmacol Ther 52(1):35–84
228. Moudi M, Go R, Yien CYS et al (2013) Vinca alkaloids. Int J Prev Med 4(11):1231–1235
229. Qian Liu Y, Yang L, Tian X (2007) Podophyllotoxin: current perspectives. Curr Bioact Compd 3(1):37–66
230. Nicolaou KC, Yang Z, Liu JJ, Ueno H, Nantermet PG, Guy RK et al (1994) Total synthesis of taxol. Nature 367(6464):630
231. Dumontet C, Jordan MA (2010) Microtubule-binding agents: a dynamic field of cancer therapeutics. Nat Rev Drug Discov 9(10):790–803
232. Lodish H, Berk A, Zipursky SL, Matsudaira P, Baltimore D, Darnell J (2000) The role of topoisomerases in DNA replication. In: Molecular Cell Biology, 4th edn. WH Freeman, New York
233. Pommier Y (2009) DNA topoisomerase I inhibitors: chemistry, biology, and interfacial inhibition. Chem Rev 109(7):2894–2902
234. Goodsell DS (2002) The molecular perspective: DNA topoisomerases. Stem Cells 20 (5):470–471. https://doi.org/10.1634/stemcells.20-5-470
235. Drugs (2017) Irinotecan hydrochloride. https://www.drugs.com/monograph/irinotecan-hydrochloride.html. Accessed 5 May 2019
236. Nitiss JL (2009) Targeting DNA topoisomerase II in cancer chemotherapy. Nat Rev Cancer 9 (5):338
237. Drugs (2018) Etoposide. https://www.drugs.com/monograph/etoposide.html. Accessed 5 May 2019
238. Collins A (1990) Topoisomerase II can relax; novobiocin is a mitochondrial poison after all. BioEssays 12(10):493–494
239. Clifford B, Beljin M, Stark GR, Taylor WR (2003) G2 arrest in response to topoisomerase II inhibitors: the role of p53. Cancer Res 63(14):4074–4081
240. Felix CA (1998) Secondary leukemias induced by topoisomerase-targeted drugs. Biochim Biophys Acta Gene Struc Expr 1400(1–3):233–255
241. Doggrell SA, Davis E, Hart J, Johnston G, Hinton T, Mullaney I (2014) Cytotoxic antibiotics. https://sites.google.com/site/pharmacologyinonesemester/24-an-introduction-to-anticancer-drugs/24-3-drugs-used-in-cancer/24-3. Accessed 5 May 2019
242. Hortobagyi GN (1997) Anthracyclines in the treatment of cancer. An overview. Drugs 54:1–7
243. Minotti G, Menna P, Salvatorelli E, Cairo G, Gianni L (2004) Anthracyclines: molecular advances and pharmacologic developments in antitumor activity and cardiotoxicity. Pharmacol Rev 56(2):185–229
244. Sobell HM (1985) Actinomycin and DNA transcription. Proc Natl Acad Sci 82 (16):5328–5331

245. Dorr RT (1992) Bleomycin pharmacology: mechanism of action and resistance, and clinical pharmacokinetics. Semin Oncol 19:3–8
246. Verweij J, Pinedo HM (1990) Mitomycin C: mechanism of action, usefulness and limitations. Anti-Cancer Drugs 1(1):5–13
247. Babiker HM, McBride A, Newton M, Boehmer LM, Drucker AG, Gowan M et al (2018) Cardiotoxic effects of chemotherapy: a review of both cytotoxic and molecular targeted oncology therapies and their effect on the cardiovascular system. Crit Rev Oncol Hematol 126:186–200
248. Parker WB (2009) Enzymology of purine and pyrimidine antimetabolites used in the treatment of cancer. Chem Rev 109(7):2880–2893
249. Peters GJ, Van der Wilt CL, Van Moorsel CJA, Kroep JR, Bergman AM, Ackland SP (2000) Basis for effective combination cancer chemotherapy with antimetabolites. Pharmacol Ther 87 (2–3):227–253
250. Lind MJ (2008) Principles of cytotoxic chemotherapy. Medicine 36(1):19–23
251. Higdon J, Drake VJ, Delage B, McNulty H (2014) Oregon State University. Folate. https://lpi.oregonstate.edu/mic/vitamins/folate. Accessed 05 May 2019
252. Wagstaff AJ, Ibbotson T, Goa KL (2003) Capecitabine: a review of its pharmacology and therapeutic efficacy in the management of advanced breast cancer. Drugs 63(2):217–236
253. Koç A, Wheeler LJ, Mathews CK, Merrill GF (2004) Hydroxyurea arrests DNA replication by a mechanism that preserves basal dNTP pools. J Biol Chem 279(1):223–230
254. Yarbro JW (1992) Mechanism of action of hydroxyurea. Semin Oncol 19:1–10
255. Encyclopaedia Britannica (2017) Antimetabolite. https://www.britannica.com/science/antimetabolite. Accessed 05 May 2019
256. Barakat K, Gajewski M, Tuszynski JA (2012) DNA repair inhibitors: the next major step to improve cancer therapy. Curr Top Med Chem 12(12):1376–1390
257. Sánchez-Pérez I (2006) DNA repair inhibitors in cancer treatment. Clin Transl Oncol 8 (9):642–646
258. Kaina B, Christmann M, Naumann S, Roos WP (2007) MGMT: key node in the battle against genotoxicity, carcinogenicity and apoptosis induced by alkylating agents. DNA Repair 6 (8):1079–1099
259. Yildiz DA, Ozkan T, Yukselten Y, Sesli NT, Ozkanca S, Gunduz M et al (2016) Lomeguatrib, a O6-Methylguanine-DNA-methyltransferase (MGMT) inhibitor, induces DNA damage induced apoptosis by targeting double-strand DNA repair in multiple myeloma. Blood 128 (22):2103
260. Reinhard J, Eichhorn U, Wiessler M, Kaina B (2001) Inactivation of O(6)-methylguanine-DNA methyltransferase by glucose-conjugated inhibitors. Int J Cancer 93(3):373–379
261. Ozkan M, Akbudak IH, Deniz K, Dikilitas M, Dogu GG, Berk V et al (2010) Prognostic value of excision repair cross-complementing gene 1 expression for cisplatin-based chemotherapy in advanced gastric cancer. Asian Pac J Cancer Prev 11(1):181–185
262. Canitrot Y, Cazaux C, Frechet M, Bouayadi K, Lesca C, Salles B et al (1998) Overexpression of DNA polymerase β in cell results in a mutator phenotype and a decreased sensitivity to anticancer drugs. Proc Natl Acad Sci U S A 95(21):12586–12590. https://doi.org/10.1073/pnas.95.21.12586
263. Husain I, Morton BS, Beard WA, Singhal RK, Prasad R, Wilson SH et al (1995) Specific inhibition of DNA polymerase β by its 14 kDa domain: role of single-and double-stranded DNA binding and 5′-phosphate recognition. Nucleic Acids Res 23(9):1597–1603
264. Barakat KH, Gajewski MM, Tuszynski JA (2012) DNA polymerase beta (pol β) inhibitors: a comprehensive overview. Drug Discov Today 17(15–16):913–920. https://doi.org/10.1016/j.drudis.2012.04.008
265. Iyer RR, Pluciennik A, Burdett V, Modrich PL (2006) DNA mismatch repair: functions and mechanisms. Chem Rev 106(2):302–323
266. Fishel R, Kolodner RD (1995) Identification of mismatch repair genes and their role in the development of cancer. Curr Opin Genet Dev 5(3):382–395

267. Martin SA, Lord CJ, Ashworth A (2010) Therapeutic targeting of the DNA mismatch repair pathway. Clin Cancer Res 16(21):5107–5113. https://doi.org/10.1158/1078-0432.CCR-10-0821
268. Takahashi M, Koi M, Balaguer F, Boland CR, Goel A (2011) MSH3 mediates sensitization of colorectal cancer cells to cisplatin, Oxaliplatin, and a poly(ADP-ribose) polymerase inhibitor. J Biol Chem 286(14):12157–12165. https://doi.org/10.1074/jbc.M110.198804
269. Pardo BG-GB, Aguilera A (2009) DNA repair in mammalian cells: DNA double-strand break repair: how to fix a broken relationship. Cell Mol Life Sci 66(6):1039–1056. https://doi.org/10.1007/s00018-009-8740-3
270. Lieber MR (2010) The mechanism of double-strand DNA break repair by the nonhomologous DNA end-joining pathway. Annu Rev Biochem 79:181–211. https://doi.org/10.1146/annurev.biochem.052308.093131
271. Bolderson E, Richard DJ, Zhou B-BS, Khanna KK (2009) Recent advances in cancer therapy targeting proteins involved in DNA double-strand break repair. Clin Cancer Res 15 (20):6314–6320. https://doi.org/10.1158/1078-0432.CCR-09-0096
272. Nijman SMB (2011) Synthetic lethality: general principles, utility and detection using genetic screens in human cells. FEBS Lett 585(1):1–6. https://doi.org/10.1016/j.febslet.2010.11.024
273. Chernikova SB, Game JC, Brown JM (2012) Inhibiting homologous recombination for cancer therapy. Cancer Biol Ther 13(2):61–68. https://doi.org/10.4161/cbt.13.2.18872
274. Oliveira NG, Castro M, Rodrigues AS, Gil OM, Toscano-Rico JM, Rueff J (2002) DNA-PK inhibitor wortmannin enhances DNA damage induced by bleomycin in V79 Chinese hamster cells. Teratog Carcinog Mutagen 22(5):343–351
275. Willmore E, de Caux S, Sunter NJ, Tilby MJ, Jackson GH, Austin CA et al (2004) A novel DNA-dependent protein kinase inhibitor, NU7026, potentiates the cytotoxicity of topoisomerase II poisons used in the treatment of leukemia. Blood 103(12):4659–4665
276. Chen X, Zhong S, Zhu X, Dziegielewska B, Ellenberger T, Wilson GM et al (2008) Rational design of human DNA ligase inhibitors that target cellular DNA replication and repair. Cancer Res 68(9):3169–3177. https://doi.org/10.1158/0008-5472.CAN-07-6636
277. Srivastava M, Nambiar M, Sharma S, Karki SS, Goldsmith G, Hegde M et al (2012) An inhibitor of nonhomologous end-joining abrogates double-strand break repair and impedes cancer progression. Cell 151(7):1474–1487. https://doi.org/10.1016/j.cell.2012.11.054
278. Srivastava M, Raghavan SC (2015) DNA double-strand break repair inhibitors as cancer therapeutics. Chem Biol 22(1):17–29. https://doi.org/10.1016/j.chembiol.2014.11.013
279. Deakyne JS, Huang F, Negri J, Tolliday N, Cocklin S, Mazin AV (2013) Analysis of the activities of RAD54, a SWI2/SNF2 protein, using a specific small-molecule inhibitor. J Biol Chem 288(44):31567–31580. https://doi.org/10.1074/jbc.M113.502195
280. Dupré A, Boyer-Chatenet L, Sattler RM, Modi AP, Lee J-H, Nicolette ML et al (2008) A forward chemical genetic screen reveals an inhibitor of the Mre11-Rad50-Nbs1 complex. Nat Chem Biol 4(2):119–125. https://doi.org/10.1038/nchembio.63
281. Zhao B, Bower MJ, McDevitt PJ, Zhao H, Davis ST, Johanson KO et al (2002) Structural basis for Chk1 inhibition by UCN-01. J Biol Chem 277(48):46609–46615
282. National Cancer Institute (2015) Chemotherapy to treat cancer. https://www.cancer.gov/about-cancer/treatment/types/chemotherapy. Accessed 05 May 2019
283. Ellis M (2014) Medical news today. Palliative chemotherapy: harms and benefits weighed in new study. https://www.medicalnewstoday.com/articles/273526.php. Accessed 05 May 2019
284. Canadian Cancer Society. Side effects of chemotherapy. http://www.cancer.ca/en/cancer-information/diagnosis-and-treatment/chemotherapy-and-other-drug-therapies/chemotherapy/side-effects-of-chemotherapy/?region=on. Accessed 05 May 2019
285. National Cancer Institute (2018) Immunotherapy to treat cancer. https://www.cancer.gov/about-cancer/treatment/types/immunotherapy#1. Accessed 05 May 2019
286. Canadian Cancer Society. Immunotherapy. http://www.cancer.ca/en/cancer-information/diagnosis-and-treatment/chemotherapy-and-other-drug-therapies/immunotherapy/?region=on. Accessed 05 May 2019

287. American Society of Clinical Oncology (2019) Understanding immunotherapy. https://www.cancer.net/navigating-cancer-care/how-cancer-treated/immunotherapy-and-vaccines/understanding-immunotherapy. Accessed 05 May 2019
288. Dance A (2017) Science. Cancer immunotherapy comes of age. https://www.sciencemag.org/features/2017/03/cancer-immunotherapy-comes-age. Accessed 05 May 2019
289. Decker WK, da Silva RF, Sanabria MH, Angelo LS, Guimarães F, Burt BM et al (2017) Cancer immunotherapy: historical perspective of a clinical revolution and emerging preclinical animal models. Front Immunol 8:829. https://doi.org/10.3389/fimmu.2017.00829
290. Targeted Oncology (2014) A brief history of immunotherapy. https://www.targetedonc.com/publications/special-reports/2014/immunotherapy-issue3/a-brief-history-of-immunotherapy. Accessed 05 May 2019
291. Coley WB (1991) The treatment of malignant tumors by repeated inoculations of erysipelas. With a report of ten original cases. 1893. Clin Orthop Relat Res 262:3–11
292. Shklar G, Schwartz JL, Trickler DP et al (1990) Prevention of experimental cancer and immunostimulation by vitamin E (immunosurveillance). J Oral Pathol Med 19(2):60–64
293. Decker WK, Safdar A (2009) Bioimmunoadjuvants for the treatment of neoplastic and infectious disease: Coley's legacy revisited. Cytokine Growth Factor Rev 20(4):271–281. https://doi.org/10.1016/j.cytogfr.2009.07.004
294. Burnet M (1957) Cancer—a biological approach: III. Viruses associated with neoplastic conditions. IV. Practical applications. Br Med J 1(5023):841
295. Friend C (1956) The isolation of a virus causing a malignant disease of the hematopoietic system in adult Swiss mice. Proc Am Assoc Cancer Res 2:106
296. Prehn RT, Main JM (1957) Immunity to methylcholanthrene-induced sarcomas. J Natl Cancer Inst 18(6):769–778
297. Kim R, Emi M, Tanabe K (2007) Cancer immunoediting from immune surveillance to immune escape. Immunology 121(1):1–14. https://doi.org/10.1111/j.1365-2567.2007.02587.x
298. Isaacs A, Lindenmann J (1957) Virus interference. I. The interferon. Proc R Soc Lond B Biol Sci 147(927):258–267. https://doi.org/10.1098/rspb.1957.0048
299. Old LJ, Clarke DA, Benacerraf B (1959) Effect of Bacillus Calmette-Guerin infection on transplanted tumours in the mouse. Nature 184(4682):291–292
300. Graham JB, Graham RM (1959) The effect of vaccine on cancer patients. Surg Gynecol Obstet 109(2):131–138
301. Miller J, Mitchell GF, Weiss NS (1967) Cellular basis of the immunological defects in thymectomized mice. Nature 214(5092):992–997
302. Steinman RM, Cohn ZA (1973) Identification of a novel cell type in peripheral lymphoid organs of mice: I. Morphology, quantitation, tissue distribution. J Exp Med 137(5):1142–1162
303. Zinkernagel RM, Doherty PC (1974) Restriction of in vitro T cell-mediated cytotoxicity in lymphocytic choriomeningitis within a syngeneic or semiallogeneic system. Nature 248:701–702
304. Kiessling R, Klein E, Pross H, Wigzell H (1975) "Natural" killer cells in the mouse. II. Cytotoxic cells with specificity for mouse Moloney leukemia cells. Characteristics of the killer cell. Eur J Immunol 5(2):117–121
305. Rodriguez V, Bodey GP, Freireich EJ et al (1978) Randomized trial of protected environment – prophylactic antibiotics in 145 adults with acute leukemia. Medicine 57(3):253–266
306. Krummel MF, Allison JP (1995) CD28 and CTLA-4 have opposing effects on the response of T cells to stimulation. J Exp Med 182(2):459–465
307. Grillo-Lopez A, White C, Dallaire B, Varns C, Shen C, Wei A, Leonard J et al (2000) Rituximab the first monoclonal antibody approved for the treatment of lymphoma. Curr Pharm Biotechnol 1(1):1–9
308. Syn NL, Teng MWL, Mok TSK, Soo RA (2017) De-novo and acquired resistance to immune checkpoint targeting. Lancet Oncol 18(12):e731–e741. https://doi.org/10.1016/S1470-2045(17)30607-1

309. Couzin-Frankel J (2013) Breakthrough of the year 2013. Cancer immunotherapy. Science 342 (6165):1432–1433. https://doi.org/10.1126/science.342.6165.1432
310. Allison J, Tasuku H (2018) British society for immunology. Nobel prize 2018: cancer immunotherapy collection. https://www.immunology.org/news/nobel-prize-2018-cancer-immunotherapy-collection. Accessed 05 May 2019
311. Sushma M (2018) Explained: the cancer therapy that got two immunologists a Nobel. https://www.downtoearth.org.in/news/health/explained-the-cancer-therapy-that-got-two-immunologists-a-nobel-61773. Accessed 05 May 2019
312. Encyclopaedia Britannica (2018) Monoclonal antibodies. https://www.britannica.com/science/monoclonal-antibody. Accessed 05 May 2019
313. Cancer Research UK (2019) Rituximab (Mabthera, Rixathon, Truxima). https://www.cancerresearchuk.org/about-cancer/cancer-in-general/treatment/cancer-drugs/drugs/rituximab. Accessed 05 May 2019
314. Chen XCH (2016) Monoclonal antibodies for Cancer therapy approved by FDA. MOJ Immunol 4(2):00120. https://doi.org/10.15406/moji.2016.04.00120
315. Romond EH, Perez EA, Bryant J et al (2005) Trastuzumab plus adjuvant chemotherapy for operable HER2-positive breast cancer. N Engl J Med 353(16):1673–1684
316. Richard S, Selle F, Lotz J-P, Khalil A, Gligorov J, Soares DG (2016) Pertuzumab and trastuzumab: the rationale way to synergy. An Acad Bras Cienc 88:565–577. https://doi.org/10.1590/0001-3765201620150178
317. Carmeliet P (2005) VEGF as a key mediator of angiogenesis in cancer. Oncology 69(Suppl. 3):4–10
318. Roviello G, Sobhani N, Generali D (2017) Bevacizumab in small cell lung cancer. Ann Transl Med 5(17):361. https://doi.org/10.21037/atm.2017.06.44
319. Pavlidis ET, Pavlidis TE (2013) Role of bevacizumab in colorectal cancer growth and its adverse effects: a review. World J Gastroenterol 19(31):5051–5060. https://doi.org/10.3748/wjg.v19.i31.5051
320. Wenger KJ, Wagner M, You SJ, Franz K, Harter PN, Burger MC et al (2017) Bevacizumab as a last-line treatment for glioblastoma following failure of radiotherapy, temozolomide and lomustine. Oncol Lett 14(1):1141–1146. https://doi.org/10.3892/ol.2017.6251
321. Hsu JY, Wakelee HA (2009) Monoclonal antibodies targeting vascular endothelial growth factor: current status and future challenges in cancer therapy. BioDrugs 23(5):289–304. https://doi.org/10.2165/11317600-000000000-00000
322. di Noia V, D'Argento E, Pilotto S, Grizzi G, Caccese M, Iacovelli R, Tortora G, Bria E (2018) Necitumumab in the treatment of non-small-cell lung cancer: clinical controversies. Expert Opin Biol Ther 18(9):937–945. https://doi.org/10.1080/14712598.2018.1508445
323. Wong SF (2005) Cetuximab: an epidermal growth factor receptor monoclonal antibody for the treatment of colorectal cancer. Clin Ther 27(6):684–694
324. Memorial Sloan Kettering Cancer Center. Checkpoint inhibitors. https://www.mskcc.org/cancer-care/diagnosis-treatment/cancer-treatments/immunotherapy/checkpoint-inhibitors. Accessed 05 May 2019
325. Pardoll DM (2012) The blockade of immune checkpoints in cancer immunotherapy. Nat Rev Cancer 12(4):252–264. https://doi.org/10.1038/nrc3239
326. Cameron F, Whiteside G, Perry C (2011) Ipilimumab: first global approval. Drugs 71 (8):1093–1104
327. Selby K (2019) Tremelimumab. https://www.asbestos.com/treatment/immunotherapy/tremelimumab/. Accessed 05 May 2019
328. Wang X, Teng F, Kong L, Yu J (2016) PD-L1 expression in human cancers and its association with clinical outcomes. Onco Targets Ther 9:5023–5039. https://doi.org/10.2147/OTT.S105862
329. Thompson RH, Gillett MD, Cheville JC, Lohse CM, Dong H, Webster WS et al (2004) Costimulatory B7-H1 in renal cell carcinoma patients: indicator of tumor aggressiveness and potential therapeutic target. Proc Natl Acad Sci U S A 101(49):17174–17179

330. Urciuoli B (2018) FDA approves Opdivo for small cell lung cancer treatment. https://www.curetoday.com/articles/fda-approves-opdivo-for-small-cell-lung-cancer-treatment. Accessed 05 May 2019
331. Bristol-Myers Squibb (2018) China national drug administration approves country's first immuno-oncology agent, Opdivo (Nivolumab injection), for previously treated non-small cell lung cancer (NSCLC). https://news.bms.com/press-release/corporatefinancial-news/china-national-drug-administration-approves-countrys-first-imm. Accessed 05 May 2019
332. Chemocare. Pembrolizumab. http://chemocare.com/chemotherapy/drug-info/Pembrolizumab.aspx. Accessed 05 May 2019
333. National Institutes of Health (2019) Study of efficacy and safety of novel spartalizumab combinations in patients with previously treated unresectable or metastatic melanoma (PLAT-forM). https://clinicaltrials.gov/ct2/show/NCT03484923. Accessed 05 May 2019
334. Shen X, Zhao B (2018) Efficacy of PD-1 or PD-L1 inhibitors and PD-L1 expression status in cancer: meta-analysis. Br Med J 362:k3529. https://doi.org/10.1136/bmj.k3529
335. Lambert JM (2005) Drug-conjugated monoclonal antibodies for the treatment of cancer. Curr Opin Pharmacol 5(5):543–549
336. Goldmacher VS, Blättler WA, Lambert JM, Chari RVJ (2002) Immunotoxins and antibody-drug conjugates for cancer treatment. In: Biomedical aspects of drug targeting. Springer, Boston, pp 291–309. https://doi.org/10.1007/978-1-4757-4627-3_15
337. Younes A, Bartlett NL, Leonard JP, Kennedy DA, Lynch CM, Sievers EL et al (2010) Brentuximab vedotin (SGN-35) for relapsed CD30-positive lymphomas. N Engl J Med 363 (19):1812–1821. https://doi.org/10.1056/NEJMoa1002965
338. Baron J, Wang ES (2018) Gemtuzumab ozogamicin for the treatment of acute myeloid leukemia. Expert Rev Clin Pharmacol 11(6):549–559. https://doi.org/10.1080/17512433.2018.1478725
339. Phillips GDL, Li G, Dugger DL, Crocker LM, Parsons KL, Mai E et al (2008) Targeting HER2-positive breast cancer with trastuzumab-DM1, an antibody–cytotoxic drug conjugate. Cancer Res 68(22):9280–9290. https://doi.org/10.1158/0008-5472.CAN-08-1776
340. Larson SM, Carrasquillo JA, Cheung N-KV, Press OW (2015) Radioimmunotherapy of human tumours. Nat Rev Cancer 15(6):347–360. https://doi.org/10.1038/nrc3925
341. Tran KQ, Zhou J, Durflinger KH, Langhan MM, Shelton TE, Wunderlich JR et al (2008) Minimally cultured tumor-infiltrating lymphocytes display optimal characteristics for adoptive cell therapy. J Immunother 31(8):742–751. https://doi.org/10.1097/CJI.0b013e31818403d5
342. Rosenberg SA, Packard BS, Aebersold PM, Solomon D, Topalian SL, Toy ST et al (1988) Use of tumor-infiltrating lymphocytes and interleukin-2 in the immunotherapy of patients with metastatic melanoma. N Engl J Med 319(25):1676–1680
343. Rosenberg SA, Restifo NP, Yang JC, Morgan RA, Dudley ME (2008) Adoptive cell transfer: a clinical path to effective cancer immunotherapy. Nat Rev Cancer 8(4):299–308. https://doi.org/10.1038/nrc2355
344. Houot R, Schultz LM, Marabelle A, Kohrt H (2015) T-cell–based immunotherapy: adoptive cell transfer and checkpoint inhibition. Cancer Immunol Res 3(10):1115–1122. https://doi.org/10.1158/2326-6066.CIR-15-0190
345. Morgan RA, Dudley ME, Wunderlich JR, Hughes MS, Yang JC, Sherry RM et al (2006) Cancer regression in patients after transfer of genetically engineered lymphocytes. Science 314 (5796):126–129
346. Robbins PF, Morgan RA, Feldman SA, Yang JC, Sherry RM, Dudley ME et al (2011) Tumor regression in patients with metastatic synovial cell sarcoma and melanoma using genetically engineered lymphocytes reactive with NY-ESO-1. J Clin Oncol 29(7):917–924. https://doi.org/10.1200/JCO.2010.32.2537
347. Morgan RA, Chinnasamy N, Abate-Daga DD, Gros A, Robbins PF, Zheng Z et al (2013) Cancer regression and neurologic toxicity following anti-MAGE-A3 TCR gene therapy. J Immunother 36(2):133–151. https://doi.org/10.1097/CJI.0b013e3182829903

348. American Cancer Society (2016) Non-specific cancer immunotherapies and adjuvants. https://www.cancer.org/treatment/treatments-and-side-effects/treatment-types/immunotherapy/nonspecific-immunotherapies.html. Accessed 06 May 2019
349. Altundag K, Altundag O, Elkiran ET, Cengiz M, Ozisik Y (2004) Addition of granulocyte-colony stimulating factor (G-CSF) to adjuvant treatment may increase survival in patients with operable breast cancer: interaction of G-CSF with dormant micrometastatic breast cancer cells. Med Hypotheses 63(1):56–58
350. Cetean S, Căinap C, Constantin A-M, Căinap S, Gherman A, Oprean L et al (2015) The importance of the granulocyte-colony stimulating factor in oncology. Clujul Med 88 (4):468–472. https://doi.org/10.15386/cjmed-531
351. American Cancer Society (2016) Cancer vaccines. https://www.cancer.org/treatment/treatments-and-side-effects/treatment-types/immunotherapy/cancer-vaccines.html. Accessed 06 May 2019
352. Takes RP, Wierzbicka M, D'Souza G, Jackowska J, Silver CE, Rodrigo JP et al (2015) HPV vaccination to prevent oropharyngeal carcinoma: what can be learned from anogenital vaccination programs? Oral Oncol 51(12):1057–1060. https://doi.org/10.1016/j.oraloncology.2015.10.011
353. Verma R, Khanna P (2013) Human papilloma virus vaccines: need to be introduced in India. Hum Vaccin Immunother 9(1):97–99. https://doi.org/10.4161/hv.22063
354. Perz JF, Armstrong GL, Farrington LA, Hutin YJF, Bell BP (2006) The contributions of hepatitis B virus and hepatitis C virus infections to cirrhosis and primary liver cancer worldwide. J Hepatol 45(4):529–538
355. Schuster SJ, Neelapu SS, Gause BL, Muggia FM, Gockerman JP, Sotomayor EM et al (2009) Idiotype vaccine therapy (BiovaxID) in follicular lymphoma in first complete remission: phase III clinical trial results. J Clin Oncol 27(18S):2–2
356. Cheever MA, Higano CS (2011) PROVENGE (Sipuleucel-T) in prostate cancer: the first FDA-approved therapeutic cancer vaccine. Clin Cancer Res 17(11):3520–3526. https://doi.org/10.1158/1078-0432.CCR-10-3126
357. Rentsch CA, Birkhäuser FD, Biot C, Gsponer JR, Bisiaux A, Wetterauer C et al (2014) Bacillus Calmette-Guérin strain differences have an impact on clinical outcome in bladder cancer immunotherapy. Eur Urol 66(4):677–688. https://doi.org/10.1016/j.eururo.2014.02.061
358. Fukuhara H, Ino Y, Todo T (2016) Oncolytic virus therapy: a new era of cancer treatment at dawn. Cancer Sci 107(10):1373–1379. https://doi.org/10.1111/cas.13027
359. Banchereau J, Steinman RM (1998) Dendritic cells and the control of immunity. Nature 392 (6673):245–252
360. Palucka K, Banchereau J (2013) Dendritic-cell-based therapeutic cancer vaccines. Immunity 39(1):38–48. https://doi.org/10.1016/j.immuni.2013.07.004
361. Quach H, Ritchie D, Stewart AK, Neeson P, Harrison S, Smyth MJ et al (2010) Mechanism of action of immunomodulatory drugs (IMiDS) in multiple myeloma. Leukemia 24(1):22–32. https://doi.org/10.1038/leu.2009.236
362. Sampaio EP, Sarno EN, Galilly R, Cohn ZA, Kaplan G (1991) Thalidomide selectively inhibits tumor necrosis factor alpha production by stimulated human monocytes. J Exp Med 173(3):699–703
363. D'Amato RJ, Loughnan MS, Flynn E, Folkman J (1994) Thalidomide is an inhibitor of angiogenesis. Proc Natl Acad Sci U S A 91(9):4082–4085
364. Haslett PAJ, Corral LG, Albert M, Kaplan G (1998) Thalidomide costimulates primary human T lymphocytes, preferentially inducing proliferation, cytokine production, and cytotoxic responses in the CD8+ subset. J Exp Med 187(11):1885–1892
365. Hardy H, Harris J, Lyon E, Beal J, Foey A (2013) Probiotics, prebiotics and immunomodulation of gut mucosal defences: homeostasis and immunopathology. Nutrients 5(6):1869–1912. https://doi.org/10.3390/nu5061869

366. Aleem E (2013) β-Glucans and their applications in cancer therapy: focus on human studies. Anti Cancer Agents Med Chem 13(5):709–719
367. Ma H-D, Deng Y-R, Tian Z, Lian Z-X (2013) Traditional Chinese medicine and immune regulation. Clin Rev Allergy Immunol 44(3):229–241. https://doi.org/10.1007/s12016-012-8332-0
368. Cancer Council Victoria. Immunotherapy. https://www.cancervic.org.au/cancer-information/treatments/treatments-types/immunotherapy. Accessed 06 May 2019
369. Cancer Research Institute. Immunotherapy by cancer type. https://www.cancerresearch.org/immunotherapy/cancer-types. Accessed 06 May 2019
370. Porter L, American Society of Clinical Oncology (2018) What you need to know about immunotherapy side effects. https://www.cancer.net/blog/2018-02/what-you-need-know-about-immunotherapy-side-effects. Accessed 06 May 2019
371. Cancer Council Victoria. Targeted therapy. https://www.cancervic.org.au/cancer-information/treatments/treatments-types/targeted-therapy. Accessed 06 May 2019
372. American Society of Clinical Oncology (2019) Understanding targeted therapy. https://www.cancer.net/navigating-cancer-care/how-cancer-treated/personalized-and-targeted-therapies/understanding-targeted-therapy. Accessed 06 May 2019
373. National Cancer Institute (2018) Targeted therapy to treat cancer. https://www.cancer.gov/about-cancer/treatment/types/targeted-therapies#2. Accessed 06 May 2019
374. Canadian Cancer Society. Targeted therapy. http://www.cancer.ca/en/cancer-information/diagnosis-and-treatment/chemotherapy-and-other-drug-therapies/targeted-therapy/?region=on. Accessed 06 May 2019
375. Joo WD, Visintin I, Mor G (2013) Targeted cancer therapy–are the days of systemic chemotherapy numbered? Maturitas 76(4):308–314
376. Lee SL (2012) Radioactive iodine therapy. Curr Opin Endocrinol Diabetes Obes 19 (5):420–428. https://doi.org/10.1097/MED.0b013e328357fa0c
377. Jordan VC (1993) A current view of tamoxifen for the treatment and prevention of breast cancer. Br J Pharmacol 110(2):507–517. https://doi.org/10.1111/j.1476-5381.1993.tb13840.x
378. Fausel C (2007) Targeted chronic myeloid leukemia therapy: seeking a cure. Am J Health Syst Pharm 64(24):S9–S15
379. Yan L, Rosen N, Arteaga C (2011) Targeted cancer therapies. Chin J Cancer 30(1):1–4
380. Ferrari M (2005) Cancer nanotechnology: opportunities and challenges. Nat Rev Cancer 5 (3):161–171
381. Bae KH, Chung HJ, Park TG (2011) Nanomaterials for cancer therapy and imaging. Mol Cells 31(4):295–302
382. Prabhu RH, Patravale VB, Joshi MD (2015) Polymeric nanoparticles for targeted treatment in oncology: current insights. Int J Nanomedicine 10:1001–1018. https://doi.org/10.2147/IJN.S56932
383. Zhou L, Huang Y, Li J, Wang Z (2010) The mTOR pathway is associated with the poor prognosis of human hepatocellular carcinoma. Med Oncol 27(2):255–261. https://doi.org/10.1007/s12032-009-9201-4
384. Zaytseva YY, Valentino JD, Gulhati P, Evers BM (2012) mTOR inhibitors in cancer therapy. Cancer Lett 319(1):1–7. https://doi.org/10.1016/j.canlet.2012.01.005
385. McDougall SR, Anderson ARA, Chaplain MAJ (2006) Mathematical modelling of dynamic adaptive tumour-induced angiogenesis: clinical implications and therapeutic targeting strategies. J Theor Biol 241(3):564–589
386. Cabebe E, Wakelee H (2006) Sunitinib: a newly approved small-molecule inhibitor of angiogenesis. Drugs Today (Barc) 42(6):387–398
387. Adams J, Kauffman M (2004) Development of the proteasome inhibitor Velcade™ (Bortezomib). Cancer Investig 22(2):304–311
388. Kabbinavar F, Hurwitz HI, Fehrenbacher L, Meropol NJ, Novotny WF, Lieberman G et al (2003) Phase II, randomized trial comparing bevacizumab plus fluorouracil (FU)/leucovorin (LV) with FU/LV alone in patients with metastatic colorectal cancer. J Clin Oncol 21(1):60–65

389. Baudino TA (2015) Targeted cancer therapy: the next generation of cancer treatment. Curr Drug Discov Technol 12(1):3–20
390. Wang J, Taylor A, Showeil R, Trivedi P, Horimoto Y, Bagwan I et al (2014) Expression profiling and significance of VEGF-A, VEGFR2, VEGFR3 and related proteins in endometrial carcinoma. Cytokine 68(2):94–100. https://doi.org/10.1016/j.cyto.2014.04.005
391. Keating GM (2014) Bevacizumab: a review of its use in advanced cancer. Drugs 74 (16):1891–1925. https://doi.org/10.1007/s40265-014-0302-9
392. Wilke H, Muro K, Van Cutsem E, Oh S-C, Bodoky G, Shimada Y et al (2014) Ramucirumab plus paclitaxel versus placebo plus paclitaxel in patients with previously treated advanced gastric or gastro-oesophageal junction adenocarcinoma (RAINBOW): a double-blind, randomised phase 3 trial. Lancet Oncol 15(11):1224–1235. https://doi.org/10.1016/S1470-2045(14)70420-6
393. Zhang H, Berezov A, Wang Q, Zhang G, Drebin J, Murali R et al (2007) ErbB receptors: from oncogenes to targeted cancer therapies. J Clin Invest 117(8):2051–2058
394. Mitri Z, Constantine T, O'Regan R (2012) The HER2 receptor in breast cancer: pathophysiology, clinical use, and new advances in therapy. Chemother Res Pract 2012:743193. https://doi.org/10.1155/2012/743193
395. Dubois EA, Cohen AF (2009) Panitumumab. Br J Clin Pharmacol 68(4):482–483. https://doi.org/10.1111/j.1365-2125.2009.03492.x
396. Ribas A, Wolchok JD (2018) Cancer immunotherapy using checkpoint blockade. Science 359 (6382):1350–1355. https://doi.org/10.1126/science.aar4060
397. Scheuermann RH, Racila E (1995) CD19 antigen in leukemia and lymphoma diagnosis and immunotherapy. Leuk Lymphoma 18(5–6):385–397
398. Nagorsen D, Kufer P, Baeuerle PA, Bargou R (2012) Blinatumomab: a historical perspective. Pharmacol Ther 136(3):334–342. https://doi.org/10.1016/j.pharmthera.2012.07.013
399. U.S. Food and Drug Administration (2017) FDA grants regular approval to blinatumomab and expands indication to include Philadelphia chromosome-positive B cell. https://www.fda.gov/drugs/resources-information-approved-drugs/fda-grants-regular-approval-blinatumomab-and-expands-indication-include-philadelphia-chromosome. Accessed 05 May 2019
400. Payandeh Z, Bahrami AA, Hoseinpoor R, Mortazavi Y, Rajabibazl M, Rahimpour A et al (2019) The applications of anti-CD20 antibodies to treat various B cells disorders. Biomed Pharmacother 109:2415–2426. https://doi.org/10.1016/j.biopha
401. Lim SH, Beers SA, French RR, Johnson PWM, Glennie MJ, Cragg MS (2010) Anti-CD20 monoclonal antibodies: historical and future perspectives. Haematologica 95(1):135–143. https://doi.org/10.3324/haematol.2008.001628
402. Van Der Weyden CA, Pileri SA, Feldman AL, Whisstock J, Prince HM (2017) Understanding CD30 biology and therapeutic targeting: a historical perspective providing insight into future directions. Blood Cancer J 7(9):e603. https://doi.org/10.1038/bcj.2017.85
403. Faramarz Naeim PNR, Song SX, Grody WW (2013) Principles of immunophenotyping. In: Atlas of Hematopathology, 2nd edn. Academic, London, pp 25–46
404. Hamann PR, Hinman LM, Hollander I, Beyer CF, Lindh D, Holcomb R, Hallett W et al (2002) Gemtuzumab ozogamicin, a potent and selective anti-CD33 antibody-calicheamicin conjugate for treatment of acute myeloid leukemia. Bioconjug Chem 13(1):47–58
405. Jen EY, Ko C-W, Lee JE, Del Valle PL, Aydanian A, Jewell C et al (2018) FDA approval: gemtuzumab ozogamicin for the treatment of adults with newly diagnosed CD33-positive acute myeloid leukemia. Clin Cancer Res 24(14):3242–3246. https://doi.org/10.1158/1078-0432
406. Hamblin TJ (2003) CD38: what is it there for? Blood 102(6):1939–1940
407. Burgler S (2015) Role of CD38 expression in diagnosis and pathogenesis of chronic lymphocytic leukemia and its potential as therapeutic target. Crit Rev Immunol 35(5):417–432
408. Lokhorst HM, Plesner T, Laubach JP, Nahi H, Gimsing P, Hansson M et al (2015) Targeting CD38 with daratumumab monotherapy in multiple myeloma. N Engl J Med 373 (13):1207–1219. https://doi.org/10.1056/NEJMoa1506348

409. Piccaluga PP, Agostinelli C, Righi S, Zinzani PL, Pileri SA (2007) Expression of CD52 in peripheral T-cell lymphoma. Haematologica 92(4):566–567
410. Malaer JD, Mathew PA (2017) CS1 (SLAMF7, CD319) is an effective immunotherapeutic target for multiple myeloma. Am J Cancer Res 7(8):1637–1641
411. Chen WC, Kanate AS, Craig M, Petros WP, Hazlehurst LA (2017) Emerging combination therapies for the management of multiple myeloma: the role of elotuzumab. Cancer Manag Res 9:307–314. https://doi.org/10.2147/CMAR.S117477
412. Wierzbicki A, Gil M, Ciesielski M, Fenstermaker RA, Kaneko Y, Rokita H et al (2008) Immunization with a mimotope of GD2 ganglioside induces CD8+ T cells that recognize cell adhesion molecules on tumor cells. J Immunol 181(9):6644–6653
413. Keyel ME, Reynolds CP (2019) Spotlight on dinutuximab in the treatment of high-risk neuroblastoma: development and place in therapy. Biologics 13:1–12. https://doi.org/10.2147/BTT.S114530
414. Nie S, Xing Y, Kim GJ, Simons JW (2007) Nanotechnology applications in cancer. Annu Rev Biomed Eng 9:257–288
415. Salvador-Morales C, Gao W, Ghatalia P, Murshed F, Aizu W, Langer R et al (2009) Multifunctional nanoparticles for prostate cancer therapy. Expert Rev Anticancer Ther 9 (2):211–221. https://doi.org/10.1586/14737140.9.2.211
416. Peer D, Karp JM, Hong S, Farokhzad OC, Margalit R, Langer R (2007) Nanocarriers as an emerging platform for cancer therapy. Nat Nanotechnol 2(12):751–760. https://doi.org/10.1038/nnano.2007.387
417. Hubbell JA (2003) Enhancing drug function. Science 300(5619):595–596. https://doi.org/10.1126/science.1083625
418. Bartlett DW, Davis ME (2007) Physicochemical and biological characterization of targeted, nucleic acid-containing nanoparticles. Bioconjug Chem 18(2):456–468. https://doi.org/10.1021/bc0603539
419. Bisht G, Rayamajhi S (2016) ZnO nanoparticles: a promising anticancer agent. Nano 3:9. https://doi.org/10.5772/63437
420. Lee Y, Lee H, Kim YB, Kim J, Hyeon T, Park H, Messersmith PB, Park TG (2008) Bioinspired surface immobilization of hyaluronic acid on monodisperse magnetite nanocrystals for targeted cancer imaging. Adv Mater 20(21):4154–4157. https://doi.org/10.1002/adma.200800756
421. Bae KH, Lee K, Kim C, Park TG (2011) Surface functionalized hollow manganese oxide nanoparticles for cancer targeted siRNA delivery and magnetic resonance imaging. Biomaterials 32(1):176–184. https://doi.org/10.1016/j.biomaterials
422. Marusyk A, Polyak K (2010) Tumor heterogeneity: causes and consequences. Biochim Biophys Acta 1805(1):105–117. https://doi.org/10.1016/j.bbcan.2009.11.002
423. Samuel N, Hudson TJ (2013) Translating genomics to the clinic: implications of cancer heterogeneity. Clin Chem 59(1):127–137. https://doi.org/10.1373/clinchem
424. Gray JW, Collins C, Henderson IC, Isola J, Kallioniemi A, Kallioniemi OP et al (1994) Molecular cytogenetics of human breast cancer. Cold Spring Harb Symp Quant Biol 59:645–652
425. Örndal C, Rydholm A, Willén H, Mitelman F, Mandahl N (1994) Cytogenetic intratumor heterogeneity in soft tissue tumors. Cancer Genet Cytogenet 78(2):127–137
426. Gorunova L, Höglund M, Andrén-Sandberg Å, Dawiskiba S, Jin Y, Mitelman F et al (1998) Cytogenetic analysis of pancreatic carcinomas: intratumor heterogeneity and nonrandom pattern of chromosome aberrations. Genes Chromosomes Cancer 23(2):81–99
427. Wright MH, Calcagno AM, Salcido CD, Carlson MD, Ambudkar SV, Varticovski L (2008) Breast tumors contain distinct CD44+/CD24-and CD133+ cells with cancer stem cell characteristics. Breast Cancer Res 10(1):R10. https://doi.org/10.1186/bcr1855
428. Rutella S, Bonanno G, Procoli A, Mariotti A, Corallo M, Prisco MG et al (2009) Cells with characteristics of cancer stem/progenitor cells express the CD133 antigen in human endometrial tumors. Clin Cancer Res 15(13):4299–4311. https://doi.org/10.1158/1078-0432

429. Lou H, Dean M (2007) Targeted therapy for cancer stem cells: the patched pathway and ABC transporters. Oncogene 26(9):1357–1360
430. Ebben JD, Treisman DM, Zorniak M, Kutty RG, Clark PA, Kuo JS (2010) The cancer stem cell paradigm: a new understanding of tumor development and treatment. Expert Opin Ther Targets 14(6):621–632. https://doi.org/10.1517/14712598.2010.485186
431. Kong D, Li Y, Wang Z, Sarkar F (2013) Cancer stem cells and epithelial-to-mesenchymal transition (EMT)-phenotypic cells: are they cousins or twins? Cancers (Basel) 3(1):716–729. https://doi.org/10.3390/cancers30100716
432. Craveiro V, Yang-Hartwich Y, Holmberg JC, Sumi NJ, Pizzonia J, Griffin B et al (2013) Phenotypic modifications in ovarian cancer stem cells following paclitaxel treatment. Cancer Med 2(6):751–762. https://doi.org/10.1002/cam4.115
433. National Institutes of Health, ClinicalTrials.gov is a database of privately and publicly funded clinical studies conducted around the world. https://www.clinicaltrials.gov/. Accessed 05 May 2019
434. Mattina J, Carlisle B, Hachem Y, Fergusson D, Kimmelman J (2017) Inefficiencies and patient burdens in the development of the targeted cancer drug sorafenib: a systematic review. PLoS Biol 15(2):e2000487. https://doi.org/10.1371/journal.pbio.2000487
435. American Cancer Society (2016) Side effects of targeted cancer therapy drugs. https://www.cancer.org/treatment/treatments-and-side-effects/treatment-types/targeted-therapy/side-effects.html. Accessed 05 May 2019
436. Rogers S, Pfuderer P (1968) Use of viruses as carriers of added genetic information. Nature 219:749–751. https://doi.org/10.1038/219749a0
437. Friedmann T, Roblin R (1972) Gene therapy for human genetic disease? Science 175:949–955. https://doi.org/10.1126/science.175.4025.949
438. Rogers S (1973) Induction of arginase activity with the Shope papilloma virus in tissue culture cells from an argininemic patient. J Exp Med 137:1091–1096. https://doi.org/10.1084/jem.137.4.1091
439. Terheggen HG, Lowenthal A, Lavinha F et al (1975) Unsuccessful trial of gene replacement in arginase deficiency. Z Kinderheilkd 119:1–3
440. Cepko CL, Roberts BE, Mulligan RC (1984) Construction and applications of a highly transmissible murine retrovirus shuttle vector. Cell 37:1053–1062. https://doi.org/10.1016/0092-8674(84)90440-9
441. Wirth T, Ylä-Herttuala S (2014) Gene therapy used in Cancer treatment. Biomedicine 2:149–162. https://doi.org/10.3390/biomedicines2020149
442. Blaese RM, Culver KW, Miller AD et al (1995) T lymphocyte-directed gene therapy for ADA-SCID: initial trial results after 4 years. Science 270:475–480. https://doi.org/10.1126/science.270.5235.475
443. Sheridan C (2011) Gene therapy finds its niche. Nat Biotechnol 29:121–128. https://doi.org/10.1038/nbt.1769
444. Amer MH (2014) Gene therapy for cancer: present status and future perspective. Mol Cell Ther 2:27. https://doi.org/10.1186/2052-8426-2-27
445. Philippidis. A 25 up-and-coming gene therapies of 2019. https://www.genengnews.com/a-lists/25-up-and-coming-gene-therapies-of-2019/. Accessed 12 Sept 2019
446. Ginn SL, Amaya AK, Alexander IE et al (2018) Gene therapy clinical trials worldwide to 2017: an update. J Gene Med 20:e3015. https://doi.org/10.1002/jgm.3015
447. Lodish H, Berk A, Zipursky SL et al (2000) Viruses: structure, function, and uses. W. H. Freeman, New York
448. Thomas CE, Ehrhardt A, Kay MA (2003) Progress and problems with the use of viral vectors for gene therapy. Nat Rev Genet 4:346–358. https://doi.org/10.1038/nrg1066
449. Zabner J, Fasbender AJ, Moninger T et al (1995) Cellular and molecular barriers to gene transfer by a cationic lipid. J Biol Chem 270:18997–19007. https://doi.org/10.1074/jbc.270.32.18997

450. Yang W, Sun T, Cao J, Liu F (2010) Survivin downregulation by siRNA/cationic liposome complex radiosensitises human hepatoma cells in vitro and in vivo. Int J Radiat Biol 86:445–457. https://doi.org/10.3109/09553001003668006
451. Baban CK, Cronin M, O'Hanlon D et al (2010) Bacteria as vectors for gene therapy of cancer. Bioengineered Bugs 1:385–394. https://doi.org/10.4161/bbug.1.6.13146
452. Pathak A, Patnaik S, Gupta KC (2009) Recent trends in non-viral vector-mediated gene delivery. Biotechnol J 4:1559–1572. https://doi.org/10.1002/biot.200900161
453. Das SK, Menezes ME, Bhatia S et al (2015) Gene therapies for Cancer: strategies, challenges and successes. J Cell Physiol 230:259–271. https://doi.org/10.1002/jcp.24791
454. Boon T (1996) Human tumor antigens recognized by T lymphocytes. J Exp Med 183:725–729. https://doi.org/10.1084/jem.183.3.725
455. Morgan RA, Dudley ME, Wunderlich JR et al (2006) Cancer regression in patients after transfer of genetically engineered lymphocytes. Science 314:126–129. https://doi.org/10.1126/science.1129003
456. Robbins PF, Morgan RA, Feldman SA et al (2011) Tumor regression in patients with metastatic synovial cell sarcoma and melanoma using genetically engineered lymphocytes reactive with NY-ESO-1. JCO 29:917–924. https://doi.org/10.1200/JCO.2010.32.2537
457. Sharpe M, Mount N (2015) Genetically modified T cells in cancer therapy: opportunities and challenges. Dis Model Mech 8:337–350. https://doi.org/10.1242/dmm.018036
458. Kochenderfer JN, Rosenberg SA (2013) Treating B-cell cancer with T cells expressing anti-CD19 chimeric antigen receptors. Nat Rev Clin Oncol 10:267–276. https://doi.org/10.1038/nrclinonc.2013.46
459. Kochenderfer JN, Dudley ME, Feldman SA et al (2012) B-cell depletion and remissions of malignancy along with cytokine-associated toxicity in a clinical trial of anti-CD19 chimeric-antigen-receptor-transduced T cells. Blood 119:2709–2720. https://doi.org/10.1182/blood-2011-10-384388
460. Boudreau JE, Bonehill A, Thielemans K, Wan Y (2011) Engineering dendritic cells to enhance Cancer immunotherapy. Mol Ther 19:841–853. https://doi.org/10.1038/mt.2011.57
461. Butterfield LH, Comin-Anduix B, Vujanovic L et al (2008) Adenovirus MART-1–engineered autologous dendritic cell vaccine for metastatic melanoma. J Immunother 31:294–309. https://doi.org/10.1097/CJI.0b013e31816a8910
462. Marshall JL, Gulley JL, Arlen PM et al (2005) Phase I study of sequential vaccinations with Fowlpox-CEA(6D)-TRICOM alone and sequentially with vaccinia-CEA(6D)-TRICOM, with and without granulocyte-macrophage Colony-stimulating factor, in patients with carcinoembryonic antigen–expressing carcinomas. JCO 23:720–731. https://doi.org/10.1200/JCO.2005.10.206
463. Norell H, Poschke I, Charo J et al (2010) Vaccination with a plasmid DNA encoding HER-2/neu together with low doses of GM-CSF and IL-2 in patients with metastatic breast carcinoma: a pilot clinical trial. J Transl Med 8:53. https://doi.org/10.1186/1479-5876-8-53
464. Sun E, Han R, Lu B (2018) Gene therapy of renal cancer using recombinant adeno-associated virus encoding human endostatin. Oncol Lett. https://doi.org/10.3892/ol.2018.9036
465. YESCARTA (axicabtagene ciloleucel) | FDA. https://www.fda.gov/vaccines-blood-biologics/cellular-gene-therapy-products/yescarta-axicabtagene-ciloleucel. Accessed 21 Sept 2019
466. KYMRIAH (tisagenlecleucel) | FDA. https://www.fda.gov/vaccines-blood-biologics/cellular-gene-therapy-products/kymriah-tisagenlecleucel. Accessed 21 Sept 2019
467. Rehman H, Silk AW, Kane MP, Kaufman HL (2016) Into the clinic: Talimogene laherparepvec (T-VEC), a first-in-class intratumoral oncolytic viral therapy. J Immunother Cancer 4:53. https://doi.org/10.1186/s40425-016-0158-5
468. IMLYGIC (talimogene laherparepvec) | FDA. https://www.fda.gov/vaccines-blood-biologics/cellular-gene-therapy-products/imlygic-talimogene-laherparepvec. Accessed 21 Sept 2019
469. Malekshah OM, Chen X, Nomani A et al (2016) Enzyme/prodrug Systems for Cancer Gene Therapy. Curr Pharmacol Rep 2:299–308. https://doi.org/10.1007/s40495-016-0073-y

470. Freeman SM, Abboud CN, Whartenby KA et al (1993) The "bystander effect": tumor regression when a fraction of the tumor mass is genetically modified. Cancer Res 53:5274
471. Zarogoulidis P, Darwiche K (2013) Suicide gene therapy for Cancer – current strategies. J Genet Syndr Gene Ther 04:pii: 16849. https://doi.org/10.4172/2157-7412.1000139
472. Chen H, Beardsley GP, Coen DM (2014) Mechanism of ganciclovir-induced chain termination revealed by resistant viral polymerase mutants with reduced exonuclease activity. Proc Natl Acad Sci U S A 111:17462–17467. https://doi.org/10.1073/pnas.1405981111
473. Kaliberov SA, Market JM, Gillespie GY et al (2007) Mutation of Escherichia coli cytosine deaminase significantly enhances molecular chemotherapy of human glioma. Gene Ther 14:1111–1119. https://doi.org/10.1038/sj.gt.3302965
474. Kaliberova LN, Della Manna DL, Krendelchtchikova V et al (2008) Molecular chemotherapy of pancreatic cancer using novel mutant bacterial cytosine deaminase gene. Mol Cancer Ther 7:2845–2854. https://doi.org/10.1158/1535-7163.MCT-08-0347
475. Deng LY (2011) Antitumor activity of mutant bacterial cytosine deaminase gene for colon cancer. WJG 17:2958. https://doi.org/10.3748/wjg.v17.i24.2958
476. Immonen A, Vapalahti M, Tyynelä K et al (2004) AdvHSV-tk gene therapy with intravenous ganciclovir improves survival in human malignant glioma: a randomised, controlled study. Mol Ther 10:967–972. https://doi.org/10.1016/j.ymthe.2004.08.002
477. Aguilar LK, Shirley LA, Chung VM et al (2015) Gene-mediated cytotoxic immunotherapy as adjuvant to surgery or chemoradiation for pancreatic adenocarcinoma. Cancer Immunol Immunother 64:727–736. https://doi.org/10.1007/s00262-015-1679-3
478. Li BJ (2007) Vascular damage and anti-angiogenic effects of tumor vessel-targeted adenovirus-mediated herpes simplex virus thymidine kinase gene. WJG 13:4006. https://doi.org/10.3748/wjg.v13.i29.4006
479. Karjoo Z, Chen X, Hatefi A (2016) Progress and problems with the use of suicide genes for targeted cancer therapy. Adv Drug Deliv Rev 99:113–128. https://doi.org/10.1016/j.addr.2015.05.009
480. Search Orphan Drug Designations and Approvals. https://www.accessdata.fda.gov/scripts/opdlisting/oopd/detailedIndex.cfm?cfgridkey=145801. Accessed 21 Sept 2019
481. Nemunaitis J, Tong AW, Nemunaitis M et al (2010) A phase I study of telomerase-specific replication competent oncolytic adenovirus (Telomelysin) for various solid tumors. Mol Ther 18:429–434. https://doi.org/10.1038/mt.2009.262
482. Clark DP, Pazdernik NJ (2016) Cancer. In: Biotechnology. Elsevier, New Jersey, pp 593–626
483. Levine AJ, Hu W, Feng Z (2008) Tumor suppressor genes. In: The molecular basis of cancer. Elsevier, New Jersey, pp 31–38
484. Cancer Genes | CancerQuest. https://www.cancerquest.org/cancer-biology/cancer-genes#table. Accessed 22 Sept 2019
485. Oncogene related genes – GeneCards Search Results. https://www.genecards.org/Search/Keyword?queryString=oncogene. Accessed 22 Sept 2019
486. Tumor suppressor gene related genes – GeneCards Search Results. https://www.genecards.org/Search/Keyword?queryString=tumor%20suppressor%20gene. Accessed 22 Sept 2019
487. Lange A, Lo H-W (2018) Inhibiting TRK proteins in clinical cancer therapy. Cancers 10:105. https://doi.org/10.3390/cancers10040105
488. Xu K, Rajagopal S, Klebba I et al (2010) The role of fibroblast Tiam1 in tumor cell invasion and metastasis. Oncogene 29:6533–6542. https://doi.org/10.1038/onc.2010.385
489. Minard ME, Kim L-S, Price JE, Gallick GE (2004) The role of the guanine nucleotide exchange factor Tiam1 in cellular migration, invasion, adhesion and tumor progression. Breast Cancer Res Treat 84:21–32. https://doi.org/10.1023/B:BREA.0000018421.31632.e6
490. Pene-Dumitrescu T, Smithgall TE (2010) Expression of a Src family kinase in chronic myelogenous leukemia cells induces resistance to Imatinib in a kinase-dependent manner. J Biol Chem 285:21446–21457. https://doi.org/10.1074/jbc.M109.090043

491. Saito S, Miyaji-Yamaguchi M, Nagata K (2004) Aberrant intracellular localization of SET-CAN fusion protein, associated with a leukemia, disorganizes nuclear export. Int J Cancer 111:501–507. https://doi.org/10.1002/ijc.20296
492. Jun HJ, Johnson H, Bronson RT et al (2012) The oncogenic lung cancer fusion kinase CD74-ROS activates a novel invasiveness pathway through E-Syt1 phosphorylation. Cancer Res 72:3764–3774. https://doi.org/10.1158/0008-5472.CAN-11-3990
493. Heider TR, Lyman S, Schoonhoven R, Behrns KE (2007) Ski promotes tumor growth through abrogation of transforming growth factor-beta signaling in pancreatic cancer. Ann Surg 246:61–68. https://doi.org/10.1097/SLA.0b013e318070cafa
494. Rice KL, de Thé H (2014) The acute promyelocytic leukaemia success story: curing leukaemia through targeted therapies. J Intern Med 276:61–70. https://doi.org/10.1111/joim.12208
495. Weekes D, Kashima TG, Zandueta C et al (2016) Regulation of osteosarcoma cell lung metastasis by the c-Fos/AP-1 target FGFR1. Oncogene 35:2852–2861. https://doi.org/10.1038/onc.2015.344
496. Muhammad N, Bhattacharya S, Steele R et al (2017) Involvement of c-Fos in the promotion of cancer stem-like cell properties in head and neck squamous cell carcinoma. Clin Cancer Res 23:3120–3128. https://doi.org/10.1158/1078-0432.CCR-16-2811
497. Greuber EK, Smith-Pearson P, Wang J, Pendergast AM (2013) Role of ABL family kinases in cancer: from leukaemia to solid tumours. Nat Rev Cancer 13:559–571. https://doi.org/10.1038/nrc3563
498. Roskoski R (2014) The ErbB/HER family of protein-tyrosine kinases and cancer. Pharmacol Res 79:34–74. https://doi.org/10.1016/j.phrs.2013.11.002
499. Ozaki T, Nakagawara A (2011) Role of p53 in cell death and human cancers. Cancers 3:994–1013. https://doi.org/10.3390/cancers3010994
500. Milella M, Falcone I, Conciatori F et al (2015) PTEN: multiple functions in human malignant tumors. Front Oncol 5. https://doi.org/10.3389/fonc.2015.00024
501. Mehta MS, Vazquez A, Kulkarni DA et al (2011) Polymorphic variants in TSC1 and TSC2 and their association with breast cancer phenotypes. Breast Cancer Res Treat 125:861–868. https://doi.org/10.1007/s10549-010-1062-1
502. McCarthy AJ, Chetty R (2018) Smad4/DPC4. J Clin Pathol 71:661–664. https://doi.org/10.1136/jclinpath-2018-205095
503. Kim WY, Kaelin WG (2004) Role of VHL gene mutation in human cancer. JCO 22:4991–5004. https://doi.org/10.1200/JCO.2004.05.061
504. Johannessen CM, Reczek EE, James MF et al (2005) The NF1 tumor suppressor critically regulates TSC2 and mTOR. PNAS 102:8573–8578. https://doi.org/10.1073/pnas.0503224102
505. Kiuru M, Busam KJ (2017) The NF1 gene in tumor syndromes and melanoma. Lab Investig 97:146–157. https://doi.org/10.1038/labinvest.2016.142
506. Dias N, Stein CA (2002) Antisense oligonucleotides: basic concepts and mechanisms. Mol Cancer Ther 1:347–355
507. Putney SD, Brown J, Cucco C et al (1999) Enhanced anti-tumor effects with microencapsulated c-myc antisense oligonucleotide. Antisense Nucleic Acid Drug Dev 9:451–458. https://doi.org/10.1089/oli.1.1999.9.451
508. Moreno PMD, Pêgo AP (2014) Therapeutic antisense oligonucleotides against cancer: hurdling to the clinic. Front Chem 2:87. https://doi.org/10.3389/fchem.2014.00087
509. Irie A, Kijima H, Ohkawa T et al (1997) Anti-oncogene ribozymes for cancer gene therapy. Adv Pharmacol 40:207–257, Elsevier
510. Scherer L, Rossi JJ (2005) Cancer therapeutic applications of ribozymes and RNAi. In: Curiel DT, Douglas JT (eds) Cancer gene therapy. Humana Press, Totowa, pp 51–63
511. Fei Q, Zhang H, Fu L et al (2008) Experimental cancer gene therapy by multiple anti-survivin hammerhead ribozymes. Acta Biochim Biophys Sin 40:466–477. https://doi.org/10.1111/j.1745-7270.2008.00430.x
512. Cai DW, Mukhopadhyay T, Roth JA (1995) Suppression of lung cancer cell growth by ribozyme-mediated modification of p53 pre-mRNA. Cancer Gene Ther 2:199–205

513. Lee S-W, Jeong J-S (2014) Use of tumor-targeting trans-splicing ribozyme for cancer treatment. In: Lafontaine D, Dubé A (eds) Therapeutic applications of ribozymes and riboswitches. Humana Press, Totowa, pp 83–95
514. Yan R, Qian X, Xin X et al (2002) Experimental study of anti-VEGF hairpin ribozyme gene inhibiting expression of VEGF and proliferation of ovarian cancer cells. Chin J Cancer 21:39–44
515. Mansoori B, Sandoghchian Shotorbani S, Baradaran B (2014) RNA interference and its role in cancer therapy. In: Advanced pharmaceutical bulletin, pp 2251–7308; eISSN. https://doi.org/10.5681/apb.2014.046
516. Agrawal N, Dasaradhi PVN, Mohmmed A et al (2003) RNA interference: biology, mechanism, and applications. MMBR 67:657–685. https://doi.org/10.1128/MMBR.67.4.657-685.2003
517. Cullen BR (2005) RNAi the natural way. Nat Genet 37:1163–1165. https://doi.org/10.1038/ng1105-1163
518. Rao DD, Vorhies JS, Senzer N, Nemunaitis J (2009) siRNA vs. shRNA: similarities and differences. Adv Drug Deliv Rev 61:746–759. https://doi.org/10.1016/j.addr.2009.04.004
519. Beheshti Zavareh R, Sukhai MA, Hurren R et al (2012) Suppression of cancer progression by MGAT1 shRNA knockdown. PLoS One 7:e43721. https://doi.org/10.1371/journal.pone.0043721
520. Nemunaitis J, Barve M, Orr D et al (2014) Summary of bi-shRNA/GM-CSF augmented autologous tumor cell immunotherapy (FANG™) in advanced cancer of the liver. Oncology 87:21–29. https://doi.org/10.1159/000360993
521. Oh J, Barve M, Matthews CM et al (2016) Phase II study of vigil® DNA engineered immunotherapy as maintenance in advanced stage ovarian cancer. Gynecol Oncol 143:504–510. https://doi.org/10.1016/j.ygyno.2016.09.018
522. Ichim TE, Li M, Qian H et al (2004) RNA interference: a potent tool for gene-specific therapeutics. Am J Transplant 4:1227–1236. https://doi.org/10.1111/j.1600-6143.2004.00530.x
523. Chakraborty C, Sharma AR, Sharma G et al (2017) Therapeutic miRNA and siRNA: moving from bench to clinic as next generation medicine. Mol Ther Nucleic Acids 8:132–143. https://doi.org/10.1016/j.omtn.2017.06.005
524. Tabernero J, Shapiro GI, LoRusso PM et al (2013) First-in-humans trial of an RNA interference therapeutic targeting VEGF and KSP in Cancer patients with liver involvement. Cancer Discov 3:406–417. https://doi.org/10.1158/2159-8290.CD-12-0429
525. Zuckerman JE, Gritli I, Tolcher A et al (2014) Correlating animal and human phase Ia/Ib clinical data with CALAA-01, a targeted, polymer-based nanoparticle containing siRNA. PNAS 111:11449–11454. https://doi.org/10.1073/pnas.1411393111
526. Aleku M, Schulz P, Keil O et al (2008) Atu027, a liposomal small interfering RNA formulation targeting protein kinase N3, inhibits Cancer progression. Cancer Res 68:9788–9798. https://doi.org/10.1158/0008-5472.CAN-08-2428
527. Salva E, Ekentok C, Özbas Turan S, Akbuga J (2016) Non-viral siRNA and shRNA delivery systems in cancer therapy. In: Abdurakhmonov IY (ed) RNA interference. InTech, London
528. Morris LGT, Chan TA (2015) Therapeutic targeting of tumor suppressor genes: therapeutic targeting of tumors. Cancer 121:1357–1368. https://doi.org/10.1002/cncr.29140
529. Kazanets A, Shorstova T, Hilmi K et al (2016) Epigenetic silencing of tumor suppressor genes: paradigms, puzzles, and potential. Biochim Biophys Acta (BBA) – Rev Cancer 1865:275–288. https://doi.org/10.1016/j.bbcan.2016.04.001
530. Liu Y, Hu X, Han C et al (2015) Targeting tumor suppressor genes for cancer therapy. BioEssays 37:1277–1286. https://doi.org/10.1002/bies.201500093
531. Ries S, Korn WM (2002) ONYX-015: mechanisms of action and clinical potential of a replication-selective adenovirus. Br J Cancer 86:5–11. https://doi.org/10.1038/sj.bjc.6600006

532. Zhang WW, Li L, Li D et al (2018) The first approved gene therapy product for cancer Ad-p53 (Gendicine): 12 years in the clinic. Hum Gene Ther 29:160–179. https://doi.org/10.1089/hum.2017.218
533. Wade M, Wahl GM (2009) Targeting Mdm2 and Mdmx in cancer therapy: better living through medicinal chemistry? Mol Cancer Res 7:1–11. https://doi.org/10.1158/1541-7786.MCR-08-0423
534. Nag S, Zhang X, Srivenugopal KS et al (2014) Targeting MDM2-p53 interaction for Cancer therapy: are we there yet? Curr Med Chem 21:553–574
535. Burgess A, Chia KM, Haupt S et al (2016) Clinical overview of MDM2/X-targeted therapies. Front Oncol 6:–7. https://doi.org/10.3389/fonc.2016.00007
536. Zak K, Pecak A, Rys B et al (2013) Mdm2 and MdmX inhibitors for the treatment of cancer: a patent review (2011-present). Expert Opin Ther Pat 23:425–448. https://doi.org/10.1517/13543776.2013.765405
537. Brown CJ, Cheok CF, Verma CS, Lane DP (2011) Reactivation of p53: from peptides to small molecules. Trends Pharmacol Sci 32:53–62. https://doi.org/10.1016/j.tips.2010.11.004
538. Zandi R, Selivanova G, Christensen CL et al (2011) PRIMA-1Met/APR-246 induces apoptosis and tumor growth delay in small cell lung cancer expressing mutant p53. Clin Cancer Res 17:2830–2841. https://doi.org/10.1158/1078-0432.CCR-10-3168
539. Zhao R, Choi BY, Lee M-H et al (2016) Implications of genetic and epigenetic alterations of CDKN2A (p16 INK4a) in Cancer. EBioMedicine 8:30–39. https://doi.org/10.1016/j.ebiom.2016.04.017
540. Tanemura A, Terando AM, Sim M-S et al (2009) CpG Island methylator phenotype predicts progression of malignant melanoma. Clin Cancer Res 15:1801–1807. https://doi.org/10.1158/1078-0432.CCR-08-1361
541. Pechalrieu D, Etievant C, Arimondo PB (2017) DNA methyltransferase inhibitors in cancer: from pharmacology to translational studies. Biochem Pharmacol 129:1–13. https://doi.org/10.1016/j.bcp.2016.12.004
542. Eckschlager T, Plch J, Stiborova M, Hrabeta J (2017) Histone deacetylase inhibitors as anticancer drugs. Int J Mol Sci 18. https://doi.org/10.3390/ijms18071414
543. Hatch SB, Yapp C, Montenegro RC et al (2017) Assessing histone demethylase inhibitors in cells: lessons learned. Epigenetics Chromatin 10:9. https://doi.org/10.1186/s13072-017-0116-6
544. Pérez-Salvia M, Esteller M (2016) Bromodomain inhibitors and cancer therapy: from structures to applications. Epigenetics 12:323–339. https://doi.org/10.1080/15592294.2016.1265710
545. Sibbald B (2001) Death but one unintended consequence of gene-therapy trial. CMAJ 164:1612
546. Wirth T, Hedman M, Mäkinen K et al (2006) Safety profile of plasmid/liposomes and virus vectors in clinical gene therapy. Curr Drug Saf 1:253–257
547. National Health Service (2018) Complementary and alternative medicine. https://www.nhs.uk/conditions/complementary-and-alternative-medicine/. Accessed 05 May 2019
548. National Cancer Institute (2015) Complementary and alternative medicine. https://www.cancer.gov/about-cancer/treatment/cam. Accessed 05 May 2019
549. Ernst E, Cohen MH, Stone J (2004) Ethical problems arising in evidence based complementary and alternative medicine. J Med Ethics 30(2):156–159. https://doi.org/10.1136/jme.2003.007021
550. Gureje O, Nortje G, Makanjuola V, Oladeji BD, Seedat S, Jenkins R (2015) The role of global traditional and complementary systems of medicine in the treatment of mental health disorders. Lancet Psychiatry 2(2):168–177. https://doi.org/10.1016/S2215-0366(15)00013-9
551. National Center for Complementary and Integrative Health (2017) Introduction. https://nccih.nih.gov/about/strategic-plans/2016/introduction. Accessed 05 May 2019

552. National Center for Complementary and Alternative Medicine (2009) What is CAM? https://web.archive.org/web/20090505211246/http://nccam.nih.gov/health/whatiscam/overview.htm. Accessed 05 May 2019
553. Kim YJ (2017) The current studies of education for a traditional and complementary medicine in Malaysia. J Evid Based Complement Altern Med 22(4):531–537. https://doi.org/10.1177/2156587217726882
554. McQuade JL, Meng Z, Chen Z, Wei Q, Zhang Y, Bei W et al (2012) Utilization of and attitudes towards traditional Chinese medicine therapies in a Chinese cancer hospital: a survey of patients and physicians. Evid Based Complement Alternat Med 2012:11
555. Berman BM (2001) Complementary medicine and medical education: teaching complementary medicine offers a way of making teaching more holistic. Br Med J 322(7279):121–122. https://doi.org/10.1136/bmj.322.7279.121
556. Robinson N (2006) Integrated traditional Chinese medicine. Complement Ther Clin Pract 12 (2):132–140
557. Chopra A, Doiphode VV (2002) Ayurvedic medicine. Core concept, therapeutic principles, and current relevance. Med Clin North Am 86(1):75–89
558. National Center for Complementary and Integrative Health (2018) Complementary, alternative, or integrative health: what's in a name? https://nccih.nih.gov/health/integrative-health?nav=gsa. Accessed 05 May 2019
559. Quan H, Lai D, Johnson D, Verhoef M, Musto R (2008) Complementary and alternative medicine use among Chinese and white Canadians. Can Fam Physician 54(11):1563–1569
560. WebMD (2018) Whole medical systems: an overview. https://www.webmd.com/balance/guide/understanding-alternative-medicine#1. Accessed 05 May 2019
561. Baars EW, Hamre HJ (2017) Whole medical systems versus the system of conventional biomedicine: a critical, narrative review of similarities, differences, and factors that promote the integration process. Evid Based Complement Alternat Med 2017:4904930. https://doi.org/10.1155/2017/4904930
562. The Guardian (2012) Integrating the methods of traditional Chinese medicine in modern healthcare. https://www.theguardian.com/world/2012/jul/10/chinese-medicine-modern-science-cooperation. Accessed 05 May 2019
563. Aichun G (1999) Huangdi Neijing Suwen Jiao Zhu Yu Yi (Yellow Emperor's Inner Classic: Plain Questions – Critically Compared, Annotated and Translated). Tianjin Kexue Jishu Chubanshe. Tianjin Science and Technology Press, Tianjin
564. Novella S (2012) What is traditional Chinese medicine? https://sciencebasedmedicine.org/what-is-traditional-chinese-medicine/. Accessed 05 May 2019
565. Liu J, Wang S, Zhang Y et al (2015) Traditional Chinese medicine and cancer: history, present situation, and development. Thorac Cancer 6(5):561–569. https://doi.org/10.1111/1759-7714.12270
566. Subhuti D Kampo medicine: the practice of Chinese Herbal Medicine in Japan. Institute for Traditional Medicine. http://www.itmonline.org/arts/kampo.htm. Accessed 05 May 2019
567. Matsuomoto M, Inoue K, Kajii E (1999) Integrating traditional medicine in Japan: the case of Kampo medicines. Complement Ther Med 4(7):254–255
568. Motoo Y, Seki T, Tsutani K (2011) Traditional Japanese medicine, Kampo: its history and current status. Chin J Integr Med 17(2):85–87. https://doi.org/10.1007/s11655-011-0653-y
569. Yamakawa J, Motoo Y, Moriya J, Ogawa M, Uenishi H, Akazawa S et al (2013) Role of Kampo medicine in integrative cancer therapy. Evid Based Complement Alternat Med 2013:570848. https://doi.org/10.1155/2013/570848
570. National Center for Complementary and Integrative Health (2019) Ayurvedic medicine: in depth. https://nccih.nih.gov/health/ayurveda/introduction.htm. Accessed 05 May 2019
571. Shah S (2019) Ayurveda: the conventional Indian Medicine System and its Global practice. Int J Innov Sci Technol 4(1):13–33. https://doi.org/10.22270/ijist.v4i1.36

572. Pandey MM, Rastogi S, Rawat AKS (2013) Indian traditional Ayurvedic system of medicine and nutritional supplementation. Evid Based Complement Alternat Med 2013:376327. https://doi.org/10.1155/2013/376327
573. Microsoft® Encarta® Online Encyclopedia (2009) Ayurveda. https://web.archive.org/web/20091028105549/http://encarta.msn.com/encyclopedia_761596196/Ayurveda.html. Accessed 05 May 2019
574. Jain R, Kosta S, Tiwari A (2010) Ayurveda and cancer. Pharm Res 2(6):393–394. https://doi.org/10.4103/0974-8490.75463
575. Rahman SZ (2001) Unani medicine in India: its origin and fundamental concepts. In: History of science, philosophy and culture in Indian civilization, vol 4 part 2. Centre for Studies in Civilizations, New Delhi, pp 298–325
576. Heyadri M, Hashempur MH, Ayati MH, Quintern D, Nimrouzi M, Mosavat SH (2015) The use of Chinese herbal drugs in Islamic medicine. J Integr Med 13(6):363–367
577. Lone AH, Ahmad T, Anwar M, Habib S, Sofi G, Imam H (2011) Leech therapy- a holistic approach of treatment in Unani (Greeko-Arab) medicine. Anc Sci Life 31(1):31
578. Qamar U, Aman U, Khalid MS, Rais UR (2015) Unani medicine for cancer care: an evidence-based review. IJAHM 5(3):1811–1825. https://doi.org/10.31142/ijahm
579. Sig AK, Guney M, Guclu AU, Ozmen E (2017) Medicinal leech therapy—an overall perspective. Integr Med Res 6(4):337–343. https://doi.org/10.1016/j.imr.2017.08.001
580. National Psoriasis Foundation, Whole Medical Systems. https://www.psoriasis.org/about-psoriasis/treatments/alternative/whole-systems. Accessed 05 May 2019
581. Smith K (2012) Homeopathy is unscientific and unethical. Bioethics 26(9):508–512. https://doi.org/10.1111/j.1467-8519.2011.01956.x
582. Homeopathy Research Institute, Homeopathy use around the world. https://www.hri-research.org/resources/essentialevidence/use-of-homeopathy-across-the-world/. Accessed 05 May 2019
583. Ernst E (2007) Homeopathy for cancer? Curr Oncol 14(4):128
584. Atwood KC (2003) Naturopathy: a critical appraisal. MedGenMed 5(4):39
585. Wahlberg A (2007) A quackery with a difference—new medical pluralism and the problem of 'dangerous practitioners' in the United Kingdom. Soc Sci Med 65(11):2307–2316
586. Ernst E (2001) Rise in popularity of complementary and alternative medicine: reasons and consequences for vaccination. Vaccine 20:S90–S93
587. Smith K. Naturopathic cancer treatment: integrative adjunctive cancer care. https://www.drsmithnd.com/naturopathic-cancer-treatment. Accessed 05 May 2019
588. Hermes BM (2016) Naturopathic cancer care – is it safe, and does it work? https://www.qualitycancertreatment.com/blog/naturopathictreatment. Accessed 05 May 2019
589. Kienle GS, Albonico HU, Baars E, Hamre HJ, Zimmermann P, Kiene H (2013) Anthroposophic medicine: an integrative medical system originating in Europe. Glob Adv Health Med 2(6):20–31. https://doi.org/10.7453/gahmj.2012.087
590. Ernst E (2006) Mistletoe as a treatment for cancer. Br Med J 333:1282. https://doi.org/10.1136/bmj.39055.493958.80
591. Olson JS (2002) Bathsheba's breast: women, cancer, and history. Press, JHU
592. National Cancer Institute (2019) Questions and answers about mistletoe. https://www.cancer.gov/about-cancer/treatment/cam/patient/mistletoe-pdq#section/_2. Accessed 05 May 2019
593. NHS Specialist Pharmacy Service (2015) What is the evidence for subcutaneous mistletoe extract in the treatment of cancer? https://www.sps.nhs.uk/articles/what-is-the-evidence-for-subcutaneous-mistletoe-extract-in-the-treatment-of-cancer/. Accessed 05 May 2019
594. Wheeler C (2010) What is mind-body medicine? https://www.psychologytoday.com/us/blog/head-toe-happiness/201006/what-is-mind-body-medicine. Accessed 05 May 2019
595. National Center for Complementary and Alternative Medicine (2009) Mind-body medicine: an overview available. https://web.archive.org/web/20090506053001/http://nccam.nih.gov/health/whatiscam/mind-body/mindbody.htm. Accessed 05 May 2019

596. Ernst E, Pittler MH, Wider B, Boddy K (2007) Mind-body therapies: are the trial data getting stronger? Altern Ther Health Med 13(5):62–64
597. Rutledge JC, Hyson DA, Garduno D, Cort DA, Paumer L, Kappagoda CT (1999) Lifestyle modification program in management of patients with coronary artery disease: the clinical experience in a tertiary care hospital. J Cardpulm Rehabil 19(4):226–234
598. Mundy EA, DuHamel KN, Montgomery GH (2003) The efficacy of behavioral interventions for cancer treatment-related side effects. Semin Clin Neuropsychiatry 8(4):253–275
599. Chaoul A, Milbury K, Sood AK, Prinsloo S, Cohen L (2014) Mind-body practices in cancer care. Curr Oncol Rep 16(12):417. https://doi.org/10.1007/s11912-014-0417-x
600. Niggemann B, Grüber C (2003) Side-effects of complementary and alternative medicine. Allergy 58(8):707–716
601. Cassileth BR, Deng G (2004) Complementary and alternative therapies for cancer. Oncologist 9(1):80–89. https://doi.org/10.1634/theoncologist.9-1-80
602. Hughes D (2010) Alternative remedies 'dangerous' for kids says report. https://www.bbc.com/news/health-12060507. Accessed 05 May 2019
603. Ajazuddin, Saraf S (2012) Legal regulations of complementary and alternative medicines in different countries. Pharmacogn Rev 6(12):154. https://doi.org/10.4103/0973-7847.99950
604. National Center for Complementary and Integrative Health (2019) NCCIH 2016 strategic plan. https://nccih.nih.gov/about/strategic-plans/2016. Accessed 05 May 2019

Chapter 7
Carbon-Based Tumour-targeted Systems

Smriti Sri, Shweta Panwar, and Pratima R. Solanki

Abstract Cancer is the second leading cause of death globally. Lung cancer and breast cancer are the most common types of cancer in men and women, respectively. Many strategies are used to target cancer such as surgery, chemotherapy, radiotherapy and immunotherapy. Immunotherapy is the latest addition to these strategies, but they also suffer from hypersensitive and allergic reactions. Lately nanoparticles have gained enormous interest in the nanomedicine due to their unique properties at the nanoscale level. These can be modified easily by biomolecules. Carbon is the extraordinary and most sought of material in the nano world. The beauty of the carbon to form covalent linkage with different orbital hybridization forms various nanoallotropes of carbon having different hybridization between C-C. Carbon is the basis of life on earth. Thus, these carbon nanomaterials are biocompatible. Earlier only diamond and graphite were the two known allotropes of carbon. But the discovery of fullerene (C_{60}) in 1985 has opened a new avenue for the discovery of carbon nanoallotropes. Soon this was followed by the discovery of carbon nanotubes (CNTs) by Iijima.

7.1 Introduction

Cancer is one of the deadliest diseases known to mankind today. It is the second leading cause of mortality. The earliest incident of cancer comes from the study by Egyptian papyrus on treatment and removal of breast cancer in 1600 BC. During the Second World War, some of the drugs known to treat cancer were discovered for the first time. However, from that time we have come so far that some of the cancers are still treatable if combined with early diagnosis, better surgery, radiotherapy, and chemotherapy. Also, the intense researches on cancer have made a better understanding towards the cellular function and immune response towards this deadly disease. Cancerous cells show different biomarkers which are either elevated or

---

S. Sri · S. Panwar · P. R. Solanki (✉)
Special Centre for Nanosciences, Jawaharlal Nehru University, New Delhi, India
e-mail: partima@mail.jnu.ac.in

R. Huang, Y. Wang (eds.), *New Nanomaterials and Techniques for Tumor-targeted Systems*, https://doi.org/10.1007/978-981-15-5159-8_7

Table 7.1 Incidents of different types of cancer in 2018 is given below

Cancer type	Prevalence (millions)
Lung	2.09
Breast	2.09
Colorectal	1.80
Prostate	1.28
Skin cancer (non-melanoma)	1.04
Stomach	1.03

downregulated in comparison to normal cells. Tumour markers are chemical substance which is released by cancer cells or normal cells in response to presence of tumour. Tumour markers are present in body fluids such as urine, blood, saliva, etc. Although these markers are proteins, recently the potential of DNA or miRNA has been explored in detail. Apart from diagnosis, the treatment perspective has also been investigated by several researchers. The nanoparticles are used as drug delivery system (DDS) to target different tumours.

Cancer is responsible for an estimated 9.6 million deaths globally in 2018. About 1 in 6 deaths is because of cancer. Low-income and middle-income countries are the worst affected with approximately 70% of global death reported in these countries. Use of tobacco is the leading cause of death accounting for a 22% cancer death. Also, limited access to early diagnosis facility worsens the scenario. Economic impact of cancer is also bothering. The total economic cost of cancer in 2010 was approximately US $1.16 trillion.[Web1] Table 7.1 shows the data of cancer person suffered in 2018 globally.

Lung cancer is one of the most common cancers in men and second most common in women. In women, breast cancer supersedes other cancers [1, 2]. The current strategies for cancer treatment include surgery, chemotherapy and radiotherapy. Surgery is the oldest known treatment to cure cancer. It may be curative, reconstructive or palliative. However, surgery in advanced stages is futile. Another treatment of cancer is radiation. In this mode of cancer treatment, high energy rays are used to destroy cancer cells and inhibit the proliferation of cancerous cells. Radiation therapy may be done internally or externally. In the external mode, high energy rays emit and pass the skin thereby targeting the underlying tissues. Internal radiation therapy also known as brachytherapy involves placing small amount of radioactive material inside the tissue. Radiation therapy suffers from selectiveness, which can cause serious damage in structure and function of surrounding tissues [3]. Chemotherapy is the most common treatment which involves use of drugs. Chemotherapeutic strategies involve the administration of high dose of drugs that affect fast-growing cells. But not only cancer but also other normal cells in the body are fast growing like skin cells, hair follicle cells, and bone marrow cells. So, this therapeutic strategy has side effects such as hair loss and bone marrow depression including anaemia, nausea, vomiting, etc.

A new therapy for cancer that evolved lately is immunotherapy. Immunotherapy aims to improve immune cells so that it can selectively kill cancer cells. Many forms of immunotherapy have been developed such as monoclonal antibodies, cytokines and vaccines. But unfortunately, immunotherapy suffers from hypersensitivity and allergic reactions. The search for an effective method to target cancer has zeroed on the use of nanoparticles for developing targeted drug delivery system (DDs) as well as diagnostic tool for early detection of cancer cells.

Researchers have come a long way in targeting cancer through the use of nanotechnology. Nanotechnology gives us the liberty to target either cell, intracellular compartments including nucleus, DNA or RNA depending on the need because of the small size of nanoparticles. Nanoparticles have high surface-to-volume ratio which gives more room for surface modification. Nanoparticles can be modified either with biotargeting ligands like monoclonal antibodies or small molecules. Another great advantage of using nanoparticles is that they can be fabricated for multiple diagnoses and can even work as therapeutic agents. In short they can function as theranostic platform for tumour targeting [4].

One of the most interesting nanomaterials is that of carbon. Carbon is the extraordinary and most sought of material in the nano world. The beauty of the carbon to form covalent linkage with different orbital hybridization forms various nanoallotropes of carbon having different hybridization between C-C. Different form of carbon has been reported in the literature (Fig. 7.1). Carbon is the basis of life on earth. Thus, these carbon nanomaterials are biocompatible, which is the best thing about these nanomaterials. For millennia there were only two known allotropes of carbon, graphite and diamond. In the recent decades, a series of new carbon nanostructures have been discovered. Fullerenes were the first one to be discovered in 1985. These are the smallest known stable carbon nanostructures which lie between the boundary of molecules and materials. The most common form of fullerene is the C_{60} molecule. Each C_{60} molecule is made up of 60 sp^2 carbon atoms arranged in a series of hexagons and pentagons to form a spherical icosahedral structure. However, other variants have been discovered C_{70}, C_{20}. The hollow cavity of fullerene can be used to load certain drugs or metal ions as well function as MRI contrast agents. C_{60} molecules bind strongly to DNA or drugs working as DDS. They can also be used as theranostic platform to target cancer. Next milestone in development of carbon nanomaterials was done by Iijima by discovery of CNTs. CNTs are formed by rolling of graphene sheets in which each carbon atom is covalently attached to other carbon by sp^2 hybridization. CNTs have unique mechanical, electrical and optical properties which allow for biological imaging and photon-assisted therapeutics. Motivated by the discovery of CNTs and its biomedical application, graphene was explored for its potential to target cancer by Hongjie Dai group at Stanford University (CA, USA) in 2008 5 years after its discovery. Graphene is an atomically thin film that consists of sp^2-hybridized hexagonally arranged carbon atoms in 2D. Graphene because of its "zero band gap semiconductor" has unique electrical properties analogous to that of metals [6]. Because of their electromagnetic and structural properties, little wonder is that they are being used as semiconductors, batteries and biosensing platform. Serendipitous discovery of carbon dots (CDs) in 2006 took the scientific community by revolution. CDs allow easy synthesis process and cheap precursors and are highly biocompatible. They are brilliantly fluorescent molecule. Further, they can be easily surface functionalized with biomolecules or drugs because of presence of ample

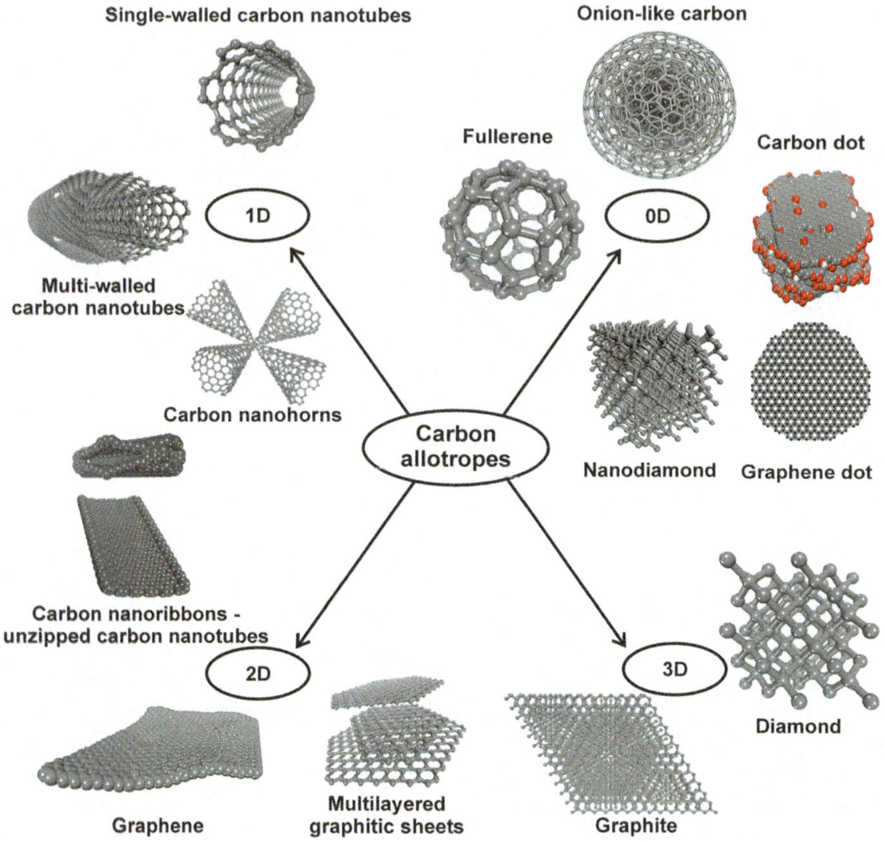

Fig. 7.1 Showing different nanostructures prepared from carbon. (Reprinted with permission from Ref. [5]. Copyright with 2015 American Chemical Society)

carboxyl moieties and amine groups at their surface. They are extensively used for bioimaging (in vitro, in vivo) and drug delivery.

Figure 7.2a shows the total number of publications of carbon nanoallotropes till date (date of search 26 April 2019, search engine web of science). The maximum number is almost half of the total research done on CNTs comprising 51.34%. It is followed by GO having 18.69%, fullerene 16.93%, CDs 9.14% and graphene quantum dots (GQDs) 3.24%. The least research is done on nanodiamonds accounting for only 0.65%. The similar trend is observed in Fig. 7.2b where the data from the last 5 years, i.e. 2015–2019, is plotted vs various carbon nanoallotropes. The higher research focus on CNTs may be explained by their various applications including diagnostics and therapy.

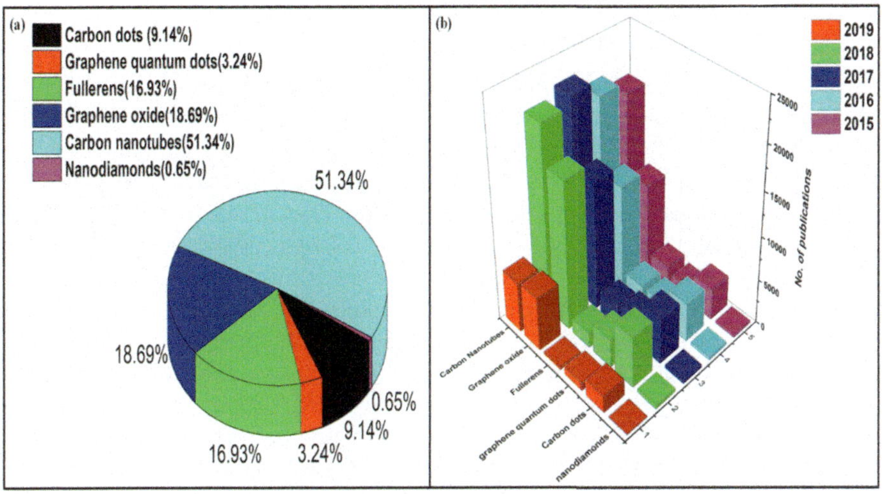

Fig. 7.2 (**a**) Showing the number of publications till date (search date 26 April 2019) on different carbon nanoallotropes. (**b**) 5-year publication trends of these carbon nanoallotropes

7.2 Scope of This Chapter

In this chapter, we have covered the recent progress in tumour targeting using different carbon nanoallotropes. Since, these nanomaterials are well explored, and there is ample work done. So, to cover each work is beyond the scope of this chapter. However, the recent trends and works for cancer targeting with major focus on nanoallotrope characteristics are the main focus of this chapter. Carbon nanoallotropes are divided into different types based on their size and thereby are of different dimensions. At the end, efforts have been tried to provide some insight and challenges for carbon-based tumour-targeting system based on our own understanding.

Carbon has the ability to transform into 0D, 1D, 2D and 3D nanomaterials with every material having different physiochemical and optical properties. The beauty of this black material is that each allotrope (diamond and graphite are both carbon allotropes but have different properties) or nanoallotrope of carbon differs from the other in their electronic or optoelectronics properties. The unusual ability of carbon to form bond with itself or with non-metallic elements makes them an extraordinary material to form a wide range of nanostructures. CDs exhibit amusing optical characteristics, inexpensive precursors ranging from grass to hair fibre [7], easy and fast synthesis approaches. All these advances make CDs a next-generation nanomaterial for biosensing and bioimaging for cancer targeting.

7.3 Classification of Carbon Nanoallotropes

Carbon nanoallotropes can be classified into two general groups on the basis of predominant type of covalent bonding between C atoms. The first group involves the graphenic nanostructures, which is mainly made up of sp^2 carbon atoms that are closely packed in a hexagonal honeycomb crystal lattice. However, they may also contain sp^3-hybridized carbon at the defect site. This group includes graphene, CNTs, CD and GQDs. Graphenic nanostructures are basically made up of graphene in which the carbon is bonded to three identical covalent bonds using sp^2 orbitals, generating a 2D lattice of densely packed hexagons. The basic unit of this classification is graphene—a 2D, single, zero band gap semiconductor having sp^2-hybridized carbon arranged in a hexagonal lattice. Thus, graphene functions as "building blocks" of the other graphenic/graphitic nanoallotropes. Graphene can be wrapped up to form 0D fullerene, rolled to give CNTs or piled above each other to form multilayered 2D carbon nanosheets. While the defects in the graphene sheets gives rise to GQDs.

7.3.1 Carbon Dots

CDs are very small, zero-dimensional nanoparticles. They were first discovered accidentally during the purification of single-walled carbon nanotubes (SWCNTs) [8]. The scientific designation as CDs was given by Sun et al. in 2006 [9]. They have received an immense interest since its discovery owing to its spectacular optical properties and high biocompatibility. Another exciting feature of CDs is that they can be synthesized from any green precursor [10]. CDs can be prepared by either top-down or bottom-up approaches. The top-down synthesis methods use bulk carbon precursors or particles that are of higher dimension to that of CD such as graphite or CNTs [11]. The top-down methods are laser ablation, arc discharge and chemical oxidation. The immediate precursors from top-down methods are not fluorescent and need surface passivation. However, the bottom-up approaches produce more fluorescent CDs immediately after synthesis. Also, there is a lot of room for the modification of CDs in situ using several bottom-up synthesis approaches such as hydrothermal synthesis and microwave pyrolysis. The most commonly used synthesis method for CD synthesis is bottom up. A wide range of precursor including natural materials can be used for CD synthesis using hydrothermal method. Microwave pyrolysis is the fastest method of CD synthesis using only a maximum of 5 min for the synthesis. Until now the highest quantum yield CDs have been synthesized mostly using citric acid as the precursor material and different surface passivating agent such as ethylenediamine and cysteine [12, 13].

The optical property of CDs is the major exciting thing about these tiny nanodots. They show size- and excitation-dependent emission, photoinduced electron transfer, upconversion luminescence, chemiluminescence and electrochemiluminescence

[14, 15]. CDs are an attractive fluorophore because of its broad absorption and narrow emission spectrum. Also, most of the CDs show excitation-dependent emission spectrum. CD emission and excitation spectrum is tunable because of quantum confinement effect by varying its size. On decreasing the size of the CDs, the exciton Bohr radius (the characteristic difference between the electron hole pair in bulk semiconductor is known as Bohr radius) decreases. Therefore, the band gap increases [16]. The small size of the CDs creates surface defects thereby causing the photoluminescence phenomenon in CDs which is their hallmark characteristics. CDs in a way have revolutionized the perception of carbon being a black material. The photoluminescence of CDs is measured in terms of quantum yield (QYs). Using various precursors and different synthetic methods the QYs are steadily increasing [16].

Graphene quantum dots (GQDs) is mostly confused with CDs. But these two are different nanodots synthesized from similar precursors. Actually, the actual difference between these two cousins is very narrow. Talking of shape, GQDs are not spherical particles but rather are small discs made by 1–3 layers of graphene stacked on each other. While CDs are mostly spherical, many variants can be found, e.g. carbon nitride structure, g-C_3N_4 CDs. GQDs consist of layers of sp^2-hybridized carbon stacked on top of each other, while in CDs the carbon core contains a mixture of sp^2 and sp^3 hybridization. As a result of this, the GQDs are crystalline in nature, while CDs can be either amorphous or crystalline. The core of CDs is made of aggregation of small molecules which are held by weak forces like π-π bond, H bonds and van der Waals forces in a quasi-spherical surface [11]. Although, the optical properties and synthesis techniques of CD and GQDs remains similar. That's why these two are considered as CDs.

CDs exhibit either upconversion or downconversion photoluminescence. The emission of UV excited CDs covers the visible spectrum ranging from blue to red, resulted in downconversion. Upconversion is because of multiphoton process, in which the absorption of two or more photons (usually in the infrared region) leads to the emitting of single photons with higher energy. The possible photoluminescence mechanism was explained by Goryacheva et al. [17].

7.3.1.1 Surface Modification of CDs

The bright emission of CDs combined with their electron-donating and electron-accepting capabilities can function as bioimaging agents, optoelectronic devices and biosensors. These calibres of CDs as cancer-targeting nanomaterials have been extensively studied and summarized in the following paragraphs.

7.3.1.2 CDs as Bioimaging Agents

CDs as bioimaging agents have been extensively explored since its serendipitous discovery. Seeing that CDs are very small nanoparticles usually between 1 and 10 nm, they can easily enter the cells after incubation. Internalization time for these can be as less as 5 min [18]. Also, the best part about CDs is they do not accumulate inside the living system and gets flushed out of the system within a span of 48–72 h [12]. Many of the cancerous cell lines such as cervical cancer (HeLa), breast cancer (MCF-7 and MDAMB-231), liver cancer (HepG245 and Huh-7 receptor negative cells NIH-3T3), lung carcinoma (A549 and neural cell line PC12 and RSC96) and oral cancer cells (FaDu and Cal-27) were explored as a potential target for bioimaging studies [12, 19, 20] (Fig. 7.3). The potential of CDs as a bioimaging agent was explored by Yang's group in live mice [21]. Till now a number of studies have been done on mice. Wang et al. fabricated 1-methyl-2-pyrrolidinone (NMP)-

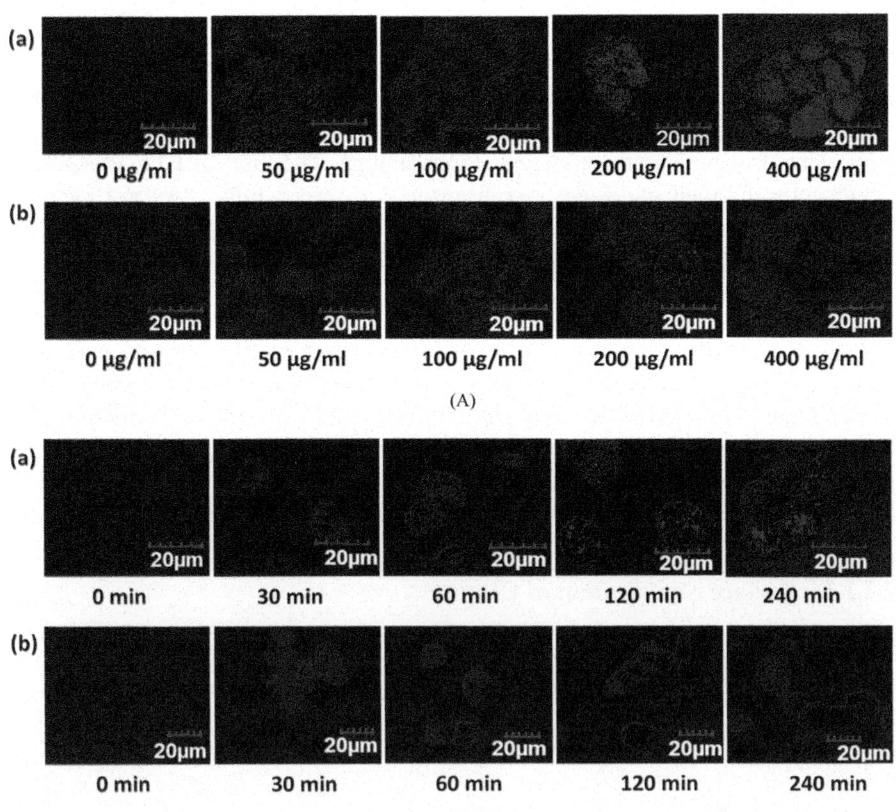

Fig. 7.3 (**A**) Concentration-dependent cellular uptake of CDs in (a) Cal 27 cells and (b) FaDu cells. (**B**) Time-dependent cellular uptake of CDs in (a) Cal 27 and (b) FaDu cells. (Reprinted with permission from Ref. [12]. Copyright 2018 American Chemical Society)

derived polymer-coated nitrogen-doped CDs. They used the excellently fluorescent CDs for glioma imaging in vitro. These also mediate glioma fluorescence imaging in vivo with good contrast via passive targeting [22].

7.3.1.3 Targeting Cancer Cells

The CDs can be easily surface functionalized with different biomolecules, thanks to the abundance of COOH and NH_2 groups as reported in CDs prepared with different compositions. The rich carboxylic group at the surface makes a lot of room for modification. CDs were functionalized with biomolecules to specifically target cancer cells. Since then numerous studies have been done. Some used bare CDs for study, while many groups functionalized CDs with cancer-targeting molecules to specifically target cancer cells. The molecules that are overexpressed in cancer cells such as folic acid, hyaluronic acid, aptamers etc. were used for functionalization. Zhao et al. modified CD surface by conjugating it with folic acid (FA-CDs) for targeted imaging of cancer cells HepG-2. The internalization of FA-CDs occurs via receptor-mediated endocytosis. The selectivity of FA-CDs to target only FA-positive cancer cells was confirmed using FA-negative cells PC-12. Also, FA-CDs were able to recognize FA-CDs could accurately recognize positive folate receptors (FR) (FR++/FR+) cancer cells in different cell mixtures ofMCF-7/HepG-2 cells and HepG-2/PC-12 cells and could distinctly indicate the expression level of FR on the membranes of the cancer cells [23]. Similar work was reported by Zhang et al. [24], Qian et al. have also done a similar work with CDs derived from aconitic acid, and these were further modified with FA for targeted imaging of cancer cells Hela, SMMC-7721, and A549 cells as models that expressed different levels of FRs on the cell surface. Also, they have successfully detected FA in pharmaceutical samples with the detection limit of 40 nM [25]. Chiu et al. synthesized gadolinium-functionalized N,S containing CDs (GdNS@CDs). These were then functionalized with FA and then loaded with anticancer drug doxorubicin (DOX), FA-GdNS@CQDs-DOX. The cancer cell were targeted and killed leaving no harm to normal cells [26]. Cancer cells overexpress CD44 cell marker. Nahain et al. anchored GQD with hyaluronic acid (HA) GQD-HA as a targeting agent. The GQD-HA efficacy to target cancer was studied in vitro in A549 cells and in tumour-bearing balb/c female mice [27].

The CDs were also modified using various aptamers to specifically target cancer. Motaghi et al. conjugated CDs with AS1411 aptamer for the spectrofluorimetric detection of cancer cells [28]. AS1411 is a nucleolin targeting aptamer. The latter protein is overexpressed in cancer cells thereby providing a suitable marker to detect cancer cells. The aptamer AS1411 initially binds to nucleolin thereby being internalized in the cell [29]. The high fluorescence intensity inside the cells marks higher the nucleolin expression thereby specifically targeting cancer cells. Li et al., recently specifically targeted desmin protein. Desmin is present in higher concentration in the colorectal cancer patients promising it to be a new colon and rectal cancer serum marker [30]. Li et al. synthesized CDs from MWCNTs (multiwalled carbon nanotubes) and were functionalized with C(O)Cl groups by $SOCl_2$ treatment.

The obtained CLCDs were conjugated onto anti-desmin, and desmin concentration was optically sensed [31]. Similarly, an on-off-on sensor was developed by Gao et al. The sensor was used to sense Fe^{3+} with a detection limit of 16 nM. Further, the CD-based sensor was able to differentiate between cancer cells from normal cells as in cancerous cells the environment is more reductive because of the presence of high level of glutathione (GSH). This reduces the Fe^{3+} therefore restoring CD fluorescence and providing a simplistic method for cancer diagnosis in vitro and in vivo [32]. CD has been also used to detect single nucleotide polymorphism (SNP), which plays a critical role in cancer. CD was used as an off sensor for detecting SNP in single-stranded DNA (ssDNA). The ssDNA was labelled with dye (6-carboxyfluorescein) (FAM). The fluorescence of the ssDNA/dye sensor system is quenched on adsorption on the CD surface [33]. Wua et al. have utilized for the detection of peroxynitrite ($ONOO^-$), reactive oxygen species produced in mitochondria [34].

7.3.1.4 CDs in Radiotherapy and Phototherapy

Radiotherapy plays an important role in curing cancer. But the success rate is limited due to the hypoxia condition of solid tumours. Increasing dose of radiosensitizer does not help as the patient develops enteritis. There is a need of radiosensitizer which should be more biocompatible and will be excreted fast. The potential of CDs as radiotherapy and phototherapeutic agent was explored by several researchers. Since CDs have ample COOH group and NH_2 groups on their surface, they can act both as electron-donating and electron-accepting groups which make them favoured material for radiotherapy [35, 36]. Hsu et al. used CDs to trigger formation of ROS which further induces apoptosis in cancerous cells [37]. Similar work was reported by Havrdova et al. in which CDs functionalized with positive-charged molecules triggered the formation of ROS in mouse fibroblast cells [37]. Ruan et al. explored the capability of GQDs on colorectal cancer for radiotherapy. GQDs in cooperation with ionizing radiations (IR) showed enhanced G2/M phase arrest of cells, inhibition of cell proliferation and enhanced apoptosis of cancerous cells [38] (Fig. 7.4). The GQDs and IR working together caused ample ROS production therefore damaging mitochondria and killing the cancer cells. They have developed a safe, cost-effective and facile nano-sensitizer for improving the tumour radiotherapy results. Kleinauskas et al. explored CDs with silver-coated shell ability as photocatalyser [39]. Qian et al. prepared a multicolour highly crystalline carbon nanocore and a hydrophilic surface. These crystalline carbon nanocores were employed for the photoacoustic and photothermal therapy with tunable fluorescence emission. The in vivo experiments carried out in U87 glioma-bearing mice suggested the CDs accumulated specifically in brain tumours and facilitate dual-modal imaging-guided photothermal therapy without any damage to healthy tissues [40].

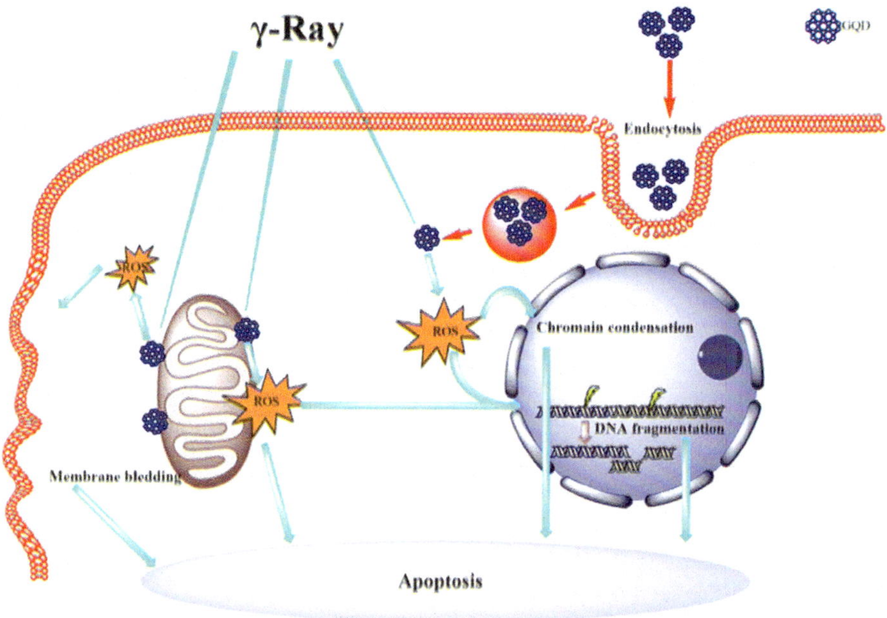

Fig. 7.4 The mechanisms of GQDs synergy with irradiation-induced cell apoptosis. (Reprinted with permission from Ref. [38]. Copyright 2018 American Chemical Society)

7.3.1.5 CDs as Theranostic Vehicle

Conventional tumour-targeting systems such as radiotherapy and chemotherapy suffer from long-lasting side effects. In the past decade, targeted drug delivery has been gaining enormous attention because of small size of drug required, enhanced delivery of drugs that are sparingly soluble in water, combination of two drugs to deliver at different locations and the efficient delivery to target. Many research groups have explored the capability of CD as a DDS because of its small size, high biocompatibility, rapid uptake and the ability to even cross blood-brain barrier due to its small size. CDs have been loaded with drug and used as a therapeutic as well as diagnostic vehicle, i.e. as theranostic platform. Jiao et al. designed a smart nanocarrier with CD entrapping the surface of mesoporous silica nanoparticles via disulphide bonds (MSNs-SS-CD) [41]. Further, Dox was loaded onto the nanocarrier. The drug-loaded nanocarrier showed considerably elevated cellular uptake and thereby more fatal effect on cancer cells. The nanocarrier showed high responsive drug release in pH 7.4 and pH 5.0. The study showed a real-time tracking of the nanocarrier for cancer imaging and killing. A very interesting study carried out by Yao et al. [42] synthesized CD from waste crab shell chitin by one-pot microwave pyrolysis. The CDs were then transition metal to explore their potential as contrast agent Gd@CDs. After that Gd@CDs was conjugated with FA to target specifically Hela cells. Further, anticancer drug DOX was loaded onto FA-Gd@CDs making nanocomposite DOX-FA-Gd@CDs. These nanocomposites showed more efficient in killing cancer cells compared to DOX only. This nanocomposite showed a potential to work as theranostic platform for cancer cells. Likewise, several other works were done on drugs like 7-(3-bromopropoxy)-2-quinolylmethyl chlorambucil (Qucbl) [43], dopamine hydrochloride (DA) for neurological disease [44] and cisplatin [45]. Chowdhuri et al. made a nano-organic framework (NMOF). Further it was encapsulated with CDs and FA to specifically target folate overexpressing cells. High drug loading was achieved by loading DOX on the NMOF-CD-FA. Apoptosis profile and pH-responsive drug release was studied in vitro in Hela cells and L929 cells [46]. Wang et al. developed a glioma-targeting theranostic vehicle using multifunctional polymer-coated carbon nanodots [47]. The CDs were used for active targeting by conjugating a peptide I_6P_7, a fragment of interleukin-6 (IL-6). The theranostic vehicle bind to IL-6 receptor (IL-6 receptors are expressed on brain capillaries, an essential key for blood-brain barrier, as well as tumour cells). Further, Dox was encapsulated onto this nanocarrier. Therefore, this nanovehicle can cross blood-brain barrier and deeply penetrate into the tumour and has a pH responsive glioma-targeted drug delivery combined with inhibition of Il-6-induced cell proliferation.

7.3.1.6 Biosensors

Biosensor is an analytical device which converts the biological signal into electrical signal using transducers. Biosensors are made up of two basic components: one is the molecular recognition system, and the other is physiochemical transducer [48]. The first biosensor developed was that of glucose by American Clark and Lyons way back in 1962. After half a century later, many variants have been tried and commercialized for glucose monitoring [49]. The major advantage of biosensor is the high specificity, low sample volume and rapid detection. Up till now there has been no report on developing a biosensor for cancer diagnosis using CDs.

7.3.1.7 Microfluidic Approach Towards Detection of Cancer

In order to develop a handheld biosensor device that can be used outside the laboratory, a microfluidic system needs to be incorporated [50]. The microfluidic platform enables us to measure electrochemical immune sensing. The microfluidic (lab-on-a chip) is advantageous over fabricated biosensor in terms of low sample volume usually in the range of picomolar to nanomolar, low sample-to-noise ratio, high specificity and sensitivity and the real-time monitoring of patient sample.

7.3.1.8 CD as Microfluidic Platform

The work towards developing a lab-on-a chip microfluidic platform is gaining gross interest for cancer detection because of its ability to detect cancerous biomarkers in a drop of blood. Recently Hu et al. developed microfluidic paper analytical devices (μPADs) for the detection of fluorescent detection of Fe^{3+} and colorimetric ELISA for ferritin [51]. In this work, they have modified the filter paper using CDs and gold nanoparticles (AuNPs). Two different regions were modified on the paper: one was masked with fluorescent CDs (F-CDs), and the next was of non-fluorescent CDs (nF-CDs). The F-CDs were used for the detection of serum ion by quenching. Further, AuNPs were deposited on nF-CDs. This region was used for the separation of serum from whole blood and sensing of serum ferritin. Ferritin is a multifunctional protein which is directly related to cancer as it affects process like cell proliferation, angiogenesis, immunosuppression and iron delivery [52]. High level of ferritin is directly linked to cancer. The whole microfluidic platform for serum iron and ferritin detection provided quick and reliable information regarding the iron metabolism and thereby the health status of the patient.

Recently, several lines of evidence have demonstrated that ferritin is a multi-functional protein with possible roles in many disease including cancer. In the context of cancer, ferritin is detected at higher levels in the sera of many cancer patients, and the higher levels correlate with aggressive disease and poor clinical outcome.

7.3.1.9 Conclusion of CDs

All the fascinating properties make CDs as an upgraded version of their popular cousins such as CNTs, graphene and fullerene [17]. They are rapidly gaining tremendous popularity because of their biocompatibility, abundant functional groups on their surface and excellent optical property. After more than a decade of these nano-light emergent CDs, these have been regarded as the most valuable and intriguing groups of carbon-based nanomaterials. Every year a large number of scientific papers are published on the CDs and its application in imaging, heath, environment and energy. There are many factors contributing to this increasing research, and the foremost is the ease of synthesizing CDs with a good yield. The synthesis routes are quite simple and economic and require less time than any other NPs. The second factor is that the surface modification is extremely easy and can be done with any known biomolecules because of more carboxyl, hydroxyl and amine groups at the surface. Third factor remains the excellent biocompatibility of CDs. Fourth factor remains the high photoluminescence of CDs and their excellent hydrophilic nature. And the last one being, there surface charge can also be tuned by varying the concentration and surface groups. These above reasons account for the gaining application of CDs in medical specially in cancer field. CDs can be used as imaging agents in vitro as well as in vivo. Taking this research to the next level,

CDs have been employed to load drug molecules at their surface. Because of their small size, they can easily be taken inside the cells and can work as cargo to load anticancer drugs. Therefore, these CDs hold a great potential especially in the cancer field.

7.3.2 Fullerenes

Fullerenes were the first discovered carbon nanoallotrope. Sir Kroto, Curl and Smalley were honoured with the Nobel Prize in Chemistry for this groundbreaking discovery in 1996. The discovery of fullerene was pretty much accidental. Scientist were using high power laser to produce hot vapours of carbon which can mimic the conditions in interstellar space. During this they found fullerene. Fullerene in general are closed circular hollow cages similar to the shape of football [53]. The carbon in fullerene is sp^2 hybridized, but since the arrangement of carbons is not planar but instead like pyramid, so the carbon may form pseudo sp^3 bond. Carbon atoms are arranged in 12 pentagons and a definite number of hexagons depending on the total number of carbons. A fullerene with 'n' hexagon will have 20 + 2n carbon atoms. The most common fullerene is C_{60}which is a spherical molecule with external diameter of 0.71 nm. All the rings in fullerene are fused and the double bonds are conjugated. Several similar structures such as C_{70}, C_{76}, C_{82} and C_{84} have been reported till now. But C_{60} is the most explored of them all. Fullerene is generally produced using a low-pressure method such as electric discharge or arc vaporization of graphite in presence of inert gases. However, there is a limitation to this tremendously 0D carbon nanoallotrope, i.e. their insolubility in aqueous medium. They form aggregates in liquid medium thereby restricting their application in nanomedicine. To overcome this barrier, several approaches have been made like (a) fullerene hydrosol preparation, (b) fullerene surface functionalization and (c) introduction of polar groups in fullerene [53].

Fullerene exhibits an exceptionally rare property. They show two opposing biological effects, antioxidant and oxidant, because of their high electron affinity (ca. 2.7–2.8 eV) and curved conjugated π bonds that can be readily attacked by radical species. The absorption of light excites the electron to excited state and an efficient intersystem crossing thereby making triplet state long-lived. For this they are also termed as radical sponges and photosensitizers [54, 55]. The powerful capacity of fullerene as antioxidants has been explored by the biologist. They have even regarded as a better antioxidant than vitamin E. Fullerenes have been used in drug delivery as well as tumour therapy. Apart from cancer other diseases targeted using fullerene are Alzheimer and Parkinson's disease and amyotrophic lateral sclerosis (Lou Gehrig's disease). They are also used being used for maintaining skin juvenility, where ROS scavenger is known as anti-ageing agents by several pharmaceutical companies. Fullerenes have used as a theranostic platform for cancer treatment. Several metal oxides are entrapped inside the hollow sphere and thereby function as MRI contrast agent.

Shi et al. synthesized a C_{60}-iron oxide nanocomposite (C_{60}-IONP). They further functionalized with PEG 2000. The formed nanocomposite showed good stability in physiological solutions. For targeting tumours they have used well-known cancer-targeting molecule FA making the conjugate C_{60}-IONP-PEG-FA. The tumour-targeting system was studied with multifunctional characteristics for cancer diagnosis photodynamic therapy (PDT), radiofrequency (RF), thermal therapy (RTT) and magnetic targeting applications, in vitro in MCF-7 cell lines and in vivo in mice models [55]. Shi et al. designed a fullerene-based DDS C_{60}@Au-PEG/DOX with stimuli-responsive controlled release property [53]. The DDS showed a promising improvement in the therapeutics and toxicity through X-ray-guided imaging-guided DOX release, photodynamic and photothermal therapies [56]. Same research group have fabricated C_{60}@Au hybrid nanocomposite by chemical deposition of Au onto C_{60} fullerene. The hybrid was then functionalized with PEG5000 via a pH-sensitive hydrazine bond. When the nanocomposite enters tumour cells because of the pH difference, the PEG dissociates from the nanocomposite system. Anticancer drug DOX is then loaded on this nanoplatform C60@Au-PEG/DOX which has the high efficacy of drug loading. By tuning the radiofrequencies, the release of DOX can be tuned. This nanocomposite C60@Au-PEG/DOX showed promising material to diagnose cancer in the early stages as well as showed chemo-RF thermal-photodynamic therapeutic efficacy, RF-controlled drug-releasing function, tumour-targeting property, tumoural acid PEG dissociating character and X-ray imaging ability. Thereby fullerene-based nanosystem can work as multifunctional DDS for cancer theranostic application [56]. Another study by the same group utilized the benefit of fullerene with an on-off state. DOX was covalently conjugated with C_{60} via a ROS-sensitive thioketal linkage. And then a hydrophilic shell (distearoyl-sn-glycero-3-phosphoethanolamine-PEG CNGRCK2HK3HK11, DSPE-PEG-NGR) was attached to the surface of C_{60}-DOX, making the nanocomposite as (C_{60}-DOX-NGR). The off-on state signifies the attachment of DOX with the nanocomposite. The on-off state can be tuned by 532 nm laser. When the nanocomposites are in off state that means the DOX is attached to the nanocomposite. However, when the laser is irradiated onto the nanocomposite, the ROS is generated which leads to the cleavage of thioketal linkage and therefore releasing the DOX to the target. Thereby a targeted tumour-specific drug delivery was done using this C_{60} nanocomposites [57].

Recent works exploits the potential of fullerene with other cancer-targeting drugs such as landomycin A [58]. Fullerene has also been used to target the signalling pathways which are key players in the cancer signalling. One of them is the calcium/calmodulin protein kinase II (CaMKIIα) activation pathway. This pathway is known to play a role in tumour progression. In this regard Xu et al. activated C60 nanosytem which promoted the ROS generation, which in turn activated CaMKIIα in osteocarcinoma cells. They explored the potential of CaMKIIα inhibition, along with the suppression of autophagic degradation for improved antitumour efficacy [59]. C60 is used for the development of photoelectrochemical (PEC) biosensor using photoactive material (PTB7-Th) as a signal indicator. This PEC biosensor provided a good linear range and good sensitivity for miRNA and thus offered a new platform for detecting cancer in the early stages [60].

7.3.3 Carbon Nanotubes(CNTs)

CNTs are 1D nanoparticles, and they were discovered way back in 1960 by Bacon [61]. However, CNTs in 1991 attracted scientific community after the pioneering discovery by Iijima [62]. CNTs are generally referred to a wide range of tubular nanostructures which have similar structures and shapes. Basically, they are made up of rolled sheets of graphene to form hexagonally arranged carbon and have a hollow tubular cylinder in nanometre size. Preferably, they are based on a hexagonal lattice of sp^2 carbon atoms such as graphene. However, the edges of graphene sheet are fused together to form a cylindrical tube with a high aspect ratio, which can be open ended or capped. CNTs are mainly categorized into two types of single-walled carbon nanotubes (SWNTs) and multiwalled carbon nanotubes (MWCNTs). SWNTs are made from a single graphene sheet and have diameter of around 0.4–2 nm to length of several micrometres, with an empty internal space. MWNTs are double-walled or multiwalled carbon nanotubes depending on the number of graphene sheets in the walls of the cylindrical structure.

Carbon nanotubes have unique physical and chemical properties such as ultra-light weight, high aspect ratio, high mechanical strength, high electrical conductivity and high thermal conductivity. The high aspect ratio, i.e. length-to-diameter ratio of carbon nanotubes, frequently increases to 10,000 because of which carbon nanotubes are considered as the most anisotropic materials. Beside this, chirality (the angle between the hexagons and the nanotube axis) is another key feature of CNTs. Depending on the chirality, the carbon atoms can be arranged in several ways around the circumference of nanotube; armchair, zigzag and chiral patterns are the most common examples. Other than these properties, CNTs also show properties which are used in biological applications such as photoluminescence, biocompatibility, a rapid uptake by cells due to anisotropy and high drug loading efficiency. CNTs present a new approach towards diagnostics and imaging capabilities. CNTs are not dispersible in organic solvents or water and are usually held strongly together in bundles by significant *van der Waals* interactions.

7.3.3.1 Classification of CNTs

CNTs are classified according to the number of cylindrical walls present in them and are mainly classified into four types (Fig. 7.5):

- *Single-walled carbon nanotubes (SWCNTs)*: A single layer of graphitic sheet wrapped into cylindrical tube structure.

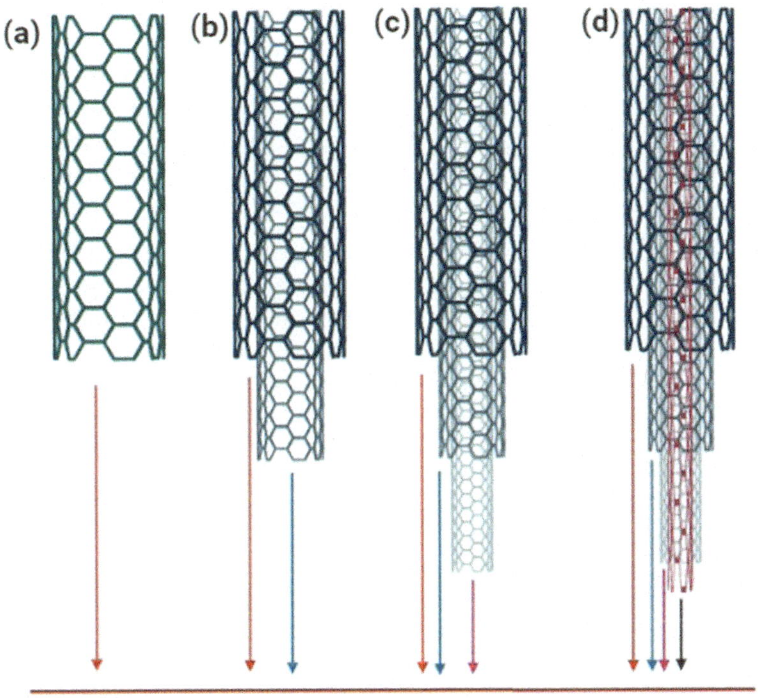

Fig. 7.5 Types of carbon nanotubes: (**a**) single-, (**b**) double-, (**c**) triple-, and (**d**) multiwalled carbon nanotubes

- *Double-walled carbon nanotubes (DWCNTs)*: Two layers of graphitic sheets wrapped over one another.
- *Triple-walled carbon nanotubes (TWCNTs)*: Three different graphitic sheets are wrapped over each.
- *Multiwalled carbon nanotubes (MWCNTs)*: Multilayers of graphitic sheets are wrapped one over other.

CNTs have been easily manufactured using laser ablation, electric arc discharge, electrolysis, high-pressure conversion, chemical vapour deposition and CoMoCat synthesis processes [63]. CNTs have exceptional nanoneedle tubular structure which facilitates their entry into cell membrane [64]. CNTs have been extensively used for cancer targeting owing to their properties such as easy availability, high aspect ratio with uniform ordered structure, photoluminescence property, good biocompatibility, non-immunogenicity and excretion from the body circulatory system through biliary pathways (clearance through urine 96%, through faeces 4%). CNTs have their both ends open which makes the inner surface accessible for drug loading and other modifications. However, CNTs suffer from certain drawback like pristine CNTs are hydrophobic in nature, they aggregate which limits their biodistribution and a fair chance of accumulation in the liver based on the in vivo studies.

7.3.3.2 Functionalization of CNTs

The pristine CNTs are hydrophobic in nature. Hydrophobic nature leads to the aggregation of CNTs in blood limiting their biodistribution. The surface alteration is done to overcome these limitations. The functionalization has been done using two main approaches: (a) covalent linkage and (b) noncovalent interaction between the surface of CNTs and biochemical functional moieties [63, 65]:

(a) *Covalent functionalization:* This is the more reliable of the two strategies as the bonding between CNTs and immobilized biomolecule is stronger. The active areas for the functionalization of CNTs are available at their terminal ends, side walls and the defect regions. Functionalizing the side walls enhance the chemical reactivity of the CNTs. However, various studies suggest ends and defects as the more favourable side for surface modification. Since this sites can be easily modified following treatment with electrochemical, mechanical and chemical routes. Also, oxygen containing groups can be incorporated following acid treatment with strong acids.
(b) *Noncovalent functionalization:* The various noncovalent functionalization like electrostatic interaction, hydrogen binding, π- π interaction and *van der Waals* force have been used to functionalize surface of CNTs, surfactants (sodium dodecyl sulphate (SDS), sodium dodecyl benzene sulfonate (SDBS), cetyltrimethylammonium bromide (CTAB), Triton-X-series and Pluronic F and E series), poly(ethylene glycol) (PEG), chitosan (CHI), porphyrin derivatives, fluorophores, polymers, lipids, nucleic acids, proteins and endohedral functionalization to increase solubility and biocompatibility, thereby making them less toxic. The PEGylating is the most propitious approach towards CNT functionalization [65].

CNTs are characterized using techniques such as X-ray diffraction (XRD), microscopy techniques (atomic force microscopy (AFM), TEM and SEM), thermogravimetric analysis (TGA), Raman spectroscopy and Fourier transform infrared spectroscopy (FTIR).

CNTs are the most sought-after material in tumour targeting and imaging. In the past, very good review has been written highlighting the surface modification, drug loading ability and tumour targeting [66–68]. CNTs are used as an attractive carrier for some of cancerous drugs like DOX, paclitaxel and carboplatin. CNTs also utilized as gene carriers for siRNAs and plasmid DNA [69]. Various studies have been done till now on CNTs on all types of cancer such as targeting breast, prostate, lung, pancreatic, colon, and gastric cancer. Also, CNT-mediated liver cancer treatment has been done. CNTs have been used to target every possible cancer including breast, prostate, lung, liver, colon and gastric, ovarian, bladder, blood, lymph node, skin, thyroid and head and neck cancers. Sheikhpour et al. have done a detailed review on CNTs as cancer diagnostics [70]. In this chapter we are only including some of the research done, as a massive amount of research has been done on CNTs accounting for 51.34% (Fig. 7.2a). Table 7.2 shows some of the cancer-targeting

Table 7.2 CNT used for different cancers targeting drugs

CNTs	Drug/vector for DNA or siRNA	Disease/model system	References
MWCNT	DOX	Breast cancer	[64, 71]
MWCNT	Betulinic acid	Drug delivery system	[72, 73]
MWCNT	Methotrexate (MTX)	Breast cancer	[74]
MWCNT	Gemcitabine	Breast cancer	[75]
SWCNT	Etoposide (drug used for chemotherapy)	Pancreatic cancer	[76]
SWCNTs, MWCNTs-	Paclitaxel	Breast cancer, lung cancer	[77–79]
MWCNT	Chelerythrine	Liver cancer	[80]
MWCNT	Camptothecin	Breast cancer	[81]
Carbon Nanofibres and CNTs	Carboplatin	Prostate cancer	[82]
Carbon Nanofibres and CNTs	Cisplatin	Prostate cancer	[82]
SWCNT	siRNA	CSC cancer cells, breast cancer, lung cancer	[83–86]
SWCNT	Aptamer (5TR1, AS1411)	Breast cancer, gastric cancer	[87]
MWCNT	Docetaxel	Lung cancer	[88]
SWCNT	EGF (epidermal growth factor)	Pancreatic cancer	[89]
SWCNT	Gambogic acid	Pancreatic cancer	[90]

drugs being used with CNT to target different cancers in the body. Apart from drugs, other cancer markers have been targeted like siRNA and aptamers sequences siRNA against EpCAM, which is overexpressed in cancer cells.

7.3.3.3 CNT-Based Electrochemical Biosensor

CNTs promote electron transfer, have high stability, low background noise and rapid electrode kinetics and are biocompatible in nature. For these reasons, CNTs have been widely used in fabricating biosensor for early detection of cancer. Table 7.3 summarizes the different cancerous biomarkers used in cancer detection using CNTs. The cancer-specific biomarkers are prostate cancer-specific antigens (PSA) and osteopontin (OPN); breast cancer biomarkers (MUC-1, HER2, HER-3); mi-RNA, CEA, CA 125 and CA 19-9; and other biomarkers such as matrix metalloprotease (MMP-3), alpha-fetoprotein (AFP) and cytokines (TNF α, IL-1β). The biosensor fabricated using CNTs achieved a very low LOD mostly in the femtomolar or picomolar range. Tian et al. achieved a very low detection limit of 1.2 fM for miRNA detection [91].

Table 7.3 CNT-based immunosensors for cancer biomarkers detection

CNTs based immunosensors	Biomarkers	References
ECL sandwich immunosensor	IL-6, PSA	[92]
DNA-linked-CNT wire motif	DNA detection	[93]
Prostate cancer biomarkers	PSA Osteopontin (OPN)	[94–99]
mi-RNA	mi-RNA	[91, 100–104]
Ovarian cancer biomarkers	CEA	[105–114]
	CA 125, CA 19-9	[115–117]
Breast cancer biomarkers	MUC-1, HER2, HER-3	[118–120]
Other cancerous antigens	TNF α, IL-1β, MMP-3, AFP	[121–123]

7.3.3.4 Limitations of CNTs

Weak *van der Waal's* forces limit capacity for drug delivery.

7.3.4 Graphene

Graphene was found in the year 2002 pursued by Nobel Prize in Physics in 2010. It is a plenteous material as it is a structure square of common graphite. It has picked up a great deal of consideration from different fields for applications in nanoelectronics, catalysis, sensing, material building and biomedicine due to its novel properties such as optical, electrical, mechanical and auxiliary properties that have been observed to be better than some other broadly found carbon nanomaterials. Graphene-based nanomaterials appear as a novel material in the field of nanomedicine for the treatment of cancer for few reasons. Firstly, the auxiliary property of these materials such as aromaticity and high surface area improves the ease of loading hydrophobic drug. Secondly, its optical retention in the near-infrared area empowers them to go about as photo agents for photodynamic and photothermal treatments of tumours. Thirdly, presence of oxygen-carrying functional groups enhances its physiological strength and goes about as site for biofunctionalization. Overall, graphene-based nanomaterials can go about as operators for creating multifunctional restorative stages for doing progressively proficient detection and treatment of malignant growths [124].

Graphene is a two-dimensional (2D) nanomaterial that consists of monoatomic layer of sp^2-hybridized carbon molecules. In graphene sheet, every carbon is associated with other three neighbouring carbon particles by covalent σ bonds, making a honeycomb cross section. As of now, graphene is the most grounded known material. The unhybridized p orbitals of carbon atom adjusted oppositely to the planar structure of graphene sheet and interconnected with each other to shape the half-full π bond that gives graphene its aromaticity.

The nature of graphene relies upon the technique by which it was delivered. The graphene sheets which have low quality regularly show poor mechanical quality or electrical conductivity because of imperfections or oxygen locales are scattered in the graphenic grid [5].

7.3.4.1 Multilayer Graphitic Nanosheets (MGNs)

MGNs comprise of between 2 and 10 graphene monolayers (Fig. 7.6). The properties of nanosheets resemble graphene. MGNs are soluble in natural solvents, framing stable straightforward suspensions [125].

7.3.4.2 Graphene Nanoribbons

At present, slender strips of graphene monolayers are accepting critical research enthusiasm as they are the elective type of graphene. Most studies concentrated on the "unzipping" of CNTs as it prompts the development of slender elongated graphene monolayer strips [126].

Graphene nanoribbons have been reported as one-dimensional sp^2-hybridized carbon segment of limited measurement with characterized edges at which carbon atoms of three non-composed nature. Depending on the edge termination [127, 128], graphene nanoribbons are divided into three types: (i) arm chair, (ii) zigzag and (iii) chiral nanoribbons. If there should be an occurrence of graphene nanoribbons with arm chair edges, the width is given by the quantity of dimer C-C lines (Na) over the nanoribbon; interestingly the width of zigzag graphene nanoribbon is expressed as the quantity of zigzag chains (Nz) over the nanoribbon. In graphene nanoribbons, edge reconstruction may happen, as the edge carbon molecules are not bound soaked. Edge reproduction can be normal in case of zigzag graphene nanoribbons at raised temperatures. The hydrogen immersion is regularly applied for the stabilization of the edge structure [127].

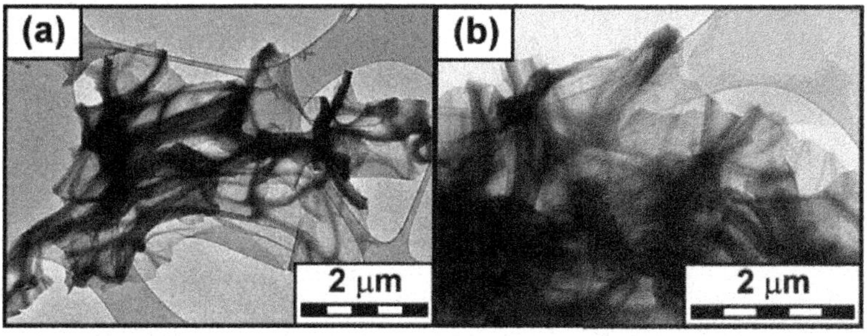

Fig. 7.6 Analysis of multilayer graphitic nanosheets by TEM. (Reprinted with permission from Ref. [125]. Copyright 2009 Springer)

The new synthesis methods have been made conceivable synthesis of subordinates of graphene like graphene oxide (GO) and reduced graphene oxide (rGO) that have further enhanced the enthusiasm for the biomedical field because of the incorporation of new highlights and novel properties in these subordinates. The monolayer of graphene and its subordinates like GO and rGO gives ultrahigh surface zone to proficient loading of helpful products like medications, fluorescent tests, nucleic acids, biomolecules and cells [129]. Beside this, these materials likewise give a circulation of delocalized π electrons superficially giving it an aromatic character, which improves the medication stacking and binding efficiency of graphene-based nanomaterials with different particles of restorative enthusiasm by π-π stacking and hydrophobic cooperation. The GO additionally gives wide assortment of oxygen-containing functional groups like ethoxy, carboxyl, carbonyl, hydroxyl, epoxy and so forth, on both the planes of GO sheets, therefore expanding its steadiness in aqueous solutions. This additionally helps in simple functionalization and derivatization of these materials through both covalent and non-covalent modifications with various biocompatible polymers like chitosan and PEG and conjugation with different target moieties such as antibodies and peptides, to produce biofunctionalized nanocomposite frameworks with upgraded biological properties [130, 131].

Other than these surface properties, GO and rGO are the photoresponsive agents for malignant growth treatments such as photothermal therapy and photodynamic therapy, as they have astonishing intrinsic properties like photoluminescence and upgraded absorbance in the near-infrared region (NIR) of electromagnetic spectrum. GO and rGO show photothermal transformation properties like plasmonic nanomaterials in response of NIR irradiation that was broadly investigated as a helpful procedure for treating diseases. In correlation with GO, rGO has been observed to be significantly more effective as a result of the nearness of more deformities and more noteworthy aromaticity in its structure that are presented in it amid decrease procedure of GO [132].

Graphene-based nanomaterials have been identified for a wide assortment of utilizations in the field of biomedical, particularly in malignant growth, researched for their phenomenal physicochemical and biocompatible properties, filling in as nanocarriers and photoagents for medication and quality conveyance to cancer cells and for photodynamic and photothermal treatments of tumour. They can be utilized both in vitro and in vivo, till its utilization as bioimaging agents for creating multifunctional theranostic stages for taking out increasingly proficient recognition and treatment of malignancies [132].

7.3.4.3 Graphene-Based Biosensors for Cancer Biomarker Detection

Graphene have high optical, electrical, mechanical and auxiliary properties that have been observed to be better than some other broadly found carbon nanomaterials, and they are biocompatible in nature. For these reasons, graphene has been widely used in fabricating biosensor for early detection of cancer. Table 7.4 summarizes the

Table 7.4 Graphene-based biosensors for cancer detection

Sensor platform/label	Analyte	LOD	References
GO/UCNP/ssDNA	PCA3	0.5 pM	[133]
FET/$(PDDA+PSS)_2(PDDA+GR)_5$/anti-PSA	PSA	11 aM	[134]
GCE/IL-rGO/anti-CEA, anti-PSA, anti-AFP/anti-CEA-Thi-CAuNP, anti-PSA-DAP-CAuNP, anti-AFP-Cd^{2+}-CAuNP	CEA, AFP, PSA	2.7 pg/ml, 3.1 pg/ml, 4.8 pg/ml	[135]
FET/GR/PtNP/scFv-anti-HER3	HER3	300 fg/ml	[136]
GCE/PEI-RGO/cDNA1/AuNP/cDNA2/TiP- Cd^{2+}/$Ru(NH_3)_6^{3+}$	miRNA-21	0.76 aM	[137]
GO-RuOMO-aptamers	Thrombin	0.76 nM	[138]
GCE/CCG/TCPP/hexapeptide	Cyclin A_2	0.32 pM	[139]
GCE/GR/ssDNA	8OHdG	0.875 nM	[140]
ITO/rGO-ZrO_2/APTES/anti-CYFRA-21-1	CYFRA-21-1	0.122 ng/ml	[141]

different cancerous biomarkers used in cancer detection using graphene. The cancer-specific biomarkers are PSA and OPN; breast cancer biomarkers (HER2, HER-3); mi-RNA; CEA; and other biomarkers such as 8-hydroxy-20-deoxyguanosine (8OHdG), alpha-fetoprotein (AFP) and cytokines (TNF α, CYYFRA-21-1, IL-1β). The biosensor fabricated using graphene achieved LOD which varies from picomolar to nanomolar range. Vilela et al. [133] produced an optical biosensor for the detection of PCA3, since they didn't participate with GO on account of target-catch DNA hybridization. Zhang et al. [134] fabricate a chemiresistor-type field-effect transistor (FET) biosensor for the detection of prostate cancer based on LbL self-assembled graphene composites. Sensor exhibitions were improved by stifling shimmer noise by suspending the graphene multilayer between gold cathodes of the sensor [134]. Xu et al. [135] discovered a triple tumour marker-based biosensor which includes CEA, PSA and AFP. Rajesh et al. [136] synthesized an immunosensor for detection of breast cancer, i.e. HER3. Cheng et al. [137] fabricate a miRNA-21 biosensor including ssDNA (DNA1) and a biotin-labelled DNA2 as target capturers. Li et al. [138] developed an immunosensor for the detection of pulmonary metastases, i.e. thrombin. By developing an optical biosensor, Feng et al. [139] develop both electrochemical and fluorescence biosensor for early-stage detection of cyclin A2 protein. Jia et al. [140] fabricated a biosensor for the detection of 8-hydroxy-20-deoxyguanosine (8OHdG) which is a cancer risk biomarker for the number of disorders such as cardiovascular and neurodegenerative diseases and diabetes. Kumar et al. [141] developed a biosensor for the detection of oral cancer via the detection of biomarker CYFRA-21-1 in saliva.

7.3.5 Nanodiamonds

Nanodiamonds are a transpiring class of new carbonaceous nanomaterials with many fascinating mechanical, chemical and optical properties (Fig. 7.7). Nanodiamonds is another allotrope of carbon-based nanomaterials. It is quite different from the various types of nanomaterials comprising graphene, fullerenes, carbon nanotubes and carbon dots. The carbon in the nanodiamond is sp^3 and consists of a diamondoid-like topology. The diameter of nanodiamonds extend from 2 to 10 nm, which are bigger than natural diamond atoms, however altogether smaller than mass diamond and diamond rough powders. They are commonly synthesized by top-down methodology, for example, jet milling or scraped area of micro-diamonds. Nanostructures of this evaluation that have diameter across of more noteworthy than 20 nm act like bulk diamonds. On the other hand, sp^3 carbon nanostructures having measurement under 1 nm are for the most part called as diamondoid and they occur naturally in petroleum deposits. The surface-bound carbon atoms of these small diamond are normally clung to hydrogen or other non-carbon particles. In this way, their properties resemble those of natural atoms as opposed to that of bulk diamond. The increase in the diameter across sp^3 carbon group, there is a significant decrease in the percentage of carbon atoms present at the surface. Yet the lower limit of nanodiamond diameter is obscure. The model nanodiamonds are produced using detonation of hazardous compounds with a negative oxygen balance, they were first found in the detonation sediment together with graphitic, nanodiamond carbon.

There are several advantages of nanodiamonds for bioimaging and therapy purposes:

1. It is a profoundly photostable fluorophore which is practically photobleachable taking into consideration long-lasting imaging without worries of single decay.
2. Nanodiamonds have surprisingly high fluorescence quantum efficiency usually ranging from 0.7 to 1.
3. They are exceptionally biocompatible nontoxic to biological tissues and organism.

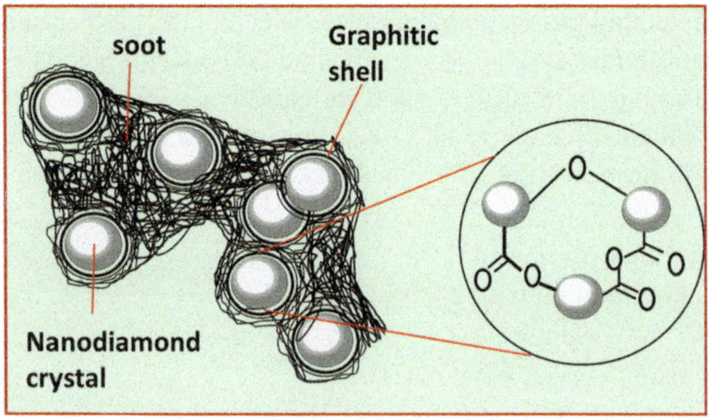

Fig. 7.7 Image shows nanodiamond clusters synthesized by the detonation method

4. There are abundant dangling bonds available for surface functionalization by means of covalent and noncovalent implies without influencing the intrinsic properties of the N-V defect centres in the centre.

With all of these advantages, nanodiamonds have been seriously utilized as fluorescent labels for in vitro imaging of subcellular structures and certain biomarkers, drug and gene delivery, in vivo whole animal imaging and tissue engineering [142–144].

7.3.5.1 Origin of Fluorescence of Nanodiamond

Due to the absence of sp^2 graphitic carbon, the fluorescent emission of nanodiamonds depends on a totally different mechanism, emerging from the presence of the defect centres. The presence of fluorescent defect centres is the exceptional element of nanodiamonds, which differentiate them from other carbon allotropic nanomaterials. Diverse deformity focuses with tunable assimilation and emission bands can be accomplished, in which the nitrogen opportunity (N-V) imperfection focus is the most widely recognized defect centre. The N-V imperfection focus can be separated into two kinds: neutral N-V° centre with fluorescence outflow at 576 nm and contrarily negatively charged $N\text{-}V^-$ centre of emanation at 638 nm [145]. Both electronic advances can be combined with phonon to give a wide band at ~700 nm. Indeed, even a solitary N-V defect centre can be imagined by confocal microscopy with the fluorescence emission in a wide range from 500 to 800 nm.

7.3.5.2 In Vitro Fluorescence Imaging with Nanodiamond

Nanodiamonds are broadly used to stain cells and tissues, and to follow the intracellular trafficking, because of their strong and stable fluorescence of every individual nanodiamond particle. In a latest work, nanodiamonds were used in 293T human kidney cells via intracellular uptake of nanodiamond which gives emission at ~700 nm which was further confirmed by confocal microscope. Likewise, nanodiamond-stained cell imaging test has been found which is also depend on the excitation dependent emission fluorescence emission Fig. 7.8, recommending conceivably a comparative reason with respect to CDs [146]. Inferable from the prevalent photostability with least photobleaching under consistent excitation, single fluorescent nanodiamond particles can be followed to give 2D in-plane directions and 3D volumetric directions, uncovering the Brownian diffusion motion inside the cell cytoplasm [145, 147]. Treussart et al. reveals that the cellular uptake of 25 nm nanodiamonds may not be endosome-associated endocytosis; however, that of 46 nm nanodiamonds is chiefly by means of clathrin-interceded endocytosis [148]. This finding has been affirmed by examining the direction of a signal nanodiamond molecule as it cooperates with the cell [149]. In a later report,

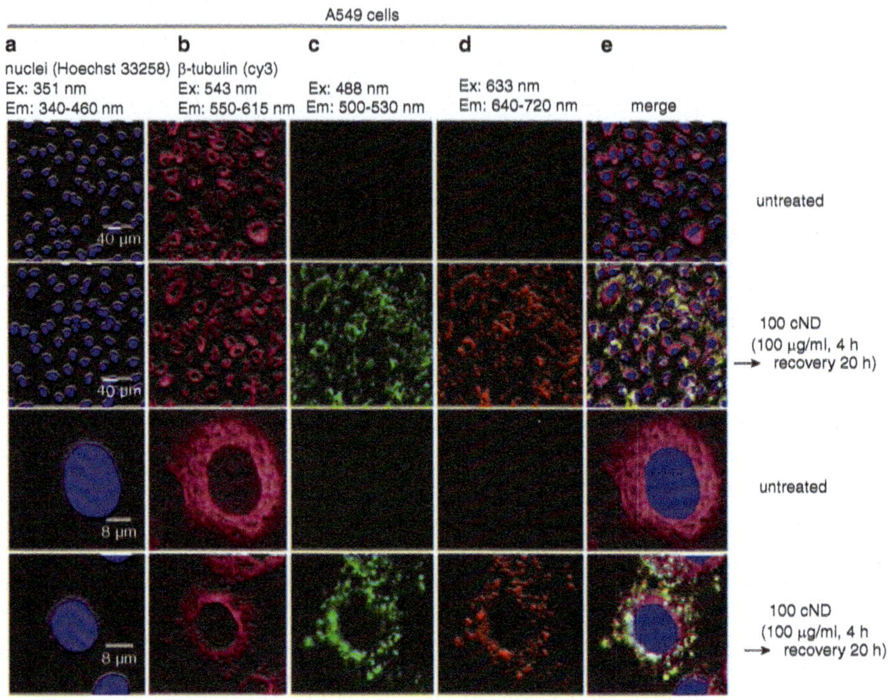

Fig. 7.8 In vitro fluorescence imaging of cells and tissues with the intrinsic photoluminescence of nanodiamonds. (Reprinted with permission from Ref. [146]. Copyright 2007 Biophysical journal)

Chang and Yu et al. have demonstrated that fluorescent nanodiamonds can be utilized for labelling, distinguishing and following transplanted lung stem/progenitor cells over a course of 7 days, misusing the photostable fluorescence of nanodiamond in the >590 nm extend under an excitation of 515–560 nm. Beside this, the fluorescence lifetime of nanodiamond is altogether more up to 20 ns than other fluorophores (<10 ns), taking into account clear recognizable proof of nanodiamond from background tissue autofluorescence and other fluorescent immunostains utilizing fluorescence lifetime microscopy (FLIM) and time-gated fluorescence (TGF) imaging. The transplanted lung stem cell named with nanodiamond is effectively recognized in profoundly differentiated lung tissues, where they are observed to be for the most part situated in the subepithelium of bronchiolar airways and structure grouped engraftment in harmed lung tissues. The equivalent TGF imaging by abusing the long fluorescence lifetime has been utilized for imaging nanodiamond-labelled malignant growth cells in both a microfluidic gadget and mouse ear blood vessels [150].

7.3.5.3 In Vivo Fluorescence Imaging with Nanodiamond

Caenorhabditis elegans (*C. elegans*) was the first multicellular organism used for the in vivo fluorescence imaging of nanodiamond instead of warm-blooded creatures, inferable from the basic and well-characterized life systems of *C. elegans*, just as the high straightforwardness of its body for a mind-blowing conveyance by fluorescence microscopy. In a study carried out by Chang et al. [151], *C. elegans* worms are sustained with fluorescent nanodiamonds, and the nanodiamonds can be found down the stomach-related framework and are tracked principally in the intestinal cells. Non-stop fluorescence imaging and following of microinjected nanodiamond into the gonad of a grown-up bisexual uncover the downstream relocation and conveyance of infused nanodiamond into the distal gonad, oocytes and early- and late-stage foetuses successively. This finding proposes that the infused nanodiamonds can be passed down from one generation to next generation without making noticeable pressure to the worms [151]. In another examination done by the equivalent group [152], fluorescent nanodiamonds were infused intraperitoneally into a live mouse, and the conveyance of the infused nanodiamonds inside the peritoneal cavity can be followed after some time by observing the fluorescence outflow in the NIR window at 780 $\pm$ 10 nm. Axillary lymph nodes light up and become obvious too, attributable to relocation of nanodiamonds from the intradermal infusion site at the front paw of a live mouse [152]. Shirakawa et al. conceived another imaging framework called specific imaging dependent on the one of a kind property of the N$-$V centre in nanodiamond that the fluorescence of the N$-$V centre exceptionally relies upon the ground-state spin configuration, which is directed by electron spin magnetic resonance, while the background fluorescence does not. By utilizing this new strategy, long-term foundation-free fluorescence imaging of *C. elegans* and mouse has been acknowledged with a more noteworthy signal-to-background proportion [153].

7.3.6 Mesoporous Carbon Nanoparticles

Recently, a great interest has been shifted towards the synthesis of mesoporous nanoparticles and its application in DDS. Previously, mesoporous silica NPs (MSN) were considered ideal as DDS due to large pore size, high surface-to-volume ratio, low cytotoxicity, easy functionalization and biodegradability. Cellular endocytosis is the preferred mechanism used for the internalization of mesoporous silica NPs below 200 nm size. In the recent years, the concern from MSN has been shifted to mesoporous carbon nanoparticles (MCNs). MCN are made through hydrothermal synthesis in the presence of silica NPs [154]. Therefore, it is important to match the polarity of the template surface with the carbon precursor. Mesoporous carbon shells are obtained after removal of the mesoporous template. The percentage weight of carbon to silica is critical for the formation of MCN [155]. MCN have rich polar functional groups which is why they are suitable for the easy functionalization with drugs or other biomolecules [156]. Gu et al. synthesized uniform MCNs using good water dispersibility using hydrothermal synthesis. The size of the prepared MCNs was below 200 nm. The MCNs were used as an effective drug carrier for camptothecin which effectively inhibited the growth of breast cancer cell MCF-7 [157]. Li et al. encapsulated aptavalve (single-stranded DNA) on the surface of DOX-loaded MCNs thus taking the MCNs to the next level of cancer targeting and imaging. The major exciting phenomenon of their cancer-targeting platform is that the fluorescent signal can be switched on and off in the presence of mucin therefore making it an on and off biosensor. Additionally, the system fabricated is capable of activating on-demand drug delivery [158]. Meng et al. developed polyethyleneimine (PEI)-grafted oxidized mesoporous carbon nanospheres. They used these biocompatible MCNs for photothermal therapy combined gene therapy platform in vitro and in vivo. The nanoplatform delivers the therapeutic gene to tumour for gene therapy and also eliminates tumour by photothermal ablation [159]. In another pioneering work by Wang et al., peptide-conjugated core-shell graphitic carbon@silica nanospheres with dual-ordered mesopores were fabricated for targeted photothermal therapy of breast cancer cells. The hydrophilic mesoporous silica shell provided good water dispersibility, while the mesoporous carbon core efficiently helps in drug loading, contact development between the drug and photothermal hotspots and high photothermal conversion efficiency [160]. Wang et al. fabricated a "four-in-one" theranostic system for photoacoustic imaging-guided synergistic targeting chemo-gene-thermo trimodal therapy for breast cancer. In this system, polyethylene glycol (PEG)-bridged PEI and a memHsp70 receptor-targeting peptide (TKD), PPT, were uniformly capped on DOX-loaded oxidized mesoporous carbon nanospheres (OMCN) to encapsulate therapeutic genes into cancer cells via active targeting accumulation. This "four-in-one" theranostic system enables distinct PA imaging visualization, NIR-/pH-sensitive drug/gene release and collective targeting therapeutic outcome which are more superior than the single therapy [161].

7.4 Future Prospect

Carbon is indeed a wonder material with the most diverse nanoallotropes ranging from 0D to 3D. A different nanoallotrope can be formed by combining sp^2- or sp^3-hybridized carbon differently. Moreover, combining these nanostructures differently give rise to a new superstructure. Also, these nanoallotropes can be easily mixed with other existing nanomaterials to give rise to composite having an altogether different properties and structure. Because of its biocompatible nature and ease of functionalization, these nanostructures have been extensively used in cancer diagnosis, therapy and therodiagnostic application. These carbon nanoallotropes hold a great potential to be used in the diagnosis and therapy of cancer. The main focus of this chapter is to summarize the use of carbon nanostructures in cancer cell imaging, drug delivery, therapy and biosensors. However, a number of applications have already been done with respect to cancer nanoallotropes. Still these are in early stages of development, and there is plenty of room for improvement. Since, these carbon nanoallotropes are quite biocompatible, and a lot more studies have been done in vitro and in vivo. There actual clinical worth can be utilized by employing them in clinical studies in humans that are much challenging. Despite having good electrical properties and a copious amount of work devoted to biosensor development based on graphene and CNTs. Carbon nanoallotropes are visualized as future of biosensors for cancer, still this field needs a substantial research to be done to develop a point-of-care testing system for cancer diagnosis involving different systems such as advanced microfluidics, lateral flow-based optical devices, device miniaturization and microarray technology [162]. CDs being such a wonderful material have not been employed for biosensor fabrication. A lot more efforts need to be devoted for electrode fabrication.

7.5 Conclusion

It is crucial to detect cancer in the early stage to increase the survival of patients. Undoubtedly, Carbon is the element with widen and most diverse range of nanoallotropes. After combining sp^2- and sp^3- hybridized carbon atoms in different ways, lead to the formation of unique variety of structures, and by exploiting the morphological transformation of sp^2 graphene sheets and the conclusions of their size distribution. Furthermore, these nanostructures can be consolidated in various approaches to make quantities of blended nanoarchitectures, new superstructures and allotropic mixtures with a wide scope of properties, dimensionalities and potential applications. In addition to the incredible assortment of composite structure and carbon nanoallotropes that have been described tentatively, there are numerous that have been anticipated to exist that are foreseen to have exceptional highlights and energizing potential applications [5, 163].

In this book chapter, we have explained the number of carbon nanoallotropes by combining different hybridized carbon atoms such as carbon quantum dots/graphene quantum dots, fullerenes, carbon nanotubes, graphene and nanodiamonds. Some researchers have reported CNTs to be biocompatible [164]. Others have proved them wrong and shown different level of biocompatibility on different cells [165, 166]. CDs which contain more graphenic layer are generally stabilized out by different functional groups that frequently influence their photoluminescent behaviour. Likewise, the photoluminescent behaviour of nanodiamonds is reliant on the dopants and surface defects, as contrast with the other pertinent elements. The chemistry, stability and properties of 1D and 2D nanoallotropes are additionally vigorously impacted by the presence of non-carbon atoms, which are generally fused during the synthesis of carbon nanoallotropes. For instance, the traces of metal catalyst were found in carbon nanotubes that influence their conducting and semi-conducting properties. So also, graphene nanoribbons are generally enriched in oxygen as carboxyl, hydroxyl and carbonyl groups.

There are such a significant number of anticipated or uncommon carbon nanoallotropes. If these materials can be synthesized, then their study can grow the comprehension of both 2D and 3D carbon materials, which will prompt the upgrade of viable chances. For instance, it is simpler to get ready compressed type of graphite as contrast with the current packed structures dependent on nanotubes and since such designs require the nanotubes to be legitimately adjusted to pressure. Although, as of now it isn't at all important to build up the direct arrangement. As of late, the Wuest group synthesizes the crystalline covalent natural materials with diamond-like system topologies which represent to a significant advance in the synthesis of new unadulterated carbon materials [163].

Finally, we state that this book chapter didn't entreat to extensively discuss about individual nanoallotropes' applications since they have been depicted finally somewhere else [167–170]. However, we do feature some especially intriguing utilizations of key nanoallotropes (fullerenes, carbon quantum dots/graphene quantum dots, nanodiamonds, carbon nanotubes and graphene) as well as some hybrid superstructures.

References

1. McGuire A, Brown J, Malone C, McLaughlin R, Kerin M (2015) Effects of age on the detection and management of breast cancer. Cancer 7(2):908–929
2. Boyle P, Levin B (2008) World cancer report 2008. IARC Press, International Agency for Research on Cancer, Lyon
3. Baskar R, Lee KA, Yeo R, Yeoh K-W (2012) Cancer and radiation therapy: current advances and future directions. Int J Med Sci 9(3):193
4. Wang MD, Shin DM, Simons JW, Nie S (2007) Nanotechnology for targeted cancer therapy. Expert Rev Anticancer Ther 7(6):833–837
5. Georgakilas V, Perman JA, Tucek J, Zboril R (2015) Broad family of carbon nanoallotropes: classification, chemistry, and applications of fullerenes, carbon dots, nanotubes, graphene, nanodiamonds, and combined superstructures. Chem Rev 115(11):4744–4822
6. Allen MJ, Tung VC, Kaner RB (2009) Honeycomb carbon: a review of graphene. Chem Rev 110(1):132–145
7. Sun D, Ban R, Zhang P-H, Wu G-H, Zhang J-R, Zhu J-J (2013) Hair fiber as a precursor for synthesizing of sulfur-and nitrogen-co-doped carbon dots with tunable luminescence properties. Carbon 64:424–434
8. Xu X, Ray R, Gu Y, Ploehn HJ, Gearheart L, Raker K, Scrivens WA (2004) Electrophoretic analysis and purification of fluorescent single-walled carbon nanotube fragments. J Am Chem Soc 126(40):12736–12737
9. Sun Y-P, Zhou B, Lin Y, Wang W, Fernando KS, Pathak P, Meziani MJ, Harruff BA, Wang X, Wang H (2006) Quantum-sized carbon dots for bright and colorful photoluminescence. J Am Chem Soc 128(24):7756–7757
10. Sharma V, Tiwari P, Mobin SM (2017) Sustainable carbon-dots: recent advances in green carbon dots for sensing and bioimaging. J Mater Chem B 5(45):8904–8924
11. Sciortino A, Cannizzo A, Messina F (2018) Carbon nanodots: a review—from the current understanding of the fundamental photophysics to the full control of the optical response. C 4 (4):67

12. Sri S, Kumar R, Panda AK, Solanki PR (2018) Highly biocompatible, fluorescence, and zwitterionic carbon dots as a novel approach for bioimaging applications in cancerous cells. ACS Appl Mater Interfaces 10(44):37835–37845
13. Zhu S, Meng Q, Wang L, Zhang J, Song Y, Jin H, Zhang K, Sun H, Wang H, Yang B (2013) Highly photoluminescent carbon dots for multicolor patterning, sensors, and bioimaging. Angew Chem 125(14):4045–4049
14. Baker SN, Baker GA (2010) Luminescent carbon nanodots: emergent nanolights. Angew Chem Int Ed 49(38):6726–6744
15. Dong Y, Wang R, Li G, Chen C, Chi Y, Chen G (2012) Polyamine-functionalized carbon quantum dots as fluorescent probes for selective and sensitive detection of copper ions. Anal Chem 84(14):6220–6224
16. Molaei MJ (2019) Carbon quantum dots and their biomedical and therapeutic applications: a review. RSC Adv 9(12):6460–6481
17. Goryacheva IY, Sapelkin AV, Sukhorukov GB (2017) Carbon nanodots: mechanisms of photoluminescence and principles of application. TrAC Trends Anal Chem 90:27–37
18. Zhou N, Zhu S, Maharjan S, Hao Z, Song Y, Zhao X, Jiang Y, Yang B, Lu L (2014) Elucidating the endocytosis, intracellular trafficking, and exocytosis of carbon dots in neural cells. RSC Adv 4(107):62086–62095
19. Song Y, Zhu S, Yang B (2014) Bioimaging based on fluorescent carbon dots. RSC Adv 4 (52):27184–27200
20. Emam A, Loutfy SA, Mostafa AA, Awad H, Mohamed MB (2017) Cyto-toxicity, biocompatibility and cellular response of carbon dots–plasmonic based nano-hybrids for bioimaging. RSC Adv 7(38):23502–23514
21. Yang S-T, Cao L, Luo PG, Lu F, Wang X, Wang H, Meziani MJ, Liu Y, Qi G, Sun Y-P (2009) Carbon dots for optical imaging in vivo. J Am Chem Soc 131(32):11308–11309
22. Wang Y, Meng Y, Wang S, Li C, Shi W, Chen J, Wang J, Huang R (2015) Direct solvent-derived polymer-coated nitrogen-doped carbon nanodots with high water solubility for targeted fluorescence imaging of glioma. Small 11(29):3575–3581
23. Zhao X, Zhang J, Shi L, Xian M, Dong C, Shuang S (2017) Folic acid-conjugated carbon dots as green fluorescent probes based on cellular targeting imaging for recognizing cancer cells. RSC Adv 7(67):42159–42167
24. Zhang J, Zhao X, Xian M, Dong C, Shuang S (2018) Folic acid-conjugated green luminescent carbon dots as a nanoprobe for identifying folate receptor-positive cancer cells. Talanta 183:39–47
25. Qian J, Quan F, Zhao F, Wu C, Wang Z, Zhou L (2018) Aconitic acid derived carbon dots: conjugated interaction for the detection of folic acid and fluorescence targeted imaging of folate receptor overexpressed cancer cells. Sensors Actuators B Chem 262:444–451
26. Chiu S-H, Gedda G, Girma WM, Chen J-K, Ling Y-C, Ghule AV, Ou K-L, Chang J-Y (2016) Rapid fabrication of carbon quantum dots as multifunctional nanovehicles for dual-modal targeted imaging and chemotherapy. Acta Biomater 46:151–164
27. Nahain A-A, Lee J-E, In I, Lee H, Lee KD, Jeong JH, Park SY (2013) Target delivery and cell imaging using hyaluronic acid-functionalized graphene quantum dots. Mol Pharm 10 (10):3736–3744
28. Motaghi H, Mehrgardi MA, Bouvet P (2017) Carbon dots-AS1411 aptamer nanoconjugate for ultrasensitive spectrofluorometric detection of cancer cells. Sci Rep 7(1):10513
29. Mongelard F, Bouvet P (2010) AS-1411, a guanosine-rich oligonucleotide aptamer targeting nucleolin for the potential treatment of cancer, including acute myeloid leukemia. Curr Opin Mol Ther 12(1):107–114
30. Ma Y, Peng J, Liu W, Zhang P, Huang L, Gao B, Shen T, Zhou Y, Chen H, Chu Z (2009) Proteomics identification of desmin as a potential oncofetal diagnostic and prognostic biomarker in colorectal cancer. Mol Cell Proteomics 8(8):1878–1890
31. Li C-F, Yan Z-K, Chen L-B, Jin J-P, Li D-D (2017) Desmin detection by facile prepared carbon quantum dots for early screening of colorectal cancer. Medicine 96(5):e5521

32. Gao G, Jiang Y-W, Jia H-R, Yang J, Wu F-G (2018) On-off-on fluorescent nanosensor for Fe3+ detection and cancer/normal cell differentiation via silicon-doped carbon quantum dots. Carbon 134:232–243
33. Li H, Zhang Y, Wang L, Tian J, Sun X (2011) Nucleic acid detection using carbon nanoparticles as a fluorescent sensing platform. Chem Commun 47(3):961–963
34. Wu X, Sun S, Wang Y, Zhu J, Jiang K, Leng Y, Shu Q, Lin H (2017) A fluorescent carbon-dots-based mitochondria-targetable nanoprobe for peroxynitrite sensing in living cells. Biosens Bioelectron 90:501–507
35. Chong Y, Ge C, Fang G, Tian X, Ma X, Wen T, Wamer WG, Chen C, Chai Z, Yin J-J (2016) Crossover between anti-and pro-oxidant activities of graphene quantum dots in the absence or presence of light. ACS Nano 10(9):8690–8699
36. Wang X, Cao L, Lu F, Meziani MJ, Li H, Qi G, Zhou B, Harruff BA, Kermarrec F, Sun Y-P (2009) Photoinduced electron transfers with carbon dots. Chem Commun 25:3774–3776
37. Hsu P-C, Chen P-C, Ou C-M, Chang H-Y, Chang H-T (2013) Extremely high inhibition activity of photoluminescent carbon nanodots toward cancer cells. J Mater Chem B 1 (13):1774–1781
38. Ruan J, Wang Y, Li F, Jia R, Zhou G, Shao C, Zhu L, Cui M, Yang D-P, Ge S (2018) Graphene quantum dots for radiotherapy. ACS Appl Mater Interfaces 10(17):14342–14355
39. Kleinauskas A, Rocha S, Sahu S, Sun Y-P, Juzenas P (2013) Carbon-core silver-shell nanodots as sensitizers for phototherapy and radiotherapy. Nanotechnology 24(32):325103
40. Qian M, Du Y, Wang S, Li C, Jiang H, Shi W, Chen J, Wang Y, Wagner E, Huang R (2018) Highly crystalline multicolor carbon nanodots for dual-modal imaging-guided photothermal therapy of glioma. ACS Appl Mater Interfaces 10(4):4031–4040
41. Jiao J, Liu C, Li X, Liu J, Di D, Zhang Y, Zhao Q, Wang S (2016) Fluorescent carbon dot modified mesoporous silica nanocarriers for redox-responsive controlled drug delivery and bioimaging. J Colloid Interface Sci 483:343–352
42. Yao Y-Y, Gedda G, Girma WM, Yen C-L, Ling Y-C, Chang J-Y (2017) Magnetofluorescent carbon dots derived from crab shell for targeted dual-modality bioimaging and drug delivery. ACS Appl Mater Interfaces 9(16):13887–13899
43. Karthik S, Saha B, Ghosh SK, Singh NP (2013) Photoresponsive quinoline tethered fluorescent carbon dots for regulated anticancer drug delivery. Chem Commun 49(89):10471–10473
44. Khan MS, Pandey S, Talib A, Bhaisare ML, Wu H-F (2015) Controlled delivery of dopamine hydrochloride using surface modified carbon dots for neuro diseases. Colloids Surf B: Biointerfaces 134:140–146
45. Feng T, Ai X, Ong H, Zhao Y (2016) Dual-responsive carbon dots for tumor extracellular microenvironment triggered targeting and enhanced anticancer drug delivery. ACS Appl Mater Interfaces 8(29):18732–18740
46. Chowdhuri AR, Singh T, Ghosh SK, Sahu SK (2016) Carbon dots embedded magnetic nanoparticles@ chitosan@ metal organic framework as a nanoprobe for pH sensitive targeted anticancer drug delivery. ACS Appl Mater Interfaces 8(26):16573–16583
47. Wang S, Li C, Qian M, Jiang H, Shi W, Chen J, Lächelt U, Wagner E, Lu W, Wang Y (2017) Augmented glioma-targeted theranostics using multifunctional polymer-coated carbon nanodots. Biomaterials 141:29–39
48. Thévenot DR, Toth K, Durst RA, Wilson GS (2001) Electrochemical biosensors: recommended definitions and classification. Anal Lett 34(5):635–659
49. Wang J (2001) Glucose biosensors: 40 years of advances and challenges. Electroanal Int J Dev Fundam Pract Asp Electroanal 13(12):983–988
50. Tothill IE (2009) Biosensors for cancer markers diagnosis. In: Seminars in cell & developmental biology. Elsevier, Amsterdam, pp 55–62
51. Hu S-W, Qiao S, Xu B-Y, Peng X, Xu J-J, Chen H-Y (2017) Dual-functional carbon dots pattern on paper chips for Fe3+ and ferritin analysis in whole blood. Anal Chem 89 (3):2131–2137

52. Alkhateeb AA, Connor JR (2013) The significance of ferritin in cancer: anti-oxidation, inflammation and tumorigenesis. Biochim Biophys Acta Rev Biomembr 1836(2):245–254
53. Chen Z, Mao R, Liu Y (2012) Fullerenes for cancer diagnosis and therapy: preparation, biological and clinical perspectives. Curr Drug Metab 13(8):1035–1045
54. Krusic P, Wasserman E, Keizer P, Morton J, Preston K (1991) Radical reactions of C60. Science 254(5035):1183–1185
55. Shi J, Wang L, Gao J, Liu Y, Zhang J, Ma R, Liu R, Zhang Z (2014) A fullerene-based multi-functional nanoplatform for cancer theranostic applications. Biomaterials 35(22):5771–5784
56. Shi J, Chen Z, Wang L, Wang B, Xu L, Hou L, Zhang Z (2016) A tumor-specific cleavable nanosystem of PEG-modified C60@ Au hybrid aggregates for radio frequency-controlled release, hyperthermia, photodynamic therapy and X-ray imaging. Acta Biomater 29:282–297
57. Shi J, Wang B, Wang L, Lu T, Fu Y, Zhang H, Zhang Z (2016) Fullerene (C60)-based tumor-targeting nanoparticles with "off-on" state for enhanced treatment of cancer. J Control Release 235:245–258
58. Bilobrov V, Sokolova V, Prylutska S, Panchuk R, Litsis O, Osetskyi V, Evstigneev M, Prylutskyy Y, Epple M, Ritter U (2019) A novel nanoconjugate of landomycin A with C 60 fullerene for cancer targeted therapy: in vitro studies. Cell Mol Bioeng 12(1):41–51
59. Xu J, Wang H, Hu Y, Zhang YS, Wen L, Yin F, Wang Z, Zhang Y, Li S, Miao Y (2019) Inhibition of CaMKIIα activity enhances antitumor effect of fullerene C60 nanocrystals by suppression of autophagic degradation. Adv Sci 6(8):1801233
60. Xia LY, Zheng YN, Liang WB, Li MJ, Hu T, Yuan R, Chai YQ (2019) [Ru (dcbpy) 2dppz] 2 +/fullerene cosensitized PTB7-Th for ultrasensitive photoelectrochemical microRNA assay. Chem Eur J 25(16):4087–4092
61. Bacon R (1960) Growth, structure, and properties of graphite whiskers. J Appl Phys 31 (2):283–290
62. Iijima S (1991) Helical microtubules of graphitic carbon. Nature 354(6348):56
63. Pastorin G, Wu W, Wieckowski S, Briand J-P, Kostarelos K, Prato M, Bianco A (2006) Double functionalisation of carbon nanotubes for multimodal drug delivery. Chem Commun 11:1182–1184
64. Mehra NK, Jain N (2013) Development, characterization and cancer targeting potential of surface engineered carbon nanotubes. J Drug Target 21(8):745–758
65. Karousis N, Tagmatarchis N, Tasis D (2010) Current progress on the chemical modification of carbon nanotubes. Chem Rev 110(9):5366–5397
66. Yang W, Thordarson P, Gooding JJ, Ringer SP, Braet F (2007) Carbon nanotubes for biological and biomedical applications. Nanotechnology 18(41):412001
67. Zhang S, Yang K, Liu Z (2010) Carbon nanotubes for in vivo cancer nanotechnology. Sci China Chem 53(11):2217–2225
68. Kesharwani P, Ghanghoria R, Jain NK (2012) Carbon nanotube exploration in cancer cell lines. Drug Discov Today 17(17–18):1023–1030
69. Kathane LL, Panchabhai SA, Thakare VM (2018) Carbon nano tubes: new techniques for cancer treatment. Trends Drug Deliv 5(1):28–39
70. Sheikhpour M, Golbabaie A, Kasaeian A (2017) Carbon nanotubes: a review of novel strategies for cancer diagnosis and treatment. Mater Sci Eng C 76:1289–1304
71. Dong X, Wei C, Liang J, Liu T, Kong D, Lv F (2017) Thermosensitive hydrogel loaded with chitosan-carbon nanotubes for near infrared light triggered drug delivery. Colloids Surf B: Biointerfaces 154:253–262
72. Tan JM, Karthivashan G, Arulselvan P, Fakurazi S, Hussein MZ (2014) Characterization and in vitro studies of the anticancer effect of oxidized carbon nanotubes functionalized with betulinic acid. Drug Des Devel Ther 8:2333
73. Tan JM, Karthivashan G, Arulselvan P, Fakurazi S, Hussein MZ (2014) Sustained release and cytotoxicity evaluation of carbon nanotube-mediated drug delivery system for betulinic acid. J Nanomater 2014:1

74. Samorì C, Ali-Boucetta H, Sainz R, Guo C, Toma FM, Fabbro C, Da Ros T, Prato M, Kostarelos K, Bianco A (2010) Enhanced anticancer activity of multi-walled carbon nanotube–methotrexate conjugates using cleavable linkers. Chem Commun 46(9):1494–1496
75. Singh R, Mehra NK, Jain V, Jain NK (2013) Gemcitabine-loaded smart carbon nanotubes for effective targeting to cancer cells. J Drug Target 21(6):581–592
76. Mahmood M, Xu Y, Dantuluri V, Mustafa T, Zhang Y, Karmakar A, Casciano D, Ali S, Biris A (2013) Carbon nanotubes enhance the internalization of drugs by cancer cells and decrease their chemoresistance to cytostatics. Nanotechnology 24(4):045102
77. Shao W, Paul A, Zhao B, Lee C, Rodes L, Prakash S (2013) Carbon nanotube lipid drug approach for targeted delivery of a chemotherapy drug in a human breast cancer xenograft animal model. Biomaterials 34(38):10109–10119
78. Arya N, Arora A, Vasu K, Sood AK, Katti DS (2013) Combination of single walled carbon nanotubes/graphene oxide with paclitaxel: a reactive oxygen species mediated synergism for treatment of lung cancer. Nanoscale 5(7):2818–2829
79. Yu B, Tan L, Zheng R, Tan H, Zheng L (2016) Targeted delivery and controlled release of Paclitaxel for the treatment of lung cancer using single-walled carbon nanotubes. Mater Sci Eng C 68:579–584
80. Cao L, Liang Y, Zhao F, Zhao X, Chen Z (2016) Chelerythrine and Fe3O4 loaded multi-walled carbon nanotubes for targeted cancer therapy. J Biomed Nanotechnol 12(6):1312–1322
81. Tian Z, Yin M, Ma H, Zhu L, Shen H, Jia N (2011) Supramolecular assembly and antitumor activity of multiwalled carbon nanotube–camptothecin complexes. J Nanosci Nanotechnol 11 (2):953–958
82. Ringel J, Erdmann K, Hampel S, Kraemer K, Maier D, Arlt M, Kunze D, Wirth MP, Fuessel S (2014) Carbon nanofibers and carbon nanotubes sensitize prostate and bladder cancer cells to platinum-based chemotherapeutics. J Biomed Nanotechnol 10(3):463–477
83. Chen H, Ma X, Li Z, Shi Q, Zheng W, Liu Y, Wang P (2012) Functionalization of single-walled carbon nanotubes enables efficient intracellular delivery of siRNA targeting MDM2 to inhibit breast cancer cells growth. Biomed Pharmacother 66(5):334–338
84. Mohammadi M, Salmasi Z, Hashemi M, Mosaffa F, Abnous K, Ramezani M (2015) Single-walled carbon nanotubes functionalized with aptamer and piperazine–polyethylenimine derivative for targeted siRNA delivery into breast cancer cells. Int J Pharm 485(1–2):50–60
85. Guo C, Al-Jamal WT, Toma FM, Bianco A, Prato M, Al-Jamal KT, Kostarelos K (2015) Design of cationic multiwalled carbon nanotubes as efficient siRNA vectors for lung cancer xenograft eradication. Bioconjug Chem 26(7):1370–1379
86. Anderson T, Hu R, Yang C, Yoon HS, Yong K-T (2014) Pancreatic cancer gene therapy using an siRNA-functionalized single walled carbon nanotubes (SWNTs) nanoplex. Biomater Sci 2 (9):1244–1253
87. Taghavi S, HashemNia A, Mosaffa F, Askarian S, Abnous K, Ramezani M (2016) Preparation and evaluation of polyethylenimine-functionalized carbon nanotubes tagged with 5TR1 aptamer for targeted delivery of Bcl-xL shRNA into breast cancer cells. Colloids Surf B: Biointerfaces 140:28–39
88. Singh RP, Sharma G, Singh S, Patne SC, Pandey BL, Koch B, Muthu MS (2016) Effects of transferrin conjugated multi-walled carbon nanotubes in lung cancer delivery. Mater Sci Eng C 67:313–325
89. Karmakar A, Iancu C, Bartos DM, Mahmood MW, Ghosh A, Xu Y, Dervishi E, Collom SL, Khodakovskaya M, Mustafa T (2012) Raman spectroscopy as a detection and analysis tool for in vitro specific targeting of pancreatic cancer cells by EGF-conjugated, single-walled carbon nanotubes. J Appl Toxicol 32(5):365–375
90. Saeed LM, Mahmood M, Pyrek SJ, Fahmi T, Xu Y, Mustafa T, Nima ZA, Bratton SM, Casciano D, Dervishi E (2014) Single-walled carbon nanotube and graphene nanodelivery of gambogic acid increases its cytotoxicity in breast and pancreatic cancer cells. J Appl Toxicol 34(11):1188–1199

91. Tian Q, Wang Y, Deng R, Lin L, Liu Y, Li J (2015) Carbon nanotube enhanced label-free detection of microRNAs based on hairpin probe triggered solid-phase rolling-circle amplification. Nanoscale 7(3):987–993
92. Sardesai NP, Barron JC, Rusling JF (2011) Carbon nanotube microwell array for sensitive electrochemiluminescent detection of cancer biomarker proteins. Anal Chem 83 (17):6698–6703
93. Weizmann Y, Chenoweth DM, Swager TM (2011) DNA– cnt nanowire networks for DNA detection. J Am Chem Soc 133(10):3238–3241
94. Okuno J, Maehashi K, Kerman K, Takamura Y, Matsumoto K, Tamiya E (2007) Label-free immunosensor for prostate-specific antigen based on single-walled carbon nanotube array-modified microelectrodes. Biosens Bioelectron 22(9–10):2377–2381
95. Tian J, Huang J, Zhao Y, Zhao S (2012) Electrochemical immunosensor for prostate-specific antigen using a glassy carbon electrode modified with a nanocomposite containing gold nanoparticles supported with starch-functionalized multi-walled carbon nanotubes. Microchim Acta 178(1–2):81–88
96. Yu X, Munge B, Patel V, Jensen G, Bhirde A, Gong JD, Kim SN, Gillespie J, Gutkind JS, Papadimitrakopoulos F (2006) Carbon nanotube amplification strategies for highly sensitive immunodetection of cancer biomarkers. J Am Chem Soc 128(34):11199–11205
97. Zou Y, Xiang C, Sun L, Xu F, Zhou H (2014) Ultrasensitive prostate specific antigen immunosensor based on gold nanoparticles functionalized polypyrrole@ carbon nanotubes. Asian J Chem 26(23):8002–8006
98. Lerner MB, D'Souza J, Pazina T, Dailey J, Goldsmith BR, Robinson MK, Johnson AC (2012) Hybrids of a genetically engineered antibody and a carbon nanotube transistor for detection of prostate cancer biomarkers. ACS Nano 6(6):5143–5149
99. Sharma A, Hong S, Singh R, Jang J (2015) Single-walled carbon nanotube based transparent immunosensor for detection of a prostate cancer biomarker osteopontin. Anal Chim Acta 869:68–73
100. Liu L, Song C, Zhang Z, Yang J, Zhou L, Zhang X, Xie G (2015) Ultrasensitive electrochemical detection of microRNA-21 combining layered nanostructure of oxidized single-walled carbon nanotubes and nanodiamonds by hybridization chain reaction. Biosens Bioelectron 70:351–357
101. Tran H, Piro B, Reisberg S, Tran L, Duc H, Pham M (2013) Label-free and reagentless electrochemical detection of microRNAs using a conducting polymer nanostructured by carbon nanotubes: application to prostate cancer biomarker miR-141. Biosens Bioelectron 49:164–169
102. Tran H, Piro B, Reisberg S, Nguyen LH, Nguyen TD, Duc H, Pham M (2014) An electrochemical ELISA-like immunosensor for miRNAs detection based on screen-printed gold electrodes modified with reduced graphene oxide and carbon nanotubes. Biosens Bioelectron 62:25–30
103. Ramnani P, Gao Y, Ozsoz M, Mulchandani A (2013) Electronic detection of microRNA at attomolar level with high specificity. Anal Chem 85(17):8061–8064
104. Li F, Peng J, Wang J, Tang H, Tan L, Xie Q, Yao S (2014) Carbon nanotube-based label-free electrochemical biosensor for sensitive detection of miRNA-24. Biosens Bioelectron 54:158–164
105. Xu H, Wang Y, Wang L, Song Y, Luo J, Cai X (2016) A label-free microelectrode array based on one-step synthesis of chitosan–multi-walled carbon nanotube–thionine for ultrasensitive detection of carcinoembryonic antigen. Nano 6(7):132
106. Kumar S, Willander M, Sharma JG, Malhotra BD (2015) A solution processed carbon nanotube modified conducting paper sensor for cancer detection. J Mater Chem B 3 (48):9305–9314
107. Feng D, Lu X, Dong X, Ling Y, Zhang Y (2013) Label-free electrochemical immunosensor for the carcinoembryonic antigen using a glassy carbon electrode modified with electrodeposited

Prussian Blue, a graphene and carbon nanotube assembly and an antibody immobilized on gold nanoparticles. Microchim Acta 180(9–10):767–774

108. Zhang Y, Chen H, Gao X, Chen Z, Lin X (2012) A novel immunosensor based on an alternate strategy of electrodeposition and self-assembly. Biosens Bioelectron 35(1):277–283
109. Gao X, Zhang Y, Chen H, Chen Z, Lin X (2011) Amperometric immunosensor for carcinoembryonic antigen detection with carbon nanotube-based film decorated with gold nanoclusters. Anal Biochem 414(1):70–76
110. Yang P, Li X, Wang L, Wu Q, Chen Z, Lin X (2014) Sandwich-type amperometric immunosensor for cancer biomarker based on signal amplification strategy of multiple enzyme-linked antibodies as probes modified with carbon nanotubes and concanavalin A. J Electroanal Chem 732:38–45
111. Cheng H, Lai G, Fu L, Zhang H, Yu A (2015) Enzymatically catalytic deposition of gold nanoparticles by glucose oxidase-functionalized gold nanoprobe for ultrasensitive electrochemical immunoassay. Biosens Bioelectron 71:353–358
112. Deng W, Liu F, Ge S, Yu J, Yan M, Song X (2014) A dual amplification strategy for ultrasensitive electrochemiluminescence immunoassay based on a Pt nanoparticles dotted graphene–carbon nanotubes composite and carbon dots functionalized mesoporous Pt/Fe. Analyst 139(7):1713–1720
113. Hu C, Zheng J, Su X, Wang J, Wu W, Hu S (2013) Ultrasensitive all-carbon photoelectrochemical bioprobes for zeptomole immunosensing of tumor markers by an inexpensive visible laser light. Anal Chem 85(21):10612–10619
114. Li N, Wang Y, Cao W, Zhang Y, Yan T, Du B, Wei Q (2015) An ultrasensitive electrochemical immunosensor for CEA using MWCNT-NH 2 supported PdPt nanocages as labels for signal amplification. J Mater Chem B 3(9):2006–2011
115. Chen S, Yuan R, Chai Y, Min L, Li W, Xu Y (2009) Electrochemical sensing platform based on tris (2, 2′-bipyridyl) cobalt (III) and multiwall carbon nanotubes–Nafion composite for immunoassay of carcinoma antigen-125. Electrochim Acta 54(28):7242–7247
116. Paul KB, Singh V, Vanjari SRK, Singh SG (2017) One step biofunctionalized electrospun multiwalled carbon nanotubes embedded zinc oxide nanowire interface for highly sensitive detection of carcinoma antigen-125. Biosens Bioelectron 88:144–152
117. Ding Y, Liu J, Jin X, Lu H, Shen G, Yu R (2008) Poly-L-lysine/hydroxyapatite/carbon nanotube hybrid nanocomposite applied for piezoelectric immunoassay of carbohydrate antigen 19-9. Analyst 133(2):184–190
118. Chen X, Zhang Q, Qian C, Hao N, Xu L, Yao C (2015) Electrochemical aptasensor for mucin 1 based on dual signal amplification of poly (o-phenylenediamine) carrier and functionalized carbon nanotubes tracing tag. Biosens Bioelectron 64:485–492
119. Arkan E, Saber R, Karimi Z, Shamsipur M (2015) A novel antibody–antigen based impedimetric immunosensor for low level detection of HER2 in serum samples of breast cancer patients via modification of a gold nanoparticles decorated multiwall carbon nanotube-ionic liquid electrode. Anal Chim Acta 874:66–74
120. Asav E, Sezgintürk MK (2014) A novel impedimetric disposable immunosensor for rapid detection of a potential cancer biomarker. Int J Biol Macromol 66:273–280
121. Sánchez-Tirado E, Salvo C, González-Cortés A, Yáñez-Sedeño P, Langa F, Pingarrón J (2017) Electrochemical immunosensor for simultaneous determination of interleukin-1 beta and tumor necrosis factor alpha in serum and saliva using dual screen printed electrodes modified with functionalized double–walled carbon nanotubes. Anal Chim Acta 959:66–73
122. Munge BS, Fisher J, Millord LN, Krause CE, Dowd RS, Rusling JF (2010) Sensitive electrochemical immunosensor for matrix metalloproteinase-3 based on single-wall carbon nanotubes. Analyst 135(6):1345–1350
123. Tu M-C, Chen H-Y, Wang Y, Moochhala SM, Alagappan P, Liedberg B (2015) Immunosensor based on carbon nanotube/manganese dioxide electrochemical tags. Anal Chim Acta 853:228–233

124. Roy S, Jaiswal A (2017) Graphene-based nanomaterials for theranostic applications. Rep Adv Phys Sci 1(04):1750011
125. Bourlinos AB, Steriotis TA, Zboril R, Georgakilas V, Stubos A (2009) Direct synthesis of carbon nanosheets by the solid-state pyrolysis of betaine. J Mater Sci 44(5):1407–1411
126. Sinitskii A, Dimiev A, Kosynkin DV, Tour JM (2010) Graphene nanoribbon devices produced by oxidative unzipping of carbon nanotubes. ACS Nano 4(9):5405–5413
127. Terrones M, Botello-Méndez AR, Campos-Delgado J, Lopez-Urias F, Vega-Cantú YI, Rodríguez-Macías FJ, Elías AL, Munoz-Sandoval E, Cano-Márquez AG, Charlier J-C (2010) Graphene and graphite nanoribbons: morphology, properties, synthesis, defects and applications. Nano Today 5(4):351–372
128. James DK, Tour JM (2012) The chemical synthesis of graphene nanoribbons—a tutorial review. Macromol Chem Phys 213(10–11):1033–1050
129. Dreyer DR, Park S, Bielawski CW, Ruoff RS (2010) The chemistry of graphene oxide. Chem Soc Rev 39(1):228–240
130. Georgakilas V, Otyepka M, Bourlinos AB, Chandra V, Kim N, Kemp KC, Hobza P, Zboril R, Kim KS (2012) Functionalization of graphene: covalent and non-covalent approaches, derivatives and applications. Chem Rev 112(11):6156–6214
131. Yang K, Feng L, Hong H, Cai W, Liu Z (2013) Preparation and functionalization of graphene nanocomposites for biomedical applications. Nat Protoc 8(12):2392
132. Patel SC, Lee S, Lalwani G, Suhrland C, Chowdhury SM, Sitharaman B (2016) Graphene-based platforms for cancer therapeutics. Ther Deliv 7(2):101–116
133. Vilela P, El-Sagheer A, Millar TM, Brown T, Muskens OL, Kanaras AG (2016) Graphene oxide-upconversion nanoparticle based optical sensors for targeted detection of mRNA biomarkers present in Alzheimer's disease and prostate cancer. ACS Sens 2(1):52–56
134. Zhang B, Li Q, Cui T (2012) Ultra-sensitive suspended graphene nanocomposite cancer sensors with strong suppression of electrical noise. Biosens Bioelectron 31(1):105–109
135. Xu T, Liu N, Yuan J, Ma Z (2015) Triple tumor markers assay based on carbon–gold nanocomposite. Biosens Bioelectron 70:161–166
136. Gao Z, Vishnubhotla R, Ducos P, Serrano MD, Ping J, Robinson MK, Johnson ATC (2016) Genetically engineered antibody functionalized platinum nanoparticles modified CVD-graphene nanohybrid transistor for the detection of breast cancer biomarker, HER3. Adv Mater Interfaces 3(17):1600124
137. Cheng F-F, He T-T, Miao H-T, Shi J-J, Jiang L-P, Zhu J-J (2015) Electron transfer mediated electrochemical biosensor for microRNAs detection based on metal ion functionalized titanium phosphate nanospheres at attomole level. ACS Appl Mater Interfaces 7(4):2979–2985
138. Li J, Hu X, Shi S, Zhang Y, Yao T (2016) Three label-free thrombin aptasensors based on aptamers and [Ru (bpy) 2 (o-mopip)] 2+. J Mater Chem B 4(7):1361–1367
139. Feng L, Wu L, Wang J, Ren J, Miyoshi D, Sugimoto N, Qu X (2012) Detection of a prognostic indicator in early-stage cancer using functionalized graphene-based peptide sensors. Adv Mater 24(1):125–131
140. Jia L-P, Liu J-F, Wang H-S (2015) Electrochemical performance and detection of 8-Hydroxy-2′-deoxyguanosine at single-stranded DNA functionalized graphene modified glassy carbon electrode. Biosens Bioelectron 67:139–145
141. Kumar S, Sharma JG, Maji S, Malhotra BD (2016) Nanostructured zirconia decorated reduced graphene oxide based efficient biosensing platform for non-invasive oral cancer detection. Biosens Bioelectron 78:497–504
142. Yu S-J, Kang M-W, Chang H-C, Chen K-M, Yu Y-C (2005) Bright fluorescent nanodiamonds: no photobleaching and low cytotoxicity. J Am Chem Soc 127 (50):17604–17605
143. Gaebel T, Popa I, Gruber A, Domhan M, Jelezko F, Wrachtrup J (2004) Stable single-photon source in the near infrared. New J Phys 6(1):98

144. Kong X, Huang LL, Liau S-CV, Han C-C, Chang H-C (2005) Polylysine-coated diamond nanocrystals for MALDI-TOF mass analysis of DNA oligonucleotides. Anal Chem 77 (13):4273–4277
145. Fu C-C, Lee H-Y, Chen K, Lim T-S, Wu H-Y, Lin P-K, Wei P-K, Tsao P-H, Chang H-C, Fann W (2007) Characterization and application of single fluorescent nanodiamonds as cellular biomarkers. Proc Natl Acad Sci 104(3):727–732
146. Chao J-I, Perevedentseva E, Chung P-H, Liu K-K, Cheng C-Y, Chang C-C, Cheng C-L (2007) Nanometer-sized diamond particle as a probe for biolabeling. Biophys J 93(6):2199–2208
147. Chang Y-R, Lee H-Y, Chen K, Chang C-C, Tsai D-S, Fu C-C, Lim T-S, Tzeng Y-K, Fang C-Y, Han C-C (2008) Mass production and dynamic imaging of fluorescent nanodiamonds. Nat Nanotechnol 3(5):284
148. Faklaris O, Joshi V, Irinopoulou T, Tauc P, Sennour M, Girard H, Gesset C, Arnault J-C, Thorel A, Boudou J-P (2009) Photoluminescent diamond nanoparticles for cell labeling: study of the uptake mechanism in mammalian cells. ACS Nano 3(12):3955–3962
149. Zhang B, Li Y, Fang CY, Chang CC, Chen CS, Chen YY, Chang HC (2009) Receptor-mediated cellular uptake of folate-conjugated fluorescent nanodiamonds: a combined ensemble and single-particle study. Small 5(23):2716–2721
150. Hui YY, Su L-J, Chen OY, Chen Y-T, Liu T-M, Chang H-C (2014) Wide-field imaging and flow cytometric analysis of cancer cells in blood by fluorescent nanodiamond labeling and time gating. Sci Rep 4:5574
151. Mohan N, Chen C-S, Hsieh H-H, Wu Y-C, Chang H-C (2010) In vivo imaging and toxicity assessments of fluorescent nanodiamonds in Caenorhabditis elegans. Nano Lett 10 (9):3692–3699
152. Vaijayanthimala V, Cheng P-Y, Yeh S-H, Liu K-K, Hsiao C-H, Chao J-I, Chang H-C (2012) The long-term stability and biocompatibility of fluorescent nanodiamond as an in vivo contrast agent. Biomaterials 33(31):7794–7802
153. Igarashi R, Yoshinari Y, Yokota H, Sugi T, Sugihara F, Ikeda K, Sumiya H, Tsuji S, Mori I, Tochio H (2012) Real-time background-free selective imaging of fluorescent nanodiamonds in vivo. Nano Lett 12(11):5726–5732
154. Titirici MM, Thomas A, Antonietti M (2007) Replication and coating of silica templates by hydrothermal carbonization. Adv Funct Mater 17(6):1010–1018
155. Hu B, Wang K, Wu L, Yu SH, Antonietti M, Titirici MM (2010) Engineering carbon materials from the hydrothermal carbonization process of biomass. Adv Mater 22(7):813–828
156. Titirici M-M, Thomas A, Antonietti M (2007) Aminated hydrophilic ordered mesoporous carbons. J Mater Chem 17(32):3412–3418
157. Gu J, Su S, Li Y, He Q, Shi J (2011) Hydrophilic mesoporous carbon nanoparticles as carriers for sustained release of hydrophobic anti-cancer drugs. Chem Commun 47(7):2101–2103
158. Li C, Qian M, Wang S, Jiang H, Du Y, Wang J, Lu W, Murthy N, Huang R (2017) Aptavalve-gated mesoporous carbon nanospheres image cellular mucin and provide on-demand targeted drug delivery. Theranostics 7(13):3319
159. Meng Y, Wang S, Li C, Qian M, Yan X, Yao S, Peng X, Wang Y, Huang R (2016) Photothermal combined gene therapy achieved by polyethyleneimine-grafted oxidized mesoporous carbon nanospheres. Biomaterials 100:134–142
160. Wang Y, Wang K, Zhang R, Liu X, Yan X, Wang J, Wagner E, Huang R (2014) Synthesis of core–shell graphitic carbon@ silica nanospheres with dual-ordered mesopores for cancer-targeted photothermochemotherapy. ACS Nano 8(8):7870–7879
161. Wang S, Li C, Meng Y, Qian M, Jiang H, Du Y, Huang R, Wang Y (2017) MemHsp70 receptor-mediated multifunctional ordered mesoporous carbon nanospheres for photoacoustic imaging-guided synergistic targeting trimodal therapy. ACS Biomater Sci Eng 3 (8):1702–1709
162. Pirsaheb M, Mohammadi S, Salimi A (2019) Current advances of carbon dots based biosensors for tumor marker detection, cancer cells analysis and bioimaging. TrAC Trends Anal Chem 115:83–99

163. Beaudoin D, Maris T, Wuest JD (2013) Constructing monocrystalline covalent organic networks by polymerization. Nat Chem 5(10):830
164. Oberdörster G, Maynard A, Donaldson K, Castranova V, Fitzpatrick J, Ausman K, Carter J, Karn B, Kreyling W, Lai D (2005) Principles for characterizing the potential human health effects from exposure to nanomaterials: elements of a screening strategy. Part Fibre Toxicol 2 (1):8
165. Kayat J, Gajbhiye V, Tekade RK, Jain NK (2011) Pulmonary toxicity of carbon nanotubes: a systematic report. Nanomedicine 7(1):40–49
166. Mehra NK, Jain AK, Lodhi N, Raj R, Dubey V, Mishra D, Nahar M, Jain NK (2008) Challenges in the use of carbon nanotubes for biomedical applications. Crit Rev Ther Drug Carrier Syst 25(2):169–206
167. Mochalin VN, Shenderova O, Ho D, Gogotsi Y (2012) The properties and applications of nanodiamonds. Nat Nanotechnol 7(1):11
168. Zeynalov EB, Allen NS, Salmanova NI (2009) Radical scavenging efficiency of different fullerenes C60–C70 and fullerene soot. Polym Degrad Stab 94(8):1183–1189
169. Liu H, Ryu S, Chen Z, Steigerwald ML, Nuckolls C, Brus LE (2009) Photochemical reactivity of graphene. J Am Chem Soc 131(47):17099–17101
170. Maggini M, Scorrano G, Prato M (1993) Addition of azomethine ylides to C60: synthesis, characterization, and functionalization of fullerene pyrrolidines. J Am Chem Soc 115 (21):9798–9799

Chapter 8
Silica-Based Tumor-targeted Systems

Wei Guo, Min Qian, Xiaoyi Zhang, and Yi Wang

Abstract In recent decades, silica-based nanomaterials have emerged as a kind of novel and multifunctional drug delivery systems (DDSs), expanding new applications in inorganic materials as well as laying a foundation for organic–inorganic hybrids tumor-targeted DDS. Silica-based delivery systems especially mesoporous ones exhibit attractive characteristics, which guarantee the high loading of diverse cargo molecules for biosensing, drug delivery, tumor imaging, nanocatalysis, and so on. Here, we categorized these silica-based DDSs according to their morphology and composition, hoping to provide a more concise and detailed insight into this potential material on current trends both in lab and clinic.

Keywords Tumor-targeted system · Silica nanosheets · Silica nanoparticles · Mesoporous nanomaterials · Medical applications

8.1 Introduction

As early as the beginning of the twenty-first century, researchers applied silica nanoparticles in solid lipid nanoparticles (SLN) for binding and transfecting plasmid DNA efficiently [1]. During the following decades, silica-based nanomaterials have emerged as a kind of novel and multifunctional drug delivery system (DDS), expanding new applications in inorganic materials as well as laying a foundation for organic–inorganic hybrids tumor-targeted DDS [2]. At the same time, the development of mesoporous technology also greatly improved the corresponding properties of

W. Guo · M. Qian · X. Zhang
Department of Pharmaceutics, School of Pharmacy, Key Laboratory of Smart Drug Delivery, Ministry of Education, Fudan University, Shanghai, China

Y. Wang (✉)
Center for Advanced Low-dimension Materials, Donghua University, Shanghai, China
e-mail: ywang@dhu.edu.cn

R. Huang, Y. Wang (eds.), *New Nanomaterials and Techniques for Tumor-targeted Systems*, https://doi.org/10.1007/978-981-15-5159-8_8

silica-based nanomaterials. Compared with those flexible inorganic materials for biomedical application, silica-based delivery system especially mesoporous ones exhibit attractive characteristics, such as easy surface functionalization, good biocompatibility, high specific surface area, stable porous structure, and tunable pore size, which guarantee the high loading of diverse cargo molecules for biosensing, drug delivery, tumor imaging, nanocatalysis, and so on [3–5].

Here, aiming to present abundant research progress of this kind of tumor-targeted delivery systems in a more orderly manner, we categorized these silica-based DDS according to their morphology and composition (nanosheets, Janus nanoparticles, nanoshells/nanocapsules, nanospheres, and organosilica nanosystems), hoping to provide a more concise and detailed insight into this potential material on current trends both in lab and clinic.

8.2 Silica Nanosheets

Compared to other inorganic nanomaterials (nanoparticles, nanotubes, nanobelts, and nanorods), the nanosheets have only one dimension (their thicknesses) in nanoscale. The most used inorganic nanosheets, the layered silicates and oxides, were nonconductor. Researchers have studied the electrical conductivity, photoelectric properties, and stability of the nanosheets, which are widely used to develop new nanocomposites.

Peng Liu researched on the electrical conductivities of the polyaniline-based nanocomposites and found that the electrical conductivities of the polyaniline/silica nanosheet composites (PANI/SNS) increased with the adding of the contents of the silica nanosheets due to the moisture absorption [6].

Silicon nanosheets have a photoelectric property very similar to graphene. However, the nanosheets are quite fragile and decompose rapidly when exposed to UV radiation, thus significantly limiting their application. Hence, Alina Lyuleeva designed novel Si-based 2D nanosheets modified with selected organic components [7]. The implementation of these silicon nanosheets in a covalent nanocomposite not only improves their stability but also facilitates subsequent device fabrication. Their research not only opens a new field of photosensitive applications but also improves the processability of these nanosheets.

Up till now, numerous researches on the properties of silica nanosheets have promoted their application in novel nanocomposites. Zheng-Ming Wang modified graphene oxide (GO) as a novel template for the self-assembly and growth of PMS films on both sides of the GO, forming the first example of a sandwich reduced graphene oxide−periodic mesoporous silica nanocomposite in which the channels are oriented perpendicularly to the surface of rGO platelets. The new nanocomposites unraveled in this study bode well for many different kinds of basic and applied research opportunities that take advantage of the synergistic integration of GO and PMS into reduced graphene oxide–PMS sandwich structures, such as sensing with molecule tunable selectivity, patterning through self-assembly,

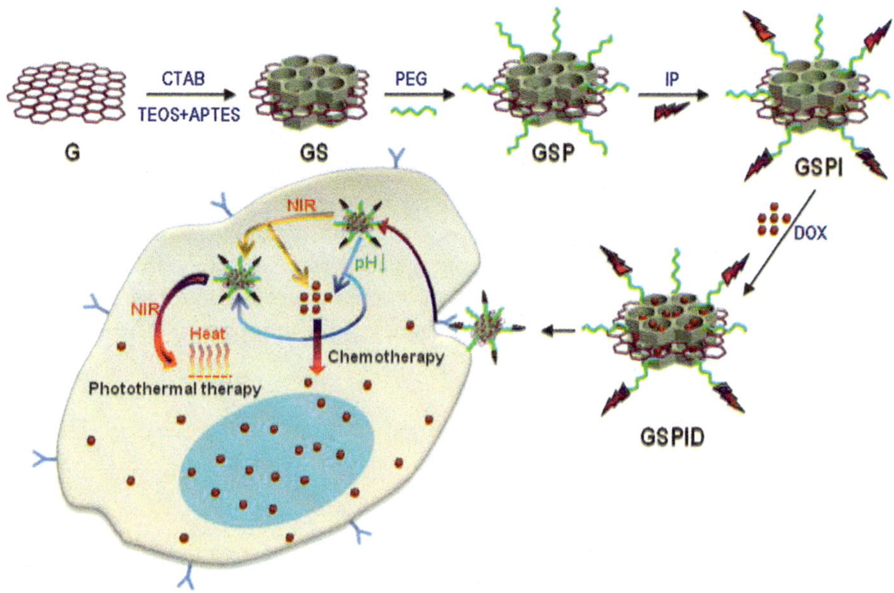

Fig. 8.1 Scheme for the design of GSPID as a multifunctional drug delivery system for combined chemo-photothermal-targeted therapy of glioma. The mesoporous silica-coated graphene nanosheet (GS) was prepared by hydrothermal reaction of CTAB, TEOS, APTES, and graphene oxide (G) nanosheet. IP, a peptide corresponding to the residues within interleukin 13, was used as a glioma-targeting ligand to modify mesoporous silica nanoparticles to form GSPI. GSPI-loaded chemotherapeutic drug doxorubicin (DOX) was named GSPID, the final pattern [9]. (Copyright 2013, American Chemical Society)

electrically stimulated drug delivery, and mesochannel membrane for electrochemical growth of short metal nanorods and barcode metal nanorods [8].

With a similar shape of sheet, silicon nanosheets have shown several superior properties to graphene nanosheets. Combining silicon nanosheets with other materials, such as graphene nanosheets, can produce composites with better properties, which has become a research hotspot. Graphene nanosheet is a photosensitizer that shows a great potential ability for photothermal therapy in vitro and in vivo. However, low drug loading efficiency, insolubility, and difficulty for interfacial interaction with the targeting matrix greatly limit its biomedical applications. Recently, Yi Wang et al. synthesized the mesoporous silica-coated graphene nanosheet (GS) to improve the interfacial properties of graphene and integrate the advantages of both materials as drug delivery vectors such as enlarging the surface area, enhancing the hydrophilicity and dispersity, being more easily covalently functionalized, and achieving high drug loading efficiency. Moreover, GS was explored as an effective bifunctional vector for chemo-photothermal therapy in their work (Fig. 8.1) [9].

Except for photothermal therapy (PTT), coating with a silica layer on nanosheet was also exploited in photodynamic therapy (PDT) and sonodynamic therapy (SDT). Covalently conjugated with red-light-excited photosensitizer of chlorin e6

(Ce6), the silica coating layer has full utilization of the markedly enhanced red emission for ROS generation. Facilely grafted the obtained PDT nanoparticles on the MoS_2 nanosheet, Jiating Xu combined PDT and PTT in one nanosystem, which provides a promising approach to induce superior cancer cell apoptosis and improve antitumor effectiveness [10].

Meanwhile, Yu-Wei Chen and his co-authors developed the medical application of silicon nanosheet composites in SDT. They strategically designed a nanosystem, mesoporous silica (MSN) grown on reduced graphene oxide nanosheet (nrGO) capped with Rose Bengal (RB)–PEG-conjugated iron oxide nanoparticles (IONs), nrGO@MSN-ION-PEG-RB, realizing the combination effect induced by focused ultrasound irradiation produces significant damage to both superficial and deep parts of the targeted tumor [11].

8.3 Silica Nanorods/Microrod

Silicon nanorods are defined herein as three-dimensional nanoparticles with rod-like structures having a certain aspect ratio, including nanorods, microrod, and nanowires.

Nanoparticles with high aspect ratio (AR: length/width) have been reported to exhibit faster rate and greater capacity to enter cells than those with a low AR [12], because elongated nanomaterials have greater surface area to interact with the cell surface. Several groups have demonstrated that nanoparticles with rod-like structures and high aspect ratios have increased internalization rates and cell uptake, high drug loading capacities, and longer circulation times.

Nanorod-based drug delivery systems could penetrate and accumulate into tumors more rapidly than spherical nanoparticles due to their greater surface area and transmembrane transporting rates. Therefore, nanorods have been shown to penetrate tumors more rapidly and cause more drugs to accumulate inside the tumor due to having greater transmembrane transport and diffusion rates than nanospheres with the same effective hydrodynamic diameter [13]. Xinglu Huang compared the biological behaviors of different MSNs and found that larger AR MSNs generally affect the particle cell uptake, cell viability, and apoptosis [14].

Rod-like mesoporous silica nanoparticles have been synthesized in different ways. Xue-Jun Wu reported a unique transformation route, using a surfactant mixture composed of zwitterionic and anionic surfactants as the templates with the assistance of a costructure-directing agent, for the synthesis of rod-like hollow mesoporous silica particles [15]. By increasing the amount of Ce6 doped inside the silica matrix, Guangbao Yang found that the morphology of MSNs changes from spheres to rod-like shapes [16].

Mesoporous silica nanorods (MSNRs) with different lengths were synthesized by Yuanyuan You and were used as nanocarriers to achieve higher drug loading and anticancer activity. As expected, MSNR-based drug delivery systems can effectively enhance the loading capacity of drugs and penetrate into tumor cells more rapidly than spherical nanoparticles due to their greater surface area and transmembrane transporting rates. Interestingly, these tailored MSNRs also enhance the cellular uptake of doxorubicin (DOX) in cancer cells, thus significantly enhancing its anticancer efficacy for hundreds of times by inducing cell apoptosis [17].

Silicon nanorods are also used to achieve effective PPT in the treatment of tumors. Shun Shen developed mesoporous silica-encapsulated gold nanorods (GNRs@mSiO_2) as NIR-light-absorbing agents and a drug delivery system for in vivo chemo-photothermal destruction of tumor. GNRs@mSiO_2 with rod-like structure and intrinsic optical properties to convert NIR light into heat can penetrate and accumulate into tumors more rapidly, as well as realize efficient photothermal therapy. The light-triggered DOX release could be acquired with NIR light as the external stimulus [18].

Silicon nanorods have extremely high drug loading capacity, which is exciting and interesting. Fei Peng produced silicon nanowires (SiNWs) with about 100 nm in diameter and about 500 nm in length for high-performance intracellular delivery of the antitumor drug DOX, that features an extremely large DOX-loading capacity (20,800 mg/g) much higher than those previously reported for nanomaterial-based drug carriers (about 120–4000 mg/g) such as mesoporous silica structure-based nanocarriers and single-walled carbon nanotubes [19].

Hollow-structured mesoporous silica materials (hMSM) which have well-defined structures, low density, huge hollow cavities, and copious mesoporous channels have aroused wide attention in many fields [20]. Xue Yang reported a straightforward and effective synthetic strategy for the synthesis of aspect-ratios-controllable mesoporous silica nanorods with hollow structure (hMSR) and its application for transcription factor (TF)-responsive drug delivery intracellular. It was demonstrated that the as-prepared hMSRs have good stability, high drug loading capacity, and fast cell uptake capability in cancer cell, which make them a potential nanocarrier for drug delivery in tumor therapy [21].

8.4 Janus Silica Nanoparticles

Owing to the combination of two distinct sides with different materials in one single unit and the asymmetry in surface chemistry, JNPs have more room to perform different functions for cancer therapy compared with the traditional core–shell NPs [22].

When dual drugs are simultaneously loaded in the single storage space, the release of each drug cannot be controlled independently. In addition, the loaded multidrugs may interact with each other leading to undesirable adverse effects, especially for drugs with different chemical properties (e.g., hydrophilicity/hydrophobicity, acidity/basicity, etc.). Therefore, the Janus nanostructure, possessing dual surface structures, is ideally suited for dual guests attaching to different domains of the Janus particles.

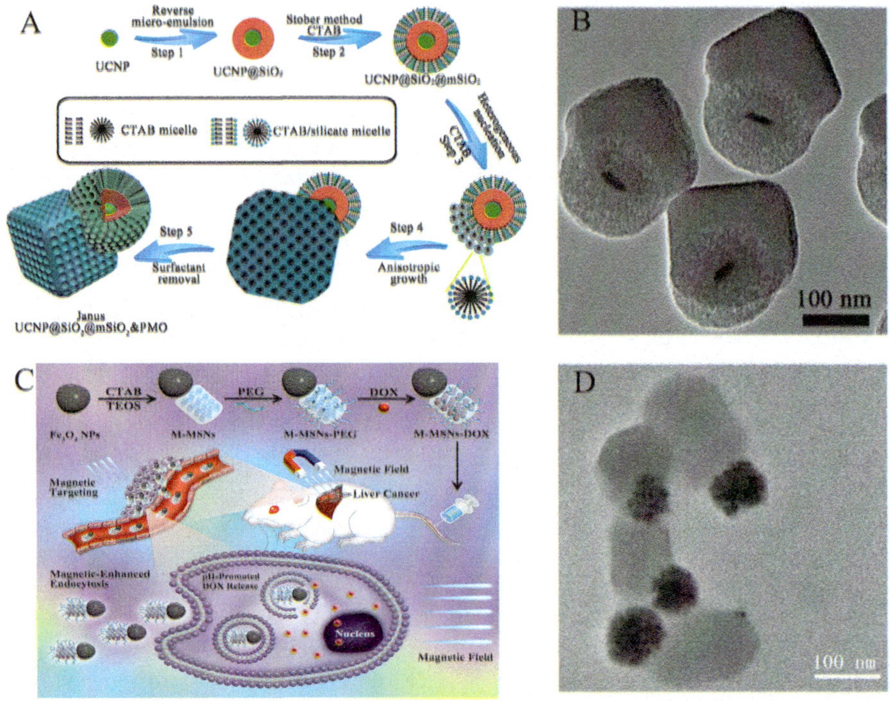

Fig. 8.2 (**a**) Schematic representation of the synthesis of UCNP@SiO2@mSiO2&PMO. (**b**) The TEM image of UCNP@SiO2@mSiO2&PMO [23] (Copyright 2014, American Chemical Society). (**c**) Schematic representation of the synthesis of magnetic Fe3O4 head with a mesoporous SiO2 body. (**d**) The TEM image of M-MSNs-DOX [24] (Copyright 2016, Elsevier Ltd)

Xiaomin Li and his co-worker synthesized multifunctional dual-compartment Janus mesoporous silica nanocomposites (UCNP@SiO2@mSiO2&PMO) based on the novel anisotropic island nucleation and growth of ordered silica mesostructures. The asymmetric Janus nanocomposites are composed by upconversion nanoparticles core@shell@shell functionalized structured UCNP@SiO2@mSiO2 nanoparticles and periodic mesoporous organosilica (PMO) single-crystal nanocubes and possess the unique dual independent mesopores and hydrophobicity/hydrophilicity domains for loading of dual guests, allowing more opportunities in multidrug delivery and combined therapy (Fig. 8.2a, b) [23].

Also, there can be only one of the two distinct parts of JNPs (usually MSN) that is used for loading drugs, while the other part (Fe3O4, Au, Ag, etc.) is used to realize target function or synergistic cancer therapy.

Dan Shao engineered multifunctional Janus nanocomposites combining a magnetic Fe3O4 head with a mesoporous SiO2 body, M-MSNs-DOX. The strong magnetism contributes to tumor-targeted delivery mediated by the enhanced permeability and retention (EPR) effect, and the antitumor agent DOX selectively inhibits cancer cell growth especially in magnetic field environment, while showing a weak influence on human normal cells in vitro (Fig. 8.2c, d) [24].

In addition to targeting function, the other side has also been developed for synergistic therapeutic effects. Lingyu Zhang obtained multifunctional octopus-type PEG-Au-PAA/mSiO2-LA Janus NPs (PEG-OJNP-LA), which were designed to integrate NIR light and pH dual stimuli-responsive properties, drug loading capacity, photothermal effect, and specific tumor targeting into one single PEG-OJNP-LA and can act as an efficient multifunctional nanoplatform for actively-targeted chemo-photothermal synergistic cancer therapy in vitro and in vivo [25]. Similarly, Zheng Wang constructed a Janus nanoplatform which could integrate indocyanine green-based photothermal therapy and silver ion-based chemotherapy in a cascade manner. In their study, ICG was released on-demand from MSN under near-infrared (NIR) irradiation to achieve photothermal therapy. And the silver ions released from the silver compartment which were triggered by light could induce efficient chemotherapy-combined photothermal therapy [26].

8.5 Silica Nanoshells/Nanocapsules

8.5.1 Hollow Mesoporous Silica (HMS)

Core–shell template@silica is first synthesized by coating the prefabricated template with a layer of sol–gel-derived silica [27]. Now there are many reported synthetic approaches, which can be classified into two major categories: hard-templating methods and soft-templating methods.

In the conventional hard-templating methods, some non-silica nanomaterials with excellent affinity with silica, such as polymer beads and oxide particles, were usually used as the templates. The desired HMS is obtained after selective removal of the templates and pore-making agents. The soft-templating methods usually employ surfactant-based micelle/microemulsion as the templates for simultaneously building a mesoporous silica shell and hollow interior by the co-assembly of the surfactant and sol–gel-derived silica. Due to the difficulty in maintaining the uniformity of the templates in solution, the size distribution for HMS made by the soft-templating methods is typically poor [28]. Interestingly, bowl-like nanostructure was reported by using the similar synthetic method of hollow nanosphere. After the removal of templates, the remaining highly porous carbon framework in N-HPCB is not rigid enough to support the hollow structure; the porous carbon shell thus collapsed into its internal cavity and finally forms the bowl-like structure [29].

After that, a number of self-templating methods have been reported to synthesize hollow nanostructures, which generally involve two steps. The first step is the synthesis of the "template" particles, and the second is the transformation of these templates into hollow structures. Unlike the conventional templating methods, the templates in these cases not only provide the scaffold for the creation of hollow structures but also directly participate in the formation of the shells. The self-templating methods can produce high-quality hollow nanostructures with uniform size, controllable shell thickness, and great convenience for large-scale production [30]. Zhang Q described a "surface-protected etching" strategy to synthesize HMS, which is useful for constructing core–shell systems where active nanomaterials are embedded in silica shell for enhanced stability against aggregation [31].

Hollow mesoporous silica materials have received intensive interest in the field of cancer treatment owing to their large drug loading capacity, controlled release property, and excellent biocompatibility. With high loading capacity of chemotherapeutic drug, hollow mesoporous structure has a brilliant application prospect in cancer treatment. The intrinsic hollow mesoporous structure endows nanoparticles with a high drug loading capacity. Jinfeng Zhang reported hollow mesoporous silicon/carbon (Si/C) NPs as a carrier for a chemotherapeutic drug DOX with a loading capacity of 31.1% [32]. Xijian Liu developed a difunctional nanoplatform based on the hollow mesoporous silica structure, which has demonstrated a good photothermal effect and excellent DOX loading capacity (as high as 49.3%) [33]. Jie Yang fabricated a multifunctional supramolecular drug delivery platform based on hollow mesoporous silica nanoparticles drug reservoir and improved loading capacity as well as effective synergistic chemo-photothermal therapy both in vitro and in vivo [34].

Hollow mesoporous materials have an interconnected bimodal pore system composed of a hollow core and a mesoporous shell and have much higher pore volume compared to the conventional mesoporous materials. Therefore, they are potentially important as a catalyst support [35]. Liu and co-workers constructed hollow manganese oxide nanoparticles (H-MnO_2) by employing monodispersed silica nanoparticles as the hard template. After further PEGylation, photosensitizer chlorine c6 (Ce6) and chemotherapeutic drug DOX were coloaded into the hollow interior. Because of the unique disintegration behavior, the coloaded therapeutic agents were released in tumor with unique responsiveness to tumor microenvironment such as acidic condition. Especially, the catalytic performance of H-MnO_2 alleviated the tumor hypoxia and enhanced PDT efficacy, inducing synergistic chemo-photodynamic tumor therapy [36].

Based on the confinement effect of mesoporous channels and the diffusion rate of the drug from the pores, HMSNs can control the release rate of drug. Yu Gao demonstrated that the pore size of HMSNs could be tuned to precisely control the drug release rate for the first time in vitro release and intracellular drug release (Fig. 8.3) [37].

In addition to the high drug storage capability and controlled release properties of HMSNs, the biological effects including biocompatibility, biodistribution, and clearance are the other major concerns from the viewpoint of clinical applications. As many of HMSNs have a poor dispersity and stability in aqueous solution due to the strong aggregation among particles, many studies have shown that the PEGylation of nanoparticles is one of the most efficient ways to enhance the blood circulation and EPR effect for enhanced therapeutic efficacy [35].

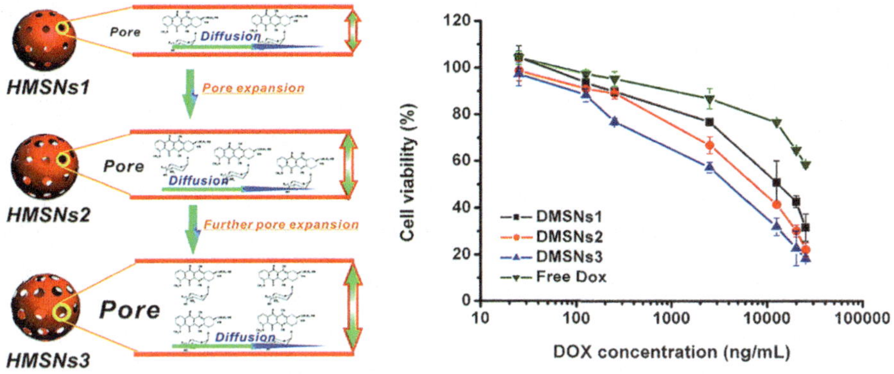

Fig. 8.3 Representative schematic illustration of controlling doxorubicin release rate by tuning the pore sizes of HMSNs to 3.2 nm (HMSNs1), 6.4 nm (HMSNs2), and 12.6 nm (HMSNs3) (left graph). It was anticipated that small pores provided limited room for the diffusion of doxorubicin, while large ones could provide enough room for the fast diffusion of drug molecules, which would lead to a higher release rate of drug molecules and enhanced toxicity (after incubating with DOX-loaded HMSNs 72 h) (right graph) [37]. (Copyright 2011, American Chemical Society)

With improved biocompatibility, the use of hollow mesoporous silicon in cancer therapy has been widely developed, particularly for pH-responsive strategies. Guangbao Yang developed pH-responsive nanoparticles which once within the slightly acidic tumor microenvironment could be converted into positively charged particles with enhanced tumor cell internalization and tumor retention [38].

Junjie Liu reported a DOX-loaded HMSNs-β-CD/Ada-PEG system that could deliver DOX in response to cascade pH stimuli in tumor microenvironment. The first pH response in tumor microenvironment could effectively resolve the "PEG dilemma" issue, and the second intracellular pH response could specifically deliver drug into cells, thus inducing cell apoptosis and inhibiting tumor growth while with minimal toxic side effects [39].

8.5.2 *Silica Coating*

Silica coating is being actively studied for metal nanoparticles to improve their properties than those without silica coating, such as higher photoacoustic signals [40, 41], enhanced fluorescence [42], and high drug payload with the large specific surface area of mesoporous silica [43].

Jingjing Liu presented a multifunctional nanoplatform based on PEGylated, MS-coated SWNTs, which could simultaneously serve as a drug loading carrier, a NIR photothermal heater, and a multimodal imaging probe, used for imaging and combination therapy of cancer. Unlike previously developed SWNT–drug complex, SWNT@MS-PEG/DOX exhibits rapid NIR light-triggered drug release behavior, which can be utilized to enhance the cancer cell killing by chemotherapeutics.

Owing to the efficient tumor accumulation as revealed by in vivo bimodal PA and MR imaging, in vivo combination therapy is then carried out, realizing an outstanding tumor growth inhibition effect by a single treatment in an animal tumor model upon systemic administration of SWNT@MS-PEG/DOX [44].

Junjie Liu reported an integrated nanocomposite based on AuNR@MSN, which could sequentially overcome biological barriers for simultaneous enhanced PDT and PTT against tumor. AuNR@MSN encapsulated photosensitizer ICG for high loading capacity and stability, while β-cyclodextrin (CD) was employed as gatekeeper to prevent premature ICG release. It exhibited no obvious toxicity, while upon NIR light irradiation, the generation of ROS and local hyperthermia led to dysfunction of mitochondria and triggered notable tumor cell apoptosis/necrosis in vitro [45].

A clinical trial was reported in which laser-excited gold–silica nanoshells (GSNs) were used in combination with magnetic resonance–ultrasound fusion imaging to focally ablate low-intermediate-grade tumors within the prostate. This current pilot device study demonstrates that GSN-directed laser excitation and ablation is a safe and technically feasible procedure for the targeted destruction of prostate tumors [46].

As Junjie Liu reported in his article, mesoporous silica nanoparticles coated with stimuli-responsive gatekeepers can provide benefits from many aspects including transportation of guest molecules (i.e., drugs or dyes) to specific locations in the body and controlled release in response to either external triggering signals or cellular stimuli, such as pH, temperature, enzymes, light, redox agents, and so on. The controlled release behavior is generally regulated by the on–off of the pore via its gatekeeper.

Cui Yanna reported a mesoporous silica nanosystem capped with disulfide-linked PEG gatekeepers for glutathione-mediated controlled release for controlled drug release in response to GSH, which is known to be high concentration in tumor microenvironment [47]. Thus, it can reduce side effects in normal tissues.

In addition to the redox response, the tumor acidic microenvironment is usually used as a response condition as well. Li Chengyi used Fe-doped silica shell as a compact inorganic cap to seal DOX into the mesoporous polymer cores, as well as a superparamagnetic agent for magnetic targeting and magnetic resonance imaging (MRI). The caps can slowly release the loaded drug under the acidic tumor environment while achieving ultralow drug leakage under normal in vivo blood circulation (physiological environment) [48]. Those strategies can efficiently avoid side effects of chemotherapeutic drug release in normal tissue.

Last but not the least, coating silica makes surface functionalization easier. In the study of Jinming Li, the mesoporous silica structure helped to establish a complex drug delivery system. The UCNPs possess a mesoporous silica-coated structure, which facilitates the conjugation of the RGD peptide and the peptide–AIEgen unit on the surface of the UCNPs to form the UCNP@SiO_2–peptide–AIEgen (UCNP–peptide–AIE). The mesoporous silica structure of UCNP–peptide–AIE also harbors the DMNPE/siRNA compound via physical absorption to form the DMNPE/siRNA–UCNP@SiO_2–peptide–AIEgen nanocomplexes (UCNP–peptide–AIE–

siRNA). The mesoporous silica structure is vital for the whole strategy in drug delivery [49].

8.5.3 Yolk–Shell Structure

A hybrid of core–shell and hollow structures, a special class of core–shell structure with a distinctive core@void@shell configuration, which are called yolk–shell or rattle-type structures, has attracted tremendous interest in recent years. In 2004, Yadong Yin and his co-worker successfully synthesized platinum–cobalt oxide yolk–shell nanostructures, which may serve as nanoscale reactors in catalytic applications. In comparison to catalysts supported on other mesoporous materials, the isolation of catalyst nanoparticles within solid shells may minimize secondary reaction of the products that degrade selectivity and product distribution [50]. With the unique properties of movable cores, interstitial hollow spaces, and the functionality of shells, yolk–shell structures have great potential for application in various fields, such as nanoreactors, biomedicine, lithium–ion batteries, and photocatalysis [51]. Here we mainly review its application in tumor treatment.

The cores of such structures can be diverse. Feng Chen reported a general synthetic strategy of chelator-free zirconium-89 (89Zr)-radiolabeled, TRC105 antibody-conjugated, silica-based yolk–shell hybrid nanoparticles for in vivo tumor vasculature targeting. Three types of inorganic nanoparticles with varying morphologies and sizes were selected as the internal cores, which were encapsulated into single hollow mesoporous silica nanoshells to form the yolk–shell-structured hybrid nanoparticles. Their strategy could be applied to the synthesis of other types of yolk–shell theranostic nanoparticles for tumor-targeted imaging and drug delivery [52].

The two parts, core and shell, are made from different materials, which can respectively take on different roles in tumor treatment, so as to realize the synergistic therapeutic effect. Jing Xu fabricated yolk–shell-structured mesoporous silica nanoparticles (YMSNs) with multifunctionalities of fluorescence imaging, photothermal therapy (PTT), and drug delivery by using a fluorescein isothiocyanate-doped silica nanoparticle partially covered by patchy gold as the core. Different from the conventional selective etching procedure, the multifunctional silica core is left intact, and the alkali etching mainly occurs in a hexadecyltrimethylammonium bromide–silica hybrid layer, which leads to the formation of the void space in YMSNs. YMSNs can load more drugs in comparison with conventional porous silica materials, resulting from the large void space inside the shell-like hollow MSNs. In vitro and in vivo antitumor experiments reveal that the resulting YMSNs can be utilized for chemo- and photothermic combination therapy as well as optical imaging (Fig. 8.4a) [53].

Specifically, yolk–shell mesoporous silica nanoparticles (YMSNs) are used as the host for loading the water-soluble anticancer drug DOX, which is a relatively high

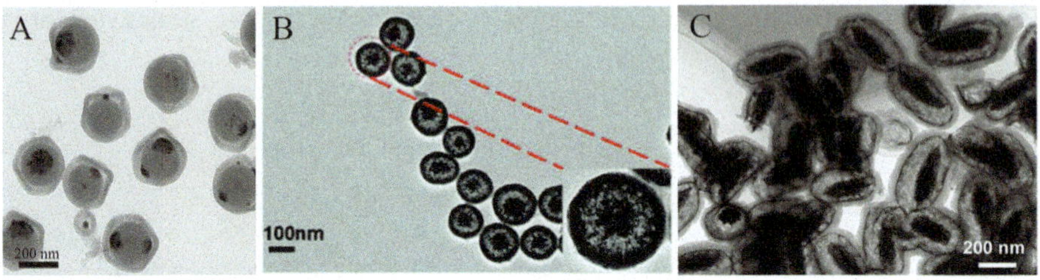

Fig. 8.4 Some TEM images of yolk–shell structure [53–55]. (**a**) Copyright 2018, American Chemical Society. (**b**) Copyright 2018, RSC Pub. (**c**) Copyright 2015, American Chemical Society

drug loading capacity and therefore potentiates additional therapeutic and safety benefits (Fig. 8.4b) [54].

The mesoporous pores of the microcapsules provide the probability to modify or store functional groups and nanoparticles. Besides, the elliptical capsules with high length\diameter ratio, yolk-like core–shell structures, may have different endocytosis. GdOF:Ln@SiO_2–ZnPc-DCs yolk-like mesoporous microcapsules using effective UCL GdOF:Ln as cores and mesoporous silica layer as shell with hollow cavities have been fabricated by a unique route and applied for both multiple imaging (CT, MRI, UCL, photothermal) and multiple therapies (PDT, PTT, and chemotherapy). The as-prepared product shows large surface area, good biocompatibility, and high ablation efficiency to cancer cells (Fig. 8.4c) [55].

In addition to inorganic silicon materials, organic silicon is also used in yolk–shell structure nanosystem. Uniform yolk–shell mesoporous nanoparticles with thioether-bridged organosilica frameworks have been successfully prepared by hydrothermal treatment of mesostructured organic–inorganic hybrid nanospheres. The yolk–shell mesoporous nanoparticles have ultrahigh condensation degree, high surface area, uniform pore size, and large pore volume [56]. Their diameter, core size, and shell thickness can be easily modulated by adjusting the reaction parameters. By using the thioether-bridged yolk–shell mesoporous nanoparticles as in situ nanoreactors, gold polyhedron-loaded yolk–shell mesoporous nanoparticles can be conveniently prepared. Furthermore, through breaking the thioether moieties to give thiol groups and then connecting with Cy5.5-maleimide, near-infrared illuminated yolk–shell mesoporous nanoparticles can be successfully synthesized, which show great promise for in vitro and in vivo imaging [57].

8.6 Silica Nanospheres

Silica nanospheres have been widely studied due to their excellent properties such as simple to synthesize and large surface area to be modified. With their tailored mesoporous structure and high surface area, MSNs as drug delivery systems (DDSs) show significant advantages over traditional drug nanocarriers [58]. The studies mainly consist of three parts: the control of size and pore diameter, surface modification, and internal doping.

The vast majority of contemporary MSN materials are synthesized by cetyltrimethylammonium bromide (CTAB)-templated silica condensation in a basic aqueous environment, which typically yields materials with pore diameters of up to 3 nm. However, it is possible to construct materials with pore sizes larger than this value, large-pore MSN (LPMSN), which exhibit supremacy over smaller pore-MSN with regard to mass transfer, diffusivity, and penetration ability of large drug molecules, proteins, or nucleic acids into or out of the pore system [59]. Nucleic acids can be effectively protected from the nucleases from the bloodstream by loading them inside the large pores of LPMSN, which is one of the important aspects in the construction of gene delivery vehicles. Gao Fei's study shows the applicability of amine-functionalized LPMSN materials for loading plasmid DNA (pDNA) and demonstrates their capability to preserve the pDNA molecules in the presence of nucleases [60].

This strategy has been used in the cancer treatment of animal model. Min et al. used siRNA-loaded LPMSN material, which was also functionalized with an imaging dye and a PEG moiety in the tumor treatment. They further demonstrated that it can efficiently (more efficiently than commercially available Lipofectamine 2000) lead to gene silencing by inhibiting the expression of green fluorescent protein (GFP) in HeLa cells. By downregulation of a therapeutically relevant vascular endothelial growth factor (VEGF), the tumor volume decreased distinctly than control group [61].

Besides, the large pore size of silicon spheres also provides favorable conditions for catalytic reactions, which can be applied in tumor treatment. Huo and co-workers integrated natural glucose oxidase (GOD, enzyme catalyst) and ultrasmall Fe_3O_4 nanoparticles (inorganic nanozyme, Fenton reaction catalyst) into the large pore-sized and biodegradable dendritic silica nanoparticles to fabricate a novel kind of sequential nanocatalyst for tumor therapeutics. The slow degradation guaranteed the full reaction of glucose oxidase with tumor glucose and Fe_3O_4 with upstream H_2O_2 for subsequent Fenton-like reaction (Fig. 8.5) [62].

MSN can be further surface functionalized easily. Tumor-specific active targeting of drug delivery carriers is one of the most promising techniques for cancer treatment, which could enhance the therapeutic efficiency and at the same time reduce the adverse effects. Till now, several ligands specifically bound to receptors overexpressed on cancer cells have been modified on MSN. For example, Yi Wang exploited an IL-13 peptide-modified MSN as a novel vehicle for delivery of hydrophobic antitumor agent DOX targeting to glioma cell and reached anti-glioma effect in vivo [63, 64].

Another research hotspot is the internal doping of silica nanosphere. Yi Wang developed a general strategy for synergistic chemo- and photothermal tumor therapy by using template semi-graphitized mesoporous silica nanoparticles (TsGMSN) as a carrier and photothermal agent. The unique structure of TsGMSN, including the semi-graphitized carbon (sGC) as hotspots on the pore wall, makes water-insoluble anticancer drug DOX directly contact the hotspots, which simultaneously achieves

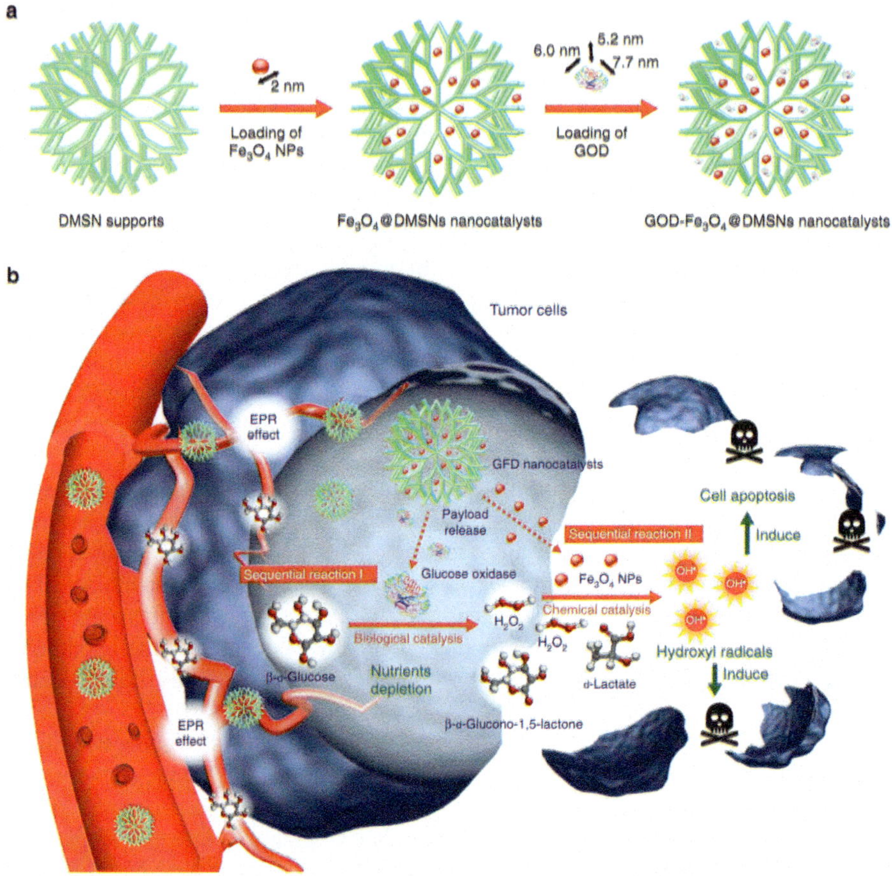

Fig. 8.5 Fabrication and catalytic-therapeutic schematics of sequential GFD NCs. (**a**) Synthetic procedure for Fe_3O_4@DMSNs nanocatalysts and GOD-Fe_3O_4@DMSNs nanocatalysts. The sizes of the prepared Fe_3O_4 nanoparticles and adopted GOD are indicated in the scheme. (**b**) The scheme of the sequential catalytic-therapeutic mechanism of GFD NCs on the generation of hydroxyl radicals for cancer therapy [62]. (Copyright 2017, Springer Nature)

efficient pH- and NIR-triggered release. Conjugating a novel DNA aptamer (HB5) which is capable of binding to HER2 receptor, the therapeutic system of TsGMSN has been demonstrated as a great platform for combined chemo-photothermal therapy targeted to HER2-positive cancers (Fig. 8.6) [65, 66].

Moreover, nanodots have been dispersed in MSN for facile preparation of nanodots [67] and improvement of properties of MSN. Yi Wang put forward a facile strategy based on solid-phase transformation of the template into polymer-coated carbonaceous nanodots (pCNDs), which was used for direct synthesis of multifunctional MSNs incorporated with dispersed fluorescent carbon nanodots. The homogeneous pCNDs incorporated into MSNs can simultaneously act as the

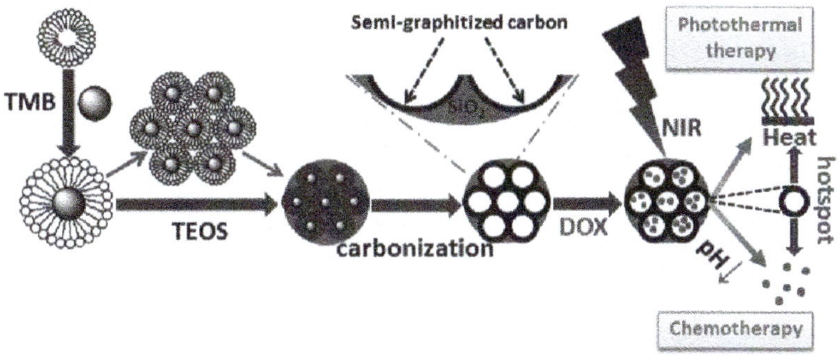

Fig. 8.6 A general strategy for synergistic chemo- and photothermal tumor therapy by using template semi-graphitized mesoporous silica nanoparticles [65]. (Copyright 2013, Wiley-VCH Verlag GmbH & Co. KGaA, Weinheim)

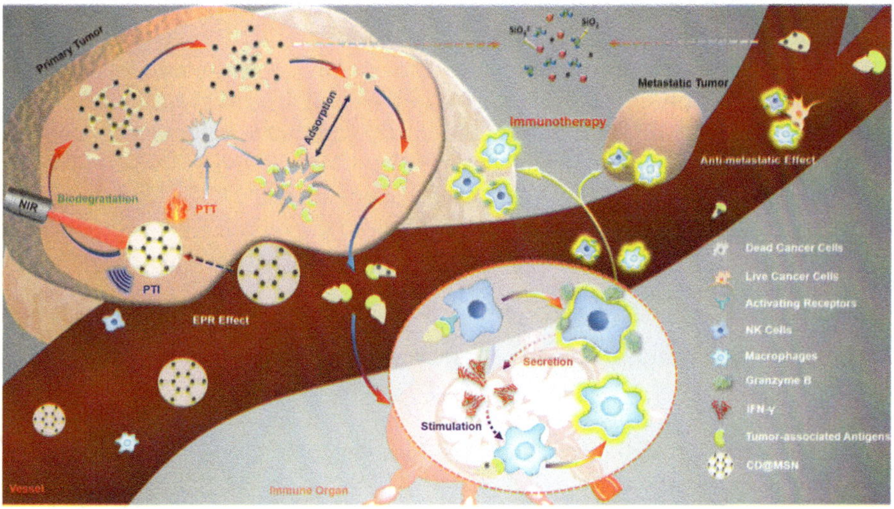

Fig. 8.7 Schematic diagram of in vivo delivery process of framework swelling-triggered biodegradable CD@MSN and its application for photothermal imaging-guided tumor-targeted PTT cooperated with debris-mediated immunotherapy against tumor metastasis [69]. (Copyright 2019, American Chemical Society)

nontoxic fluorescence probe for highly efficient fluorescence imaging and also anchoring sites for anticancer drug loading [68].

More interestingly, the incorporation of carbon nanodots can improve the degradation of MSN. Besides, biodegradation into small debris with large surface area for elevated protein adsorption was suggested to be beneficial for carrying more antigens to activate the cancer immunity (Fig. 8.7) [69].

In addition to fluorescent probes, doping other ingredients with therapeutic properties can achieve synergistic therapy with the loading drug. Zhao et al. explored a facile and unique outside-in approach for the synthesis of high dispersive mesoporous silica nanoparticle (MSN)/MoS_2-PEG nanoparticles (SMPs) based on the differences of precursor solubility in different solvents, for the first time, exhibiting both higher drug loading efficiency of 64.7% owing to the pore structure of mesoporous silica and excellent photothermal performance with photothermal efficiency of 41.5%. The results in vivo also demonstrated the great synergistic effect of SMPs, i.e., the heat produced during PTT can increase the chemotherapy sensitivity of DOX and synergistically improve its therapeutic effect [70].

8.7 Organosilica Nanosystems

Inspired by the well-established synthetic mechanism for MSNs, a simple but efficient "chemical homology" mechanism is put forward to clarify the formation procedure of the pure –Si–O–Si– framework and the organic R group-bridged –Si–R–Si– framework for the synthesis of mesoporous organosilica nanoparticles (MONs). The hydrolysis and condensation of the bissilylated organosilica precursor are similar to that of TEOS where different hydrolyzed/condensed bissilylated organosilica molecules are linked through –Si–O–Si– bonds. Several types of MONs have been successfully constructed including solid MONs, hollow MONs (HMONs), and core–shell-structured MONs. Due to the versatility of incorporated R organic moieties, such a composition tailoring of the mesopore wall is expected to bring abundant specific functions and performances for various applications in nanotechnology [71].

The biorelated degradability and clearance of siliceous nanomaterials have been questioned worldwide, since they are crucial prerequisites for the successful translation in clinics [72]. This hybridization strategy can endow hybrid nanosystems with tunable biodegradation and improved biocompatibility.

Kuroda and co-workers synthesized MONs with an ethenylene-bridged silsesquioxane framework by using bis(triethoxysilyl)ethenylene as the organosilica precursor. It was found that such ethenylene-bridged MONs were resistant to biodegradation, compared to traditional colloidal MSNs. Long-term biodegradation assay showed 90% biodegradation percentage for MSNs. However, only less than 2% MONs were biodegraded in a 15-day assay [73]. It provides a new strategy for its clinical application in tumor treatment.

In contrast, the organic–inorganic hybrid composition of MONs can also lead to enhanced biodegradability of silica in the presence of a high content (ca. 40–60%) of physiologically active organic fragments within the framework [74]. For incorporating physiologically active organic groups within the framework, bis(propyl)disulfide and ethylene groups were successfully co-incorporated into the framework of MONs. It was found that these MONs could be quickly degraded only after incubation in physiological conditions for 48 h [75]. The incorporation of organic groups into the MON framework can substantially reduce the surface-exposed silanol groups, fortunately leading to extremely low hemolytic effects of MONs against red blood cells (RBCs) [76]. This can reduce the long-term toxicity problems caused by the accumulation of drug delivery carriers in other tissues.

Organosilica with a silsesquioxane framework can also be used as the capping material for stimuli-responsive drug release. For instance, disulfide-bond-bridged silsesquioxane was coated onto the surface of poly (lactic-co-glycolic acid) (PLGA) NPs for redox-responsive cargo release [77]. Thus, it can reduce the release of drugs in normal tissues and therefore reduce side effects.

Above all, the biodegradation essence and rates of MONs could be practically controlled by optimizing the incorporated organic groups within the framework, which is difficult to realize based on traditional MSNs; biodegradation can be resisted, whereas the incorporation of biologically active organic groups can accelerate the biodegradation rates in response to some bioactive triggers, guaranteeing the rapid release of antitumor drug in targeted sites and the biosafety of silica-based nanosystems in vivo.

MONs with molecularly incorporated organic groups are also beneficial for stimuli-responsive drug release based on the new mechanism of hydrophobic–hydrophobic interaction between the drug molecules and frameworks of MONs. M. Wu synthesized phenylene-bridged HMONs for the delivery of the anticancer drugs DOX and bleomycin (BLM). In the presence of hydrophobic–hydrophobic interactions and supramolecular π–π stacking between the aromatic DOX/BLM molecules and the phenylene-bridged framework of HMONs, the release of DOX/BLM from the carrier was low (less than 8% within 24 h) under normal physiological condition but could be easily interfered with in mild acidic condition, which means MONs can be used for pH-responsive drug release [78].

The combination of organosilica with biocatalysis and responsive release strategies sparks novel nanosystems. Li et al. successfully combined reactive oxygen species (ROS)-involved photodynamic therapy (PDT) and chemodynamic therapy (CDT) with an in situ polymerized HMON biocatalysis nanoreactor, providing a good paradigm for enhancing ROS-mediated antitumor efficacy and offering a new method to expand the application of silica-based tumor-targeted DDS [79].

Furthermore, Chen et al. developed molecularly organic–inorganic hybrid HMON with silsesquioxane framework achieving unique intrinsic reducing/acidic and external high-intensity focused ultrasound (HIFU)-responsive drug-releasing performances, improved biological effects (e.g., lowered hemolytic effect and improved histocompatibility), and enhanced ultrasonography behavior. The doxorubicin-loaded HMONs with concurrent thioether and phenylene hybridization exhibit drastically enhanced therapeutic efficiency against cancer growth and metastasis, as demonstrated both in vitro and in vivo [80].

Besides, organosilica is also used in the assembly of deformable nanosystems. Teng et al. synthesized deformable thioether-, benzene-, and ethane-bridged hollow periodic mesoporous organosilica (HPMO) nanocapsules successfully by a preferential etching approach, which show more intake by breast cancer MCF-7 cells

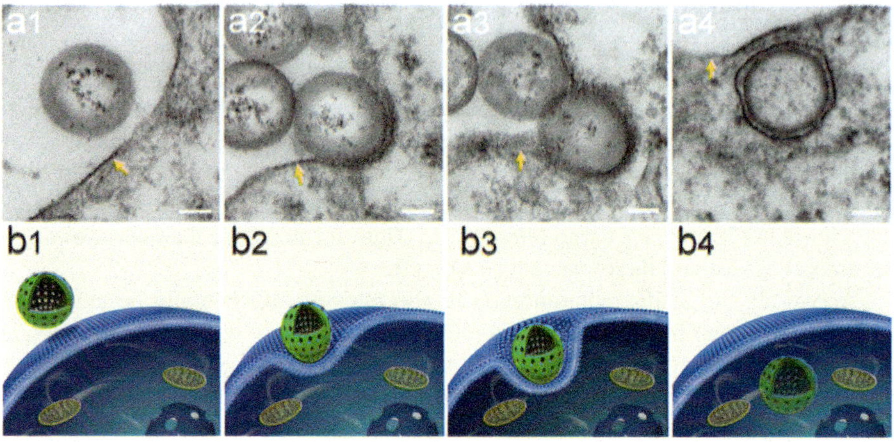

Fig. 8.8 (**a1–a4**) TEM images and (**b1–b4**) schematic illustration of the uptake processes of HPMO-Cy5.5-PEG by human breast cancer MCF-7 cells at 37 °C. The arrows indicate the cellular membranes. Scale bars, 100 nm [81]. (Copyright 2018, American Chemical Society)

compared with solid counterparts. The deformable HPMO nanocapsules were further loaded with anticancer drug DOX, which shows high killing effects for MCF-7 cells, demonstrating the promise for biomedical applications (Fig. 8.8) [81].

8.8 Conclusion

Although numerous cutting-edge researches on silica-based DDS are mentioned in this chapter, the application of these well-biocompatible systems in clinical practice remains difficult. Indeed, there are plenty of major challenges in translating silica-based nanoparticles from laboratory to clinic, but we believe that, by exploring related biosafety issues and designing proper structures, these could be certainly overcome with improved nanodelivery systems exerting their varieties of advantages. As co-treating with drugs, these systems will be no longer just excipients but developed as "adjuvants" delivering drugs at the same time. Meanwhile, if researchers continue to focus too much on therapeutic effects of silica-based DDS rather than their side effects such as off-target, their successful clinical application will remain out of reach. In addition, with the attention on immune system in recent years, researchers have gradually focused on the interaction between silica-based DDS and the immune system in vivo. Future researches on silica-based DDS must also pay more attention to this field to ensure this kind of nanomedicine can fully realize its potential in clinic and help patients. Although current clinical application and research on silica-based nano-DDS are still few in number, it is believed that their huge potential in clinical practice can exhibit amazing results quickly in near future.

References

1. Carsten O et al (2001) Cationic solid-lipid nanoparticles can efficiently bind and transfect plasmid DNA. J Control Release 77(3):345–355
2. Barbe C et al (2004) Silica particles: a novel drug-delivery system. Adv Mater 128:1959–1966
3. Jing W et al (2011) Solvent evaporation induced aggregating assembly approach to three-dimensional ordered mesoporous silica with ultralarge accessible mesopores. J Am Chem Soc 133:20369–20377
4. Rui L et al (2010) pH-responsive nanogated ensemble based on gold-capped mesoporous silica through an acid-labile acetal linker. J Am Chem Soc 132:1500–1501
5. Roman AP et al (2017) Silica-based multifunctional nanodelivery systems toward regenerative medicine. Mater Horiz 4:772–799
6. Liu P et al (2008) Preparation and characterization of conducting polyaniline/silica nanosheet composites. Curr Opinion Solid State Mater Sci 12(1):9–13
7. Lyuleeva A et al (2017) Lewis acid induced functionalization of photoluminescent 2D silicon nanosheets for the fabrication of functional hybrid films. Adv Funct Mater 27:6711–6718
8. Wang ZM et al (2010) Graphene oxide-periodic mesoporous silica sandwich nanocomposites with vertically oriented channels. ACS Nano 4(12):7437–7450
9. Yi W et al (2013) Multifunctional mesoporous silica-coated graphene nanosheet used for chemo-photothermal synergistic targeted therapy of glioma. J Am Chem Soc 135 (12):4799–4804
10. Xu J et al (2017) Integration of IR-808 sensitized upconversion nanostructure and MoS2 nanosheet for 808 nm NIR light triggered phototherapy and bioimaging. Small 13(36):6090864
11. Chen YW et al (2016) A theranostic nrGO@MSN-ION nanocarrier developed to enhance the combination effect of sonodynamic therapy and ultrasound hyperthermia for treating tumor. Nanoscale 8(25):12648–12657
12. Petros RA et al (2010) Strategies in the design of nanoparticles for therapeutic applications. Nat Rev Drug Discov 9(8):615–627
13. Vikash P et al (2011) Fluorescent nanorods and nanospheres for real time in vivo probing of nanoparticle shape-dependent tumor penetration. Angew Chem Int Ed Eng 50 (48):11417–11420
14. Huang X et al (2010) The effect of the shape of mesoporous silica nanoparticles on cellular uptake and cell function. Biomaterials 31(3):438–448
15. Wu XJ et al (2011) A unique transformation route for synthesis of rodlike hollow mesoporous silica particles. J Phys Chem C 115(23):11342–11347
16. Yang G et al (2014) Mesoporous silica nanorods intrinsically doped with photosensitizers as a multifunctional drug carrier for combination therapy of cancer. Nano Res 8(3):751–764
17. You Y et al (2017) High-drug-loading mesoporous silica nanorods with reduced toxicity for precise cancer therapy against nasopharyngeal carcinoma. Adv Funct Mater 27(42):1703313
18. Shun S et al (2013) Targeting mesoporous silica-encapsulated gold nanorods for chemo-photothermal therapy with near-infrared radiation. Biomaterials 12:3150–3158
19. Peng F et al (2013) Silicon-nanowire-based nanocarriers with ultrahigh drug-loading capacity for in vitro and in vivo cancer therapy. Angew Chem 125(5):1497–1501
20. Li YS et al (2014) Hollow-structured mesoporous materials: chemical synthesis, functionalization and applications. Adv Mater 20:3176–3205
21. Yang X et al (2016) Synthesis of hollow mesoporous silica nanorods with controllable aspect ratios for intracellular triggered drug release in cancer cells. ACS Appl Mater Interfaces 8 (32):20558–20569
22. Hu J et al (2012) Fabrication, properties and applications of Janus particles. Chem Soc Rev 11:4356–4378
23. Li X et al (2014) Anisotropic growth-induced synthesis of dual-compartment Janus mesoporous silica nanoparticles for bimodal triggered drugs delivery. J Am Chem Soc 136 (42):15086–15092

24. Shao D et al (2016) Janus "nano-bullets" for magnetic targeting liver cancer chemotherapy. Biomaterials 100:118–133
25. Zhang L et al (2016) Tailored synthesis of octopus-type Janus nanoparticles for synergistic actively-targeted and chemo-photothermal therapy. Angew Chem Int Ed Eng 55(6):2118–2121
26. Wang Z et al (2017) Janus silver/silica nanoplatforms for light-activated liver cancer chemo/photothermal therapy. ACS Appl Mater Interfaces 9(36):30306–30317
27. Parrill TM et al (1992) Transmission infrared study of Acie-catalyzed catalyzed sol-gel silica coatings during room ambient drying. J Biomed Mater Res A 7:2230–2239
28. Fang X et al (2013) Self-templating synthesis of hollow mesoporous silica and their applications in catalysis and drug delivery. Nanoscale 5(6):2205–2218
29. Fei P et al (2016) From hollow carbon spheres to N-doped hollow porous carbon bowls: rational design of hollow carbon host for Li-S batteries. Adv Energy Mater 6:1502539
30. Qiao Z et al (2009) Self-templated synthesis of hollow nanostructures. NanoToday 6:494–507
31. Zhang J et al (2008) Permeable silica shell through surface-protected etching. Nano Lett 8:2867–2871
32. Zhang J et al (2017) Degradable hollow mesoporous silicon/carbon nanoparticles for photoacoustic imaging-guided highly effective chemo-thermal tumor therapy in vitro and in vivo. Theranostics 7(12):3007–3020
33. Liu XJ et al (2014) Folic acid-conjugated hollow mesoporous silica/CuS nanocomposites as a difunctional nanoplatform for targeted chemo-photothermal therapy of cancer cells. J Mater Chem B 2:5358–5367
34. Yang J et al (2020) Supramolecular nanomaterials based on hollow mesoporous drug carriers and macrocycle-capped CuS nanogates for synergistic chemo-photothermal therapy. Theranostics 10(2):615–629
35. Li Y et al (2014) Hollow-structured mesoporous materials: chemical synthesis, functionalization and applications. Adv Mater 26(20):3176–3205
36. Qian XQ et al (2019) Manganese-based functional nanoplatforms: nanosynthetic construction, physiochemical property, and theranostic applicability. Adv Funct Mater 30(9):1907066
37. Gao Y et al (2011) Controlled intracellular release of doxorubicin in multidrug-resistant cancer cells by tuning the shell-pore sizes of mesoporous silica nanoparticles. ACS Nano 5 (12):9788–9798
38. Yang G et al (2018) Smart nanoreactors for pH-responsive tumor homing, mitochondria-targeting, and enhanced photodynamic-immunotherapy of cancer. Nano Lett 18(4):2475–2484
39. Liu J et al (2016) Hollow mesoporous silica nanoparticles facilitated drug delivery via cascade pH stimuli in tumor microenvironment for tumor therapy. Biomaterials 83:51–65
40. Chen YS et al (2011) Silica-coated gold nanorods as photoacoustic signal nanoamplifiers. Nano Lett 11(2):348–354
41. Chen YS et al (2010) Enhanced thermal stability of silica-coated gold nanorods for photoacoustic imaging and image-guided therapy. Opt Express 18(9):8867–8878
42. Ming T et al (2009) Strong polarization dependence of plasmon-enhanced fluorescence on single gold nanorods. Nano Lett 9(11):3896–3903
43. Zhang Z et al (2012) Mesoporous silica-coated gold nanorods as a light-mediated multifunctional theranostic platform for cancer treatment. Adv Mater 24(11):1418–1423
44. Liu J et al (2015) Mesoporous silica coated single-walled carbon nanotubes as a multifunctional light-responsive platform for cancer combination therapy. Adv Funct Mater 25(3):384–392
45. Liu J et al (2018) Tumor acidity activating multifunctional nanoplatform for NIR-mediated multiple enhanced photodynamic and photothermal tumor therapy. Biomaterials 157:107–124
46. Rastinehad AR et al (2019) Gold nanoshell-localized photothermal ablation of prostate tumors in a clinical pilot device study. PNAS 116(37):18590–18596
47. Cui Y et al (2012) Mesoporous silica nanoparticles capped with disulfide-linked PEG gate-keepers for glutathione-mediated controlled release. ACS Appl Mater Interfaces 4 (6):3177–3183

48. Li C et al (2018) Side effects-avoided theranostics achieved by biodegradable magnetic silica-sealed mesoporous polymer-drug with ultralow leakage. Biomaterials 186:1–7
49. Li J et al (2019) Photocontrolled SiRNA delivery and biomarker-triggered luminogens of aggregation-induced emission by up-conversion NaYF4:Yb3+Tm3+@SiO_2 nanoparticles for inducing and monitoring stem-cell differentiation. ACS Appl Mater Interfaces 11 (25):22074–22084
50. Yadong Y et al (2004) Formation of hollow nanocrystals through the nanoscale Kirkendall effect. Science 304:711–714
51. Liu J et al (2010) Monodisperse yolk-shell nanoparticles with a hierarchical porous structure for delivery vehicles and nanoreactors. Angew Chem Int Ed 49:4981–4985
52. Chen F et al (2018) General synthesis of silica-based yolk/shell hybrid nanomaterials and in vivo tumor vasculature targeting. Nano Res 11:4890–4904
53. Xu J et al (2018) Multifunctional yolk-shell mesoporous silica obtained via selectively etching the shell: a therapeutic nanoplatform for cancer therapy. ACS Appl Mater Interfaces 10 (29):24440–24449
54. Pei Y et al (2018) An autonomous tumor-targeted nanoprodrug for reactive oxygen species-activatable dual cytochrome c/doxorubicin antitumor therapy. Nanoscale 10:11418–11429
55. Lv R et al (2015) A yolk-like multifunctional platform for multimodal imaging and synergistic therapy triggered by a single near-infrared light. ACS Nano 9(2):1630–1647
56. Liu J et al (2011) Yolk/shell nanoparticles: new platforms for nanoreactors, drug delivery and lithium-ion batteries. Chem Commun 47:12578–12591
57. Teng Z et al (2014) Yolk−shell structured mesoporous nanoparticles with thioether-bridged organosilica frameworks. Chem Mater 26(20):5980–5987
58. Tang FQ et al (2012) Mesoporous silica nanoparticles: synthesis, biocompatibility and drug delivery. Adv Mater 12:1504–1534
59. Nikola Ž et al (2015) Large pore mesoporous silica nanomaterials for application in delivery of biomolecules. Nanoscale 7:2199–2209
60. Gao F et al (2009) Monodispersed mesoporous silica nanoparticles with very large pores for enhanced adsorption and release of DNA. J Phys Chem B 113(6):1796–1804
61. Na HK et al (2012) Efficient functional delivery of siRNA using mesoporous silica nanoparticles with ultralarge pores. Small 8(11):1752–1761
62. Huo M et al (2017) Tumor-selective catalytic nanomedicine by nanocatalyst delivery. Nat Commun 8:357
63. Wang Y et al (2012) Tumor cell targeted delivery by specific peptide-modified mesoporous silica nanoparticles. J Mater Chem 22:14608–14616
64. Shi J et al (2017) An MSN-PEG-IP drug delivery system and IL13Rα2 as targeted therapy for glioma. Nanoscale 9:8970–8981
65. Wang Y et al (2014) A general strategy for dual-triggered combined tumor therapy based on template semi-graphitized mesoporous silica nanoparticles. Adv Healthc Mater 3(4):485–489
66. Wang K et al (2015) Specific aptamer-conjugated mesoporous silica–carbon nanoparticles for HER2-targeted chemo-photothermal combined therapy. Acta Biomater 16:196–205
67. Wang Y et al (2016) Facile growth of well-dispersed and ultra-small MoS2 nanodots in ordered mesoporous silica nanoparticles. Chem Commun 52:10217–10220
68. Wang Y et al (2016) Facile incorporation of dispersed fluorescent carbon nanodots into mesoporous silica nanosphere for pH-triggered drug delivery and imaging. Carbon 108:146–153
69. Qian M et al (2019) Biodegradable mesoporous silica achieved via carbon nanodots-incorporated framework swelling for debris-mediated photothermal synergistic immunotherapy. Nano Lett 19(12):8409–8417
70. Zhao J et al (2018) Outside-in synthesis of mesoporous silica/molybdenum disulfide nanoparticles for antitumor application. Chem Eng J 351(1):157–168
71. Chen Y et al (2016) Chemistry of mesoporous organosilica in nanotechnology: molecularly organic-inorganic hybridization into frameworks. Adv Mater 28(17):3235–3272

72. Jonas GC et al (2017) Degradability and clearance of silicon, organosilica, silsesquioxane, silica mixed oxide, and mesoporous silica nanoparticles. Adv Mater 29(9):1604634
73. Chihiro U et al (2011) Aqueous colloidal mesoporous nanoparticles with ethenylene-bridged silsesquioxane frameworks. J Am Chem Soc 133(21):8102–8105
74. Fatieiev Y et al (2015) Enzymatically degradable hybrid organic–inorganic bridged silsesquioxane nanoparticles for in vitro imaging. Nanoscale 7:15046–15050
75. Jonas C et al (2015) One-pot construction of multipodal hybrid periodic mesoporous organosilica nanoparticles with crystal-like architectures. Adv Mater 27:145–149
76. Igor IS et al (2009) Mesoporous silica nanoparticles for reducing hemolytic activity towards mammalian red blood cells. Small 5(1):57–62
77. Manuel Q, Carlos M, Pablo B (2013) Hybrid PLGA-organosilica nanoparticles with redox-sensitive molecular gates. Chem Mater 25(13):2597–2602
78. Wu M et al (2015) A salt-assisted acid etching strategy for hollow mesoporous silica/organosilica for pH-responsive drug and gene co-delivery. J Mater Chem B 3:766–775
79. Li L et al (2019) In situ polymerized hollow mesoporous organosilica biocatalysis nanoreactor for enhancing ROS-mediated anticancer therapy. Adv Funct Mater 30(4):1907716
80. Chen Y et al (2014) Hollow mesoporous organosilica nanoparticles: a generic intelligent framework-hybridization approach for biomedicine. J Am Chem Soc 136(46):16326–16334
81. Teng Z et al (2018) Deformable hollow periodic mesoporous organosilica nanocapsules for significantly improved cellular uptake. J Am Chem Soc 140(4):1385–1393

Chapter 9
Lipid-Based Tumor-targeted Systems

Yaxi Li, Chen Zhang, Tianliang Min, Yuan Ping, and Kai Li

Abstract With the booming development of bionanotechnology, a variety of multifunctional nanoparticle platforms have been explored to facilitate drug delivery in cancer treatment. Thanks to their good biocompatibility and versatile structures, the lipid-based nanoparticles can serve as ideal candidates to carry therapeutic reagents to cancer tissues. Additionally, the natural polymeric coating of liposomes could easily fuse with cellular membrane and minimize clearance by blood circulation. To date, their composition, properties, targeting strategies, reasonable design methods, and pharmacokinetics have been extensively studied to facilitate preclinical and clinical translations. In this chapter, we will introduce the structure and function of the lipid-based nanoparticle system, and a comprehensive review of the lipid-based nanoparticle drug delivery system will be presented.

9.1 Introduction

The emergence of nanotechnology sheds new lights on the cancer research and offers new opportunities to improve the current treatment modalities in terms of cancer detection, diagnosis, and therapy. With the rapid development of nanomaterials for the delivery of cancer therapeutics, the controlled drug delivery system that was first introduced into pharmaceutical industry makes it possible for nanoparticles to selectively target tumor tissue after the intravenous injection [1]. Although many different types of nanocarriers have been developed, lipid-based carriers have shown numerous promises in pharmaceutical industry and represent the first generation of drug delivery systems for cancer nanomedicine [2–5].Owing to the excellent biocompatibility and flexible chemical structure,

Y. Li · C. Zhang · T. Min · K. Li (✉)
Department of Biomedical Engineering, Southern University of Science and Technology, Shenzhen, Guangdong, China
e-mail: lik@sustech.edu.cn

Y. Ping
College of Pharmaceutical Sciences, Zhejiang University, Hangzhou, China

R. Huang, Y. Wang (eds.), *New Nanomaterials and Techniques for Tumor-targeted Systems*, https://doi.org/10.1007/978-981-15-5159-8_9

Bangham and co-workers found that liposomes could encapsulate and release solutes, attracting intensive studies on their structures and function [6, 7]. In addition, Gregoriadis and co-workers also explored liposomes as drug carrier for enzymes, anticancer, and antimicrobial drugs, suggesting their potentials in a wide range of drug delivery applications [8].

Through different delivery strategies, the lipid-based drug delivery systems could encapsulate hydrophobic or hydrophilic molecules into lipid membrane bilayer and achieve both passive and active targeting ability to tumors [9, 10]. Recently, the liposomal formulations are not only actively studied in laboratories but also are readily to meet the standard of Food and Drug Administration (FDA) as the excipient for chemotherapeutic drugs. So far, there are a number of lipid formulation being currently under the clinical trials, and some approved by FDA have been commercially available. With further functionalization, stimuli-responsive liposomes have also been developed, such as pH-responsive liposomes, temperature-sensitive, magnetic liposomes, etc. Furthermore, liposomes integrated with theranostic features have also been intensively investigated. In order to improve the tissue specificity during delivery process, the modification with targeting ligands over the liposomal surface significantly reshapes the pharmacokinetic profiles of the parental liposomes.

9.2 Lipid-Based Drug Delivery Systems

9.2.1 Liposome

Liposomes can be described as micro-vesicles which self-assemble into concentrated lipid bilayers, in which an aqueous core is located [11, 12]. As shown in Fig. 9.1, the major constituents of liposome are natural or synthetic phospholipids, such as phosphatidylcholine (PC) and phosphatidylethanolamine (PE) [13]. Cholesterol is also a major component in liposomes, which can align among phospholipids with hydroxyl group orienting towards the aqueous phase. The presence of cholesterol can prevent liposomes from destabilization by blood proteins, which intend to interact strongly with liposomes [13].To improve the performance of liposomes, some polymers with hydrophilic functional groups, such as polyethylene glycol (PEG), are attached to liposome surface through different strategies. As a result, the PEG coating on lipid bilayers forms steric hindrance to reduce the interaction between liposomes and plasma proteins, such as opsonin [13].

Conventional liposomes are composed of the same phospholipids and cholesterol that form biological membrane, thus showing amphipathic properties [14].Compared with other drug delivery systems, conventional liposomes are vulnerable to elimination by circulatory system and reticuloendothelial system (RES) [15]. In addition, conventional liposomes usually tend to accumulate in the tissues of mononuclear phagocyte system (MPS), particularly the liver and spleen [16]. As shown by previous studies, the conventional liposomes can be adsorbed by cells of the RES within 15–30 min after the intravenous injection [17].To address these

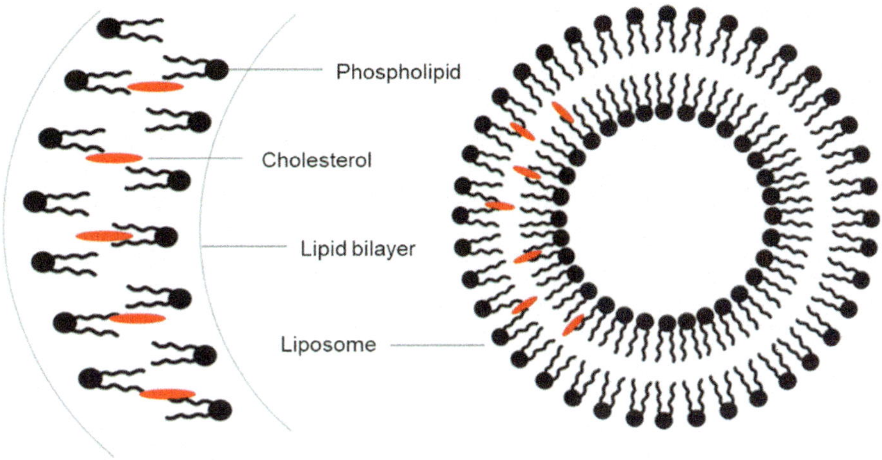

Fig. 9.1 The schematic illustration of liposome structure. (Redraw with permission from Ref. [13]. Copyright 2017 Elsevier B.V)

concerns, researchers have developed various strategies to modify the surface of liposomes and regulate the size in a narrow distribution range [18]. Currently, the attachment of antifouling polymers, such as PEG segments, on the liposome surface is the most common approach. Depending on the nature of liposome surface, PEG segments can be decorated by covalent combination, physical coating, or physical adsorption [19].

Besides conventional liposomes, researchers have also developed a variety of functional liposomes to cater to the versatile biological applications. Through nanoengineering, the functional liposomes can be responsive to external stimuli or act as carriers of biomolecules. For instance, the pH-sensitive liposomes mainly composed of dioleoylphosphatidylethanolamine (DOPE) can form hexagonal structure with improved fusogenic behavior [20]. The weak acidic amphiphiles such as cholesteryl hemisuccinate (CHEMS) with carboxyl groups have also been employed to provide functionality for further conjugation. At neutral pH, the mixture of DOPE and amphiphiles forms a bilayer structure by hydration. At mild acidic environment, the protonation of carboxyl groups results in a rapid transition of DOPE to hexagonal structure, thus displaying the membrane fusion capability or triggering the release of loaded drugs [21, 22].As a result of increased fermentative metabolism in the acid tumor microenvironment, the pH-sensitive nature can promote the drug release in the tumor tissue.

Liposomes can also shuttle biomolecules to the disease sites for therapeutic treatment. Especially, positively charged liposome can be used as a carrier to load negatively charged proteins, peptides, DNA, and RNA, especially [23]. A major challenge of positively charged liposomes for in vivo applications is their poor blood circulation performance, due to complement activation and macrophage uptake [24]. Two major types of positively charged liposomes, namely, DOTMA and

DOTAP, have been widely used in the liposomal formulation for nucleic acid delivery [23]. They usually contain positively charged groups in the molecular structures such as quaternary amines to facilitate the interaction with polyanionic nucleic acid to load these biomacromolecules [25]. For instance, the positively charged liposomes were used to encapsulate macrophage-specific promoter-driven Cas9 expression plasmids pM330 and pM458 for specific in vivo gene-editing applications [26]. Moreover, lipid-like nanoparticles can be used to deliver Cas9 messenger RNA (mRNA) into the livers, kidneys, and lungs for genome editing. In a recent study, positively charged liposome adjuvant CAF01 stabilized with cholesterol was optimized for the prolonged delivery of the antigen by migratory antigen-presenting cells to the preconditioned lymph node [27]. This formulation induced an antigen-specific intestinal IgA response upon subcutaneous administration, indicating its great potential for vaccine applications.

Another example of functional liposome is the sterically stabilized liposome. An ideal liposome formulation should possess prolonged half-life to facilitate high drug delivery efficiency. Sterically stabilized liposomes are a class of liposomes with polymeric matrices (such as palmitoyl glucoside acid or PEG lipid derivatives) on the surface, which are able to yield tunable surface property for extended half-life in the circulatory system [28]. The mechanism is that they can effectively prevent the nonspecific interaction with various components in the blood, such as opsonin, due to the high degree of hydration of hydrophilic groups on the surface that can reduce the affinity with mononuclear phagocyte system [29]. Moreover, sterically stabilized liposomes can delay drug release and improve the selectivity of specific target tissues, making it more suitable for therapeutic application in practice [30]. With prolonged circulation half-life, sterically stabilized liposomes can passively target solid tumors through newly formed vascular pores in tumor tissue, which can greatly improve the efficacy of encapsulated drugs [30]. However, this targeting strategy is not particularly effective in some cases. For example, when treating leukemia cells in the circulatory system or eradicating cancer cells that have already metastasized to the blood or lymphatic system, the targeting effect of sterically stabilized liposomes is poor [31]. To address this concern, the sterically stabilized immune-liposome with specific antibody functionalization was developed for specific binding to target sites, which is a promising solution.

In general, liposomes can be categorized into different types according to various preparation approaches, bilayers numbers, or vesicle sizes. Based on their sizes as the most common method, liposomes can be generally classified into multilamellar vesicles, large unilamellar vesicles, and small unilamellar vesicles (Fig. 9.2).

Multilamellar vesicles usually consist of multiple phospholipid bilayers. The size of multilamellar vesicles falls into a size ranging from 0.05 to 10 μm [32]. The thin film hydration approach is a most widely used to obtain multilamellar vesicles [33]. To prepare the multilamellar vesicles, a thin film of dry lipids is usually added into the buffer solution at above lipid phase transition temperature to form multilamellar vesicle aqueous suspensions. Thin film hydration is an ideal way for drug encapsulation (either hydrophilic drugs or lipophilic drugs), because hydrophilic drugs can be dissolved in aqueous hydration buffer while lipophilic drugs can

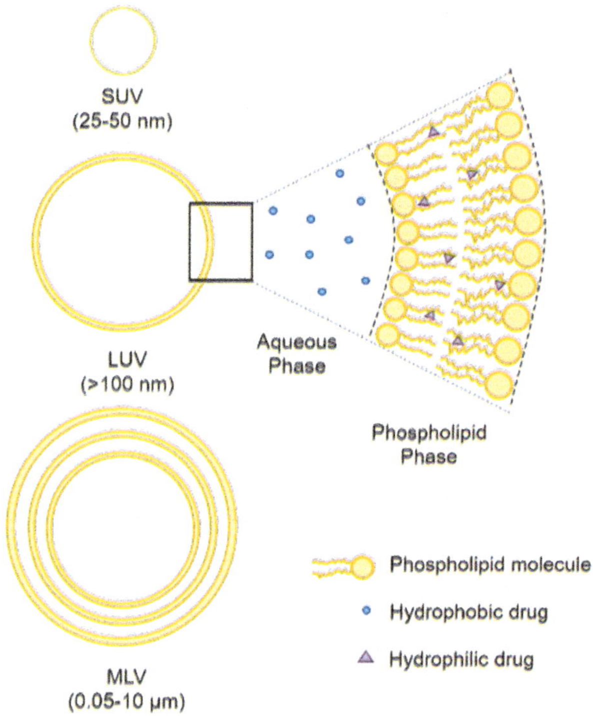

Fig. 9.2 Liposome types: illustration of small unilamellar vesicles (SUVs), large unilamellar vesicles (LUVs), and multilamellar vesicles (MLVs). (Redraw with permission from Ref. [32]. Copyright 2016 the American Society for Pharmacology and Experimental Therapeutics)

be dispersed in the lipid film. However, this method requires delicate control of preparation conditions and advanced characterization techniques, since minor changes may result in bench-to-bench variation of liposomal characteristics and behaviors. As a result, thin film hydration is limited by the low drug encapsulation efficiency, usually from 5% to 15% [32]. Several methods have been explored to improve the drug encapsulation efficiency, which is of high importance to reduce the dosage and enhance the therapeutic efficacy. Olson et al. reported that the hydrated ratio of lipid film to aqueous or drug solution could affect the number of drug molecules encapsulated in final lipid vesicles. Using the lipid films with similar compositions, they successfully achieved a high drug encapsulation ratio from 11.3% to 22.6% due to slower hydration and gentler mingling [34]. In another study, Vemuri and co-workers added a small amount of charged lipids (10–20 mol %) into bilayers before liposome formulation to improve the encapsulation capacity of multilamellar vesicles. The charged lipids are used to increase the interlamellar distance among successive bilayers and reduce vesicular aggregation when multilamellar vesicles synthesis [35].

Large unilamellar vesicles consist of a single phospholipid bilayer, and their sizes are usually larger than 100 nm [32]. Due to the unique structural signature, they are able to encapsulate a mass of solution in their cavity with high encapsulation efficiency. Therefore, one can use a small number of lipids to prepare unilamellar vesicles for encapsulating a considerable quantity of drugs [36]. Furthermore, large unilamellar vesicles have the property of replicable drug release rate, which is a key factor in controlled drug delivery systems [37]. Large unilamellar vesicles can be prepared by the reverse-phase evaporation technique [38]. In brief, a water/oil (W/O) emulsion, in which the water phase contains drugs and organic phase consists of lipids, can be obtained under mechanical stirring or sonication. Upon organic solvent evaporation, the system transforms into a gel matrix and forms a liposome aqueous suspension. Using this technique, the drug-loading ratio can be significantly improved (~60–65%). However, this process may have negative impact on drug properties because protein or DNA can undergo conformational changes in presence of mechanical force and organic solvents. Another technique for preparing large unilamellar vesicles is called detergent removal technique [39]. In this technique, a detergent and phospholipids are mixed to form micelles. The detergent is subsequently removed to obtain micelles with a large number of phospholipids, and these micelles ultimately form single bilayer vesicle. The detergents that are used in this technique usually have high critical micelle concentrations (CMCs), including odium cholate, sodium deoxycholate, and octyl glucoside. To remove the detergent, several techniques have been employed, including dialysis, column chromatography, and Bio-Beads adsorption. For instance, detergents can be removed after passing through a dialysis cell, while single bilayer liposomal vesicles with the sizes of 50–100 nm are still left. On the other hand, in column chromatography technique, detergent (e.g., deoxycholate) can be separated by a G-25 column, and single layer phospholipid vesicles with diameters of ~100 nm are collected [40]. In Bio-Beads adsorption technique, the detergent can bind to Bio-Beads, and phospholipid vesicles are thus separated from the phospholipid/detergent mixture. However, this technique is only suitable for the separation of nonionic detergent in nature [41].

Small unilamellar vesicles consist of single or multiple phospholipid bilayers. The hydrodynamic diameters of small unilamellar vesicles in aqueous phase typically range from 25 to 50 nm [42]. Two main techniques have been employed to prepare small unilamellar vesicles: sonication or extrusion from multiple lamellar vesicles or large unilamellar vesicles and direct injection from diethyl ether or ethanol. In the sonication technique, an aqueous suspension of multiple lamellar vesicles or large unilamellar vesicles is sonicated in a bath or probe sonicator under nitrogen or argon gas protection, resulting in smaller vesicles with reduced sizes [42]. Compared with probe sonication, bath sonication can provide an aseptic environment with well-controlled operating temperature. In extrusion technique, multiple lamellar vesicles or large unilamellar vesicles are extruded at 4 °C under high pressure through a narrow orifice, which can lead to reduced vesicle diameters of 20 nm. However, the processing temperature cannot be accurately controlled, which may affect the packing of phospholipid in liposomes and result in inconsistency. In direct injection technique, the lipids are first dissolved in organic solvent, followed by injection into excessive water by a syringe infusion pump. After organic solvent is removed, small unilamellar vesicles (50–200 nm) suspended in water are obtained. The main limitation of this technique is that the organic solvent cannot be removed completely [43].

9.2.2 Lipid Core Micelles

Lipid core micelles stand for a major type of lipid-based nanoparticles [44]. When polyethylene glycol-phosphatidylethanolamine (PEG-PE) conjugates exceed a critical micelle concentration, the conjugated mixtures tend to form lipid core micelles rather than PEGylated liposomes [45]. The size of nanoparticles mainly depends on the molecular weight of PEG segment. The driving force of micelle formation can be attributed to the hydrogen bonds between hydrophilic segments and aqueous phase as well as the van der Waals force among hydrophobic segments. With both hydrophilic and hydrophobic components, lipid core micelles have the ability to carry a variety of drugs as effective drug delivery systems [46]. The presence of PEG helps lipid core micelles prolong their circulation time in blood, while the hydrophobic phospholipid segments endow the micelles with high stability and excellent drug encapsulation in the phospholipid single layer.

In general, the majority of PEG-PE conjugates have lower critical micelle concentration (CMC) values as compared to conventional detergents [47]. It is a great advantage in micelle formulation since PEG-PE conjugates can maintain micelle formation even at low concentrations, which means the micelles have excellent stability in aqueous environment. For instance, the lipid core micelles prepared from DSPE-PEG_{2000} and DSPE-PEG_{5000} can remain stable in blood plasma within 48 hours, indicating that the blood plasma can barely affect the stability of DSPE-PEG lipid core micelles after systemic administration [48].

Lipid core micelles can be prepared from a facile strategy. The solvent in mixture of PEG-PE and drugs is evaporated to form a dry lipid film, following which an aqueous buffer is poured into the dry lipid film under fierce shaking to obtain lipid core micelles [46]. Due to the nature of micelle formulation, the drug-loading rate of lipid core micelles can be regulated by the hydrophobicity of PEG-PE conjugates or addition of another micelle-forming compound. For instance, Gao and colleagues used L-α-phosphatidylcholine (egg PC) with PEG-PE to achieve a loading efficiency with 33 mg of pacilitaxel (PTX)/g of PEG-PE-egg PC micelles [49], compared with 15 mg/g in the case of pure PEG-PE micelles. Additionally, the PTX-loaded PEG-PE-egg PC hybrid micelles have comparable cytotoxicity to targeted tumor cells as the free PTX.

9.2.3 Solid Lipid Nanoparticles

Solid lipid nanoparticles with sizes from 50 to 100 nm are also called lipospheres or solid lipid nanospheres [50]. The components of solid lipid include glycerides, lipid acids, and glyceride mixtures. The solid lipids as matrix materials can be stabilized by a variety of biocompatible surfactants and used for drug encapsulation with good physical stability [51, 52]. Apart from that, solid lipid nanoparticles also have many other advantages, such as economical preparation and large-scale production, easy and convenient storage, low toxicity and acidity, controlled release, and protection of labile drug degeneration. There are mainly three techniques for the preparation of solid lipid nanoparticles, including high-pressure homogenization, microemulsion, and solvent evaporation.

In high-pressure homogenization, lipids are heated above their melting point so that drugs are dissolved in the molten lipid. Subsequently, either hot or cold homogenization techniques are applied to fabricate solid lipid nanoparticles. In the hot homogenization method [53], the molten mixture of lipids and drugs forms pre-emulsion in an aqueous surfactant solution at the same temperature and then homogenized by a piston-gap homogenizer. When the emulsion is cooled down to room temperature, the solid lipid nanoparticles appear due to the recrystallization of lipids. However, the hot homogenization technique is not suitable for high-temperature-sensitive drugs, while the cold homogenization technique is a better choice. In cold homogenization technique [54], the molten mixture of lipids and drugs is cooled down to form solid lipids particles. These particles are ground into smaller sizes in the range of 50–100 nm, which are dispersed in a cold surfactant solution as a suspension. Finally, the suspension is homogenized at a lower temperature and disintegrated with the high-pressure homogenizer to obtain nanoparticles. To avoid melting of lipids in homogenizer, it is critical to have a large difference between lipid melting point and homogenization temperature.

In microemulsion technique, water, co-surfactants, and surfactants are mixed and added into the molten lipids under mild stirring to form microemulsion at a temperature above melting point of lipids [55]. The microemulsion is further dispersed in 2–3 °C water with successively stirring to obtain smaller particles [56]. The success of this technique relies on three key factors: (1) the lipids used should be in solid state at room temperature, (2) the ratio of surfactants to lipids should be appropriate, and to obtain a thermodynamically stable and transparent microemulsion system, (3) the whole process of microemulsion must be under a temperature above the lipid melting point.

Another important technique is called solvent evaporation; lipids are dissolved into organic solvent and then emulsified in an aqueous phase to form emulsion. The solid lipid nanoparticles are then precipitated from the emulsion with organic solvent evaporating [57]. This technique is similar with the preparation of polymeric nanoparticles through micro/nanoemulsion. As compared to the abovementioned two techniques, solvent evaporation technique exits a weakness that involves organic solvent removal and the potential toxicity from organic solvent [54].

Several key factors will determine the drug-loading efficiency in solid lipid nanoparticles [53]. In general, the solubility and miscibility of drugs drop with decreased temperature. As a result, solubilizers such as glycerides can be added into lipid matrices to improve the miscibility and solubility of drugs in molten lipids, leading to high loading efficiency. On the other hand, the physical and chemical properties of solid lipids also play a critical role. Usually, the lipid with imperfect crystalline lattice matrix is capable of encapsulating more drugs, while that with perfect crystalline lattice has a tendency to expulse drugs. In practice, a mixture of lipids with different fatty acid chains can induce imperfect crystalline formation, which helps improve the drug-loading efficiency [58].

Due to the fact that hydrophilic drugs have limited solubility in molten lipids, solid lipid nanoparticles usually have poor encapsulation ability for carrying hydrophilic drugs [54]. This problem can be solved by addition of organic counterions, which can facilitate the formation of ionic pairs with charged drug molecules for better encapsulation [59]. Another shortcoming of solid lipid nanoparticles is the burst effect, which is the rapid initial release of a high-dose loading drugs followed by a slow incomplete release, resulting in nonuniform drug release [60]. The burst effect is mainly caused by a large aggradation of drugs in surface layers of nanoparticles and fast diffusion into surrounding medium. In addition, the burst release of anticancer drugs will cause a series of complications [52]. Some feasible methods have been proposed to ease the burst effect to achieve smooth drug release, including reduction of the surfactant concentration, rapidly cooling of the lipid emulsion, and using lipid compositions without perfect crystal lattices.

9.2.4 Lipid-Polymer Hybrid Nanoparticles

Lipid-polymer hybrid nanoparticles represent an emerging type of lipid nanoparticles, which constitute three parts, a polymeric core, a lipid layer, and a lipid-PEG stealth layer [61]. Loaded drugs are encapsulated by polymer chains as a core, while the lipid layer is able to avoid the drug leakage and decrease the degradation rate of polymer. In outer layer, the hydrophobic lipid tail can entangle with the lipid layer and the hydrophilic PEG segments expose to the aqueous environment to stabilize the nanoparticles (Fig. 9.3). Therefore, the lipid-PEG stealth layer can prolong the blood circulation of nanoparticles in in vivo applications by reducing opsonization and providing steric stabilization [62].

Lipid-polymer hybrid nanoparticles can be prepared by two- or one-step methods. Two-step preparation involves preforming both polymeric nanoparticles and lipid vesicles, followed by mixing them through electrostatic interaction [61]. The polymeric nanoparticles can be prepared through precipitation [63], high-pressure homogenization [64], and emulsion-solvent-evaporation [65]. On the other hand, the preformation of lipid vesicles involves solvent evaporation, followed by hydrating to obtain a thin lipid film. Upon vortex or ultrasonication of the mixture, the lipid-polymer hybrid nanoparticles are separated from the polymer/lipid hybrid suspension. In addition, other two-step approaches have also been employed. For example, the polymeric nanoparticles can be dispersed in a dichloromethane solvent with lipids, and lipid-polymer nanoparticles are then fabricated by spray drying [66]. Another example is particle replication in non-wetting templates, in which the polymeric nanoparticles are first prepared and stabilized by poly (vinyl alcohol) coating. Subsequently, the poly (vinyl alcohol) is replaced by lipids to form lipid-polymer particles [67].

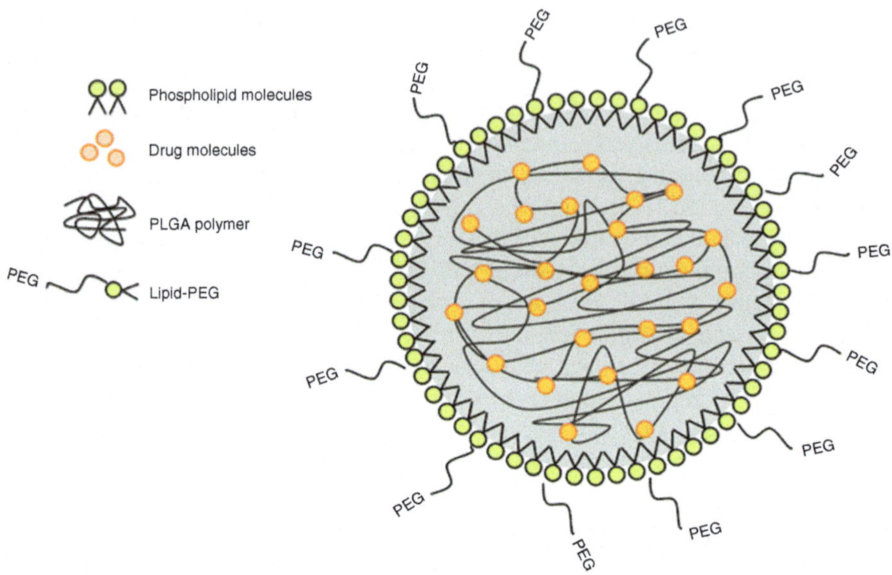

Fig. 9.3 Lipid-polymer hybrid nanoparticles. (Redraw with permission from Ref. [62]. Copyright 2014 Royal Society of Chemistry)

There are several variables in two-step preparation that may have impact on the properties of final products [68]. For example, one can obtain small and homogeneous lipid vesicles using extrusion through a membrane, which will lead to lipid-polymer nanoparticles with small sizes. In practice, the dispersibility of lipid-polymer nanoparticles is of high importance, which is determined mainly by the net charge of lipids. Using lipids with the same charge can successfully minimize the nanoparticle aggregation. In addition, the ratio of lipid vesicle to polymeric nanoparticle (Av/Ap) is another important factor to affect the colloidal stability of lipid-polymer nanoparticles. The lipid vesicle can act as a stabilizer to minimize aggregation at a high Av/Ap ratio, ensuring higher stability and less aggregation. As the lipid-PEG can act as a colloidal stabilizer, using lipid-PEG with longer PEG chains can afford lipid-polymer nanoparticles with high colloidal stability [63].

As compared to two-step method, one-step method is considered to be more efficient, which can be realized by nanoprecipitation and emulsification-solvent-evaporation techniques [61]. In either approach, polymer and lipid solutions are mixed and self-assemble to form lipid-polymer nanoparticles with a polymer core. In nanoprecipitation, polymer and drug are dissolved in water-miscible organic solvent to form a homogenous solution, which is subsequently dropped in lipid-PEG aqueous under continuous stirring to form lipid-polymer nanoparticles. The mechanism is that polymer precipitates due to the sudden drop of solubility in a poor solvent while the lipids self-assemble due to the hydrophobic interaction with polymer chains. In 2010, Fang and colleagues further developed a rapid nanoprecipitation method [69]. They used bath sonication with high and continuous energy input to accelerate self-assembly of nanoparticles. This method significantly boosts the productivity of nanoprecipitation by 20-fold as compared to conventional nanoprecipitation approach. In addition, using microchannel in nanoprecipitation process is able to improve the size homogeneity of lipid-polymer nanoparticles. On the other hand, Valencia and colleagues introduced microchannel into nanoprecipitation process to realize homogeneous polymer nucleation to produce lipid-polymer nanoparticles with uniform sizes [70].

In nanoprecipitation, high lipid-to-polymer (L/P) ratio can ensure excellent colloidal stability, encapsulation rate, and releasing kinetics [71, 72]. When the lipid concentration is higher than the critical micelle concentration, it can lead to formation of liposomes apart from lipid-polymer nanoparticles [71]. Noteworthy is that the efficiency of lipid coating plays an important role in nanoprecipitation [72]. The lipid coating acts as a barrier that can maintain the encapsulated drugs in nanoparticles during self-assembly, which can promote the drug encapsulation ratio and protect drug from sudden release. Similar to its role in two-step procedure, the presence of lipid-PEG can also improve the colloidal stability of final products obtained from nanoprecipitation [71].

Besides nanoprecipitation, both single and double emulsification-solvent evaporation can be used for synthesis of lipid-polymer nanoparticles [73]. In single emulsification, polymers and drugs are dissolved in organic solvent, while lipids are dissolved in water. Subsequently, the organic phase is added to aqueous phase under strong stirring or ultrasonication to facilitate emulsion formation. After organic solvent evaporation, the polymer cores are formed, and the lipids self-assemble around the cores to afford lipid-polymer nanoparticles. The difference between single and double emulsification is that the former is suitable for encapsulating hydrophobic drugs, while the latter is used for carrying hydrophilic drugs. In double emulsification-solvent evaporation, encapsulated drugs cannot be added in organic solvent with polymer simultaneously [74]. Instead, the hydrophilic drugs have to be dissolved in aqueous phase and then emulsified in organic solution containing polymers and lipids to obtain water-in-oil (W/O) emulsion. Subsequently, the emulsion was emulsified again in aqueous solution with lipid-PEG to form a W/O/W emulsion. After removal of organic solvent, lipid-polymer nanoparticles are obtained with an aqueous core surrounded by an inner lipid layer.

In emulsification-solvent evaporation, increasing the lipid to polymer ratio will reduce the nanoparticle sizes with enhanced colloidal stability but induce lower drug encapsulation efficiency owing to the high surface pressure of small lipid layer [73–76]. In addition, the ionic interaction between encapsulated drug and lipid molecules also has inevitable impact on the physical properties of the final product [74]. For instance, stronger hydrophobic interaction among them can provide higher encapsulation efficiency and slower release of lipophilic drugs.

9.2.5 Characteristics and Properties of Lipid-Based Nanoparticles

9.2.5.1 Morphology of Lipid-Based Nanoparticles

In general, most lipid-based nanoparticles are in spherical shapes [77]. However, the morphology of lipid-based nanoparticles can be rationally designed by altering the building blocks. For example, Miyata and Hotani successfully demonstrated that the spherical liposomes can be induced into dumbbell shape and disk shape by initiating polymerization of encapsulated G-actin in liposomes [78]. They first encapsulated G-actin monomers in spherical liposomes, followed by polymerization into actin filament to realize the subsequent shape changes of liposomes, which was confirmed under microscopy. In addition, using different fabrication technique is also a promising solution to design distinct nanoparticle morphologies. For instance, non-wetting template technology has been applied to prepare needle-like lipid-polymer hybrid nanoparticles [79].

The morphology of lipid-based nanoparticles, especially their lamellarity, has significant effect on a variety of physical and physiological properties of the nanoparticles, including drug encapsulation efficiency, drug release profile, cellular uptake, and in vivo circulation time in blood. More layers can provide better stability to liposome vesicles because high lamellarity can effectively prevent deformation and sharp temperature changes [80]. In addition, lamellarity affects the drug encapsulation efficiency. In general, MLVs with multiple phospholipid bilayers are capable of encapsulating more drugs as compared to LUVs and SUVs. As reported in literature, MLVs that are larger than 2 mm in size can load 78.8–81.4% of luteinizing hormone-releasing hormone, while LUVs with average sizes of 241–269.5 nm can encapsulate 66.7–77.9% of the same hormone [81]. Moreover, MLVs have shown faster release of ketoprofen-cyclodextrin complex than LUVs at the same conditions, indicating the impact of lamellarity on drug release profiles [82].

9.2.5.2 Size and Size Distribution

The size and distribution of lipid-based nanoparticles play a crucial role in their in vivo biodistribution and excretion route from bodies. Most of lipid-based nanoparticles cannot be filtered out by the capillary bed of the lungs upon intravenous injection. When their size is below 5.5 nm, renal excretion of nanoparticles becomes dominant [83]. For those with sizes beyond 200 nm, the clearance mainly occurs in the spleen and liver due to the recognition and uptake by reticuloendothelial system (RES) [84]. For example, Awasthi and colleagues demonstrated that the liposome uptake in the spleen of a rabbit model was enhanced with increased liposome size from 136 to 318 nm, suggesting a direct correlation between spleen uptake and the size [85]. On the contrary, the relationship between liver uptake and nanoparticle size is ambiguous and requires further exploration. Theoretically, the hepatic sinusoidal endothelial cells in a healthy human liver contain fenestrae with a size of approximately 107 nm, meaning that the nanoparticles with size beyond 100 nm cannot pass through [86]. In fact, liposomes internalized by the liver usually tend to localize into Kupffer cells. For example, the liposomes prepared from phosphatidylserine with an average size between 200 and 400 nm can be internalized into Kupffer cells and hepatic parenchymal cells [87].

9.2.5.3 Surface Charge

When colloidal particles carry net charge, iron arrangement around the particles changes accordingly and form a double electrical layer including Stern layer and diffuse layer. The electric potential of a theoretical boundary in the diffuse layer is called zeta potential that indicates the potential difference between bulk fluid and fluid layer with opposite charged ions. It is of high importance in assessment of the surface characteristics of nanoparticles, revealing their colloidal stability [88, 89]. If the nanoparticle surface is neutral in nature, the colloidal system cannot be stably dispersed in aqueous phase and intends to form aggregates. To improve the stability, one can introduce net surface charges to the nanoparticles for increased repulsive force. Additionally, the surface charge also has a profound effect on cellular uptake of nanoparticles [90, 91]. With increased surface charge, the uptake of chitosan-based polymeric nanoparticles by murine macrophages can be improved [91]. The surface charge of nanoparticles can interact with cells. However, it is still not well understood which type of charge is best for nanoparticles delivery due to the controversial results from different groups. However, it has been demonstrated that slightly charge of nanoparticles, no matter negative or positive, can be beneficial for nanoparticle delivery [92–95].

9.2.5.4 Phase Transition Temperature

Temperature insensitivity is an important factor in liposomal drugs delivery system [32]. Different lipids have their unique phase transition temperatures (PTTs) and exhibit varied physical states and properties above or below their PPTs. In general, lipids are in highly ordered structure at a temperature below the PTT and in a liquid crystalline state above the PTT. For example, the fluidity of liposomal bilayers can be fine-tuned by using different combination of phospholipids to regulate drug release [96]. When the bilayers contain more lipids with high PTT, the drug release can be reduced. On the contrary, the lipid bilayers will release more drugs when low PTT lipids are used, due to the increase of fluidity, allowing drugs to escape from the liposome interior [96].

9.2.5.5 Plasma Protein Interaction: Particle Stability and Clearance

In complicated biological environment, especially for in vivo studies, the interaction between plasma proteins and nanoparticles determines their fate (e.g., stability, circulation, and excretion) [97]. This interaction can be investigated by hemolysis assays. Szebeni and colleagues found that the hemolytic complement activity of rats showed a significant reduction upon injection of hemoglobin with liposomes, while no significant change was observed in the rats after injection of free hemoglobin [98]. In live bodies, lipoprotein and its apolipoprotein can disrupt lipid-based nanocarriers to trigger undesired release of encapsulated drugs [99]. Other plasma proteins, including immunoglobulins and albumin, also have impact on the in vivo fate of the lipid-based nanoparticles [100].

9.3 Tumor-targeted

In order to shuttle therapeutic reagents to tumor tissues with high efficiency, the drug delivery systems used should possess excellent biocompatibility, high drug-loading capacity, and good tumor-targeted ability. Among a variety of nanocarriers, lipid-based nanoparticles such as liposomes represent one of the most widely used systems that can promote targeted uptake by tumor tissues with reduced unnecessary toxicity to normal tissues. In addition, they can help extend the circulation time in blood to facilitate enhanced permeability and retention (EPR) effect, which can be attributed to passive targeting. Thanks to the tunable functionality of the nanoparticle surface, immobilization of targeting moieties such as ligands or antibodies will endow them with active targeting ability. In this section, we will briefly discuss the application of lipid-based nanoparticles in both in vitro and in vivo targeting studies.

9.3.1 Passive Targeting

It was believed that no matter the particles are mediated with or without ligands, they both can reach the target sites via passive targeting (Fig. 9.4) [101, 102]. The tumor tissues have distinct vascular characteristics from normal tissues, such as larger interstitial spaces, which can significantly influence the drug delivery efficiency at tumor sites. As compared to normal tissues, tumor vessels are leaky, chaotic, and non-homogeneously distributed. In addition, the non-functional lymphatic system in tumor tissues could not achieve effective drainage of nanoparticle with appropriate size. In this contribution, researchers have made great efforts to design lipid nano-particle drug delivery strategies relying on the EPR effect [103–105]. In order to achieve better passive targeting ability, a prolonged blood circulation time is critical,

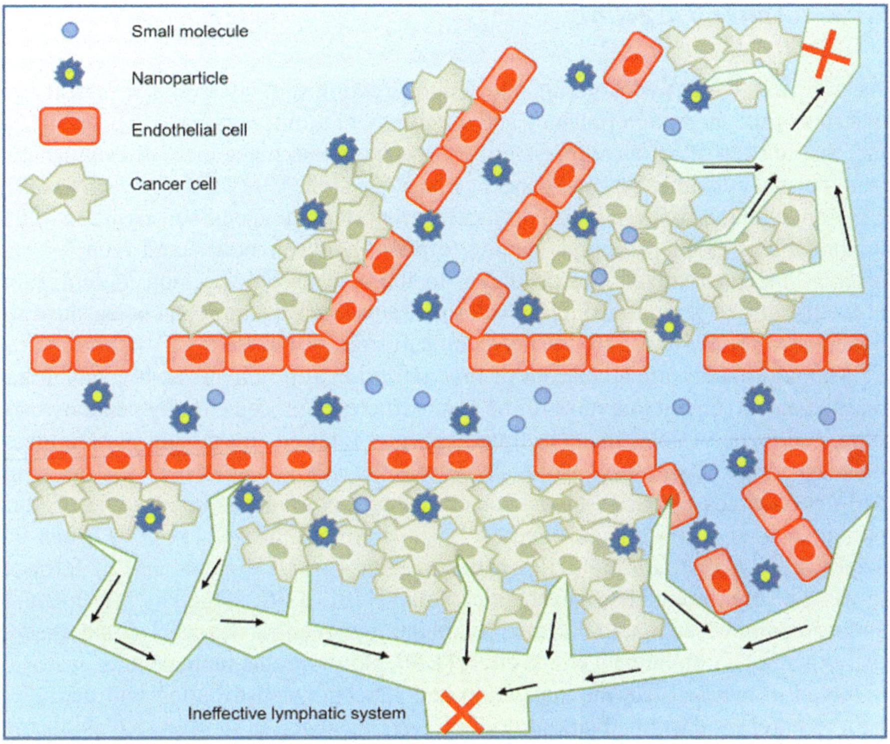

Fig. 9.4 Passive targeting of nanoparticles. (Redraw with permission from Ref. [32]. Copyright 2016 the American Society for Pharmacology and Experimental Therapeutics)

which can be simply realized through modification on lipid nanoparticle surface [106].

In the arena of biomedicine, liposomes exhibit great advantages when used as drug delivery system by encapsulating drugs with short biological half-lives into lipid-based carriers with prolonged half-life in blood [107]. Normally, the lipid-based nanoparticles are coated with PEG layers to enhance their stability and blood circulation time. For instance, under appropriate circumstances, PEGylated liposomal oxaliplatin could be stored for up to 12 months and achieve long circulation half-lives in vivo, leading to increased suppression of tumor growth as well as reduced side effects [108]. In another case, solid lipid nanoparticles (SLNs) loaded with *sclareol* could inhibit A549 growth more efficiently than free drugs [109]. Similar studies also showed that SLN-formulated temozolomide exerted larger effects towards melanoma using EPR effect [110], proving that passive targeting strategy is reasonable in building lipid-based drug delivery system.

9.3.2 Active Targeting

As compared to passive targeting, the active targeting strategy enjoys several merits in terms of the targeting efficiency and specificity to tumor regions [111]. Due to the fact that tumors often overexpress unique receptors that make them distinguishable from normal and other types of cancer cells, it is a straightforward approach to build a targeting ligand-functionalized drug delivery system for specific recognition of the tumor targets [112]. The high binding affinity between ligand and receptor can promote the internalization efficiency into tumor cells [113]. Upon identification of the specific overexpressed tumor cell surface biomarkers, one can choose the right ligand accordingly to satisfy the needs in active tumor-targeted.

Although the exact mechanism of internalization still remains ambiguous, it has reached an agreement that most of the nanoparticles enter cells through endocytosis. First, endocytic vesicles made of ligand-receptor complex inside and plasma membrane envelops outside are formed. These vesicles will then enter cells and fuse with early endosomes [114]. With the maturing process of endosomes, some protons pump, such as ATPase, would recruit H^+ ions around them [115–117]. When the acidity is increased and the pH of endosomes drops to 5, ligands will be released from the conformation-changed receptors [118, 119]. Finally, the matured endosomes and used ligand-receptor combination will either be recycled or degraded by lysosomes, completing a whole circle [120]. However, the more detailed pathway involved in nanoparticle internalization and subcellular distribution will definitely help researchers design advanced drug delivery systems. In in vivo studies, factors such as chemical properties of the drugs [121] (amphiphilicity, ion charge, and log P value), tumor permeability and penetrability [122], binding density and affinity [123, 124], as well as binding site microenvironment will affect their uptake and targeting efficiency to tumors [125]. Thus, it is of importance that the nanoplatform could strike the balance between those challenges.

To date, a number of receptors have been studied extensively. In practice, folate, transferrin, epidermal growth factor receptor (EGFR), and ssDNA and RNA aptamers are the most widely used targeting ligands. Among them, folate receptor (FR) is a well-known cellular surface biomarker for cancer because of its overexpression in tumor cells as compared to normal cells. *Via* conjugation of folate on the PEG chains of the lipid-based nanocarriers, researchers demonstrated ~31-fold higher cellular uptake than the non-functionalized one [126]. In addition, folate modification could enhance the life span of nanoparticles in nude mice xenografted with KB tumors, proving that the specific ligand-receptor binding can benefit the long-term circulation of nanoparticles in vivo. Recent studies from several research groups have successfully demonstrated that folate-functionalized drug-loaded liposomes have significant better ability to inhibit tumor growth than bare liposomes [127–130]. More details about the strategies for linking folate receptor onto liposomes could be found in a recently published review [131].

Transferrin receptor is another appealing cell marker that is always overexpressed at the tumor cell surface (~100-fold higher) as compared to the normal ones [132–137]. As long as the coated nanoparticles are functionalized with transferrin, they can be endowed with high binding affinity to transferrin receptors and enter into the cells subsequently. This process has been thoroughly investigated by previous studies [138, 139]. In liposomal formulations, it has been demonstrated that the conjugation of transferrin receptors with drugs (e.g., doxorubicin and cisplatin) can promote intracellular accumulation than free drugs or conventional liposomes [140–144]. Moreover, transferrin receptor-conjugated liposomes encapsulating DOX are more toxic to cancer cells than free DOX, suggesting that this strategy can overcome multidrug resistance of tumor cells [145]. The ability of such similar formulation to overcome blood-brain barrier was also demonstrated by isotope labeling [146].

Tumor cells have the potential of limitless proliferation, so the cytokines related to growth and progression stimulation are of vital importance to their cell cycles. One of them is a tyrosine kinase receptor, epidermal growth factor receptor (EGFR), which is typically overexpressed in cancer cells, especially the most invasive breast carcinomas [147–149]. Although the modification with anti-EGFR antibody will not significantly affect the circulation halftime of the liposomes, anti-EGFR liposomes can reach 92% of internalization efficiency while the non-targeted liposomes are only 5% [150]. Details of the composition and approach to build anti-EGFR liposomes could be searched in recent patents [151]. Usually, anti-EGFR liposomes are designed towards breast cancer, since EGFR expression and overexpression could occur in 14–91% patients suffering [152]. By tailoring the lipid composition, the cellular uptake of the modified liposomes could be modulated against triple negative breast cancer [153]. Anti-EGFR-liposomes have also been reported in targeting other kinds of tumor, e.g., EGFR-mutant non-small cell lung cancer cells [154]. Biological evaluation has been set up for EGFR-targeting liposomes on tumor-bearing mouse model [155].

Aptamers now have become a promising targeting moiety other than antibodies. As ssDNA or RNA oligonucleotides, aptamers can take the advantage of their small and stable structures to facilitate their immobilization on liposomes [156]. Via electrostatic interaction, hydrophobic interactions, and hydrogen bonding, aptamers can also modulate the overall physical properties of nanocarriers. For example, one aptamer named sgc8 could target to protein tyrosine kinase 7 (PTK7), which is a specific receptor that usually upregulates on the surface of some cancer cells. After simple incubation for 30 minutes, the aptamer-conjugated liposomes could bind to the hematologic cancer cell targets, such as leukemia CCRF-CEM cells [157]. The same principle is also applied in DNA aptamer conjugation. After incubation with human breast cancer cells (MCF-7), AS1411-liposomes showed higher killing efficiency in both in vitro and in vivo as compared to free drugs and non-functionalized liposomes [158]. As researchers have discovered a number of unique antigens that only exist in hematologic cancer cells, aptamers can be potential candidates for liposome functionalization to achieve targeted delivery [159–164].

Besides direct targeting of tumors, indirect targeting of tumoral endothelium is also a well-known strategy for designing nanomedicines [165]. By depriving the oxygen and nutrients that cancer cells need through targeting endothelium in newly formed blood vessels, this strategy could not only inhibit the growth of existing tumors but also restrict the development of new neoplastic tissues to prevent their metastasis [166, 167]. To date, a number of targets in tumoral endothelium microenvironment have been identified for cancer treatment, including vascular endothelial growth factor (VEGF), integrin, and matrix metalloproteinases (MMPs).

VEGF and its receptors play major roles in the progress of neovascularization and angiogenesis in cancer [168]. Usually, the tumor microenvironment is hypoxic, which would in turn trigger the upregulation of VEGF expression in tumor endothelium cells. Taking advantage of the versatile functionalities of liposomes as nanocarriers, they can be decorated with anti-VEGF antibody while delivering siRNA sequences to downregulate the expression of VEGF through RNA interference. For instance, VEGF-siRNA co-loaded liposomes could enhance the downregulation of VEGF and induce early cell apoptosis, being more effective than free siRNA and single-component loaded carrier [169].

$\alpha_v\beta_3$ integrin is the most wildly distributed protein-type endothelial cell receptor that could distinguish tumor and angiogenic endothelial cells from normal cells [170]. Arg-Gly-Asp (RGD) oligopeptide is a well-known targeting moiety to $\alpha_v\beta_3$ integrin [171, 172]. Aside from targeting tumor related tissues, the RGD-functionalized liposomes have been actively used in killing target solid tumors and prevention of cancer metastasis and angiogenesis [173]. For instance, studies on resistant glioma cells and glioma-bearing mice have shown that the RGD-decorated liposomes could significantly promote the transportation across BBB and enhance the efficacy for killing brain glioma [174].

MMPs are a family of endopeptidases that depend on zinc, which can mediate endothelial cells invasion and migration, assist formation of capillary tubes, and recruit accessory cells [175, 176]. One of the most famous MMPs is membrane type 1 matrix metalloproteinase (MT1-MMP), which could be found in endothelial cells across a variety of solid tumors. As an example, lysolipid-containing thermosensitive liposomes delivering MMP inhibitor are found to be able to cause a sevenfold decrease in metastatic lung nodules and a sixfold reduction in micro-vessels inside the tumor tissues [177].

To promote the tumoral endothelium targeting ability, modification of the liposome surface with ions and charges is also a promising strategy for increased accumulation in the interstitial space of tumor vessels [178–180]. Because anionic glycoproteins are often overexpressed on the surface of tumor endothelium, cationic liposomes will have higher affinity in targeted delivery than anionic or neutral ones [181]. Studies have shown that the micro-vessel density and tumor perfusion could also be greatly reduced by more than 50% in mouse models treated by cationic liposomes [182].

9.4 Pharmacological Properties of Lipid-Based Nanoparticles

As lipid-based nanoparticles are complexes consisting of versatile coats and cores, their pharmacological behaviors are very different from free drugs and much more complicated. These behaviors and properties are highly related to the nanoplatforms' size, dose, charge, membrane packing, steric stabilization, and liposomal formulations, as well as their metabolism, kinetics, and distribution in vivo [183]. Additionally, when surface modification happens, the pharmacological properties of such nanoplatforms should be reconsidered as a whole new system [184]. Besides the in vivo behaviors of the nanocarriers, the release profile of drugs encapsulated in lipid-based nanoparticles should be considered as well to reveal the overall pharmacokinetics. Therefore, several parameters should be carefully considered from various aspects when investigating the pharmacological properties of these nanoplatforms.

First, it is clear that liposomes should be able to encapsulate sufficient amounts of drugs and escape from mononuclear phagocyte system with prolonged circulation time in blood. To achieve this goal, small sizes are always preferred, and the impact of sizes on in vivo fate of nanoplatforms has been intensively studied. When investigating the size effect, the basic composition of the liposomes must remain the same because the surface properties could also generate significant impact on the nanoparticle clearance rate and pathway in bodies [185]. Also, the aggregation of unstable neutral liposomes or presence of protein coronas around the liposomes can increase the exact sizes of the liposomes [186]. These factors should be carefully examined.

Second, it is possible that high lipid dose can saturate the reticuloendothelial system (RES) clearance effect, so that more lipid-based nanoparticles will enter the cells [187, 188]. Besides, the phagocytic capacity of RES macrophages towards high lipid dose is limited, resulting in longer circulation time of nanoparticles [189]. However, in the presence of DSPE-PEG on nanoparticle surface, the clearance kinetics could be less dependent on lipid dose [190–192].

Third, the surface charge of lipid-based nanoparticles is of high importance. The modification with charged group can change liposomal steric hindrance, further affecting their clearance by RES [193]. Previous studies have shown that cationic modification and incorporation of high-phase transition anionic lipids in nanoparticles are beneficial for prolonging their half-life in blood [194, 195]. Actually, even the absorption kinetics of those phospholipid head groups with similar charges may vary because of distinct surface signatures [196].

Fourth, membrane packing of the lipid system could influence the blood circulation clearance and drug release profile [197]. The bilayer fluidity of the liposomes can make a difference in protein absorption as well as in in vivo biodistribution. The high-phase transitional liposomes usually have better stability and thus avoid the premature leakage of interior drugs [198]. Subsequently, high doses of liposomes can be delivered to solid tumors.

Fifth, as mentioned above, steric hindrance, such as PEGylated formulation, is able to reduce their clearance by RES and promote the circulation time [199]. Occupied by flexible hydrophilic polymers such as PEG, the surface of nanoparticles is hindered from direct interaction with macrophages from RES to achieve long circulation time and abundant accumulation in tumor tissues [200–208].

The detailed pharmacological properties of lipid-based nanoparticles in in vivo study have been described in published reviews, which has discussed the impact of size, surface coating, and composition on the performance of versatile nanoparticles [13, 25, 32, 209, 210].

9.5 Lipid-Based Nanocarriers for Cancer Treatment

Cancer is among the leading causes of human death. To date, the treatment of cancer still faces a number of challenges. Among them, precise delivery of drugs to tumor lesions for targeted and effective therapy is a major scientific problem that needs to be solved [211, 212]. With the rapid development of nanobiotechnology, antitumor nanodrugs hold great promises in improving cancer treatment efficiencies and minimizing side effects to normal tissues. Several studies have shown that lipid-based nanocarriers can not only promote the drug accumulation in tumor tissues to inhibit tumor growth by changing the pharmacokinetics of drug molecules [213] but also reduce the toxicity of drugs to normal tissues and organs [214]. In recent years, lipid-based nanocarriers have shown great potential in antitumor drug delivery and improvement of cancer treatment for better prognosis and overall patient outcomes [214, 215]. Here, we mainly introduce the lipid-based nanodrugs that have been approved for clinical cancer therapy and the advanced lipid-based drug delivery systems that are undergoing clinical trials.

9.5.1 Clinically Approved Lipid-Based Nanodrugs

At present, several lipid-based nanodrugs have been approved by Food and Drug Administration (FDA) or European Medicines Agency (EMA) for clinical use in cancer treatment [216–222] (Table 9.1). Compared with the newer nanodrug undergoing exploration, the approved lipid-based nanocarrier systems have experienced a more direct clinical approval approach [218]. This is due to the fact that the approved lipid-based nanodrug has been practiced on patients for a long time and proven to be safe and effective in humans [219–221]. In this section, we will discuss the approved lipid-based nanodrugs and the types of cancer they can treat.

Many clinically approved nanopharmaceutical formulations are used to treat different cancer types at various stages. Liposomes are one of the most promising nanoplatforms in nanomedication for cancer treatment. Phospholipid bilayer is the core component in cell membrane structure, and the interest in study of phospholipid

Table 9.1 Clinically approved lipid-based nanodrugs for cancer treatment

Name	Company/ Institution	Particle type	Application/Indication	Indication approval (year)
Doxil/ Caelyx	Ortho Biotech	Liposomal doxorubicin	Ovarian cancer	FDA (1995)
	Schering-Plough	(PEGylated)	HIV-associated Kaposi's sarcoma	EMA (1996)
			Multiple myeloma	
DaunoXome	Galen	Liposomal daunorubicin	HIV-associated Kaposi's sarcoma	FDA (1996)
	NeXstar	(non-PEGylated)	(primary)	
Myocet	Teva UK	Liposomal doxo-rubicin (non-PEGylated)	Treatment of metastatic breast cancer (primary)	EMA (2000)
Depocyt	SkyePharma	Liposome/lipid NPs	Acute myeloid leukaemia	FDA (1999)
				EMA (2001)
Lipusu	Luye Pharma	Liposomal paclitaxel	Multiple cancers	CFDA (2003)
Marqibo	Spectrum	Liposomal vin-cristine (non-PEGylated)	Philadelphia chromosome-negative acute lymphoblastic leukemia (tertiary)	FDA (2012)
MEPACT	Millennium	Liposomal mifamurtide (non-PEGylated)	Treatment for osteosarcoma (primary following surgery)	EMA (2009)
Onivyde MM-398	Merrimack	Liposomal irinotecan (PEGylated)	Metastatic pancreatic cancer (secondary)	FDA (2015)

membrane system started earlier than that in the field of nanotechnology. So the study of liposomes is relatively mature. Noteworthy is that all these clinically approved formulations, except Abraxane (albumin-bound paclitaxel nanoparticles), are liposomal systems that encapsulate anticancer drugs. Doxil, polyethylene glycol (PEG)-functionalized liposome doxorubicin, was the first clinically approved (FDA 1995) cancer nanodrug [222]. It was first developed by ALZA, after which ALZA was acquired by Johnson & Johnson, and the product was transferred to Johnson & Johnson [223]. Doxorubicin is a member of anthracycline-based chemotherapeutic drugs and is the first one used in liposome-based drug delivery systems. As one of the earliest cancer chemotherapeutic nanodrugs, Doxil was used to treat a variety of cancers, including breast cancer, lung cancer, stomach cancer, ovarian cancer, sarcoma, myeloma, leukemia, and lymphoma.

Later, other liposome formulas were successively approved. For example, liposomal doxorubicin (Myocet) and liposomal mifamurtide (MEPACT) were approved by EMA [224, 225]. Liposomal daunorubicin (DaunoXome), liposomal vincristine l (Marqibo), and liposome irinotecan (Onivyde) were approved by FDA [226]. In addition, the paclitaxel liposome developed and sold by Nanjing Green Leaf Company of China, named as Lipusu, was approved for marketing in 2003. Lipusu successfully solved the problem of low solubility of paclitaxel, relieving the risk of hypersensitivity caused by solvent. It also significantly reduced the side effects of paclitaxel, achieving great success in the market [227, 228]. These approved particles exhibit better functions than their free counterparts, which can preferentially accumulate in tumor sites through enhanced penetration and retention (EPR) effects to reduce the side effects to normal tissues and organs.

9.5.2 Lipid-Based Drug Delivery Systems in Clinical Trials

Cancer nanodrug currently undergoing clinical trials has received great attention. On one hand, the fiery heat of approved nanopharmaceuticals such as Doxil and DaunoXome has encouraged the development of nanodrugs for cancer therapy [229, 230]. On the other hand, these older generations of drugs become less capable of meeting the growing needs in the mediation market [231]. Here, we will introduce the nanodrugs in current clinical trials (Table 9.2).

9.5.2.1 Chemotherapeutics and Anticancer Drugs

In clinical trials, many types of chemotherapeutic reagents and anticancer drugs are encapsulated and delivered by lipid-based nanoparticles, most of which are liposomes [232]. Many of these liposomes have similar features in design to approved liposome systems, such as non-targeting functionalization and single drug component. Here, we will introduce some novel formulations.

One of the unique features and advantages of liposomes is their long circulation time and slow drug release. Clinical transformation of liposomal cytarabine (DepoCyt, DepoTech Corp.) takes full advantage of these merits of liposome. Cytarabine, also known as arabinofuranosyl cytidine (ara-C), is a chemotherapeutic agent commonly used in leukemia and lymphoma. One of the complications of invasive lymphoma and leukemia is the meningeal involvement [233]. In this case, not many treatments are effective. Intrathecal injection of ara-C (delivering the drug to the cerebrospinal fluid (CSF)) is one of the options, but it is toxic and usually requires frequent administration [234]. Liposomal formulations can be a promising strategy to solve these problems. In addition, versatile design features of liposome systems have been introduced for clinical research. For example, VYEXOS/CPX-351 is a synergistic therapeutic formula that contains a certain ratio of two anticancer drugs (cytarabine and daunorubicin), and the recommended dose of each drug has been determined by early clinical results [235, 236].

Table 9.2 Lipid-based drug delivery systems for cancer treatment in clinical trials [211–241]

Name	Company/ Institution	Particle type	Application/Indication	ClinicalTrials. gov Identifier	Status
Chemotherapeutics and anticancer drugs					
LE-DT	INSYS Therapeutics Inc	Liposome Encapsulated Docetaxel	Solid Tumors	NCT01151384	Phase 1
IHL-305	Yakult Honsha Co., LTD	Irinotecan liposome	Advanced Solid Tumors	NCT00364143	Phase 1
SN-38	Alliance for Clinical Trials in Oncology	Liposomal SN-38	Metastatic Colorectal Cancer	NCT00311610	Phase 2
Oncoprex	Genprex	FUS1 (TUSC2) encapsulated liposome	Lung cancer	NCT01455389	Phase 1/2
PROMITIL	Lipomedix Pharmaceuticals	Pegylated liposomal mitomycin-C	Solid tumors	NCT01705002	Phase 1
Halaven E7389-LF	Eisai	Liposomal eribulin mesylate	Solid tumors	NCT01945710	Phase 1
ThermoDox	Celsion	Lyso-thermosensitive liposomal doxorubicin	Temperature-triggered doxorubicin release:	NCT00826085	Phase 1/2
				NCT02112656	Phase 3
			Breast cancer recurrence at chest wall (microwave hypothermia)	NCT02181075	Phase 1
				NCT02536183	Phase 1
			Hepatocellular carcinoma (radiofrequency ablation)		
			Liver tumors (mild hypothermia)		
			Refractory solid tumors (magnetic resonance high intensity focused ultrasound)		

(continued)

Table 9.2 (continued)

Name	Company/ Institution	Particle type	Application/Indication	ClinicalTrials. gov Identifier	Status
VYEXOS CPX-351	Celator Pharmaceuticals	Liposomal formulation of cytarabine: daunorubicin (5:1 molar ratio)	Leukemias	NCT01804101	Not Applicable
				NCT02286726	Phase 2
				NCT01943682	Phase 1
				NCT02269579	Phase 2
				NCT01696084	Phase 3
MM-302	Merrimack Pharmaceuticals	HER2-targeted liposomal doxorubicin (PEGylated)	Breast cancer	NCT01304797	Phase 1
				NCT02213744	Phase 2/3
188Re-BMEDA-liposome	Institute of Nuclear Energy Research	188Re-N,N-bis (2-mercaptoethyl)-N0,N0-diethylethylenediamine pegylated liposome	Advanced solid tumors	NCT02271516	Phase 1
Mitoxantrone Hydrochloride		Mitoxantrone liposome	Lymphoma and breast cancer	NCT02131688	Phase 1
Liposome	CSPC ZhongQi Pharmaceutical Technology			NCT02596373	Phase 2
				NCT02595242	Phase 1
				NCT02597387	Phase 2
				NCT02597153	Phase 2
JVRS-100	Milton S. Hershey Medical Center	Cationic liposome incorporating plasmid DNA complex for immune system stimulation	Leukemia	NCT00860522	Phase 1
Lipocurc	SignPath Pharma	Liposomal curcumin	Solid tumorsvNCT02138955	Phase 1/2	
LiPlaCis	LiPlasome Pharma	Liposomal formulated cisplatin with specific degradation-controlled drug release via phospholipase A2 (PLA2)	Advanced or refractory tumors	NCT01861496	Phase 1/2

Nanosomal Docetaxel Lipid Suspension	Jina Pharmaceuticals Inc.	Nanosomal Docetaxel Lipid Suspension (75 mg/m2)	Triple Negative Breast Cancer	NCT03671044	Phase 3
		Nanosomal Docetaxel Lipid Suspension (100 mg/m2)			
LIPUSU	Nanjing Luye Sike Pharmaceutical Co.,Ltd.	Paclitaxel Liposome	Advanced solid tumors, or gastric, breast cancer	NCT01994031	Phase 4
				NCT02142790	Phase 4
				NCT02163291	Phase 2
				NCT02142010	Not Applicable
Gene therapy					
TKM-080301	Arbutus Biopharma	Lipid particle targeting polo-like kinase 1 (PLK1) for delivery of siRNA	Hepatocellular carcinoma	NCT02191878	Phase 1/2
siRNA-EphA2-DOPC	ProNAi Therapeutics	siRNA liposome for EphA2 knockdown	Solid tumors	NCT01591356	Phase 1
PNT2258	Sierra Oncology	Proprietary single-stranded DNAi (PNT100) encapsulated in lipid nanoparticles	Lymphomas	NCT02378038	Phase 2
				NCT02226965	Phase 2
				NCT01733238	Phase 2
BP1001	Bio-Path Holdings	Growth factor receptor bound protein-2 (Grb-2) antisense oligonucleotide encapsulated in neutral liposomes	Leukemias	NCT01159028	Phase 1
DCR-MYC	Dicerna Pharmaceuticals	DsiRNA lipid nanoparticle for NYC oncogene silencing	Solid tumors, multiple myeloma, lymphoma, or hepatocellular carcinoma	NCT02110563	Phase 1
				NCT02314052	Phase 1/2

(continued)

Table 9.2 (continued)

Name	Company/ Institution	Particle type	Application/Indication	ClinicalTrials.gov Identifier	Status
Atu027	Silence Therapeutics GmbH	AtuRNAi liposomal formulation for PKN3 knockdown in vascular endothelium	Pancreatic cancer	NCT01808638	Phase 1/2
SGT-53	SynerGene Therapeutics	Cationic liposome with anti-transferrin receptor antibody, encapsulating	Glioblastoma, solid tumors, or pancreatic cancer	NCT02354547	Phase 1
				NCT00470613	Phase 1
		Wildtype p53 sequence		NCT02354547	Phase 1
				NCT02340156	Phase 2
SGT-94	SynerGene Therapeutics	RB94 plasmid DNA in a liposome with anti-transferrin receptor antibody	Solid tumors	NCT01517464	Phase 1
MRX34	Mirna Therapeutics	Double-stranded RNA mimic of miR-34 encapsulated in liposomes	Liver cancer	NCT01829971	Phase 1
TargomiRs	EnGeneIC	Anti-EGFR bispecific antibody minicells (bacteria derived nanoparticles) with a miR-16 based microRNA payload	Mesothelioma and nonsmall cell lung cancer	NCT02369198	Phase 1

9.5.2.2 Lipid-Based Nanocarriers for Gene Therapy

Gene therapy refers to the introduction of foreign normal genes into target cells to correct or compensate diseases caused by defective and abnormal genes for therapeutic purposes, which can usually be achieved by packaging and delivering siRNA or RNA [237, 238]. In cancer treatment, gene therapy has shown promising performance, and lipid-based nanoparticles as gene carriers make it easier to realize [239]. There are not many examples of these systems involved in current clinical trials. For example, the SGT-53 system has succeeded in restoring the function of the human suppressor p53 through delivering a plasmid containing the wild-type p53 sequence [240]. The significance of this technique is enormous because p53 dysfunction is found in most cancers and considered as a necessary condition for tumor growth. Therefore, the technique of restoring the normal anticancer function of the body can potentially be used to treat many cancers with less side effects.

In addition, engineered siRNAs of current clinical studies show enhanced performance as compared to traditional siRNAs. For example, the liposome siRNA formula, Atu027, utilizes proprietary AtuRNAi to target and knock down PKN3, a well-known gene that causes malignant cell growth. Early clinical results indicated limited cytokine activation, which showed great compliance to patients and can achieve disease stabilization in 41% of pancreatic cancer patients. DCR-MYC is a lipid-based therapeutic system that inhibits the expression of key oncoproteins (MYC), outperforming the existing standard therapies [241]. DCR-MYC not only can be used alone for therapeutic treatment but also can play a significant role in combination with other standard therapies. To date, many gene therapies using lipid-based nanocarriers are undergoing clinical trials and have shown great potential in clinical treatment.

9.5.3 Tumor-targeted Lipid-Based Drug Delivery Systems in Clinical Trials

Many previous studies have pointed out the disadvantages of clinically approved lipid-based drug delivery systems. Researchers have dedicated to explore other new drug delivery systems that are more advanced in several respects than the approved ones, and some are already in clinical trials (Table 9.3). For instance, some advanced drug delivery systems in clinical trials employ active targeting mechanisms to improve biodistribution or stimulus response mechanisms to achieve controlled drug release in specific tissues of the body [242–245]. None of these features are available from any currently approved lipid-based drug delivery systems.

Table 9.3 Tumor-targeted lipid-based drug delivery systems in clinical trials [211, 212]

Name	Company/ Institution	Particle type	Application/Indication	ClinicalTrials. gov Identifier	Status
Active tumor-targeting Systems					
Anti-EGFR immunoliposomes	University Hospital, Basel, Switzerland	Anti-EGFR immunoliposomes loaded with doxorubicin	Solid Tumors	NCT01702129	Phase 1
ND-L02-s0201	Nitto Denko	siRNA lipid nanoparticle conjugated to Vitamin A	Hepatic fibrosis	NCT02227459	Phase 1
DCR-MYC	Dicerna Pharmaceuticals, Inc.	Stable lipid particle suspension that targets the oncogene MYC	Solid Tumors or Multiple Myeloma	NCT02110563	Phase 1
MM-302	Merrimack Pharmaceuticals	HER2-targeted liposomal doxorubicin (PEGylated)	Breast cancer	NCT01304797	Phase 1
				NCT02213744	Phase 2/3
WXFL10030390	Shanghai Jiatan Pharmatech Co., Ltd	Targeting lipid particle	Advanced Solid Tumors or Lymphoma	NCT03730142	Phase 1
SGT-53	SynerGene Therapeutics	Cationic liposome with anti-transferrin receptor antibody, encapsulating	Glioblastoma, solid tumors pancreatic cancer	NCT02354547	Phase 1
				NCT00470613	Phase 1
		Wildtype p53 sequence		NCT02354547	Phase 1
				NCT02340156	Phase 2
SGT-94	SynerGene Therapeutics	RB94 plasmid DNA in a liposome with anti-transferrin receptor antibody	Solid tumors	NCT01517464	Phase 1

Stimuli-responsive Drug Delivery Systems					
ThermoDox	Celsion	Lyso-thermosensitive liposomal doxorubicin	Temperature-triggered doxorubicin release:	NCT00826085	Phase 1/2
				NCT02112656	Phase 3
			Breast cancer recurrence at chest wall (microwave hypothermia)	NCT02181075	Phase 1
				NCT02536183	Phase 1
			Hepatocellular carcinoma (radiofrequency ablation)		
			Liver tumors (mild hypothermia)		
			Refractory solid tumors (magnetic resonance high intensity focused ultrasound)		
LiPlaCis	LiPlasome Pharma	Liposomal formulated cisplatin with specific degradation-controlled drug release via phospholipase A2 (PLA2)	Advanced or refractory tumors	NCT01861496	Phase 1/2

9.5.3.1 Active Tumor-targeted Systems

Interestingly, although some nanoparticles and targeting antibodies have been approved for clinical use, integrated systems that combine these two technologies are rare in approved products. This can be mainly attributed to the lack of significant variability in clinical trial results. Efforts have been made and some delivery systems show promising outcomes. For instance, BIND-014, a nanocarrier loaded with docetaxel for specific targeting of prostate specific membrane antigen (PSMA) [242], have shown good inhibitory effect on a variety of tumor models (e.g., prostate cancer and non-small cell lung cancer). Additionally, its pharmacokinetic behavior is significantly better than free docetaxel with lower systemic toxicity. ND-L02-s0201 is a vitamin A-bound liposome for target siRNA delivery to treat liver fibrosis.

SynerGene therapeutics (SGT-53) using a similar strategy with transferrin targeting can provide plasmids containing wild-type p53 sequences [243, 244]. Early clinical data have shown that the active targeting system has been successful in restoring the function of human suppressor p53 in the tumor tissues. This same targeting approach was applied to another SynerGene Therapeutics formulation (SGT-94), RB94-loaded liposomes. Other examples include HER-2-targeted PEGylated liposomal doxorubicin (MM-302), which is under development for the treatment of breast cancer. The concept is based on the FDA-approved HER-2 targeting monoclonal antibody Herceptin (trastuzumab) and the recently approved antibody drug conjugate version (Kadcyla) [245].

9.5.3.2 Stimuli-Responsive Drug Delivery Systems

Stimuli-responsive design can also be introduced to lipid-based drug delivery systems for better performance in cancer killing accuracy. Following Doxil's success, Celsion Corporation is currently developing ThermoDox as a temperature-sensitive version of liposomal doxorubicin. ThermoDox is a thermosensitive liposome that releases doxorubicin at high temperatures (about 42 °C). Clinical trials of ThermoDox present tolerability in patients with breast cancer [246, 247]. And more clinical studies are ongoing for the treatment of liver metastases, chest wall breast cancer, and hepatocellular carcinoma. In addition, some drug delivery systems that can respond to some biosignals are in clinical trials. For example, the presence of phospholipase A2 will induce faster the liposome degradation. Phospholipase A2 is more abundant in tumor tissues, so that the cisplatin payload is only released at the target tumor sites [248, 249].

9.6 Challenges in Translational Research

Although we are very optimistic about the future of nanomedicine, the clinical translation of nanomedicine products still faces serious challenges. It is true that each nanodrug formula will face unique challenges in its clinical transformation. On the other hand, most nanodrugs will encounter the same challenges from many aspects. Here, we will focus on the main challenges in common from a clinical and transformation perspective [250, 251].

9.6.1 *Biological Evaluation Model*

The biological evaluation of nanomedicine can be divided into in vitro cytological assessment and in vivo assessment using animal models. In vitro experiments can help in deeper understanding of nanoparticle-cell interactions, and it is of high importance to conduct in vitro evaluation of nanomedicines to evaluate their biocompatibility before performing animal studies. However, the porous plate environment used in conventional cell culture processes lacks the complexity of living tissue and blood fluidics and does not fully mimic the complex physiological barriers between the organism and the nanomedicine [252]. A variety of tumor models are currently available in nanomedicine research, including subcutaneous and orthotopic tumor models based on tumor cell lines, human tumor xenograft models, and genetically engineered mice. However, none of the above tumor models can accurately reflect the symptoms of human malignancies. Fortunately, as the number of clinical trials increases, the understanding of correlation between human and small animal data is increasing.

9.6.2 *Design and Large-Scale Synthesis of Nanodrug*

Approaches to address the key technical challenges of nanoparticles, such as amplification synthesis and performance optimization/prediction, are critical to ensuring the clinical success of novel nanoparticle formulations. Clinical conversion and marketing highly rely on the consistent and repeatable production. With a few exceptions, most of the nanoparticles used in preclinical studies are exclusively synthesized in small batches and even for clinical studies. The ability for large-scale synthesis is one of the most challenging stages in practice [253, 254]. Due to the complexity of human disease, the manufacture of nanodrugs must be consistent and highly reproducible before bringing them into the clinical trial phase. In fact, the transformation from laboratory development to clinical use is always accompanied by sophisticated optimization of nanodrug formulation parameters and manufacturing process. Therefore, it is particularly important for the early nanodrug design to make forward-looking considerations for its subsequent large-scale preparation.

9.6.3 Pharmacokinetic Assessment

Another main challenge faced in the clinical transformation of nanodrugs is to control their biological fate (e.g., increasing accumulation at the target site and reducing off-target site accumulation). Noteworthy is that many currently approved and clinically investigated nanoparticles are PEGylated or PEG terminated, which can effectively reduce their interaction with immune cells and rapid clearance [255].

The two main strategies for clinically approved lipid-based nanodrugs include the use of particles of a certain size to increase their accumulation in tumor tissues due to EPR effect and the use of liposome systems that exhibit significantly enhanced binding affinity to target cells. As studies have shown that drug payloads are more effective when distributed through pathological tissues, researchers should still consider the ability to further promote the barrier destruction and tissue penetration of nanoparticles [256–258].

9.6.4 Study-Design Challenges

Many approved drugs do not provide benefits for all patients who receive the same treatment. This problem stems from the original clinical research design, which often overlooks different characteristics that may affect a population of particular patients. It is important to collect and analyze data using advanced technologies, such as artificial intelligence, on how the patients respond in clinical trials. In addition, efforts should be made from the management departments to propose guidelines and standards to specify the quality control and characterization of nanomedicine, encouraging more nanodrugs for clinical trials [259, 260].

9.7 Conclusion

In this chapter, we discuss the conventional lipid-based nanoparticle systems for drug delivery and tumor-targeted applications. Compared to the small therapeutic molecules, liposomes enjoy the advantage in terms of biocompatibility and biodegradability in in vivo tests. The utilization of therapeutic small molecules or liposomes are not contradictory, because nearly all molecular anticancer drugs can be encapsulated into liposomal platform with appropriate size, charge, structure and physicochemical properties. Although the mechanism and pharmacokinetic behaviors are difficult to comprehensively understand, the drugs combined with lipid coats could always present more cytotoxicity and specificity for the tumoral targets than drug molecules.

To date, more and more studies have been focused on novel nanocarrier design to meet the increasing demands in highly specific and effective cancer treatment. Particularly, the smart construction of the lipid-based tumor-targeted system should strike the balance between effective uptake and stimuli-sensitive release. Apart from surface modification of ligand and/or antibody receptor-mediated strategy, targeting tumor vessels and microenvironment is also a promising approach that is worth the investigation. More and more hybrid multifunctional lipid-based drug delivery systems such as micelles, solid nanoparticles, and lipid nanocarriers have emerged for optimization towards the next-generation clinical applications. Importantly, the complex pharmacokinetics of these lipid-based drug delivery systems must be carefully considered case by case to promote their transformation to clinical applications. With the dedication from the whole society, we believe more lipid-based drug delivery systems will finally turn into clinical practice and benefit more patients.

Acknowledgments We are grateful to the funding support from the National Natural Science Foundation of China (31870991) and Thousand Young Talents Program.

References

1. Kreuter J (2007) Nanoparticles – a historical perspective. Int J Pharm 331(1):1–10
2. Gregoriadis G (1976) The carrier potential of liposomes in biology and medicine. N Engl J Med 295(13):704–710
3. Tran MA, Watts RJ, Robertson GP (2009) Use of liposomes as drug delivery vehicles for treatment of melanoma. Pigm Cell Melanoma Res 22(4):388–399
4. Torchilin VP (2005) Recent advances with liposomes as pharmaceutical carriers. Nat Rev Drug Discov 4(2):145–160
5. Bangham AD, Standish MM, Watkins JC (1965) Diffusion of univalent ions across the lamellae of swollen phospholipids. J Mol Biol 13(1):238–252
6. Sessa G, Weissmann G (1968) Phospholipid spherules (liposomes) as a model for biological membranes. J Lipid Res 9(3):310–318
7. Klausner RD, Kleinfeld AM, Hoover RL, Karnovsky MJ (1980) Lipid domains in membranes. Evidence derived from structural perturbations induced by free fatty acids and lifetime heterogeneity analysis. J Biol Chem 255(4):1286–1295
8. Gregoriadis G (1995) Engineering liposomes for drug delivery: progress and problems. Trends Biotechnol 13(12):527–537
9. Allen TM (1998) Liposomal drug formulations. Rationale for development and what we can expect for the future. Drugs 56(5):747–756
10. Gubernator J (2011) Active methods of drug loading into liposomes: recent strategies for stable drug entrapment and increased in vivo activity. Expert Opin Drug Deliv 8(5):565–580
11. Allen TM*, Hansen CB, de Menezes DEL (1995) Pharmacokinetics of long-circulating liposomes. Adv Drug Deliv Rev 16(1–2):267–284
12. Lian T, Ho RJY (2011) Trends and developments in liposome drug delivery systems. J Pharm Sci 90(6):667–680

13. Mu LM, Ju RJ, Liu R et al (2017) Dual-functional drug liposomes in treatment of resistant cancers. Adv Drug Deliv Rev 115:46–56
14. Michelia MR, Bovab R, Maginib A, Emilianib C∗ (2012) Lipid-based nanocarriers for CNS-targeted drug delivery. Recent Pat CNS Drug Discov 7(1):71–86
15. Senior JH (1987) Fate and behavior of liposomes in vivo: a review of controlling factors. Crit Rev Ther Drug Carrier Syst 3(2):123–193
16. Mazzacuva F, Isacchi B, Bergonzi M et al (2011) Development and evaluation of conventional and PEGylated curcumin liposomes, absorption and tissue distribution studies in mice. Planta Med 77(12):87–120
17. Laverman P, Carstens MG, Boerman OC et al (2001) Factors affecting the accelerated blood clearance of polyethylene glycol-liposomes upon repeated injection. J Pharmacol Exp Ther 298(2):607–612
18. Lv Z, Yang Y, Wang J et al (2018) Optimization of the preparation conditions of borneol-modified ginkgolide liposomes by response surface methodology and study of their blood brain barrier permeability. Molecules 23(2):303–317
19. Yoshioka H, Goto H (1998) Inhibition adsorption of proteins on the liposome surface. US patent 5, 846, 458, 1998
20. Hope MJ, Mui B, Ansell S et al (1998) Cationic lipids, phosphatidylethanolamine and the intracellular delivery of polymeric, nucleic acid-based drugs. Membr Biochem 15(1):1–14
21. Ellens H, Bentz J, Szoka FC (1984) PH-induced destabilization of phosphatidylethanolamine-containing liposomes: role of bilayer contact. Biochemistry 23(7):1532–1538
22. Miyazaki M, Yuba E∗, Hayashi H et al (2018) Hyaluronic acid-based pH-sensitive polymer-modified liposomes for cell-specific intracellular drug delivery systems. Bioconjug Chem 29 (1):44–55
23. Xia Y, Tian J, Chen X∗ (2016) Effect of surface properties on liposomal siRNA delivery. Biomaterials 79:56–68
24. Dakwar GR, Braeckmans K, Demeester J et al (2015) Disregarded effect of biological fluids in siRNA delivery: human ascites fluid severely restricts cellular uptake of nanoparticles. ACS Appl Mater Interfaces 7(43):24322–24329
25. Wang Y∗, Miao L, Satterlee A et al (2015) Delivery of oligonucleotides with lipid nanoparticles. Adv Drug Deliv Rev 87:68–80
26. Luo YL, Xu CF, Li HJ et al (2018) Macrophage-specific in vivo gene editing using cationic lipid-assisted polymeric nanoparticles. ACS Nano 12(2):994–1005
27. Christensen D, Bøllehuus Hansen L, Leboux R et al (2019) A liposome-based adjuvant containing two delivery systems with the ability to induce mucosal immunoglobulin a following a parenteral immunization. ACS Nano 13(2):1116–1126
28. Bume G, Cevc G (1990) Liposomes for the sustained drug release in vivo. Biochim Biophys Acta 1029(1):91–97
29. Lasic DD, Martin FJ, Gabizon A et al (1991) Sterically stabilized liposomes: a hypothesis on the molecular origin of the extended circulation times. Biochim Biophys Acta 1070 (1):187–192
30. ElBayoumi TA, Torchilin VP (2010) Liposomes methods and protocols volume 1: pharmaceutical nanocarriers. In: Tamer AE, Vladimir PT (eds) Current trends in liposome research. Springer, London, pp 1–28
31. Pattni BS, Chupin VV, Torchilin VP∗ (2015) New developments in liposomal drug delivery. Chem Rev 115(19):10938–10966
32. Yingchoncharoen P, Kalinowski DS, Richardson DR (2016) Lipid-based drug delivery systems in cancer therapy: what is available and what is yet to come. Pharm Rev 68(3):701–787
33. Bangham AD (1982) Preparation of liposomes and methods for measuring their permeabilities. In: Hesketh TR, Kornberg HL, Metcalfe JC et al (eds) Technique in life science – technique in lipid and membrane biochemistry. Elsevier, Amsterdam, pp 1–25
34. Olson F, Hunt CA, Szoka FC et al (1979) Preparation of liposomes of defined size distribution by extrusion through polycarbonate membranes. Biochim Biophys Acta 557(1):9–23

35. Vemuri S, Rhodes CT (1995) Preparation and characterization of liposomes as therapeutic delivery systems: a review. Pharm Acta Helv 70(2):95–111
36. Hauser H (1982) Methods of preparation of lipid vesicles: assessment of their suitability for drug encapsulation. Trends Pharmacol Sci 3:274–277
37. Tyrrell DA, Heath TD, Colley CM et al (1976) New aspects of liposomes. Biochim Biophys Acta 457(3–4):259–302
38. Szoka F, Papahadjopoulos D (1978) Procedure for preparation of liposomes with large internal aqueous space and high capture by reverse-phase evaporation. Proc Natl Acad Sci U S A 75 (9):4194–4198
39. Milsmann MHW, Schwendener RA, Weder HG (1978) The preparation of large single bilayer liposomes by a fast and controlled dialysis. Biochim Biophys Acta 512(1):147–155
40. Enoch HG, Strittmatter P (1979) Formation and properties of 1000-A-diameter, single-bilayer phospholipid vesicles. Proc Natl Acad Sci U S A 76(1):145–149
41. Gerritsen WJ, Verkley AJ, Zwaal RF et al (1978) Freeze-fracture appearance and disposition of band 3 protein from the human erythrocyte membrane in lipid vesicles. Eur J Biochem 85 (1):255–261
42. Huang CH (1969) Phosphatidylcholine vesicles. Formation and physical characteristics. Biochemistry 8(1):344–352
43. Deamer D, Bangham AD (1976) Large volume liposomes by an ether vaporization method. Biochim Biophys Acta 443(3):629–634
44. Torchilin VP (2007) Micellar nanocarriers: pharmaceutical perspectives. Pharm Res 24 (1):1–16
45. Edwards K, Johnsson M, Karlsson G et al (1997) Effect of polyethylene glycol-phospholipids on aggregate structure in preparations of small unilamellar liposomes. Biophys J 73 (1):258–266
46. Lukyanov AN, Torchilin VP (2004) Micelles from lipid derivatives of water soluble polymers as delivery systems for poorly soluble drugs. Adv Drug Deliv Rev 56(9):1273–1289
47. Rosen MJ, Kunjappu JT (2012) Surfactants and interfacial phenomena, 4th edn. Wiley, Hoboken
48. Lukyanov AN, Gao Z, Mazzola L et al (2002) Polyethylene glycol diacyl lipid micelles demonstrate increased accumulation in subcutaneous tumors in mice. Pharm Res 19 (10):1424–1429
49. Gao Z, Lukyanov AN, Chakilam AR et al (2003) Diacyl lipid-polymer micelles as nanocarriers for poorly soluble anticancer drugs. J Drug Target 11(2):87–92
50. Wong HL, Bendayan R, Rauth AM et al (2007) Chemotherapy with anticancer drugs encapsulated in solid lipid nanoparticles. Adv Drug Deliv Rev 59(6):491–504
51. Schwarz C, Mehnert W, Lucks JS et al (1994) Solid lipid nanoparticles (SLN) for controlled drug delivery. I. Production, characterization and sterilization. J Control Release 30(1):83–96
52. Wissing SA, Kayser O, Müller RH (2004) Solid lipid nanoparticles for parenteral drug delivery. Adv Drug Deliv Rev 56(9):1257–1272
53. Müller RH, Mäder K, Gohla S (2000) Solid lipid nanoparticles (SLN) for controlled drug delivery – a review of the state of the art. Eur J Pharm Biopharm 50(1):161–177
54. Mehnert W, Mäder K (2001) Solid lipid nanoparticles: production, characterization and applications. Adv Drug Deliv Rev 47:165–196
55. Gasco MR (1993) Method for producing solid lipid microspheres having a narrow size distribution. US Patent US5250236. 1991 Aug 2
56. Boltri L, Canal T, Esposito PA et al (1993) Lipid nanoparticles: evaluation of some critical formulation parameters. Proc Int Symp Control Release Bioact Mater 20:346–347
57. Sjöström B, Westesen K, Bergenståhl B (1993) Preparation of submicron drug particles in lecithin-stabilized o/w emulsions: II. Characterization of cholesteryl acetate particles. Int J Pharm 94(1–3):89–101
58. Westesen K, Siekmann B, Koch MHJ (1993) Investigations on the physical state of lipid nanoparticles by synchrotron radiation X-ray diffraction. Int J Pharm 93(1–3):189–199

59. Cavalli R, Caputo O, Gasco MR (1993) Solid lipospheres of doxorubicin and idarubicin. Int J Pharm 89(1):9–12
60. zurMühlen A, Schwarz C, Mehnert W (1998) Solid lipid nanoparticles (SLN) for controlled drug delivery–drug release and release mechanism. Eur J Pharm Biopharm 45(2):149–155
61. Hadinoto K, Sundaresan A, Cheow WS (2013) Lipid-polymer hybrid nanoparticles as a new generation therapeutic delivery platform: a review. Eur J Pharm Biopharm 85(3):427–443
62. Raemdonck K, Braeckmans K, Demeester J et al (2014) Merging the best of both worlds: hybrid lipid-enveloped matrix nanocomposites in drug delivery. Chem Soc Rev 43 (1):444–472
63. Thevenot J, Troutier AL, David L et al (2007) Steric stabilization of lipid/polymer particle assemblies by poly(ethylene glycol)-lipids. Biomacromolecules 8(11):3651–3660
64. Fenart L, Casanova A, Dehouck B et al (1999) Evaluation of effect of charge and lipid coating on ability of 60-nm nanoparticles to cross an in vitro model of the blood-brain barrier. J Pharmacol Exp Ther 291(3):1017–1022
65. Mieszawska AJ, Gianella A, Cormode DP et al (2012) Engineering of lipid-coated PLGA nanoparticles with a tunable payload of diagnostically active nanocrystals for medical imaging. Chem Commun 48(47):5835–5837
66. Messerschmidt SK, Musyanovych A, Altvater M et al (2009) Targeted lipid-coated nanoparticles: delivery of tumor necrosis factor-functionalized particles to tumor cells. J Control Release 137(1):69–77
67. Hasan W, Chu K, Gullapalli A et al (2012) Delivery of multiples iRNAs using lipid coated PLGA nanoparticles for treatment of prostate cancer. Nano Lett 12(1):287–292
68. Troutier AL, Delair T, Pichot C et al (2005) Physicochemical and interfacial investigation of lipid/polymer particle assemblies. Langmuir 21(4):1305–1313
69. Fang RH, Aryal S, Hu CMJ et al (2010) Quick synthesis of lipid-polymer hybrid nanoparticles with low polydispersity using a single-step sonication method. Langmuir 26 (22):16958–16962
70. Valencia PM, Basto PA, Zhang L et al (2010) Single-step assembly of homogenous lipid-polymeric and lipid-quantum dot nanoparticles enabled by microfluidic rapid mixing. ACS Nano 4(3):1671–1679
71. Zhang L, Chan JM, Gu FX et al (2008) Self-assembled lipid–polymer hybrid nanoparticles: a robust drug delivery platform. ACS Nano 2(8):1696–1702
72. Chan JM, Zhang L, Yuet KP et al (2009) PLGA-lecithin-PEG core-shell nanoparticles for controlled drug delivery. Biomaterials 30(8):1627–1634
73. Bershteyn A, Chaparro J, Yau R et al (2008) Polymer-supported lipid shells, onions, and flowers. Soft Matter 4(9):1787–1791
74. Cheow WS, Hadinoto K (2011) Factors affecting drug encapsulation and stability of lipid-polymer hybrid nanoparticles. Colloids Surf B Biointerfaces 85(2):214–220
75. Liu Y, Pan J, Feng SS (2010) Nanoparticles of lipid monolayer shell and biodegradable polymer core for controlled release of paclitaxel: effects of surfactants on particles size, characteristics and in vitro performance. Int J Pharm 395(1–2):243–250
76. Chu CH, Wang YC, Huang HY et al (2011) Ultrafine PEG coated poly(lactic-co-glycolic acid) nanoparticles formulated by hydrophobic surfactant-assisted one-pot synthesis for biomedical applications. Nanotechnology 22(18):185601
77. Wacker M (2013) Nanocarriers for intravenous injection–the long hard road to the market. Int J Pharm 457(1):50–62
78. Pastorino F, Brignole C, Marimpietri D et al (2003) Vascular damage and anti-angiogenic effects of tumor vessel-targeted liposomal chemotherapy. Cancer Res 63(21):7400–7409
79. Kaur S, Banerjee R (2012) Lipid-coated PLGA nanoparticles as robust siRNA delivery vehicles. Nanomedicine 7(6):803
80. Tayebi L, Vashaee D, Parikh AN (2012) Stability of uni- and multillamellar spherical vesicles. Chem Phys Chem 13(1):314–322

81. Pignatello R, Musumeci T, Graziano AC et al (2016) A study on liposomal encapsulation of a lipophilic prodrug of LHRH. Pharm Dev Technol 21(6):664–671
82. Maestrelli F, González-Rodríguez ML, Rabasco AM et al (2006) Effect of preparation technique on the properties of liposomes encapsulating ketoprofen-cyclodextrin complexes aimed for transdermal delivery. Int J Pharm 312(1–2):53–60
83. Choi HS, Liu W, Misra P et al (2007) Renal clearance of quantum dots. Nat Biotechnol 25 (10):1165–1170
84. Owens DE III, Peppas NA (2006) Opsonization, biodistribution, and pharmacokinetics of polymeric nanoparticles. Int J Pharm 307(1):93–102
85. Awasthi VD, Garcia D, Goins BA et al (2003) Circulation and biodistribution profiles of long-circulating PEG-liposomes of various sizes in rabbits. Int J Pharm 253(1–2):121–132
86. Wisse E, Jacobs F, Topal B et al (2008) The size of endothelial fenestrae in human liver sinusoids: implications for hepatocyte-directed gene transfer. Gene Ther 15(17):1193–1199
87. Daemen T, Velinova M, Regts J et al (1997) Different intrahepatic distribution of phosphatidylglycerol and phosphatidylserine liposomes in the rat. Hepatology 26(2):416–423
88. Shaw DJ, Costello B (1993) Introduction to colloid and surface chemistry: Butterworth-Heinemann. Elsevier, Oxford
89. Hunter RJ (1981) Zeta potential in colloid science: principles and applications, 1st edn. Academic, London
90. Brandhonneur N, Chevanne F, Vié V et al (2009) Specific and non-specific phagocytosis of ligand-grafted PLGA microspheres by macrophages. Eur J Pharm Sci 36(4–5):474–485
91. He C, Hu Y, Yin L et al (2010) Effects of particle size and surface charge on cellular uptake and biodistribution of polymeric nanoparticles. Biomaterials 31(13):3657–3666
92. Davis ME, Chen ZG, Shin DM (2008) Nanoparticle therapeutics: an emerging treatment modality for cancer. Nat Rev Drug Discov 7:771–782
93. Liu Y, Li K, Pan J et al (2010) Folic acid conjugated nanoparticles of mixed lipid monolayer shell and biodegradable polymer core for targeted delivery of Docetaxel. Biomaterials 31 (2):330–338
94. Ge J, Li K, Ding D et al (2012) Lipid-PEG-Folate encapsulated nanoparticles with aggregation induced emission characteristics: cellular uptake mechanism and two-photon fluorescence imaging. Small 8(23):3655–3663
95. Li K, Jiang Y, Ding D et al (2011) Folic acid-functionalized two-photon absorbing nanoparticles for MCF-7 cancer cell imaging. Chem Commun 47(26):7323–7325
96. Gabizon A, Papahadjopoulos D (1988) Liposome formulations with prolonged circulation time in blood and enhanced uptake by tumors. Proc Natl Acad Sci U S A 85(18):6949–6953
97. Williams KJ, Phillips MC, Rodrigueza WV (1998) Structural and metabolic consequences of liposome-lipoprotein interactions. Adv Drug Deliv Rev 32(1–2):31–43
98. Szebeni J, Wassef NM, Spielberg H et al (1994) Complement activation in rats by liposomes and liposome-encapsulated hemoglobin: evidence for anti-lipid antibodies and alternative pathway activation. Biochem Biophys Res Commun 205(1):255–263
99. Tall AR, Tabas I, Williams KJ (1986) Lipoprotein-liposome interactions. Methods Enzymol 128(4):647–657
100. Sabín J, Prieto G, Ruso JM et al (2009) Interactions between DMPC liposomes and the serum blood proteins HSA and IgG. J Phys Chem B 113(6):1655–1661
101. Goren D, Horowitz AT, Zalipsky S et al (1996) Targeting of stealth liposomes to erB-2 (Her/2) receptor: in vitro and in vivo studies. Br J Cancer 74(11):1749–1756
102. Riviere K, Huang Z, Jerger K et al (2011) Antitumor effect of folate-targeted liposomal doxorubicin in KB tumor-bearing mice after intravenous administration. J Drug Target 19 (1):14–24
103. Matsumura Y, Maeda H (1986) A new concept for macromolecular therapeutics in cancer-chemotherapy – mechanism of tumor itropic accumulation of proteins and the antitumor agent Smancs. Cancer Res 46:6387–6392

104. Ganta S, Devalapally H, Shahiwala A et al (2008) A review of stimuli responsive nanocarriers for drug and gene delivery. J Control Release 126(3):187–204
105. van Vlerken LE, Duan Z, Seiden MV et al (2007) Modulation of intracellular ceramide using polymeric nanoparticles to overcome multidrug resistance in cancer. Cancer Res 67 (10):4843–4850
106. Noble CO, Kirpotin DB, Hayes ME et al (2004) Development of ligand-targeted liposomes for cancer therapy. Expert Opin Ther Targets 8(4):335–353
107. El-Hammadi MM, Arias JL (2019) An update on liposomes in drug delivery: a patent review (2014–2018). Expert Opin Ther Pat 29(11):891–907
108. Doi Y, Shimizu T, Ishima Y et al (2019) Long-term storage of PEGylated liposomal oxaliplatin with improved stability and long circulation times in vivo. Int J Pharm 564:237–243
109. Hamishehkar H, Bahadori MB, Vandghanooni S et al (2018) Preparation, characterization and anti-proliferative effects of sclareol loaded solid lipid nanoparticles on A549 human lung epithelial cancer cells. J Drug Deliv Sci Tec 45:272–280
110. Nausicaa C, Benedetta F, Casimiro G et al (2018) Solid lipid nanoparticles carrying temozolomide for melanoma treatment. Preliminary in vitro and in vivo studies. Int J Mol Sci 19(2):255
111. Bae YH (2009) Drug targeting and tumor heterogeneity. J Control Release 133(1):2–3
112. Cho K, Wang X, Nie S et al (2008) Therapeutic nanoparticles for drug delivery in cancer. Clin Cancer Res 14(5):1310–1316
113. Danhier F, Feron O, Préat V (2010) To exploit the tumor microenvironment: passive and active tumor-targeted of nanocarriers for anti-cancer drug delivery. J Control Release 148 (2):35–146
114. Bareford LM, Swaan PW (2007) Endocytic mechanisms for targeted drug delivery. Adv Drug Deliv Rev 59(8):748–758
115. Clague MJ, Urbé S, Aniento F et al (1994) Vacuolar ATPase activity is required for endosomal carrier vesicle formation. J Biol Chem 269(1):21–24
116. Lee RJ, Wang S, Low PS (1996) Measurement of endosome pH following folate receptor-mediated endocytosis. Biochim Biophys Acta 1312(3):237–242
117. Nishi T, Forgac M (2002) The vacuolar (H+)-ATPases–nature's most versatile proton pumps. Nat Rev Mol Cell Biol 3(2):94–103
118. Rudenko G, Henry L, Henderson K et al (2002) Structure of the LDL receptor extracellular domain at endosomal pH. Science 298(5602):2353–2358
119. Kamen BA, Smith AK (2004) A review of folate receptor alpha cycling and 5-methyltetrahydrofolate accumulation with an emphasis on cell models in vitro. Adv Drug Deliv Rev 56(8):1085–1097
120. Lakadamyali M, Rust MJ, Zhuang X (2006) Ligands for clathrin-mediated endocytosis are differentially sorted into distinct populations of early endosomes. Cell 124(5):997–1009
121. Weijer R, Broekgaarden M, Kos M et al (2015) Enhancing photodynamic therapy of refractory solid cancers: combining second-generation photosensitizers with multi-targeted liposomal delivery. J Photochem Photobiol Chem 23:103–131
122. Jain RK, Stylianopoulos T (2010) Delivering nanomedicine to solid tumors. Nat Rev Clin Oncol 7(11):653–664
123. Park JW, Hong K, Kirpotin DB et al (2002) Anti-HER2 immunoliposomes: enhanced efficacy attributable to targeted delivery. Clin Cancer Res 8(4):1172–1181
124. Orcutt KD, Rhoden JJ, Ruiz-Yi B et al (2012) Effect of small molecule binding affinity on tumor uptake in vivo. Mol Cancer Ther 11(6):1365–1372
125. Juweid M, Neumann R, Paik C et al (1992) Micropharmacology of monoclonal antibodies in solid tumors: direct experimental evidence for a binding site barrier. Cancer Res 52 (19):5144–5153
126. Yang G, Yang T, Zhang W et al (2014) In vitro and in vivo antitumor effects of folate-targeted ursolic acid stealth liposome. J Agric Food Chem 62(10):2207–2215

127. Min HK, Kim CS, Han J et al (2019) Folate receptor-targeted liposomal nanocomplex for effective synergistic photothermal-chemotherapy of breast cancer in vivo. Colloids Surf B 173:539–548
128. Hattori Y, Shimizu S, Ozaki K et al (2019) Effect of cationic lipid type in folate-PEG-modified cationic liposomes on folate receptor-mediated siRNA transfection in tumor cells. Pharmaceutics 11(4):181
129. Silindir-Gunay M, Karpuz M, Ozturk N et al (2019) Radiolabeled, folate-conjugated liposomes as tumor imaging agents: formulation and in vitro evaluation. J Drug Deliv Sci Technol 50:321–328
130. Handali S, Moghimipour E, Kouchak M et al (2019) New folate receptor targeted nano liposomes for delivery of 5-fluorouracil to cancer cells: strong implication for enhanced potency and safety. Life Sci 227:39–50
131. Kumar P, Huo P, Liu B (2019) Formulation strategies for folate-targeted liposomes and their biomedical applications. Pharmaceutics 11(8):381
132. Trinder D, Zak O, Aisen P (1996) Transferrin receptor-independent uptake of differic transferrin by human hepatoma cells with antisense inhibition of receptor expression. Hepatology 23(6):1512–1520
133. Kalinowski D, Richardson DR (2005) The evolution of iron chelators for the treatment of iron overload disease and cancer. Pharmacol Rev 57(4):547–583
134. Prost AC, Ménégaux F, Langlois P et al (1998) Differential transferrin receptor density in human colorectal cancer: a potential probe for diagnosis and therapy. Int J Oncol 13 (4):871–875
135. Shinohara H, Fan D, Ozawa S et al (2000) Site-specific expression of transferrin receptor by human colon cancer cells directly correlates with eradication by antitransferrin recombinant immunotoxin. Int J Oncol 17(4):643–651
136. Gomme PT, McCann KB, Bertolini J (2005) Transferrin: structure, function and potential therapeutic actions. Drug Discov Today 10(4):267–273
137. Daniels TR, Delgado T, Helguera G et al (2006) The transferrin receptor part II: targeted delivery of therapeutic agents into cancer cells. Clin Immunol 121(2):159–176
138. Chang J, Jallouli Y, Kroubi M et al (2009) Characterization of endocytosis of transferrin-coated PLGA nanoparticles by the blood-brain barrier. Int J Pharm 379(2):285–292
139. Ulbrich K, Hekmatara T, Herbert E et al (2009) Transferrin- and transferrin-receptor-antibody-modified nanoparticles enable drug delivery across the blood-brain barrier (BBB). Eur J Pharm Biopharm 71(2):251–256
140. Tang J, Wang Q, Yu Q et al (2019) A stabilized retro-inverso peptide ligand of transferrin receptor for enhanced liposome-based hepatocellular carcinoma-targeted drug delivery. Acta Biomater 83:379–389
141. Fu J, Li W, Xin X et al (2019) Transferrin modified nano-liposome co-delivery strategies for enhancing the cancer therapy. J Pharm Sci
142. Riaz MK, Zhang X, Wong KH et al (2019) Pulmonary delivery of transferrin receptors targeting peptide surface-functionalized liposomes augments the chemotherapeutic effect of quercetin in lung cancer therapy. Int J Nanomedicine 14:2879–2902
143. Jhaveri A, Luther E, Torchilin V (2019) The effect of transferrin-targeted, resveratrol-loaded liposomes on neurosphere cultures of glioblastoma: implications for targeting tumour-initiating cells. J Drug Targeting 27(5–6):601–613
144. Wang Y, Yang Y, Yu Y et al (2020) Transferrin modified dioscin loaded PEGylated liposomes: characterization and in vitro antitumor effect. J Nanosci Nanotechnol 20(3):1321–1331
145. Li S, Zhao H, Mao X et al (2019) Transferrin receptor targeted cellular delivery of doxorubicin via a reduction-responsive peptide-drug conjugate. Pharm Res 36(12):168
146. dos Santos Rodrigues B, Kanekiyo T, Singh J (2019) ApoE-2 brain-targeted gene therapy through transferrin and penetratin tagged liposomal nanoparticles. Pharm Res 36(11):161
147. Witton CJ, Reeves JR, Going JJ et al (2003) Expression of the HER1-4 family of receptor tyrosine kinases in breast cancer. J Pathol 200(3):290–297

148. Abd El-Rehim DM, Pinder SE, Paish CE et al (2004) Expression and co-expression of the members of the epidermal growth factor receptor (EGFR) family in invasive breast carcinoma. Br J Cancer 91(8):1532–1542
149. Bossuyt V, Fadare O, Martel M et al (2005) Remarkably high frequency of EGFR expression in breast carcinomas with squamous differentiation. Int J Surg Pathol 13(4):319–327
150. Mamot C, Drummond DC, Noble CO et al (2005) Epidermal growth factor receptor-targeted immunoliposomes significantly enhance the efficacy of multiple anticancer drugs in vivo. Cancer Res 65(24):11631–11638
151. Zu Kim Y, hee Park Y, Choi HJ et al (2018) Compositions and methods related to anti-EGFR antibody drug conjugates: U.S. Patent 10,118,965[P]. 2018-11-6
152. Guo P, Yang J, Liu D et al (2019) Dual complementary liposomes inhibit triple-negative breast tumor progression and metastasis. Sci Adv 5(3):eaav5010
153. Abumanhal-Masarweh H, da Silva D, Poley M et al (2019) Tailoring the lipid composition of nanoparticles modulates their cellular uptake and affects the viability of triple negative breast cancer cells. J Control Release 307:331–341
154. Takenaka T, Nakai S, Katayama M et al (2019) Effects of gefitinib treatment on cellular uptake of extracellular vesicles in EGFR-mutant non-small cell lung cancer cells. Int J Pharm 572:118762
155. Li JJ, Feng GY, Chen CC et al (2019) Biological evaluation of an EGFR targeting liposomal drug in a peritoneal tumor-bearing mouse model. J Nucl Med 60(supplement 1):1041
156. Deshpande PP, Biswas S, Torchilin VP (2013) Current trends in the use of liposomes for tumor-targeted. Nanomedicine (Lond) 8(9):1509–1528
157. Kang H, O'Donoghue MB, Liu H et al (2010) A liposome-based nanostructure for aptamer directed delivery. Chem Commun 46(2):249–251
158. Xing H, Tang L, Yang X et al (2013) Selective delivery of an anticancer drug with aptamer-functionalized liposomes to breast cancer cells in vitro and in vivo. J Mater Chem B Mater Biol Med 1(39):5288–5297
159. Moosavian SA, Sahebkar A (2019) Aptamer-functionalized liposomes for targeted cancer therapy. Cancer Lett 448:144–154
160. Hong S, Ding P, Luo Y et al (2019) Aptamer-integrated α-Gal liposomes as bispecific agents to trigger immune response for killing tumor cells. J Biomed Mater Res Part A 107 (6):1176–1183
161. Frohnmeyer E, Tuschel N, Sitz T et al (2019) Aptamer lateral flow assays for rapid and sensitive detection of cholera toxin. Analyst 144(5):1840–1849
162. Yang X, Zhao J, Duan S et al (2019) Enhanced cytotoxic T lymphocytes recruitment targeting tumor vasculatures by endoglin aptamer and IP-10 plasmid presenting liposome-based nanocarriers. Theranostics 9(14):4066–4083
163. Guo X, Dong C, Liu Q et al (2019) The sustained and targeted treatment of hemangiomas by propranolol-loaded CD133 aptamers conjugated liposomes-in-microspheres. Biomed Pharmacother 114:108823
164. Cheng Y, Ou Z, Li Q et al (2019) Cabazitaxel liposomes with aptamer modification enhance tumor-targeting efficacy in nude mice. Mol Med Rep 19(1):490–498
165. Lammers T, Hennink WE, Storm G (2008) Tumour-targeted nanomedicines: principles and practice. Br J Cancer 99(3):392–397
166. Denekamp J, Hobson B (1982) Endothelial-cell proliferation in experimental tumours. Br J Cancer 46(5):711–720
167. Denekamp J (1984) Vasculature as a target for tumour therapy. Prog Appl Microcirc 4:28–38
168. Carmeliet P (2005) VEGF as a key mediator of angiogenesis in cancer. Oncology 69(Suppl 3):4–10
169. Yao Y, Wang T, Liu Y et al (2019) Co-delivery of sorafenib and VEGF-siRNA via pH-sensitive liposomes for the synergistic treatment of hepatocellular carcinoma. Artif Cells Nanomed Biotechnol 47(1):1374–1383

170. Byrne JD, Betancourt T, Brannon-Peppas L (2008) Active targeting schemes for nanoparticle systems in cancer therapeutics. Adv Drug Deliv Rev 60(15):1615–1626
171. Hood JD, Bednarski M, Frausto R et al (2002) Tumor regression by targeted gene delivery to the neovasculature. Science 296(5577):2404–2407
172. Desgrosellier JS, Cheresh DA (2010) Integrins in cancer: biological implications and therapeutic opportunities. Nat Rev Cancer 10(1):9–22
173. Tang Z, Feng W, Yang Y et al (2019) Gemcitabine-loaded RGD modified liposome for ovarian cancer: preparation, characterization and pharmacodynamic studies. Drug Des Dev Ther 13:3281–3290
174. Li XT, Tang W, Xie HJ et al (2019) The efficacy of RGD modified liposomes loaded with vinorelbine plus tetrandrine in treating resistant brain glioma. J Liposome Res 29(1):21–34
175. Vihinen P, Ala-aho R, Kähäri VM (2005) Matrix metalloproteinases as therapeutic targets in cancer. Curr Cancer Drug Targets 5(3):203–220
176. Genís L, Gálvez BG, Gonzalo P et al (2006) MT1-MMP: universal or particular player in angiogenesis? Cancer Metastasis Rev 25(1):77–86
177. Lyu Y, Xiao Q, Yin L et al (2019) Potent delivery of an MMP inhibitor to the tumor microenvironment with thermosensitive liposomes for the suppression of metastasis and angiogenesis. Signal Transduction Targeted Ther 4(1):1–9
178. Andresen T L, Jensen S S, Henriksen JR et al (2019) Cationic liposomes: WIPO Patent 2019012107[P]
179. Seraj S, Lee J, Ahn HJ (2019) Systemic delivery of Eg5 shRNA-expressing plasmids using PEGylated DC-Chol/DOPE cationic liposome: long-term silencing and anticancer effects in vivo. Biochem Pharmacol 166:192–202
180. Ichihara H, Motomura M, Matsumoto Y (2019) Therapeutic effects and anti-metastasis effects of cationic liposomes against pancreatic cancer metastasis in vitro and in vivo. Biochem Biophys Res Commun 511(3):504–509
181. Zhang M, Lemay SG (2019) Interaction of anionic bulk nanobubbles with cationic liposomes: evidence for reentrant condensation. Langmuir 35(11):4146–4151
182. Wui SR, Kim KS, Ryu JI et al (2019) Efficient induction of cell-mediated immunity to varicella-zoster virus glycoprotein E co-lyophilized with a cationic liposome-based adjuvant in mice. Vaccine 37(15):2131–2141
183. Drummond DC, Meyer O, Hong K et al (1999) Optimizing liposomes for delivery of chemotherapeutic agents to solid tumors. Pharmacol Rev 51(4):691–743
184. Allen TM, Hansen CB, Menezes DLD (1995) Pharmacokinetics of long-circulating liposomes. Adv Drug Deliv Rev 16(2):267–284
185. Senior J, Crawley JCW, Gregoriadis G (1985) Tissue distribution of liposomes exhibiting long half-lives in the circulation after intravenous injection. Biochim Biophys Acta 839:1–8
186. Ahl PL, Bhatia SK, Meers P et al (1997) Enhancement of the in vivo circulation lifetime of L-a-distearoylphosphatidylcholine liposomes: importance of liposomal aggregation versus complement opsonization. Biochim Biophys Acta 1329(2):370–382
187. Gregoriadis G, Senior J (1980) The phospholipid component of small unilamellar liposomes controls the rate of clearance of entrapped solutes from the circulation. FEBS Lett 119 (1):43–46
188. Abra RM, Hunt CA (1981) Liposome disposition in vivo. III. Dose and vesicle-size effects. Biochim Biophys Acta 666(3):493–503
189. Hwang KJ (1987) Liposome pharmacokinetics. In: Ostro MJ (ed) Liposomes: from biophysics to therapeutics. Marcel Dekker Inc, New York, pp 109–156
190. Allen TM, Hansen C (1991) Pharmacokinetics of stealth versus conventional liposomes: effect of dose. Biochim Biophys Acta 1068(2):133–141
191. Huang SK, Mayhew E, Gilani S et al (1992) Pharmacokinetics and therapeutics of sterically stabilized liposomes in mice bearing C-26 colon carcinoma. Cancer Res 52(24):6774–6781
192. Woodle MC, Lasic DD (1992) Sterically stabilized liposomes. Biochim Biophys Acta 1113 (2):171–199

193. Papahadjopoulos D, Allen TM, Gabizon A et al (1991) Sterically stabilized liposomes: improvements in pharmacokinetics and antitumor therapeutic efficacy. Proc Natl Acad Sci U S A 88(24):11460–11464
194. Gabizon A, Papahadjopoulos D (1988) Liposome formulations with prolonged circulation time in blood and enhanced uptake in tumors. Proc Natl Acad Sci U S A 85(18):6949–6953
195. Gabizon A, Price DC, Huberty J et al (1990) Effect of liposome composition and other factors on the targeting of liposomes to experimental tumors: biodistribution and imaging studies. Cancer Res 50(19):6371–6378
196. Fielding RM, Mukwaya G, Sandhaus RA (1998) Long circulating liposomes: old drugs, new therapeutics. J Control Release 44(1):1–9
197. Papahadjopoulos D, Jacobson K, Nir S et al (1973) Phase transitions in phospholipid vesicles: fluorescence polarization and permeability measurements concerning the effect of temperature and cholesterol. Biochim Biophys Acta 311(3):330–348
198. Bally MB, Nayar R, Masin D et al (1990) Liposomes with entrapped doxorubicin exhibit extended blood residence times. Biochim Biophys Acta 1023(1):133–139
199. van Etten EWM, van Vianen W, Tijhuis RHG et al (1995) Sterically stabilized amphotericin B-liposomes: toxicity and biodistribution in mice. J Control Release 37(1):123–129
200. Nosova AS, Koloskova OO, Nikonova AA et al (2019) Diversity of PEGylation methods of liposomes and their influence on RNA delivery. Med Chem Comm 10(3):369–377
201. Li B, Cai M, Lin L et al (2019) MRI-visible and pH-sensitive micelles loaded with doxorubicin for hepatoma treatment. Biomater Sci-UK 7(4):1529–1542
202. Kanamala M, Palmer BD, Ghandehari H et al (2018) PEG-benzaldehyde-hydrazone-lipid based PEG-sheddable pH-sensitive liposomes: abilities for endosomal escape and long circulation. Pharm Res-DORDR 35(8):154
203. Shen Z, Ye H, Kröger M et al (2018) Aggregation of polyethylene glycol polymers suppresses receptor-mediated endocytosis of PEGylated liposomes. Nanoscale 10(9):4545–4560
204. Wei H, Zhao Y, Guo Y et al (2018) PEGylated self-assembled nano-bacitracin A: probing the antibacterial mechanism and real-time tracing of target delivery in vivo. ACS Appl Mater Interfaces 10(13):10688–10705
205. Ma K, Fu D, Liu Y et al (2018) Cancer cell targeting, controlled drug release and intracellular fate of biomimetic membrane-encapsulated drug-loaded nano-graphene oxide nanohybrids. J Mater Chem B 6(31):5080–5090
206. Kang X, Chen H, Li S et al (2018) Magnesium lithospermate B loaded PEGylated solid lipid nanoparticles for improved oral bioavailability. Colloids Surf B Biointerfaces 161:597–605
207. Shimizu T, Abu Lila AS, Fujita R et al (2018) A hydroxyl PEG version of PEGylated liposomes and its impact on anti-PEG IgM induction and on the accelerated clearance of PEGylated liposomes. Eur J Pharm Biopharm 127:142–149
208. Kuang Y, Zhang K, Cao Y et al (2017) Hydrophobic IR-780 dye encapsulated in cRGD-conjugated solid lipid nanoparticles for NIR imaging-guided photothermal therapy. ACS Appl Mater Interfaces 9(14):12217–12226
209. Porter CJH, Trevaskis NL, Charman WN (2007) Lipids and lipid-based formulations: optimizing the oral delivery of lipophilic drugs. Nat Rev Drug Discov 6(3):231–248
210. Williams H D, Ford L, Igonin A et al (2019) Unlocking the full potential of lipid-based formulations using lipophilic salt/ionic liquid forms. Adv Drug Deliv Rev
211. Anselmo AC, Mitragotri S (2016) Nanoparticles in the clinic. Bioeng Transl Med 1(1):10–29
212. Beloqui A, Solinís MÁ, Rodríguez-Gascón A et al (2016) Nanostructured lipid carriers: promising drug delivery systems for future clinics. Nanomedicine 12(1):143–161
213. Desai N (2012) Challenges in development of nanoparticle-based therapeutics. AAPS J 14 (2):282–295
214. Susan H (2015) Lipid-based nano-delivery systems for skin delivery of drugs and bioactives. Front Pharmacol 6:219
215. Wagner V, Dullaart A, Bock AK et al (2006) The emerging nanomedicine landscape. Nat Biotechnol 24(10):1211–1217

216. Van DVFM (2014) The emerging landscape for nanomedicine in atherosclerosis. Endocrine BioScientifica 35
217. Schütz CA, Juillerat-Jeanneret L, Mueller H et al (2013) Therapeutic nanoparticles in clinics and under clinical evaluation. Nanomedicine 8(3):449–467
218. Mishra DK, Ruchita S, Mishra PK (2018) Lipid based nanocarriers: a translational perspective. Nanomedicine 14(7):2023–2050
219. Bobo D, Robinson KJ, Islam J et al (2016) Nanoparticle-based medicines: a review of FDA-approved materials and clinical trials to date. Pharm Res 33(10):2373–2387
220. Eifler AC, Thaxton CS (2011) Nanoparticle therapeutics: FDA approval, clinical trials, regulatory pathways, and case study. Methods Mol Biol 726(726):325
221. Vivot A, Jacot J, Zeitoun JD et al (2017) Clinical benefit, price and approval characteristics of FDA-approved new drugs for treating advanced solid cancer, 2000–2015. Ann Oncol 28 (5):1111–1116
222. James ND, Coker RJ, Tomlinson D et al (1994) Liposomal doxorubicin (Doxil): an effective new treatment for Kaposi's sarcoma in aids. Clin Oncol 6(5):294–296
223. Gabizon A et al (1994) Prolonged circulation time and enhanced accumulation in malignant exudates of doxorubicin encapsulated in polyethylene-glycol coated liposomes. Cancer Res 54 (4):987–992
224. Karthik V, Yi L, Dennis N et al (2014) Pharmacokinetics and pharmacodynamics of liposomal mifamurtide in adult volunteers with mild or moderate hepatic impairment. Br J Clin Pharmacol 77(6):998–1010
225. Gabizon A, Shmeeda H, Barenholz Y (2003) Pharmacokinetics of pegylated liposomal doxorubicin. Clin Pharmacokinet 42(5):419
226. Hattori Y, Shi L, Ding W et al (2009) Novel irinotecan-loaded liposome using phytic acid with high therapeutic efficacy for colon tumors. J Control Release 136(1):30–37
227. Yingchoncharoen P, Kalinowski DS, Richardson DR (2016) Lipid-based drug delivery systems in cancer therapy: what is available and what is yet to come. Pharmacol Rev 68 (3):701–787
228. Babu A, Templeton AK, Munshi A et al (2014) Nanodrug delivery systems: a promising technology for detection, diagnosis, and treatment of cancer. AapsPharmscitech 15 (3):709–721
229. Etheridge ML, Campbell SA, Erdman AG et al (2013) The big picture on nanomedicine: the state of investigational and approved nanomedicine products. Nanomedicine 9(1):1–14
230. Ventola CL (2017) Progress in nanomedicine: approved and investigational nanodrugs. Pharm Ther 42(12):742
231. Min Y, Caster JM, Eblan MJ et al (2015) Clinical translation of nanomedicine. Chem Rev 115 (19):11147–11190
232. Lytton-Jean AKR, Kauffman KJ, Kaczmarek JC et al (2015) Cancer nanotherapeutics in clinical trials. Cancer Treat Res 166:293–322
233. Bolwell BJ, Cassileth PA, Gale RP (1988) High dose cytarabine: a review. Leukemia 2(5):253
234. Lengfelder E, Haferlach C, Saussele S et al (2009) High dose ara-C in the treatment of newly diagnosed acute promyelocytic leukemia: long-term results of the German AMLCG. Leukemia 23(12):2248–2258
235. Wiernik PH, Banks PL, Case CD et al (1992) Cytarabine plus idarubicin or daunorubicin as induction and consolidation therapy for previously untreated adult patients with acute myeloid leukemia. Blood 79(2):313–319
236. Vogler WR, Velezgarcia E, Weiner RS et al (1992) A phase III trial comparing idarubicin and daunorubicin in combination with cytarabine in acute myelogenous leukemia: a southeastern cancer study group study. J Clin Oncol 10(7):1103–1111
237. Muhammad A, Champeimont J, Mayr UB et al (2012) Bacterial ghosts as carriers of protein subunit and DNA-encoded antigens for vaccine applications. Expert Rev Vaccines 11 (1):97–116

238. Kanasty R, Dorkin JR, Vegas A et al (2013) Delivery materials for sirna therapeutics. Nat Mater 12(11):967–977
239. Jin GR et al (2016) Multifunctional organic nanoparticles with aggregation induced emission (AIE) characteristics for targeted photodynamic therapy and RNA interference therapy. Chem Commun 52(13):2752–2755
240. Senzer N, Nemunaitis J, Nemunaitis D et al (2013) Phase I study of a systemically delivered p53 nanoparticle in advanced solid tumors. Mol Ther 21(5):1096–1103
241. Lane DP (1995) On the regulation of the p53 tumour suppressor, and its role in the cellular response to DNA damage. Philos Trans R Soc Lond Biol 347(1319):83–87
242. Yokoyama M (2005) Drug targeting with nano-sized carrier systems. J Artif Organs 8 (2):77–84
243. Bottini M, Sacchetti C, Pietroiusti A et al (2014) Targeted nanodrugs for cancer therapy: prospects and challenges. J Nanosci Nanotechnol 14(1):98–114
244. Yang HX, Qin XL et al (2019) An in vivo miRNA delivery system for restoring infarcted myocardium. ACS Nano
245. Nahta R, Hung MC, Esteva FJ (2004) The her-2-targeting antibodies trastuzumab and pertuzumab synergistically inhibit the survival of breast cancer cells. Cancer Res 64 (7):2343–2346
246. Mikhail AS, Negussie AH, Pritchard WF et al (2017) Lyso-thermosensitive liposomal doxorubicin for treatment of bladder cancer. Int J Hyperth 33(7):1–28
247. Yang J (2012) Stimuli-responsive drug delivery systems. Adv Drug Deliv Rev 64 (11):965–966
248. Yin Q, Shen J, Zhang Z et al (2013) Reversal of multidrug resistance by stimuli-responsive drug delivery systems for therapy of tumor. Adv Drug Deliv Rev 65(13–14):1699–1715
249. Gu M, Wang X, Toh TB et al (2017) Applications of stimuli-responsive nanoscale drug delivery systems in translational research. Drug Discov Today 23(5):1043–1052
250. Hare JI, Lammers T, Ashford MB et al (2016) Challenges and strategies in anti-cancer nanomedicine development: an industry perspective. Adv Drug Deliv Rev 108:25–38
251. Moore R (2014) Challenges to nanomedicine. Springer, New York
252. Eliasof S, Lazarus D, Peters CG et al (2013) Correlating preclinical animal studies and human clinical trials of a multifunctional, polymeric nanoparticle. Proc Natl Acad Sci U S A 110 (37):15127–15132
253. Malinoski FJ (2014) The nanomedicines alliance: an industry perspective on nanomedicines. Nanomedicine 10(8):1819–1820
254. Bregoli L, Movia D, Gavigan-Imedio JD et al (2016) Nanomedicine applied to translational oncology: a future perspective on cancer treatment. Nanomed Nanotechnol Biol Med 12 (1):81–103
255. Sharma A, Madhunapantula SRV, Robertson GP (2012) Toxicological considerations when creating nanoparticle based drugs and drug delivery systems? Expert Opin Drug Metab Toxicol 8(1):47–69
256. Adiseshaiah PP, Crist RM, Hook SS et al (2016) Nanomedicine strategies to overcome the pathophysiological barriers of pancreatic cancer. Nat Rev Clin Oncol 13(12):750–765
257. Nie S (2010) Understanding and overcoming major barriers in cancer nanomedicine. Nanomedicine 5(4):523–528
258. Maity AR, Stepensky D (2016) Pharmacokinetics and pharmacodynamics of nano-drug delivery systems. In: Intracellular delivery III. Springer, Cham, pp 341–362
259. Sahakyan N, Haddad A, Richardson S et al (2017) Personalized nanoparticles for cancer therapy: a call for greater precision. Anti Cancer Agents Med Chem 17(8):1033–1039
260. Onoue S, Yamada S, Chan K (2014) Nanodrugs: pharmacokinetics and safety. Int J Nanomedicine 9:1025

Chapter 10
Dendrimer-Based Tumor-targeted Systems

Zhijun Ouyang, Du Li, Mingwu Shen, and Xiangyang Shi

Abstract Dendrimers have a unique three-dimensional structure, highly branched macromolecular features, and abundant terminal functional groups. These features of dendrimers support their uses as universal nanoplatforms to prepare multifunctional nanodevices for a myriad of biomedical applications, particularly for targeted imaging and therapy of different biosystems. The periphery of the dendrimer enables the attachment of targeting ligands and imaging agents, while the internal cavity allows the embedment of metal nanoparticles (NPs), anticancer drugs, and other inorganic NPs. Meanwhile, the versatile dendrimer nanotechnology enables different types of integration with varied inorganic components. The resulting hybrid dendrimer-based particles are useful in a variety of imaging, therapy, or theranostic applications. This chapter focuses on the latest advances in the dendrimer-based targeting systems for the imaging, therapy, and theranostics of cancer.

Keywords Dendrimers · Targeting · Contrast agents · Therapy · Theranostic agents

Abbreviations

3D	3-dimensional
^{99m}Tc	Technetium-99m
α-TOS	α-tocopheryl succinate
AuNPs	Gold nanoparticles
AFM	Atomic force microscopy
Au DENPs-FA	Folic acid-modified dendrimer-entrapped gold nanoparticles
ASGPR	Asialoglycoprotein receptors
BBB	Blood brain barrier
bis-MPA	2,2-bis(hydroxymethyl)propionic acid

Z. Ouyang · D. Li · M. Shen · X. Shi (✉)
State Key Laboratory for Modification of Chemical Fibers and Polymer Materials, International Joint Laboratory for Advanced Fiber and Low-dimension Materials, College of Chemistry, Chemical Engineering and Biotechnology, Donghua University, Shanghai, People's Republic of China
e-mail: xshi@dhu.edu.cn

R. Huang, Y. Wang (eds.), *New Nanomaterials and Techniques for Tumor-targeted Systems*, https://doi.org/10.1007/978-981-15-5159-8_10

CT	Computer tomography
DOTA	Tetraazacyclododecane-1,4,7,10-tetraacetic acid
DOTA-NHS	(2,2′,2″-(10-(2-(2,5-dioxopyrrolidin-1-yloxy)-2-oxoethyl)-1,4,7,10-tetraazacyclododecane-1,4,7-triyl) triacetic acid
DTPA	Diethylenetriaminepentaacetic acid
DENPs/DSNPs	Dendrimer-entrapped nanoparticles/dendrimer-stabilized nanoparticles
DOX	Doxorubicin
EPR	Enhanced permeability and retention
EGFP	Enhanced green fluorescent protein
FA	Folic acid
FAR	FA receptors
FA-PEG-COOH	PEG-modified FA with carboxyl end group
FI	Fluorescein isothiocyanate
FDG	2-18F-fluoro-2-deoxy-D-glucose
FAM	Fluorescein amidite
FPP	FA-PEG-PAMAM
Gd-Au DENPs	Gd-loaded Au DENPs
G	Generation
Gal	Galactose
HA	Hyaluronic acid
HPAO	3-(4’-hydroxyphenyl) propionic acid-OSu
hMSCs	Human mesenchymal stem cells
HER2	Human epidermal growth factor receptor-2
HEGFR	Human epidermal growth factor receptor
HCC	Hepatocellular carcinoma
HUVEC	Human umbilical vein endothelial cells
IO	Iron oxide
IL-6	Interleukin-6
LA	Lactobionic acid
LA-Au DENPs	LA-modified dendrimer-entrapped gold nanoparticles
Luc	Luciferase
MR	Magnetic resonance
MWCNTs	Multi-walled carbon nanotubes
*m*PEG-COOH	PEG monomethyl ether with carboxyl end group
NCPs	Nanocomposite particles
PEG	Polyethylene glycol
NAcGal	N-acetylgalactosamine
PAMAM	Poly(amidoamine)
PLL	Poly(l-lysine)
PPI	poly(propyleneimine)
PSMA	Prostate specific membrane antigen
PET	Positron emission computed tomography
PGA	Poly(γ-glutamic acid)

QDs	Quantum dots
RGD	Arg-Gly-Asp
RGD-4C	Cyclized RGD
RT	Radiotherapy
SPECT	Single-photon emission computed tomography
SMVT	Sodium-dependent multivitamin transporter
US	Ultrasound

10.1 Introduction

During the past decades, the arising flourish field of nanomedicine has captured the attention and ignited the imagination of some of the most renowned scientific researchers and physicians worldwide. Nanomedicine, the integration of nanotechnologies with biology, chemistry, and medicine, is expanding its applications in disease diagnosis, temporal and spatial site-specific drug delivery, and in situ therapy [1]. As an outstanding achievement in nanomedicine, nanotheranostics has been widely applied since they enable real-time visualization, diagnosis and recognition of the disease, and early-stage therapy. More recently, a myriad of nanoplatforms have been introduced in the field of nanomedicine. Among these platforms, dendrimers, a class of highly branched, monodispersed, synthetic macromolecules with defined conformation and architecture, have drawn a great deal of attention [1–3].

Dendrimers are discrete nanostructures composed of onion skin-like branched layers [4–6]. Starting with a core, the branching units radically grow with concentric layers to produce unique three-dimensional architectures similar to some ball-shaped proteins. These branched tree-like concentric layers are defined as generations (denoted as G, Fig. 10.1) [7]. The outer generations of each dendrimer display with a precise number of functional groups [7]. The periphery groups could be connected with different theranostic components to proceed favorable nanoparticle (NP)-drug and NP-tissue interactions. These features render dendrimers with great advantages in cancer detection and treatment as nanocarriers for traditional medical drugs, proteins, and imaging agents [8–16]. It is notable that the size and shape are two most important physical properties of dendrimers, determining various types of events such as cellular uptake, transport, and tissue accumulation in vivo. For example, the effects of dendrimer generation (from G1 to G7) and dendrimer architecture (three-dimensional (3D) dendrimers, dendrons, lipid-bearing dendrons, or Janus dendrimers) could determine the nanoparticular interactions with biological membranes and organelles, which in turn directly influences their cellular uptake, metabolism, excretion, bioavailability, tissue accumulation, and organ toxicity [17–21]. Among the most studied dendrimer species in nanomedicine, poly(amidoamine) (PAMAM), poly(l-lysine) (PLL), poly(propyleneimine) (PPI), and 2,2-bis (hydroxymethyl)propionic acid (bis-MPA) are all commercially available [1].

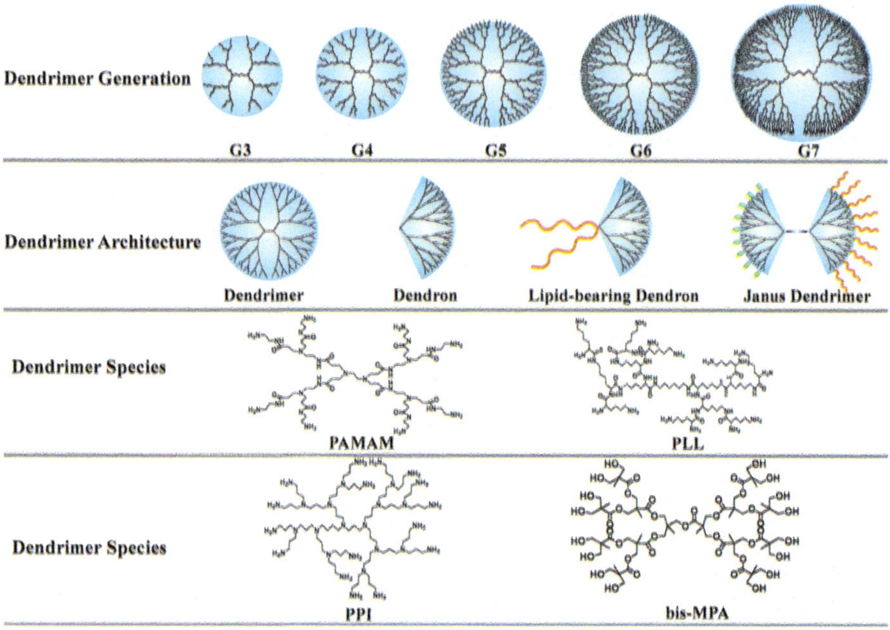

Fig. 10.1 The dendrimer generation, species, architecture, and surface functionality play important roles in dendrimer-mediated theranostic systems [25]. (Reproduced with permission from Ref. 25, Copyright 2016 Elsevier)

After surface modification and conjugation with targeting ligands and imaging agents, the dendrimer-based platform could be further entrapped with metal NPs or used to stabilize or assemble magnetic NPs [22–24] for various biomedical applications. In particular, dendrimer-based contrast agents, as well as imaging and therapeutic agents, are emerging as promising candidates for many nanomedicinal applications.

This chapter aims to give a general literature review covering up-to-date dendrimer-based targeted systems. Followed by a brief introduction toward the targeting strategies, recent advances in the dendrimer-based theranostic platforms will be systematically elaborated. Several key developments in the molecular imaging as well as therapeutic treatment of cancer will be focused and discussed in detail. Finally, challenges and future perspectives relating to this field will be briefly addressed. It should be noted that this chapter is not a comprehensive literature review but rather discusses some prominent developments of dendrimer-based targeting systems for cancer diagnosis and treatment.

10.2 Dendrimer-Based Targeting Delivery System

Due to the unique macromolecular properties and well-defined coordination with a wide range of theranostic components, and their adaptability to a variety of drug targeting applications in nanomedicine, dendrimers have attracted massive attention in developing different targeting systems. Many of these platforms have been designed for in vivo targeted delivery of contrast agents or drugs for disease diagnosis and treatment. Nowadays, two types of nanocarrier targeting mechanisms, passive targeting and active targeting, have been widely used to deliver theranostic agents (Fig. 10.2).

10.2.1 Dendrimer-Mediated Passive Targeting Delivery System

Passive targeting is usually based on the enhanced permeability and retention (EPR) effect of tumors to deliver NPs by penetrating tumor vasculature [26–28]. EPR effect, depending on the pathophysiological differences between tumor

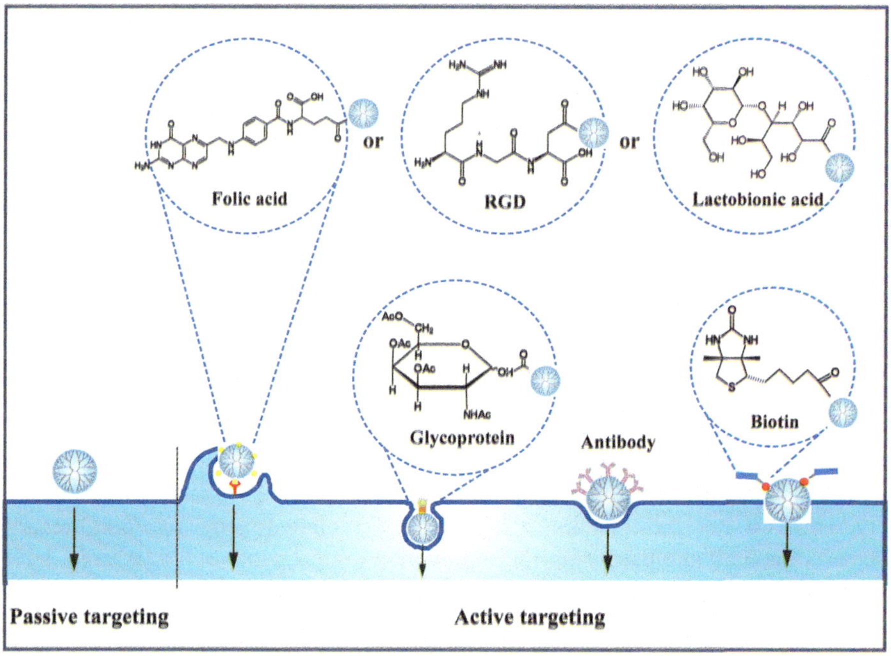

Fig. 10.2 Schematic illustration of passive targeting and active targeting mechanism of dendrimer-based nanoplatforms

microenvironments and normal tissues, could improve the delivery efficacy of therapeutic agents to tumors [29]. By taking advantage of tissue physiological properties and NP physicochemical parameters, the concentration of dendrimer-based nanomaterials is normally 10–30 times higher in the tumor than that in the blood. This is ascribed to the homing effect of lesion site, driven by sizes of NPs rather than specific recognition.

Cancer cells divide and reproduce at an exponential speed and construct complex networks of blood vessel, forming a highly disordered aberrant and leaky vascular system. Meanwhile, cancer cells produce excessive vascular permeable mediators that facilitate dilation of blood vessels. Moreover, solid tumors always have dysfunctional lymphatic clearance problems. The anatomical and pathophysiological abnormalities in tumor areas in conjunction with overproduction of permeability mediators lead to extensive leakage of blood plasma components, macromolecules, and NPs including dendrimers into the tumor interstitium. As a result, dendrimers possessing multiple functional groups or dendrimer-based nanohybrids can be used as passive targeting agents to be enriched in tumor region for theranostic applications.

10.2.2 Dendrimer-Mediated Active Targeting Delivery System

Active targeting, on the other hand, involves the therapeutic delivery of a nanosystem to a specific site via molecular recognition. Specific interactions between the ligands on the surface of dendrimers and receptors expressed on the cancer cells may facilitate NP internalization by triggering receptor-mediated endocytosis.

Due to the unique macromolecular properties and well-defined coordination with a wide range of theranostic agents, dendrimers have attracted increasing attention. Many dendrimer-mediated strategies are proposed for in vitro or in vivo targeted delivery of imaging contrast agents or drugs for diagnosis and treatment of certain diseases [30–33]. Covalent conjugation of specific targeting moieties, such as folic acid (FA) [34, 35], peptides [36–38], monoclonal antibodies [39], glycoprotein [40, 41], and sugar [42] to nanocarriers enables receptor-mediated active targeting of nanosystems to tumor tissues. More specifically, dendrimer-based targeted drug delivery can enhance chemotherapeutic effects and prevent damage to normal tissues. The dendrimer-drug conjugates may be engineered for targeted drug delivery to disease sites, thus avoiding off-target delivery of toxic drugs to normal organs. Herein, this section presents a brief introduction of dendrimer-based targeted delivery systems.

10.2.2.1 FA-Targeted Delivery System

FA is one of the most studied targeting ligands for tumor targeting due to its high affinity to FA receptors (FAR) overexpressed in many types of cancer cell lines [43, 44]. The multivalent FA-modified dendrimers [45] have facilitated active targeting to tumor cells due to the fact that the multivalent characters of dendrimers enable the attachment of various diagnostic and therapeutic payloads. Since FAR expression mostly occurs on the surface of cancer cells, instead of normal tissues [46], FA has been considered as a promising targeting ligand for dendrimer-based nanodevices to actively target tumors.

10.2.2.2 Peptide-Targeted Delivery System

Peptides have also been used to target cancer cells overexpressing specific receptors. A well-known example is Arg-Gly-Asp (RGD) peptide, which has a high affinity to $a_v\beta_3$ integrins overexpressed on the surface of tumor microvasculature [36, 47–49]. PAMAM dendrimers modified with RGD peptide on their surface could be used as a vehicle for various targeted drug and gene delivery applications [37, 50]. Earlier work by Baker et al. has demonstrated the use of G5 PAMAM dendrimers modified with a doubly cyclized RGD (RGD-4C) peptide and Alexa Fluor 488 dye for tumor neovasculature targeting [36].

Similarly, Lesniak et al. [51] and Hill et al. [50, 52] conjugated other types of RGD peptides onto the surface of G5 PAMAM dendrimers, aiming to target tumor microvasculature or odontoblast-like cells. For tumor microvasculature targeting, the binding characteristics were assessed both in vitro and in vivo, using human SK-RC-52 cells and the mouse xenografted SK-RC-52 tumor model. The results indicated that the tetrameric RGD-dendrimers exhibited the highest level of tumor targeting. Furthermore, PAMAM dendrimer-RGD conjugates can be used to mediate cellular binding and adhesion. In general, these probes are considered to be nontoxic to normal cells and possess the ability to deliver imaging agents to tumor tissues overexpressing $a_v\beta_3$ integrins via a receptor-mediated manner [53].

10.2.2.3 Monoclonal Antibody-Targeted Delivery System

Monoclonal antibodies, when conjugated to dendrimers, could function against specific antigens and further selectively target cancer cells with the minimal damage to normal cells. An example illustrated in prostate-specific membrane antigen (PSMA) J591 antibody conjugated to G5 PAMAM. The in vitro studies showed that the designed probe could specifically bind cells expressing PSMA antigen [54]. Other studies demonstrated that the binding and internalization of interleukin-6 (IL-6) monoclonal antibody-conjugated dendrimers could target human epidermal growth factor receptor-2 (HER2)-expressing cancer cells

[55]. Through further confirmation with internalization and competitive experiments using free antibody, the rapid and efficient cellular internalization of the G4.5 PAMAM/IL-6 conjugate was revealed [56]. IL-6 is an important multifunctional cytokine that plays a crucial role in angiogenesis. Due to the rapid neovascularization of tumors, IL-6 is overexpressed in cancer cells. Studies have shown that the high affinity of IL-6 to human epidermal growth factor receptor (HEGFR) could lead to significant internalization of dendrimer-modified IL-6 conjugates into HeLa cells through the receptor-mediated endocytosis pathway [57]. Furthermore, the drug release of IL-6-conjugated dendrimer conjugates was significantly higher in the nucleus than native PAMAM dendrimers [57]. Dendrimers have also been conjugated to anti-EGFR antibodies, such as cetuximab, for enhanced selective cellular uptake of antitumor drugs to treat brain tumors [58].

10.2.2.4 Glycoprotein-Targeted Delivery System

The presence of lectin receptors on different cell surfaces enables glycosylated carriers to be used for targeted drug delivery. N-Acetylgalactosamine (NAcGal) sugar molecules conjugated to G5 PAMAM dendrimers via peptide and thiourea linkages could selectively bind asialoglycoprotein receptors (ASGPR), which are highly expressed on the surface of hepatic cancer cells, and then triggered the receptor-mediated endocytosis of the dendrimers in hepatic cancer cells [41]. Similarly, Huang et al. reported that G2.5 PAMAM dendrimers modified with polyethylene glycol (PEG) and galactose (Gal) could mediate liver-targeted drug delivery. The release of doxorubicin (DOX) from Gal-PEG-G2.5-DOX dendrimers was accelerated with the decreasing of pH from 8.0 to 5.6. Results showed that the cytotoxicity of the conjugates against BEL-7402 cells (human hepatoma cell line) was weaker than that of free DOX likely due to the gradual release of the drug [40].

10.2.2.5 Biotin-Targeted Delivery System

Biotin is one important clinical target for nanomedicine. It is an essential micronutrient in normal cellular functions, such as fatty acid biosynthesis, gluconeogenesis growth, and catabolism. Substantial evidence suggests that the level of biotin has a connection with the rapid proliferating speed of cancer cells. As a consequence, specific interaction between biotin and its receptor may be exploited for targeted therapeutic delivery [59]. It has been reported that partially acetylated G5 PAMAM conjugated with biotin can serve as a nanodevice for in vitro cancer targeting [60]. A similar investigation has extended this work by reporting that biotinylated PAMAM could be internalized by cancer cells via receptor-mediated endocytosis and charge-mediated adsorptive endocytosis. Langmuir-Blodgett monolayer technique, atomic force microscopy (AFM), and lactate dehydrogenase assays were used to investigate the interaction of biotinylated G4 PAMAM with the blood-brain barrier (BBB) in vitro. According to cellular uptake results, these biotinylated dendrimers displayed both charge- and sodium-dependent multivitamin transporter (SMVT)-mediated uptake [61]. It is noteworthy that SMVT is responsible for biotin uptake and highly expressed in human keratinocytes and peripheral blood mononuclear, intestinal, liver, and renal epithelial cells [62–64]. The biotinylation of dendrimers could act as an attractive strategy for targeted delivery of chemotherapeutic agents to cancer cells.

10.2.2.6 Sugar-Targeted Delivery System

Dendrimers modified with sugar moieties such as glucose, mannose, galactose, or disaccharide are referred as to glycodendrimers. The sugar and gluconic acid are linked by an ether bond and can be hydrolyzed into galactose and gluconic acid under the catalysis of an enzyme. These dendrimers are expected to be delivered to the lectin-rich organs. Selime et al. [42] studied the targeting effect of the carrier on hepatocytes by modifying the magnetic NPs with the glycodendrimers and proved that the specific binding of the ASGPR on the surface of the hepatocytes to the lactobionic acid (LA) galactosyl group can significantly improve the targeting of the glycodendrimer devices.

10.3 Dendrimer-Based Targeted Imaging Systems

Cancer treatment efficiency can be improved by developing accurate diagnostics toward specific cancer types and monitoring therapy in real time. Accurate diagnosis is essential for effective treatment of cancer. Early detection of cancer provides more effective treatment options and improved chance of patient recovery. Several imaging methods, such as computer tomography (CT), magnetic resonance (MR) imaging, positron emission computed tomography (PET), single-photon emission computed tomography (SPECT), and ultrasound (US), have been developed to achieve this purpose. However, these imaging techniques are subject to some limitations including poor resolution, inherent radiation damage, and non-specificity due to the use of conventional contrast agents.

Current research has indicated that dendrimer-based nanodevices enable enhanced and precision cancer diagnostics. In this section, we will introduce the dendrimer-based devices for targeted tumor imaging.

10.3.1 Targeted Tumor CT Imaging

CT enables 3D anatomical imaging with high spatial and density resolution. Compared to other clinical imaging modalities such as PET and MR imaging, CT has deep tissue penetration capability, cost-effectiveness, and simple 3D image reconstruction ability [65]. In order to distinguish target soft tissues from the surrounding ones in CT scanning, X-ray contrast agents play an important role to provide detailed anatomical and functional information. However, commercially available iodinated small molecular CT contrast agents (e.g., Omnipaque) have disadvantages of rapid clearance speed in blood, non-specificity, and renal toxicity at a relatively high concentration. Therefore, novel CT contrast agents with longer blood circulation time and enhanced contrast ability have been emerging, focusing on critical nanomedicine requirements such as targeted cancer imaging, prolonged blood circulation time, and improved X-ray attenuation effect [52, 66–72]. Among these contrast agents, metallic NPs have become an eminent choice for their excellent X-ray attenuation effect and low toxicity characteristics [73].

It has been found that dendrimers are suitable macromolecular CT contrast agents. Similar to other macromolecular contrast agents developed for CT imaging, dendrimers have been bound with iodine compounds because they inherently lack radio-opaque elements [74]. Recently, gold NPs (AuNPs) have received much attention to be used for CT imaging owing to their biocompatibility [75, 76], higher X-ray absorption coefficient than traditional iodine-based contrast agent [77], and extended blood circulation time due to their nanoscale sizes. It has been reported that amine-terminated G5 PAMAM dendrimers entrapped with AuNPs could be applied as contrast agents for CT imaging of tumors [77]. In another study, silver NPs stabilized with G5 PAMAM dendrimers with an appropriate core size were designed to act as a potential CT imaging contrast agent, even though the atomic number of Ag is lower than that of iodine [78]. However, these dendrimer-based contrast agents do not have high affinity to specific cancer cells, and they are only able to be delivered to tumor site via passive EPR effect. Recently, dendrimer-entrapped/dendrimer-stabilized NPs (DENPs/DSNPs) have been conjugated with various targeting ligands (e.g., FA, RGD peptide, and antibodies) for targeted CT imaging of cancer cells. For instance, Wang et al. [79] reported the use of FA-modified Au DENPs (Au DENPs-FA) as a nanoplatform for in vitro and in vivo targeted human lung adenocarcinoma CT imaging. In this study, the formed Au DENPs-FA were used to image human lung adenocarcinoma cells (SPC-A1 cells) and the xenograft mouse tumor model, which expresses FAR. Micro-CT imaging results showed that SPC-A1 cells could display CT contrast enhancement after treated with the Au DENPs-FA in vitro, and the xenograft tumor model could be diagnosed after intravenous, intratumoral, or intraperitoneal injection of the particles.

To enable high content of Au loading within the dendrimer internal cavity, PEGylation of dendrimer periphery is essential. Peng et al. designed PEGylated FA-targeted Au DENPs for specific CT imaging of FAR-overexpressing tumors in vivo [80]. In this work, G5.NH_2 PAMAM dendrimers modified with both PEG-modified FA with carboxyl end groups (FA-PEG-COOH) and PEG monomethyl ether with carboxyl end group (*m*PEG-COOH) were used as templates to entrap Au NPs. The formed FA-modified PEGylated Au DENPs enabled targeted CT imaging of the high FAR-expressing model cancer cells (KB cells) in vitro (Fig. 10.3a) and the corresponding tumor xenografts in vivo (Fig. 10.3b). Technically, the FA-modified PEGylated Au DENPs performed better than the

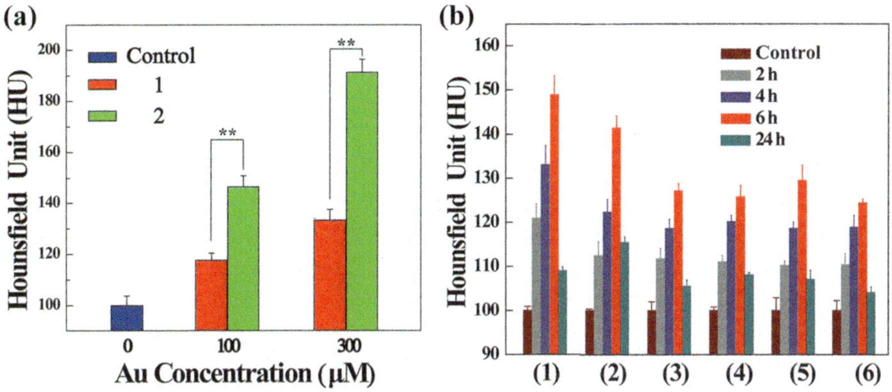

Fig. 10.3 (**a**) The CT values of the KB-HFAR cells incubated with[$(Au^0)_{300}$-G5.NHAc-*m*PEG] (1) and [$(Au^0)_{300}$-G5.NHAc-(PEG-FA)-*m*PEG] (2) DENPs at different Au concentrations for 4 h ($n = 3$); (**b**) CT values of the KB tumor in nude mice before (0 h) and after injected with [$(Au^0)_{300}$-G5.NHAc-(PEG-FA)-*m*PEG] DENPs (1–4; 1, intravenous injection; 2, intraperitoneal injection; 3, intravenous injection of FA-blocked tumor; and 4, intraperitoneal injection of FA-blocked tumor) and [$(Au^0)_{300}$-G5.NHAc-*m*PEG] DENPs (5 and 6; 5, intravenous injection; 6, intraperitoneal injection) by different injection routes at 2, 4, 6, and 24 h post injection [80]. (Reproduced with permission from Ref. 80, Copyright Royal Society of Chemistry)

FA-modified acetylated Au DENPs [79] in the aspects of the improved biocompatibility rendered by PEGylation and CT imaging sensitivity due to the improved Au loading within the dendrimer internal cavities.

Moreover, LA has also been modified onto the dendrimer nanoplatform to target cancer cells. In a recent study, Liu et al. [81] described a novel approach to prepare LA-modified Au DENPs (LA-Au DENPs) for in vitro and in vivo targeted CT imaging of human hepatocellular carcinoma. In this work, amine-terminated G5 PAMAM dendrimers pre-modified with fluorescein isothiocyanate (FI) and PEG-linked LA were adopted as templates to prepare Au NPs. The leftover dendrimer amine termini were acetylated to create LA-Au DENPs. Similarly, in order to developed cost-effective nanoscale contrast agents for targeted tumor CT imaging, Cao et al. [82] synthesized Au DENPs using G2 PAMAM dendrimers pre-modified with FI and LA via a PEG spacer as templates. The formed Au DENPs were used as a nanoprobe for targeted CT imaging of hepatocellular carcinoma (HCC) overexpressing ASGPR.

10.3.2 Targeted Tumor Fluorescence Imaging

Although dendrimers, especially PAMAM or PPI dendrimers, have been demonstrated to have intrinsic blue fluorescence emission property, the quantum yield is quite low, and hence they cannot be effectively used for fluorescence imaging of tumors. To generate a dendrimer-based hybrid system for tumor fluorescence imaging, quantum dots (QDs) have been combined with dendrimers to synthesize DENPs or DSNPs that can be decorated with PEG to improve their biocompatibility and be further modified with targeting ligands for specific fluorescence imaging of tumors. Zhao et al. synthesized FA-PEG-PAMAM (FPP)-decorated CdSe/ZnS quantum dots, and the FA-PEG chains linked to PAMAM dendrimers were used to target FAR-overexpressing tumor cells and render the QDs with improved water solubility [44]. In vitro imaging results showed that cellular uptake of FPP-coated QDs was more significant than that of non-targeted ones. In another work, Cui and coworkers conjugated a DNA aptamer (GBI-10) on the surface of dendrimer-modified CdSe QDs to target the extracellular matrix protein tenascin-C on the surface of human glioblastoma cells [83]. Similarly, RGD peptide conjugated to dendrimer-modified CdSe QDs was used to target tumor vasculature. The prepared nanoprobe was injected into nude mice with melanoma (A375) tumor xenografts via tail vein. In vivo imaging system was used to disclose the biodistribution of the nanoprobe. RGD-conjugated QDs could target human umbilical vein endothelial cells (HUVEC) and A375 melanoma cells, as well as xenografted A375 melanoma tumors on nude mice [84]. Therefore, high-performance targeting molecule-conjugated dendrimer-modified QDs may serve as a new generation of nanoprobes for fluorescence imaging of tumors.

10.3.3 Targeted Tumor MR Imaging

MR imaging is a powerful, noninvasive imaging technique for disease diagnosis [85–88], and it possesses various attractive advantages, including a nonionizing radiation source, tomographic capabilities, and high spatial resolution between different soft tissues [22, 73, 89–93]. At present, MR imaging has been widely used for clinical purposes and has been generally considered as one of the most convenient and noninvasive imaging modalities.

With the versatile dendrimer nanotechnology, dendrimers can be combined with different MR contrast agents for improved imaging applications [94]. Paramagnetic Gd(III) [95] and Mn(II) [96], as well as superparamagnetic iron oxide (IO) NPs [97, 98] have been integrated with dendrimers to generate multifunctional dendrimer nanohybrids for different MR imaging applications. In general, clinically used low molecular weight Gd(III) chelates, such as Gd-diethylenetriaminepentaacetic acid (DTPA) and Gd-tetraazacyclododecane-1,4,7,10-tetraacetic acid (DOTA) can be rapidly extravasate from blood circulation and eliminated via renal filtration, displaying non-specificity and renal toxicity at a relatively high concentration. Macromolecular Gd(III) complexes instead show prolonged blood circulation and preferential accumulation in solid tumors due to the hyperpermeability of tumor vasculature [99]. The loading of Gd(III) onto dendrimers as MR contrast agents has been described by several groups [95, 96, 100] for T_1-weighted MR imaging applications.

For targeted tumor MR imaging, modification of specific targeting ligands on the surface of the dendrimer-Gd nanocomplexes is necessary [88]. The commonly used targeting ligands include monoclonal antibodies [101, 102], peptides [103, 104], FA [22, 105], polysaccharides [106], proteins [107], and oligonucleotides [108]. FA is one of the most studied cancer-targeting ligands for targeted MR imaging applications [109]. Swanson et al. [105] prepared an FA-targeted dendrimer-based MR imaging contrast agent by conjugating of FA and Gd-DOTA onto G5 PAMAM dendrimers. The targeted device displayed a significant high T_1 MR contrast enhancement in the KB (a human epithelial carcinoma cell line) xenografted tumor than the non-targeted ones. In a similar work, Chen et al. [110] loaded FA and Gd-DTPA onto PEG-cored dendrimers with 16 functional hydroxyl groups. In vitro cellular uptake results revealed that the KB cells had a more significantly enhanced uptake of the targeted particles than that of the non-targeted ones. MR imaging data demonstrated that KB tumor had a more enhanced contrast after injection of the FA-targeted device than the FAR-negative HT-1080 tumor. G4 PAMAM dendrimers have also been developed as a target-specific T_1-weighted MR contrast agent by Konda et al. [111].

Dendrimers can also be modified with peptide ligands [103] for targeted cancer MR imaging. Tan and coworkers [104] synthesized and evaluated a cyclic decapeptide CGLIIQKNEC (CLT1)-targeted dendrimeric MR imaging agent via linking Gd-DOTA and the CLT1 peptide onto the surface of G2 and G3 nanoglobular lysine dendrimers having a cubic silsesquioxane core. The CLT1 peptide modification rendered the nanoglobular agents with greater contrast enhancement in MDA-MB-231-bearing female athymic mice than non-targeted agents. The advantages to employ targeted dendrimeric contrast agents for specific MR imaging of different tumor types have been demonstrated by many other studies [112, 113].

Besides dendrimer-based Gd complexes used as positive T_1-weighted contrast agents for MR imaging, IO NPs can also be modified with dendrimers for targeted T_2-weighted MR imaging of tumors. Earlier work has shown that IO NPs synthesized via controlled coprecipitation of Fe(II) and Fe(III) ions in aqueous phase can be assembled with FA-functionalized dendrimers for targeted MR imaging of FAR-expressing cancer cells [114] and a xenografted tumor model [43].

Recently, we have shown that ultrasmall IO NPs with an r_1 relaxivity of 0.66 mM^{-1} s^{-1} can be assembled onto the surface of dendrimers for targeted T_2-weighted MR imaging of tumors [115]. In this work, ultrasmall citric acid-stabilized IO NPs were conjugated with RGD peptide-modified G5 PAMAM dendrimers via amide bond formation. Due to the close packing of the IO NPs onto the dendrimer surface, the formed G5.NHAc-RGD-Fe_3O_4 NPs displayed an r_2 relaxivity of 5.899 mM^{-1} s^{-1}, rendering their uses as a platform for targeted MR imaging of integrin $\alpha_v\beta_3$-expressing C6 glioma cells in vitro and the xenografted tumor model in vivo. The results showed that the tumor MR signal intensity from the

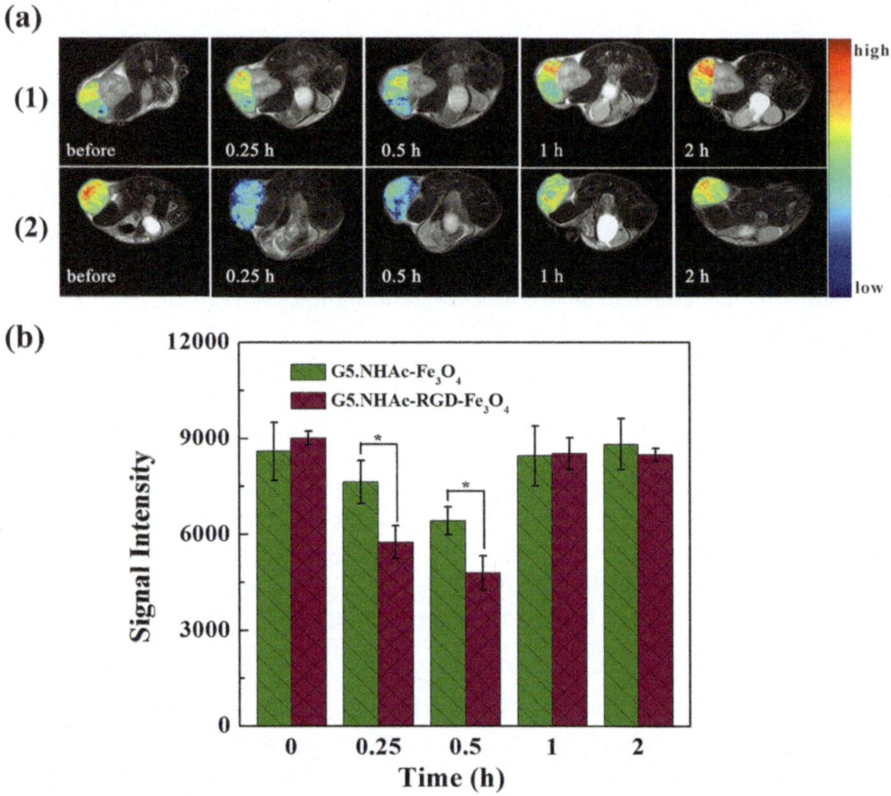

Fig. 10.4 In vivo T_2-weighted MR images of tumors (**a**) and the signal intensity (**b**) after intravenous injections of 0.1 mL of PBS solution containing G5.NHAc-Fe_3O_4 NPs (1) or G5.NHAc-RGD-Fe_3O_4 NPs (2) at different time points post injection (600 μg Fe for each mouse) [115]. (Reproduced with permission from Ref. 115, Copyright 2015 American Chemical Society)

mice injected with G5.NHAc-RGD-Fe_3O_4 NPs decreased more significantly than that injected with the G5.NHAc-Fe_3O_4NPs from 0.25 to 0.5 h (Fig. 10.4a), which was further validated through quantitative MR signal intensity measurements (Fig. 10.4b).

10.3.4 Targeted Tumor PET Imaging

PET is a functional imaging technique observing the metabolic processes in the body. PET system detects pairs of gamma rays emitted indirectly by a positron-emitting radionuclide (tracer), which is introduced into the body. Molecular imaging techniques, such as PET and SPECT, utilize positron and gamma to emit radionuclide signals that allow the whole-body scan in a single examination. The noninvasive evaluation of physiology and pathology merges with external and internal radiotherapy (RT), allowing for possible realization of theranostics and personalized therapy. Compared with SPECT, PET is relatively more sensitive and exhibits higher spatial resolution.

The PET tracer 2-18F-fluoro-2-deoxy-D-glucose (FDG) has been most widely adopted in clinical oncology. Besides, L-methyl-11C-methionine (11C-methionine) has also been successfully adopted for PET imaging of brain and lung tumors, non-Hodgkin's lymphoma, breast cancer, and head and neck cancer. Among all the clinical PET radionuclides used, gallium-68 (half-life, $t_{1/2} = 68$ min) exhibits positron emission intensity >87%. Recently, Ghai et al. radiolabeled G4 PAMAM dendrimers modified with (2,2′,2″-(10-(2-(2,5-dioxopyrrolidin-1-yloxy)-2-oxoethyl)-1,4,7,10-tetraazacyclododecane-1,4,7-triyl) triacetic acid (DOTA-NHS) with^{68}Ga via DOTA chelation to provide a conjugate used for PET imaging [116].

In a recent report [117], multifunctional dendrimers were chelated and readily labeled with^{64}Cu. This system served as a potential platform for PET imaging. In another work, Ma et al. [118] reported a^{64}Cu-radiolabeled multifunctional FA-targeted dendrimers for PET imaging of FAR-expressing tumors. In this work, amine-terminated G5 PAMAM dendrimers modified with FI, FA, and DOTA and with leftover amine being acetylated were labeled with^{64}Cu via the DOTA chelation. The FA modification rendered the dendrimers with targeting specificity for specific micro-PET imaging of the FAR-expressing tumors.

10.4 Dendrimer-Based Dual-Mode Contrast Agents for Targeted Imaging of Tumors

10.4.1 Dual-Mode CT/MR Imaging

MR imaging provides prominent 3D soft tissue details and functional information of the lesions, while CT supplies high-resolution 3D tomography information of the anatomic structure. It is attractive to develop multifunctional contrast agents that enable dual-mode CT/MR imaging of different biological systems. In a recent study [119], we showed that G5 dendrimers pre-modified with DOTA-NHS and PEG monomethyl ether could be used for templated synthesis of AuNPs and for subsequent Gd(III) complexation and dendrimer remaining terminal amine acetylation. Due to the coexistence of two radio-dense imaging elements of Gd(III) and AuNPs, the prepared multifunctional Gd-loaded Au DENPs (Gd-Au DENPs) could be used as a dual-mode CT/MR contrast agent for CT imaging of the heart, kidney, liver, and bladder of a rat within 45 min and MR imaging of the liver, kidney, and bladder of mice with extended circulation time and major organ clearance.

To achieve targeted dual-mode CT/MR imaging of tumors, we prepared an FA-targeted multifunctional dendrimeric nanoprobe [120] using G5 dendrimers as a platform to be sequentially modified with DOTA-NHS, PEGylated FA, and PEG monomethyl ether. The dendrimers were then entrapped with AuNPs in the interiors, complexed Gd(III) ions, and acetylated to neutralize the leftover dendrimer amine termini. The generated Gd-AuDENPs-FA probe enabled targeted CT/MR dual-mode imaging of the cancer cells in vitro and the xenografted tumor model in vivo. In another work, by replacing FA with RGD peptide, Chen et al. [121] reported the use of multifunctional Au DENPs loaded with Gd chelator/Gd(III) complexes and surface-modified with thiolated RGD peptide for targeted dual-mode CT/MR imaging of tiny tumors expressing $\alpha_v\beta_3$ integrin. Using a similar strategy, Wang et al. [122] presented the preparation and characterization of hyaluronic acid (HA)-modified multifunctional Au DENPs for targeted dual-mode CT/MR imaging of CD44 receptor-expressing cancer cells. Instead of chelating Gd (III), the authors prepared {$(Au^0)_{100}$-G5.NH_2-FI-DOTA(Mn)-HA}NPs with Mn (II) chelated to have a high X-ray attenuation intensity and favorable r_1 relaxivity (5.42 mM^{-1} s^{-1}) for targeted CT/MR dual-mode imaging of hepatocellular carcinoma overexpressing CD44 receptors.

To generate CT/T_2 MR dual-mode contrast agents, the combination of IO NPs and AuNPs can be realized via the dendrimer nanotechnology. Cai et al. [123] reported a unique dendrimer-assisted approach to create Fe_3O_4/Au nanocomposite particles (NCPs) for targeted CT/MR imaging of tumors (Fig. 10.5a). In this approach, preformed Fe_3O_4 NPs were layer-by-layer self-assembled with multilayers of poly(γ-glutamic acid) (PGA)/PLL/PGA/FA-modified Au DENPs that were crosslinked via 1-ethyl-3-(3-dimethylaminopropyl)carbodiimide chemistry. The assembled Au core NPs were then adopted as seed particles to mediate the growth of Au shells via repeated Au salt reduction process. After the remaining amines of dendrimers were subsequently acetylated, the final Fe_3O_4/Au_n.Ac-FA NCPs were formed. At an optimized Au/Fe_3O_4 molar ratio of 2.02, the Fe_3O_4/Au_n.Ac-FA NCPs showed a relatively high r_2 relaxivity (92.67 mM^{-1} s^{-1}) and good X-ray attenuation property. The FA-mediated targeting enables the Fe_3O_4/Au_n.Ac-FA NCPs to be specifically uptaken by FAR-overexpressing cancer cells for targeted dual-mode CT/MR imaging of tumors (Fig. 10.5b and c).

10.4.2 Dual-Mode T_1/T_2 MR Imaging

Dual-mode T_1/T_2 MR imaging [124] makes it possible to study soft tissue anatomy and pathological phenomenon, leading to potential differentiation between normal and tumor tissues, early tumor detection, and prediction of cancer stages [125]. Very recently, Haribabu et al. [125] demonstrated an effective approach to encapsulate manganese ferrite NPs [126, 127] within dendrimers that were later tagged with FA ligands. In this report, IO element could serve as a T_2 MR contrast agent, while manganese oxide composition served as T_1 MR contrast agent. The authors characterized different ratios of Fe-Mn conjugates and found that when the Mn/Fe molar ratio was at 0.5, the NPs displayed an r_2/r_1 ratio of 4.6 at 1.5 T, enabling targeted dual-mode T_1/T_2 MR imaging of FAR-expressing cancer cells.

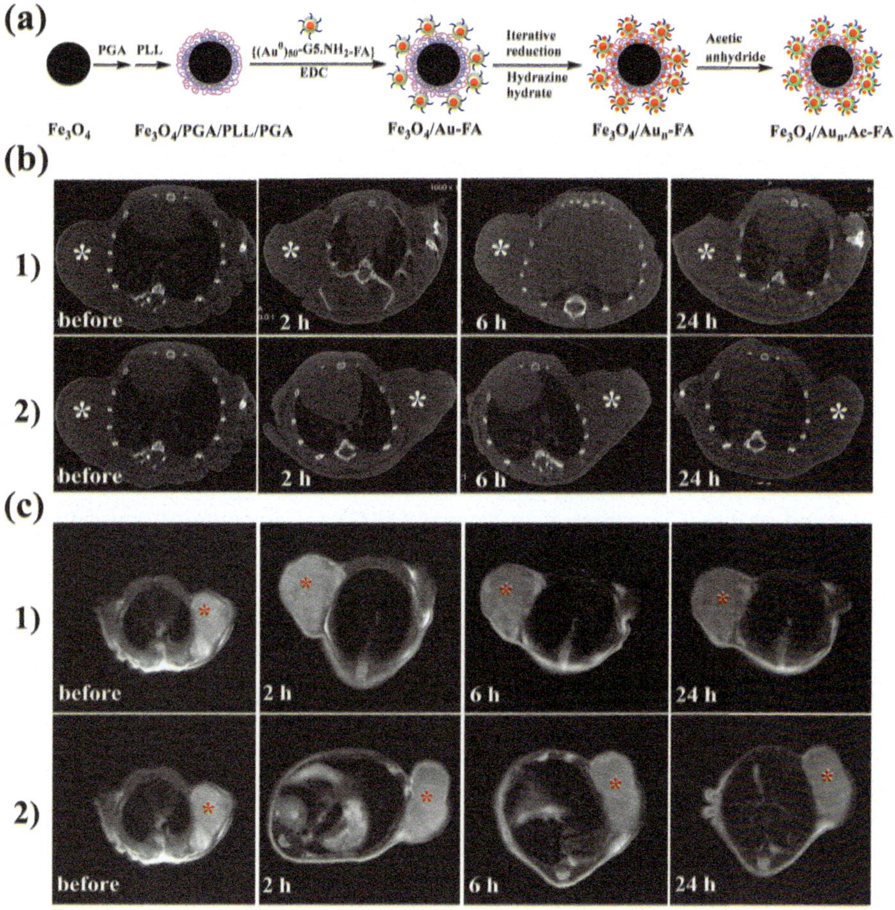

Fig. 10.5 (**a**) Schematic illustration of the fabrication of the Fe_3O_4/Au_n.Ac-FA NCPs. (**b**) is the CT images and (**c**) is the T_2-weighted MR images of nude mice bearing transplanted KB tumors before and at different time points post intravenous injection of the (1) targeted Fe_3O_4/Au_n.Ac-FA (Au/Fe_3O_4 molar ratio = 2.02) and (2) non-targeted Fe_3O_4/Au_n.Ac (Au/Fe_3O_4 molar ratio = 1.59) NCPs [123]. (Reproduced with permission from Ref. 123, Copyright 2015 WILEY-VCH Verlag GmbH & Co. KGaA)

10.4.3 Dual-Mode SPECT/MR Imaging

PET and SPECT can deliver functional information with high sensitivity, but yield low anatomical resolution. MR offers high-resolution anatomical images, especially when specific contrast agents are applied. PET and SPECT in combination with MR have been integrated through the dendrimer platform and applied to improve the precision and accuracy of imaging and diagnosis [128]. For example, Luo et al. [129] have demonstrated the preparation of FA-conjugated dendrimers loaded with Mn

(II) and Technetium-99 m (^{99m}Tc) both via DOTA chelation for targeted dual-mode SPECT/MR imaging of FAR-expressing cancer cells and a xenografted tumor model after intravenous injection.

10.4.4 Dual-Mode SPECT/CT Imaging

Via the dendrimer nanotechnology, both the SPECT and CT imaging elements can be incorporated [131], allowing for targeted dual-mode SPECT/CT imaging of tumors. Due to the high cost of high-generation dendrimers (G4 or above), cost-effective low-generation PAMAM dendrimers have been attractive to be used to construct dual-mode SPECT/CT imaging contrast agents. In a work delivered by Xu et al. [132], low-generation Au DENPs could be used for targeted tumor SPECT/CT dual-mode imaging. The formed RGD-modified Au DENPs labeled with^{99m}Tc possessed a uniform size distribution, good colloidal stability, and biocompatibility and could be used for targeted dual-mode SPECT/CT imaging of $\alpha_v\beta_3$ integrin-expressing tumors. In another work, Li et al. [130] used G2 PAMAM dendrimers to prepare multifunctional Au DENPs modified with FA and labeled with^{99m}Tc for targeted dual-mode SPECT/CTimaging of tumors. The formed particles with an average Au core diameter of 1.6 nm could be dispersed in water, displayed stability under different conditions, and were cytocompatible. Experiments demonstrated that the multifunctional nanoprobe enabled targeted SPECT/CT dual-mode imaging of FAR-overexpressing cancer cells and tumors (Fig.10.6).

10.4.5 Tri-mode Imaging

As dual modality is better than single modality, and tri-modal imaging could work even better compared to dual-mode imaging in terms of imaging precision. Recently, Chen et al. [133] have shown that dendrimers incorporated with AuNPs, Gd (III), and cyanine dye (Cy5.5) could serve as a tri-modal imaging agent for CT, MR, and fluorescence imaging, respectively, to provide better spatial and density resolutions with a high sensitivity. This is important for future translational precision nanomedicine.

10.5 Dendrimer-Based Targeting Therapy Systems

Most cancer therapeutics are small drug molecules ingested or injected into the bloodstream and then diffuse through vascular pores or the extracellular matrix to reach tumors. Under these circumstances, incremental nanometric size differences become extremely important in complexed therapeutics that involves drug and

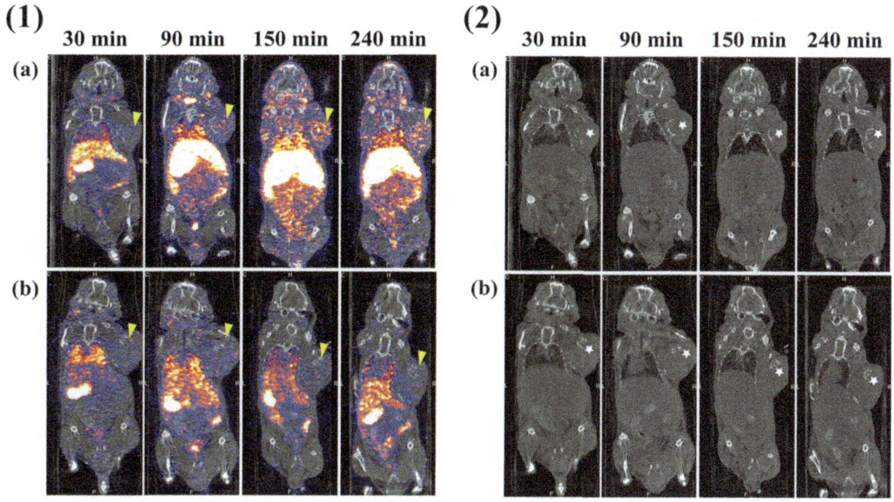

Fig. 10.6 In vivo SPECT/CT (1) and CT (2) images of tumors after intravenous injection of the {$(Au^0)_6$-G2-DTPA(^{99m}Tc)-PEG-FA} DENPs (a) or {$(Au^0)_6$-G2-DTPA(^{99m}Tc)-*m*PEG} DENPs (b) ([^{99m}Tc] = 740 MBq·mL^{-1}, [Au] = 0.08 M, in 100 μL PBS for each mouse) at different time points post injection. The dashed red circles indicate the tumor sites [130]. (Reproduced with permission from Ref. 130, Copyright 2016 American Chemical Society)

imaging moieties. However, the exact size of molecules that can easily transverse vascular pores from the bloodstream and reach tumor tissue is unclear. It could be probably limited to protein size (< 20 nm). Studies have shown that NPs larger than 100 nm could not make their way to go across the vascular endothelium [134] unless cell endothelium is traumatized by radiation or heating [135]. As such, smaller-sized NPs with complex and multifunctional structures are required to enter the vasculature and ultimately eliminate cancer cells.

Another feature concerned is the specific pathophysiological changes in cancer cells. In order to achieve successful treatment of tumors, it is crucial to combine with one or more therapeutic agents that can be effectively delivered to specifically kill abnormal cells without side effect. However, it still remains limitations, such as most chemotherapeutics agents will stop cell growth or induce apoptosis if activated accidentally in normal cells. Since tumors have developed mechanisms to avoid anticancer drugs, such as molecular pumps [136, 137], higher doses of therapeutic agents are often required for tumors therapy. To prevent tumors from developing resistance mechanism, therapeutic agents often require several different mechanisms of action to work simultaneously. Finally, it is critical for the therapeutic agents to monitor the response of the residual disease to treatment immediately after treatment. This is important since some of remaining cells may regrow, leading to drug-resistant tumors. It will help eliminate the few remaining tumor cells by identifying residual disease at the end of treatment (rather than after tumor regeneration). Therefore, an ideal therapeutic agent should possess the ability of target cancer cells, image the tumor range and sense its characteristics, deliver a therapeutic, and monitor the cancer cells' response to treatment.

10.5.1 Targeted Gene Therapy

Dendrimers have been used as one of the most promising vectors for non-vial gene delivery applications due to their high-density of surface amine groups and well-defined molecular architecture. In our previous studies, Xiao et al. [138] used FA-modified Au DENPs for targeted gene delivery to cancer cells overexpressing FAR. The formed Au DENPs-FA vector could effectively compact pDNA for gene delivery applications. The transfection assays using two different pDNAs encoding luciferase (Luc) and enhanced green fluorescent protein (EGFP) demonstrated that the FA modification of Au DENPs enabled the vector to specifically deliver genes to FAR-overexpressing cancer cells. The formed Au DENPs -FA displayed a low cytotoxicity and could be used for targeted gene therapy to FAR-overexpressing cancer cells of different types. In another work, RGD-targeted Au DENPs have been used for targeted siRNA delivery to cancer cells to silence the Bcl-2 and VEGF protein expression [139]. Similarly, we developed the use of RGD-modified Au DENPs for highly efficient and specific gene delivery to stem cells [140]. The results showed that all the Au DENP vectors were capable of transfecting the human mesenchymal stem cells (hMSCs) with both pDNAs encoding luciferase (Luc) and enhanced green fluorescent protein (EGFP) and pDNAs encoding the human bone morphogenetic protein-2 gene for osteogenic differentiation of stem cells.

10.5.2 Targeted Chemotherapy

Doxorubicin (DOX) is one of the most widely used drugs in cancer chemotherapy [141]. Wang and coworkers [142] reported a general approach to using multifunctional PAMAM dendrimer-based platform to encapsulate DOX for targeted cancer therapy. In this study, G5 PAMAM dendrimers were covalently modified with FI and FA and acetylated to neutralize their remaining terminal amines (G5.NHAc-FI-FA), followed by encapsulation of DOX for targeted delivery to FAR-overexpressing cancer cells. The formed G5.NHAc-FI-FA dendrimer could encapsulate approximately one DOX molecule per dendrimer with good water solubility and stability and released the DOX in a sustained manner. The G5. NHAc-FI-FA/DOX complexes could target and display specific therapeutic efficacy to FAR-overexpressing cancer cells. Using the same dendrimer carrier, other anti-cancer drugs such as 2-methoxyestradiol [143] and combretastatin A4 [144] can be encapsulated and sustainably released for targeted chemotherapy of FAR-expressing cancer cells. To further improve the targeting specificity, dendrimers were covalently linked with a targeting ligand LA via a PEG spacer [145]. For brain tumors, a dendrimer-based dual-targeting drug delivery system was designed by conjugating both transferrin and wheat germ agglutinin on the periphery of PEGylated PAMAM dendrimers for DOX loading. As opposed to free drug or the single-targeting delivery system, the dual-targeting system displayed an enhanced transportation ratio across the BBB and improved accumulation of DOX in the tumor sites [146].

In order to improve the drug loading capacity, by taking unique structural characteristics of the dendrimers into consideration, Wen et al. [23] reported the use of multifunctional dendrimer-modified multi-walled carbon nanotubes (MWCNTs) for targeted and pH-responsive delivery of DOX into cancer cells. In this work, amine-terminated G5 PAMAM dendrimers modified with FI and FA were covalently linked to acid-treated MWCNTs, followed by acetylation of the remaining dendrimer terminal amines to neutralize the positive surface potential. The MWCNT/G5.NHAc-FI-FA had a high drug payload and encapsulation efficiency both up to 97.8%, displayed an acidic pH-triggered fast DOX release profile, and were able to target to FAR-overexpressing cancer cells to inhibit the cancer cell growth.

Besides the physical encapsulation of drugs within dendrimer interior, drugs can also be covalently conjugated onto the surface of dendrimers for stimuli-responsive delivery to cancer cells. Recently [147], partially acetylated and FA-modified G5 PAMAM dendrimers were covalently conjugated with DOX onto their periphery through a pH-sensitive cis-aconityl linkage to create the G5.NHAc-FA-DOX conjugates. The DOX release from the G5.NHAc-FA-DOX conjugates followed an acid-triggered manner with a higher release rate under an acidic pH condition. In vitro cytotoxicity evaluation demonstrated that the therapeutic activity of dendrimer-DOX conjugates against cancer cells was solely related to the DOX drug released, and the FA conjugation onto the dendrimers allowed specific targeted inhibition of cancer cells. In another work [40], polymeric prodrug of DOX with potential liver-targeting ability and with a pH-triggered drug release profile was designed. Linear dendritic block copolymer composed of PAMAM dendrimer and galactose-modified PEG chain was synthesized, followed by DOX coupling via an acid-labile hydrazone linker. The constructed conjugates showed rapid acid pH-triggered DOX release and higher cytotoxicity against human hepatoma cells (Bel-7402) than the non-targeted conjugates and displayed better in vivo antitumor efficacy than free DOX due to the galactose receptor-mediated enhanced tumor uptake. The conjugates could target hepatoma tissue as demonstrated by MR imaging to have prolonged retention time due to the galactose receptor-mediated liver targeting.

10.6 Dendrimer-Based Theranostic System

The earlier dendrimer-based drug delivery systems generally include simple association, complexation, or encapsulation of small molecular pharmaceuticals with dendrimers. The primary consideration was to increase the residence time and adjust biodistribution profiles and targeting therapy to diseased sites compared to free drug administration. Nowadays, attention has been paid to more advanced complexes with dedicate design that enable real-time diagnosis, imaging, and targeting using dendrimer nanotechnology. These multifunctional integrated nanoplatforms can be readily applied for theranostics of cancer after merging radio-dense elements for molecular imaging applications and anticancer drugs for therapeutic applications into a dendrimer-based platform [143, 148, 149].

Recently, G5 PAMAM dendrimers pre-functionalized with both targeting ligand FA and anticancer drug methotrexate (G5-FA-MTX) were utilized to synthesize AuNPs for CT imaging and chemotherapy of cancer cells [150]. The formed Au DENPs with an Au core size of 1.7 nm possessed good colloidal stability and better X-ray attenuation property than Omnipaque for CT imaging of cancer cells. Further, the FA-mediated targeting afforded the Au DENPs with specific targeting to FAR-overexpressing cancer cells for specific therapy of the target cells. To fully demonstrate the potential of multifunctional Au DENPs for cancer theranostics, an Au DENP-based platform was developed by our group [149] for targeted cancer CT imaging and therapy in vitro and in vivo. In this work, G5.NH_2 dendrimers were covalently conjugated with FI, PEGylated α-tocopheryl succinate (α-TOS), and PEGylated FA, entrapped with AuNPs and acetylated to shield their remaining terminal amines. The formed multifunctional Au DENPs had an Au core size of 3.3 nm, carried 9.8 α-TOS molecules per dendrimer, and displayed good water dispersibility and stability, and high X-ray attenuation intensity. The FA-mediated targeting enabled the multifunctional Au DENPs to be specifically uptaken by FAR-overexpressing cancer cells for effective targeted CT imaging of the cancer cells and the xenografted tumor model (Fig.10.7a). Likewise, the antitumor therapeutic efficacy of α-TOS was proven not to be compromised after covalent attachment onto the dendrimer surface. Instead, the formed multifunctional Au DENPs could exert their specific therapeutic efficacy of α-TOS to the FAR-overexpressing cancer cells in vitro and the xenografted tumor model in vivo (Fig.10.7b).

Similarly, in another work [151], we covalently linked G5 PAMAM dendrimers with FA and DOX through acid-sensitive cis-aconityl linkage to form G5·NHAc-FA-DOX conjugates and using the dendrimers as templates to entrap AuNPs. The prepared DOX-Au DENPs possessed an 2.8 nm Au core, 9.0 DOX moieties per dendrimer and great colloid stability under different conditions. The formed DOX-Au DENPs exhibited a pH-responsive DOX release profile with a faster DOX release rate under a slightly acidic pH condition than physiological one. The multifunctional DOX-conjugated Au DENPs afforded specific chemotherapy and CT imaging of FAR-overexpressing cancer cells in vitro.

Besides the combination of CT imaging agents and chemotherapeutic drugs for CT imaging-guided chemotherapy, Gd chelate-modified dendrimers can be used to encapsulate anticancer drug (e.g., DOX) for MR imaging-guided chemotherapy [152]. Gd(III)-based dendritic NPs have also been developed recently for specific targeted gene therapy under MR imaging guidance [153].

Nuclear medicine, utilizing radionuclides and radiolabeled compounds to diagnose and treat diseases, plays an increasingly important role in medical sciences. In a recent study, Zhu et al. [154] reported the preparation of dendrimer-based

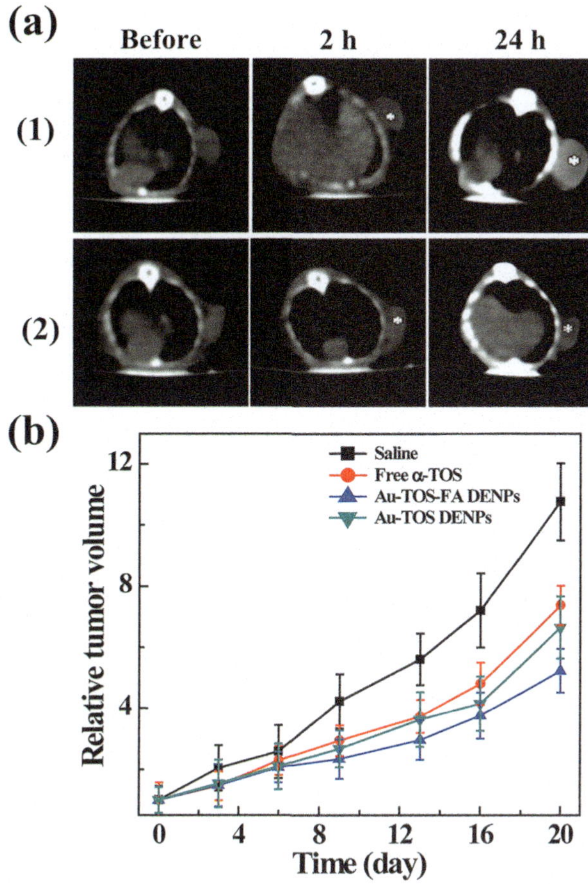

Fig. 10.7 (**a**) Representative transverse CT images of the U87MG tumor xenografts in nude mice before and after intravenous injection of Au-TOS-FA (1) and Au-TOS (2) DENPs for 2 h and 24 h, respectively. (**b**) The growth of U87MG xenografted tumors after various treatments [149]. (Reproduced with permission from Ref. 149, Copyright 2014 Elsevier)

nanodevices labeled with ^{131}I for SPECT imaging and targeted RT of tumors. Specifically, amine-terminated G5 PAMAM dendrimers were modified with 3-(4-′-hydroxyphenyl)propionic acid-OSu (HPAO) and PEGylated FA, followed by acetylation of the dendrimer leftover surface amines and labeling of ^{131}I. Prior to^{131}I labeling, the G5·NHAc-HPAO-PEG-FA dendrimers conjugated with approximately 9.4 HPAO moieties per dendrimer were proven to be noncytotoxic at a concentration up to 20 μM and to target FAR-overexpressing cancer cells through the FA ligands. Radioactive^{131}I could be efficiently labeled onto the dendrimer platform through the phenol group for targeted SPECT imaging and RT of a FAR-overexpressing xenografted tumor model in vivo. Using the same approach, Zhao et al. [155] and Cheng et al. [156] reported the synthesis of chlorotoxin- and *Buthus martensii* Karsch chlorotoxin-conjugated multifunctional dendrimers labeled with^{131}I for targeted SPECT imaging and RT of gliomas overexpressing matrix metallopeptidase 2.

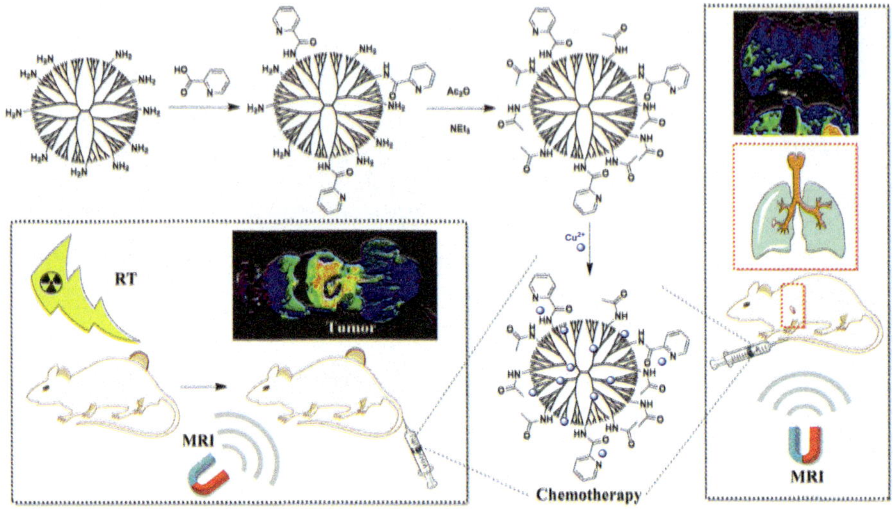

Fig. 10.8 Schematic diagram illustrating the synthesis of Cu(II) complexes with Pyr-functionalized PAMAM dendrimers for the RT-enhanced T_1 MR imaging and chemotherapy of tumors and tumor metastasis [157]. (Reproduced with permission from Ref. 157, Copyright 2019 American Chemical Society)

To minimize the components of the therapeutic and imaging elements, it is essential to create the dendrimer-based nanodevices with the same component that can be used for both therapy and diagnosis functionalities. In our recent study [157], we described the use of G5 PAMAM dendrimers modified with pyridine for Cu (II) complexation, and the developed PAMAM dendrimer-coordinated Cu (II) complexes were able to be used as a nanoplatform for RT-enhanced MR imaging and chemotherapy of tumors (Fig. 10.8). The dendrimer-copper complexes could effectively inhibit the proliferation of different tumor cell lines, promote the production of reactive oxygen species in the cells, and further lead to cell apoptosis. Since the copper ions can shorten the T_1 relaxation time, the complexes displayed an r_1 of 0.7024 mM^{-1} s^{-1}. Compared with copper salt ($CuCl_2$) and small molecule complex copper-pyridine, dendrimer-copper complexes exhibited better MR imaging effect due to its good water solubility and nanometer size. Furthermore, low-dose RT could enhance the MR imaging sensitivity as well as chemotherapy effects in a mouse breast cancer subcutaneous xenograft model and a lung tumor metastasis model.

10.7 Conclusions and Outlooks

In summary, this chapter has reviewed the recent advances in dendrimer-based nanodevices as tumor targeting systems in nanomedicine. Dendrimers conjugated with various imaging and therapeutic agents were used for different types of targeting delivery. As described, passive and active targeting strategies may be systematically engineered to provide innovative ideas for optimizing theranostic platform and improving the delivery efficiency. After incorporated with several imaging agents, dendrimers could be used in molecular imaging, including single-mode CT, MR, and PET and dual-mode CT/MR, T_1/T_2MR, SPECT/MR, SPECT/CT, as well as tri-modal targeted imaging applications. There appears to be considerable promise and benefits for the dendrimer-based platforms used in current nanomedicine and nanotherapeutics, including enhanced therapeutic results and reduced off-target side effects, compared with the whole-body exposure to traditional pharmaceuticals.

Despite substantial progresses that have been made in the development of dendrimer-based nanosystems, great challenges still remain in many important aspects of nanomedicine. For instance, some of the stimuli-responsive dendrimer-based drug delivery systems lack cancer targeting specificity and in vivo applications. In addition, the dendrimer-based organic-inorganic hybrid systems have not been integrated well for simultaneous imaging and therapy of various types of cancer. Further, the clinical translation of dendrimer-based systems has not yet been well achieved. Such challenges include the following aspects: (i) establish well-defined and better characterized NPs that may be synthesized reproducibly with standardized methodologies; (ii) mask the dendrimer-based particles via surface chemistry beyond PEGylation to reduce side effects and avoid unnecessary activation; (iii) develop smart vehicles, such as pH-responsive, temperature-sensitive, and light-triggered ones, to be responsive to tumor microenvironment; (iv) control the size of the particle that enables ideal kidney excretion and effective tumor penetration; and (v) establish intrinsic features and surface chemistry to ensure longer circulation time. All in all, these challenges could drive effective collaborations between scientists, engineers, and physicians to develop multifunctional dendrimer-based tumor targeting systems for different biomedical applications. We hope that dendrimer-based nanoplatforms could serve as optimized therapy agents as well as quantitative imaging agents to meet the global medical needs with cost-effectiveness.

Acknowledgments This research is financially supported by the Shanghai Leading Talents Program, National Natural Science Foundation of China (21773026 and 81761148028), the Science and Technology Commission of Shanghai Municipality (17540712000, 19XD1400100, 19410740200 and 18520750400), and the 111 project (BP0719035).

References

1. Kannan RM, Nance E, Kannan S, Tomalia DA (2014) Emerging concepts in dendrimer-based nanomedicine: from design principles to clinical applications. J Intern Med 276(6):579–617
2. Svenson S, Tomalia DA (2012) Dendrimers in biomedical applications-reflections on the field. Adv Drug Deliv Rev 64:102–115

3. Hu JJ, Xu TW, Cheng YY (2012) NMR insights into Dendrimer-based host–guest systems. Chem Rev 112(7):3856–3891
4. Cheng YY, Zhao LB, Li YW, Xu TW (2011) Design of biocompatible dendrimers for cancer diagnosis and therapy: current status and future perspectives. Chem Soc Rev 40(5):2673–2703
5. Tomalia DA (2012) Interview: An architectural journey: from trees, dendrons/dendrimers to nanomedicine. Nanomedicine 7(7):953–956
6. Tomalia DA, Christensen JB, Boas U (2012) Dendrimers, dendrons, and dendritic polymers: discovery, applications, and the future. Cambridge University Press, Cambridge
7. Tomalia DA (2005) Birth of a new macromolecular architecture: dendrimers as quantized building blocks for nanoscale synthetic polymer chemistry. Prog Polym Sci 30(3):294–324
8. Lee CC, MacKay JA, Fréchet JMJ, Szoka FC (2005) Designing dendrimers for biological applications. Nat Biotechnol 23:1517–1526
9. Somani S, Dufès C (2014) Applications of dendrimers for brain delivery and cancer therapy. Nanomedicine 9(15):2403–2414
10. Sun QH, Sun XR, Ma XP, Zhou ZX, Jin E, Zhang B, Shen YQ, Van Kirk EA, Murdoch WJ, Lott JR, Lodge TP, Radosz M, Zhao YL (2014) Integration of nanoassembly functions for an effective delivery cascade for cancer drugs. Adv Mater 26(45):7615–7621
11. Zhou ZX, Ma XP, Murphy CJ, Jin E, Sun QH, Shen YQ, Van Kirk EA, Murdoch WJ (2014) Molecularly precise dendrimer–drug conjugates with tunable drug release for cancer therapy. Angew Chem Int Ed 53(41):10949–10955
12. Ye MZ, Qian Y, Tang JB, Hu HJ, Sui MH, Shen YQ (2013) Targeted biodegradable dendritic MRI contrast agent for enhanced tumor imaging. J Control Release 169(3):239–245
13. Tian WD, Ma YQ (2013) Theoretical and computational studies of dendrimers as delivery vectors. Chem Soc Rev 42(2):705–727
14. Lim J, Simanek EE (2012) Triazine dendrimers as drug delivery systems: from synthesis to therapy. Adv Drug Deliv Rev 64(9):826–835
15. Wang Y, Zhao Q, Zhang H, Yang S, Jia XR (2014) A novel poly(amido amine)-dendrimer-based hydrogel as a mimic for the extracellular matrix. Adv Mater 26(24):4163–4167
16. Ghobril C, Charoen K, Rodriguez EK, Nazarian A, Grinstaff MW (2013) A dendritic thioester hydrogel based on thiol–thioester exchange as a dissolvable sealant system for wound closure. Angew Chem Int Ed 52(52):14070–14074
17. Dufès C, Uchegbu IF, Schätzlein AG (2005) Dendrimers in gene delivery. Adv Drug Deliv Rev 57(15):2177–2202
18. Yang JP, Zhang Q, Chang H, Cheng YY (2015) Surface-engineered dendrimers in gene delivery. Chem Rev 115(11):5274–5300
19. Haensler J, Szoka FC Jr (1993) Polyamidoamine cascade polymers mediate efficient transfection of cells in culture. Bioconjug Chem 4(5):372–379
20. Arima H, Motoyama K, Higashi T (2013) Sugar-appended polyamidoamine dendrimer conjugates with cyclodextrins as cell-specific non-viral vectors. Adv Drug Deliv Rev 65 (9):1204–1214
21. Gray WD, Wu RJ, Yin X, Zhou JH, Davis ME, Luo Y (2013) Dendrimeric bowties featuring hemispheric-selective decoration of ligands for microRNA-based therapy. Biomacromolecules 14(1):101–109
22. Zhang WL, Li N, Huang J, Yu JH, Wang DX, Li YP, Liu SY (2010) Gadolinium-conjugated FA-PEG-PAMAM-COOH nanoparticles as potential tumor-targeted circulation-prolonged macromolecular MRI contrast agents. J Appl Polym Sci 118(3):1805–1814
23. Wen SH, Liu H, Cai HD, Shen MW, Shi XY (2013) Targeted and pH-responsive delivery of doxorubicin to cancer cells using multifunctional dendrimer-modified multi-walled carbon nanotubes. Adv Healthc Mater 2(9):1267–1276
24. Shi XY, Wang SH, Meshinchi S, Van Antwerp ME, Bi XD, Lee I, Baker JR (2007) Dendrimer-entrapped gold nanoparticles as a platform for cancer-cell targeting and imaging. Small 3(7):1245–1252

25. Hu JJ, Hu K, Cheng YY (2016) Tailoring the dendrimer core for efficient gene delivery. Acta Biomater 35:1–11
26. Xu XY, Ho W, Zhang XQ, Bertrand N, Farokhzad O (2015) Cancer nanomedicine: from targeted delivery to combination therapy. Trends Mol Med 21(4):223–232
27. Bamrungsap S, Zhao ZL, Chen T, Wang L, Li CM, Fu T, Tan WH (2012) Nanotechnology in therapeutics: a focus on nanoparticles as a drug delivery system. Nanomedicine 7 (8):1253–1271
28. Cole AJ, Yang VC, David AE (2011) Cancer theranostics: the rise of targeted magnetic nanoparticles. Trends Biotechnol 29(7):323–332
29. Danhier F, Feron O, Preat V (2010) To exploit the tumor microenvironment: passive and active tumor targeting of nanocarriers for anti-cancer drug delivery. J Control Release 148 (2):135–146
30. Zhu JY, Shi XY (2013) Dendrimer-based nanodevices for targeted drug delivery applications. J Mater Chem B 1(34):4199–4211
31. Kukowska-Latallo JF, Candido KA, Cao ZY, Nigavekar SS, Majoros IJ, Thomas TP, Balogh LP, Khan MK, Baker JR (2005) Nanoparticle targeting of anticancer drug improves therapeutic response in animal model of human epithelial cancer. Cancer Res 65(12):5317–5324
32. Majoros IJ, Myc A, Thomas T, Mehta CB, Baker JR (2006) PAMAM dendrimer-based multifunctional conjugate for cancer therapy: synthesis, characterization, and functionality. Biomacromolecules 7(2):572–579
33. Majoros IJ, Thomas TP, Mehta CB, Baker JR (2005) Poly(amidoamine) dendrimer-based multifunctional engineered nanodevice for cancer therapy. J Med Chem 48(19):5892–5899
34. Singh P, Gupta U, Asthana A, Jain NK (2008) Folate and folate-PEG-PAMAM dendrimers: synthesis, characterization, and targeted anticancer drug delivery potential in tumor bearing mice. Bioconjug Chem 19(11):2239–2252
35. Thomas TP, Huang BH, Choi SK, Silpe JE, Kotlyar A, Desai AM, Zong H, Gam J, Joice M, Baker JRJ (2012) Polyvalent dendrimer-methotrexate as a folate receptor-targeted cancer therapeutic. Mol Pharm 9(9):2669–2676
36. Shukla R, Thomas TP, Peters J, Kotlyar A, Myc A, Baker JR (2005) Tumor angiogenic vasculature targeting with PAMAM dendrimer-RGD conjugates. Chem Commun 14 (46):5739–5741
37. Liu J, Gray WD, Davis ME, Luo Y (2012) Peptide- and saccharide-conjugated dendrimers for targeted drug delivery: a concise review. Interface Focus 2(3):307–324
38. Li ZM, Huang P, Zhang XJ, Lin J, Yang S, Liu B, Gao F, Xi P, Ren QS, Cui DX (2010) RGD-conjugated dendrimer-modified gold nanorods for in vivo tumor targeting and photothermal therapy. Mol Pharm 7(1):94–104
39. Thomas TP, Patri AK, Myc A, Myaing MT, Ye JY, Norris TB, Baker JR (2004) In vitro targeting of synthesized anti body-conjugated dendrimer nanoparticles. Biomacromolecules 5 (6):2269–2274
40. Huang J, Gao F, Tang XX, Yu JH, Wang DX, Liu SY, Li YP (2010) Liver-targeting doxorubicin-conjugated polymeric prodrug with pH-triggered drug release profile. Polym Int 59(10):1390–1396
41. Medina SH, Tekumalla V, Chevliakov MV, Shewach DS, Ensminger WD, El-Sayed MEH (2011) N-acetylgalactosamine-functionalized dendrimers as hepatic cancer cell-targeted carriers. Biomaterials 32(17):4118–4129
42. Selim KMK, Ha YS, Kim SJ, Chang YM, Kim TJ, Lee GH, Kang IK (2007) Surface modification of magnetite nanoparticles using lactobionic acid and their interaction with hepatocytes. Biomaterials 28(4):710–716
43. Shi XY, Wang SH, Swanson SD, Ge S, Cao ZY, Van Antwerp ME, Landmark KJ, Baker JR (2008) Dendrimer-functionalized shell-crosslinked iron oxide nanoparticles for in-vivo magnetic resonance imaging of tumors. Adv Mater 20(9):1671–1678

44. Zhao YL, Liu S, Li YP, Jiang W, Chang YL, Pan S, Fang XX, Wang YA, Wang JY (2010) Synthesis and grafting of folate-PEG-PAMAM conjugates onto quantum dots for selective targeting of folate-receptor-positive tumor cells. J Colloid Interface Sci 350(1):44–50
45. Kono K, Liu MJ, Frechet JMJ (1999) Design of dendritic macromolecules containing folate or methotrexate residues. Bioconjug Chem 10(6):1115–1121
46. Sudimack J, Lee RJ (2000) Targeted drug delivery via the folate receptor. Adv Drug Deliv Rev 41(2):147–162
47. Majoros IJ, Williams CR, Becker A, Baker JRJ (2009) Methotrexate delivery via folate targeted dendrimer-based nanotherapeutic platform. Wiley Interdiscip Rev Nanomed Nanobiotechnol 1(5):502–510
48. Choi SK, Thomas T, Li MH, Kotlyar A, Desai A, Baker JRJ (2010) Light-controlled release of caged doxorubicin from folate receptor-targeting PAMAM dendrimer nanoconjugate. Chem Commun 46(15):2632–2634
49. Myc A, Kukowska-Latallo J, Cao P, Swanson B, Battista J, Dunham T, Baker JR (2010) Targeting the efficacy of a dendrimer-based nanotherapeutic in heterogeneous xenograft tumors in vivo. Anti-Cancer Drugs 21(2):186–192
50. Shukla R, Hill E, Shi XY, Kim J, Muniz MC, Sun K, Baker JRJ (2008) Tumor microvasculature targeting with dendrimer-entrapped gold nanoparticles. Soft Matter 4(11):2160–2163
51. Lesniak WG, Kariapper MST, Nair BM, Tan W, Hutson A, Balogh LP, Khan MK (2007) Synthesis and characterization of PAMAM dendrimer-based multifunctional nanodevices for targeting alpha(v)beta(3) integrins. Bioconjug Chem 18(4):1148–1154
52. Hill E, Shukla R, Park SS, Baker JR (2007) Synthetic PAMAM-RGD conjugates target and bind to odontoblast-like MDPC 23 cells and the predentin in tooth organ cultures. Bioconjug Chem 18(6):1756–1762
53. Waite CL, Roth CM (2009) PAMAM-RGD conjugates enhance siRNA delivery through a multicellular spheroid model of malignant glioma. Bioconjug Chem 20(10):1908–1916
54. Patri AK, Myc A, Beals J, Thomas TP, Bander NH, Baker JR (2004) Synthesis and in vitro testing of J591 antibody-dendrimer conjugates for targeted prostate cancer therapy. Bioconjug Chem 15(6):1174–1181
55. Shukla R, Thomas TP, Peters JL, Desai AM, Kukowska-Latallo J, Patri AK, Kotlyar A, Baker JR (2006) HER2 specific tumor targeting with dendrimer conjugated anti-HER2 mAb. Bioconjug Chem 17(5):1109–1115
56. Mekuria SL, Tsai HC (2015) Preparation of self-assembled core-shell nano structure of conjugated generation 4.5 poly (amidoamine) dendrimer and monoclonal Anti-IL-6 antibody as bioimaging probe. Colloid Surf B: Biointerfaces 135:253–260
57. Mekuria SL, Debele TA, Chou HY, Tsai HC (2016) IL-6 antibody and RGD peptide conjugated poly(amidoamine) dendrimer for targeted drug delivery of HeLa cells. J Phys Chem B 120(1):123–130
58. Xu LY, Zhang H, Wu YL (2014) Dendrimer advances for the central nervous system delivery of therapeutics. ACS Chem Neurosci 5(1):2–13
59. Yellepeddi VK, Kumar A, Palakurthi S (2009) Biotinylated poly(amido)amine (PAMAM) dendrimers as carriers for drug delivery to ovarian cancer cells in vitro. Anticancer Res 29 (8):2933–2943
60. Yang WJ, Cheng YY, Xu TW, Wang XY, Wen LP (2009) Targeting cancer cells with biotin-dendrimer conjugates. Eur J Med Chem 44(2):862–868
61. Yellepeddi VK, Kumar A, Maher DM, Chauhan SC, Vangara KK, Palakurthi S (2011) Biotinylated PAMAM dendrimers for intracellular delivery of cisplatin to ovarian cancer: role of SMVT. Anticancer Res 31(3):897–906
62. Luo SH, Kansara VS, Zhu XD, Mandava NK, Pal D, Mitra AK (2006) Functional characterization of sodium-dependent multivitamin transporter in MDCK-MDR1 cells and its utilization as a target for drug delivery. Mol Pharm 3(3):329–339
63. Vadlapudi AD, Vadlapatla RK, Mitra AK (2012) Sodium dependent multivitamin transporter (SMVT): a potential target for drug delivery. Curr Drug Targets 13(7):994–1003

64. Janoria KG, Hariharan S, Paturi D, Pal D, Mitra AK (2006) Biotin uptake by rabbit corneal epithelial cells: role of sodium-dependent multivitamin transporter (SMVT). Curr Eye Res 31 (10):797–809
65. Lusic H, Grinstaff MW (2013) X-ray-computed tomography contrast agents. Chem Rev 113 (3):1641–1666
66. Rabin O, Perez JM, Grimm J, Wojtkiewicz G, Weissleder R (2006) An X-ray computed tomography imaging agent based on long-circulating bismuth sulphide nanoparticles. Nat Mater 5(2):118–122
67. Kim D, Park SS, Lee JH, Jeong YY, Jon S (2007) Antibiofouling polymer-coated gold nanoparticles as a contrast agent for in vivo x-ray computed tomography imaging. J Am Chem Soc 129(24):7661–7665
68. Popovtzer R, Agrawal A, Kotov NA, Popovtzer A, Balter J, Carey TE, Kopelman R (2008) Targeted gold nanoparticles enable molecular CT imaging of cancer. Nano Lett 8 (12):4593–4596
69. Aviv H, Bartling S, Kieslling F, Margel S (2009) Radiopaque iodinated copolymeric nanoparticles for X-ray imaging applications. Biomaterials 30(29):5610–5616
70. de Vries A, Custers E, Lub J, van den Bosch S, Nicolay K, Grull H (2010) Block-copolymer-stabilized iodinated emulsions for use as CT contrast agents. Biomaterials 31(25):6537–6544
71. Hallouard F, Anton N, Choquet P, Constantinesco A, Vandamme T (2010) Iodinated blood pool contrast media for preclinical X-ray imaging applications – a review. Biomaterials 31 (24):6249–6268
72. Kojima C, Umeda Y, Ogawa M, Harada A, Magata Y, Kono K (2010) X-ray computed tomography contrast agents prepared by seeded growth of gold nanoparticles in PEGylated dendrimer. Nanotechnology 21(24):245104
73. Nune SK, Gunda P, Thallapally PK, Lin YY, Forrest ML, Berkland CJ (2009) Nanoparticles for biomedical imaging. Expert Opin Drug Deliv 6(11):1175–1194
74. Barrett T, Ravizzini G, Choyke PL, Kobayashi H (2009) Dendrimers in medical nanotechnology application of dendrimer molecules in bioimaging and cancer treatment. IEEE Eng Med Biol 28(1):12–22
75. Hainfeld JF, Slatkin DN, Focella TM, Smilowitz HM (2006) Gold nanoparticles: a new X-ray contrast agent. Br J Radiol 79(939):248–253
76. Chien CC, Chen HH, Lai SF, Wu KC, Cai XQ, Hwu YK, Petibois C, Chu Y, Margaritondo G (2012) Gold nanoparticles as high-resolution X-ray imaging contrast agents for the analysis of tumor-related micro-vasculature. J Nanobiotechnol 10:10
77. Wang H, Zheng LF, Guo R, Peng C, Shen MW, Shi XY, Zhang GX (2012) Dendrimer-entrapped gold nanoparticles as potential CT contrast agents for blood pool imaging. Nanoscale Res Lett 7:190
78. Liu H, Wang H, Guo R, Cao XY, Zhao JL, Luo Y, Shen MW, Zhang GX, Shi XY (2010) Size-controlled synthesis of dendrimer-stabilized silver nanoparticles for X-ray computed tomography imaging applications. Polym Chem 1(10):1677–1683
79. Wang H, Zheng LF, Peng C, Shen MW, Shi XY, Zhang GX (2013) Folic acid-modified dendrimer-entrapped gold nanoparticles as nanoprobes for targeted CT imaging of human lung adenocarcinoma. Biomaterials 34(2):470–480
80. Peng C, Qin JB, Zhou BQ, Chen Q, Shen MW, Zhu MF, Lu XW, Shi XY (2013) Targeted tumor CT imaging using folic acid-modified PEGylated dendrimer-entrapped gold nanoparticles. Polym Chem 4(16):4412–4424
81. Liu H, Wang H, Xu YH, Guo R, Wen SH, Huang YP, Liu WN, Shen MW, Zhao JL, Zhang GX, Shi XY (2014) Lactobionic acid-modified dendrimer-entrapped gold nanoparticles for targeted Computed Tomography imaging of human hepatocellular carcinoma. ACS Appl Mater Interfaces 6(9):6944–6953
82. Cao YY, He Y, Liu H, Luo Y, Shen MW, Xia JD, Shi XY (2015) Targeted CT imaging of human hepatocellular carcinoma using low-generation dendrimer-entrapped gold nanoparticles modified with lactobionic acid. J Mater Chem B 249:286–295

83. Li ZM, Huang P, He R, Lin J, Yang S, Zhang XJ, Ren QS, Cui DX (2010) Aptamer-conjugated dendrimer-modified quantum dots for cancer cell targeting and imaging. Mater Lett 64(3):375–378
84. Li ZM, Huang P, Lin J, He R, Liu B, Zhang XM, Yang S, Xi P, Zhang XJ, Ren QS, Cui DX (2010) Arginine-Glycine-Aspartic acid-conjugated dendrimer-modified quantum dots for targeting and imaging melanoma. J Nanosci Nanotechnol 10(8):4859–4867
85. Laurent S, Forge D, Port M, Roch A, Robic C, Vander Elst L, Muller RN (2008) Magnetic iron oxide nanoparticles: synthesis, stabilization, vectorization, physicochemical characterizations, and biological applications. Chem Rev 108(6):2064–2110
86. Schiffmann R, Van der Knaap MS (2009) Invited article: an MRI-based approach to the diagnosis of white matter disorders. Neurology 72(8):750–759
87. Serres S, Soto MS, Hamilton A, McAteer MA, Carbonell W, Robson MD, Ansorge O, Khrapitchev A, Bristow C, Balathasan L (2012) Molecular MRI enables early and sensitive detection of brain metastases. Proc Natl Acad Sci U S A 109(17):6674–6679
88. Langereis S, Dirksen A, Hackeng TM, Van Genderen MH, Meijer E (2007) Dendrimers and magnetic resonance imaging. New J Chem 31(7):1152–1160
89. Cai HD, An X, Cui J, Li JC, Wen SH, Li KA, Shen MW, Zheng LF, Zhang GX, Shi XY (2013) Facile hydrothermal synthesis and surface functionalization of polyethyleneimine-coated iron oxide nanoparticles for biomedical applications. ACS Appl Mater Interfaces 5(5):1722–1731
90. Alexiou C, Jurgons R, Seliger C, Iro H (2006) Medical applications of magnetic nanoparticles. J Nanosci Nanotechnol 6(9–10):2762–2768
91. Sosnovik DE, Weissleder R (2007) Emerging concepts in molecular MRI. Curr Opin Biotechnol 18(1):4–10
92. Na HB, Song IC, Hyeon T (2009) Inorganic nanoparticles for MRI contrast agents. Adv Mater 21(21):2133–2148
93. Mahajan S, Koul V, Choudhary V, Shishodia G, Bharti AC (2013) Preparation and *in vitro* evaluation of folate-receptor-targeted spion-polymer micelle hybrids for MRI contrast enhancement in cancer imaging. Nanotechnology 24(1):015603
94. McMahon MT, Bulte JWM (2018) Two decades of dendrimers as versatile MRI agents: a tale with and without metals. Wiley Interdiscip Rev Nanomed Nanobiotechnol 10(3):e1496
95. Wiener EC, Brechbiel MW, Brothers H, Magin RL, Gansow OA, Tomalia DA, Lauterbur PC (1994) Dendrimer-based metal-chelates – a new class of magnetic-resonance-imaging contrast agents. Magn Reson Med 31(1):1–8
96. Zhu J, Gale EM, Atanasova I, Rietz TA, Caravan P (2014) Hexameric Mn-II dendrimer as MRI contrast agent. Chem Eur J 20(44):14507–14513
97. Aime S, Caravan P (2009) Biodistribution of gadolinium-based contrast agents, including gadolinium deposition. J Magn Reson Imaging 30(6):1259–1267
98. Kanal E, Tweedle MF (2015) Residual or retained gadolinium: practical implications for radiologists and our patients. Radiology 275(3):630–634
99. Xu RZ, Wang YL, Wang XL, Jeong EK, Parker DL, Lu ZR (2007) In vivo evaluation of a PAMAM-Cystamine-(Gd-DO3A) conjugate as a biodegradable macromolecular MRI contrast agent. Exp Biol Med 232(8):1081–1089
100. Artemov D (2003) Molecular magnetic resonance imaging with targeted contrast agents. J Cell Biochem 90(3):518–524
101. Kobayashi H, Sato N, Saga T, Nakamoto Y, Ishimori T, Toyama S, Togashi K, Konishi J, Brechbiel MW (2000) Monoclonal antibody-dendrimer conjugates enable radiolabeling of antibody with markedly high specific activity with minimal loss of immunoreactivity. Eur J Nucl Med 27(9):1334–1339
102. Xu H, Regino CA, Koyama Y, Hama Y, Gunn AJ, Bernardo M, Kobayashi H, Choyke PL, Brechbiel MW (2007) Preparation and preliminary evaluation of a biotin-targeted, lectin-targeted dendrimer-based probe for dual-modality magnetic resonance and fluorescence imaging. Bioconjug Chem 18(5):1474–1482

103. Han L, Li JF, Huang SX, Huang RQ, Liu SH, Hu X, Yi PW, Shan D, Wang XX, Lei H (2011) Peptide-conjugated polyamidoamine dendrimer as a nanoscale tumor-targeted T_1 magnetic resonance imaging contrast agent. Biomaterials 32(11):2989–2998
104. Tan MQ, Wu XM, Jeong EK, Chen QJ, Lu ZR (2010) Peptide-targeted nanoglobular Gd-DOTA monoamide conjugates for magnetic resonance cancer molecular imaging. Biomacromolecules 11(3):754–761
105. Swanson SD, Kukowska-Latallo JF, Patri AK, Chen CY, Ge S, Cao ZY, Kotlyar A, East AT, Baker JR (2008) Targeted gadolinium-loaded dendrimer nanoparticles for tumor-specific magnetic resonance contrast enhancement. Int J Nanomed 3(2):201–210
106. Wolfenden ML, Cloninger MJ (2005) Mannose/glucose-functionalized dendrimers to investigate the predictable tunability of multivalent interactions. J Am Chem Soc 127 (35):12168–12169
107. van Baal I, Malda H, Synowsky SA, van Dongen JL, Hackeng TM, Merkx M, Meijer E (2005) Multivalent peptide and protein dendrimers using native chemical ligation. Angew Chem, Int Ed 44(32):5052–5057
108. Choi Y, Mecke A, Orr BG, Banaszak Holl MM, Baker JR (2004) DNA-directed synthesis of generation 7 and 5 PAMAM dendrimer nanoclusters. Nano Lett 4(3):391–397
109. Antony AC (1992) The biological chemistry of folate receptors. Blood 79(11):2807–2820
110. Chen WT, Thirumala R, Shih TF, Chen RC, Tu SY, Lin CI, Yang PC (2010) Dynamic contrast-enhanced folate-receptor-targeted MR imaging using a Gd-loaded PEG-dendrimer-folate conjugate in a mouse xenograft tumor model. Mol Imaging Biol 12(2):145–154
111. Konda SD, Aref M, Wang S, Brechbiel M, Wiener EC (2001) Specific targeting of folate-dendrimer MRI contrast agents to the high affinity fo late receptor expressed in ovarian tumor xenografts. Magn Reson Mater Phys Biol Med 12(2–3):104–113
112. Kobayashi H, Kawamoto S, Saga T, Sato N, Ishimori T, Konishi J, Ono K, Togashi K, Brechbiel MW (2001) Avidin-dendrimer-(1B4M-Gd) 254: a tumor-targeting therapeutic agent for gadolinium neutron capture therapy of intraperitoneal disseminated tumor which can be monitored by MRI. Bioconjug Chem 12(4):587–593
113. Park J, Lee JJ, Jung JC, Yu DY, Oh C, Ha S, Kim TJ, Chang YM (2008) Gd-DOTA conjugate of RGD as a potential tumor-targeting MRI contrast agent. Chembiochem 9(17):2811–2813
114. Wang SH, Shi XY, Van Antwerp ME, Cao ZY, Swanson SD, Bi XD, Baker JR (2007) Dendrimer-functionalized iron oxide nanoparticles for specific targeting and imaging of cancer cells. Adv Funct Mater 17(16):3043–3050
115. Yang J, Luo Y, Xu YH, Li JC, Zhang ZX, Wang H, Shen MW, Shi XY, Zhang GX (2015) Conjugation of iron oxide nanoparticles with RGD-modified dendrimers for targeted tumor MR imaging. ACS Appl Mater Interfaces 7(9):5420–5428
116. Ghai A, Singh B, Hazari PP, Schultz MK, Parmar A, Kumar P, Sharma S, Dhawan D, Mishra AK (2015) Radiolabeling optimization and characterization of Ga-68 labeled DOTA-polyamido-amine dendrimer conjugate – animal biodistribution and PET imaging results. Appl Radiat Isot 105:40–46
117. Seo JW, Baek H, Mahakian LM, Kusunose J, Hamzah J, Ruoslahti E, Ferrara KW (2014) Cu-64-labeled LyP-1-Dendrimer for PET-CT imaging of atherosclerotic plaque. Bioconjug Chem 25(2):231–239
118. Ma WH, Fu FF, Zhu JY, Huang R, Zhu YZ, Liu ZW, Wang J, Conti PS, Shi XY, Chen K (2018) Cu-64-Labeled multifunctional dendrimers for targeted tumor PET imaging. Nanoscale 10(13):6113–6124
119. Chen H, Viel S, Ziarelli F, Peng L (2013) F-19 NMR: a valuable tool for studying biological events. Chem Soc Rev 42(20):7971–7982
120. Tirotta I, Dichiarante V, Pigliacelli C, Cavallo G, Terraneo G, Bombelli FB, Metrangolo P, Resnati G (2015) F-19 magnetic resonance imaging (MRI): from Design of Materials to clinical applications. Chem Rev 115(2):1106–1129

121. Chen Q, Wang H, Liu H, Wen S, Peng C, Shen M, Zhang G, Shi X (2015) Multifunctional dendrimer-entrapped gold nanoparticles modified with RGD peptide for targeted computed tomography/magnetic resonance dug-modal imaging of tumors. Anal Chem 87(7):3949–3956
122. Wang RZ, Luo Y, Yang SH, Lin J, Gao DM, Zhao Y, Liu JG, Shi XY, Wang XL (2016) Hyaluronic acid-modified manganese-chelated dendrimer-entrapped gold nanoparticles for the targeted CT/MR dual-mode imaging of hepatocellular carcinoma. Sci Rep 6:33844
123. Cai HD, Li KG, Li JC, Wen SH, Chen Q, Shen MW, Zheng LF, Zhang GX, Shi XY (2015) Dendrimer-assisted formation of Fe_3O_4/Au nanocomposite particles for targeted dual mode CT/MR imaging of tumors. Small 11(35):4584–4593
124. Yang H, Qin C, Yu C, Lu Y, Zhang H, Xue F, Wu D, Zhou Z, Yang S (2014) RGD-conjugated nanoscale coordination polymers for targeted T-1- and T-2-weighted magnetic resonance imaging of tumors in vivo. Adv Funct Mater 24(12):1738–1747
125. Haribabu V, Farook AS, Goswami N, Murugesan R, Girigoswami A (2016) Optimized Mn-doped iron oxide nanoparticles entrapped in dendrimer for dual contrasting role in MRI. J Biomed Mater Res Part B Appl Biomater 104(4):817–824
126. Pradhan P, Giri J, Banerjee R, Bellare J, Bahadur D (2007) Preparation and characterization of manganese ferrite-based magnetic liposomes for hyperthermia treatment of cancer. J Magn Magn Mater 311(1):208–215
127. Tang ZX, Sorensen CM, Klabunde KJ, Hadjipanayis GC (1991) Preparation of manganese ferrite fine particles from aqueous-solution. J Colloid Interface Sci 146(1):38–52
128. Qiao Z, Shi XY (2015) Dendrimer-based molecular imaging contrast agents. Prog Polym Sci 44:1–27
129. Luo Y, Zhao L, Li X, Yang J, Guo L, Zhang G, Shen M, Zhao J, Shi X (2016) The design of a multifunctional dendrimer-based nanoplatform for targeted dual mode SPECT/MR imaging of tumors. J Mater Chem B 4(45):7220–7225
130. Li X, Xiong ZG, Xu XY, Luo Y, Peng C, Shen MW, Shi XY (2016) Tc-99m-labeled multifunctional low-generation dendrimer-entrapped gold nanoparticles for targeted SPECT/CT Dual-Mode imaging of tumors. ACS Appl Mater Interfaces 8(31):19883–19891
131. Wen SH, Zhao LZ, Zhao QH, Li D, Liu CC, Yu ZB, Shen MW, Majoral JP, Mignani S, Zhao JH, Shi XY (2017) A promising dual mode SPECT/CT imaging platform based on Tc-99m-labeled multifunctional dendrimer-entrapped gold nanoparticles. J Mater Chem B 5 (21):3810–3815
132. Xu XY, Zhao LZ, Li X, Wang P, Zhao JH, Shi XY, Shen MW (2017) Targeted tumor SPECT/CT dual mode imaging using multifunctional RGD-modified low generation dendrimer-entrapped gold nanoparticles. Biomater Sci 5(12):2393–2397
133. Chen JW, Sun YQ, Chen Q, Wang L, Wang SH, Tang YQ, Shi XY, Wang H (2016) Multifunctional gold nanocomposites designed for targeted CT/MR/optical trimodal imaging of human non-small cell lung cancer cells. Nanoscale 8(28):13568–13573
134. Kong G, Braun RD, Dewhirst MW (2001) Characterization of the effect of hyperthermia on nanoparticle extravasation from tumor vasculature. Cancer Res 61(7):3027–3032
135. Kong G, Braun RD, Dewhirst MW (2000) Hyperthermia enables tumor-specific nanoparticle delivery: effect of particle size. Cancer Res 60(16):4440–4445
136. Culver KW (1994) Clinical-applications of gene-therapy for cancer. Clin Chem 40 (4):510–512
137. Rosenberg SA (1992) The immunotherapy and gene-therapy of cancer. J Clin Oncol 10 (2):180–199
138. Xiao TY, Hou WX, Cao XY, Wen SH, Shen MW, Shi XY (2013) Dendrimer-entrapped gold nanoparticles modified with folic acid for targeted gene delivery applications. Biomater Sci 1 (11):1172–1180
139. Kong LD, Wu YL, Alves CS, Shi XY (2016) Efficient delivery of therapeutic siRNA into glioblastoma cells using multifunctional dendrimer-entrapped gold nanoparticles. Nanomedicine 11(23):3103–3115
140. Kong LD, Alves CS, Hou WX, Qiu JR, Moehwald H, Tomas H, Shi XY (2015) RGD peptide-modified dendrimer-entrapped gold nanoparticles enable highly efficient and specific gene delivery to stem cells. ACS Appl Mater Interfaces 7(8):4833–4843

141. Wang YJ, Bansal V, Zelikin AN, Caruso F (2008) Templated synthesis of single-component polymer capsules and their application in drug delivery. Nano Lett 8(6):1741–1745
142. Wang Y, Cao XY, Guo R, Shen MW, Zhang ME, Zhu MF, Shi XY (2011) Targeted delivery of doxorubicin into cancer cells using a folic acid-dendrimer conjugate. Polym Chem 2 (8):1754–1760
143. Wang Y, Guo R, Cao XY, Shen MW, Shi XY (2011) Encapsulation of 2-methoxyestradiol within multifunctional poly(amidoamine) dendrimers for targeted cancer therapy. Biomaterials 32(12):3322–3329
144. Zhang ME, Guo R, Wang Y, Cao XY, Shen MW, Shi XY (2011) Multifunctional dendrimer/combretastatin A4 inclusion complexes enable in vitro targeted cancer therapy. Int J Nanomed 6:2337–2349
145. Fu FF, Wu YL, Zhu JY, Wen SH, Shen MW, Shi XY (2014) Multifunctional lactobionic acid-modified dendrimers for targeted drug delivery to liver cancer cells: investigating the role played by peg spacer. ACS Appl Mater Interfaces 6(18):16416–16425
146. He H, Li Y, Jia XR, Du J, Ying X, Lu WL, Lou JN, Wei Y (2011) PEGylated poly (amidoamine) dendrimer-based dual-targeting carrier for treating brain tumors. Biomaterials 32(2):478–487
147. Zhang ME, Zhu JY, Zheng Y, Guo R, Wang SG, Mignani S, Caminade AM, Majoral JP, Shi XY (2018) Doxorubicin-conjugated PAMAM dendrimers for ph-responsive drug release and folic acid-targeted cancer therapy. Pharmaceutics 10(3):162
148. Zheng Y, Fu FF, Zhang MG, Shen MW, Zhu MF, Shi XY (2014) Multifunctional dendrimers modified with alpha-tocopheryl succinate for targeted cancer therapy. MedChemComm 5 (7):879–885
149. Zhu JY, Zheng LF, Wen SH, Tang YQ, Shen MW, Zhang GX, Shi XY (2014) Targeted cancer theranostics using alpha-tocopheryl succinate-conjugated multifunctional dendrimer-entrapped gold nanoparticles. Biomaterials 35(26):7635–7646
150. Zheng L, Zhu J, Shen M, Chen X, Baker JR Jr, Wang SH, Zhang G, Shi X (2013) Targeted cancer cell inhibition using multifunctional dendrimer-entrapped gold nanoparticles. MedChemComm 4(6):1001–1005
151. Zhu JY, Wang GY, Alves CS, Tomas H, Long ZJ, Shen MW, Rodrigues J, Shi XY (2018) Multifunctional dendrimer-entrapped gold nanoparticles conjugated with doxorubicin for ph-responsive drug delivery and targeted computed tomography imaging. Langmuir 34 (41):12428–12435
152. Zhu J, Xiong Z, Shen M, Shi X (2015) Encapsulation of doxorubicin within multifunctional gadolinium-loaded dendrimer nanocomplexes for targeted theranostics of cancer cells. RSC Adv 5(38):30286–30296
153. Wang Q, Li J, An S, Chen Y, Jiang C, Wang X (2015) Magnetic resonance-guided regional gene delivery strategy using a tumor stroma-permeable nanocarrier for pancreatic cancer. Int J Nanomed 10:4479–4490
154. Zhu J, Zhao L, Cheng Y, Xiong Z, Tang Y, Shen M, Zhao J, Shi X (2015) Radionuclide I-131-labeled multifunctional dendrimers for targeted SPECT imaging and radiotherapy of tumors. Nanoscale 7(43):18169–18178
155. Zhao L, Zhu J, Cheng Y, Xiong Z, Tang Y, Guo L, Shi X, Zhao J (2015) Chlorotoxin-conjugated multifunctional dendrimers labeled with radionuclide I-131 for single photon emission computed tomography imaging and radiotherapy of gliomas. ACS Appl Mater Interfaces 7(35):19798–19808
156. Cheng YJ, Zhu JY, Zhao LZ, Xiong ZJ, Tang YQ, Liu CC, Guo LL, Qiao WL, Shi XY, Zhao JH (2016) I-131-labeled multifunctional dendrimers modified with BmK CT for targeted SPECT imaging and radiotherapy of gliomas. Nanomedicine 11(10):1253–1266
157. Fan Y, Zhang JL, Shi MH, Li D, Lu CH, Cao XY, Peng C, Mignani SG, Majoral JP, Shi XY (2019) Poly(amidoamine) dendrimer-coordinated copper(ii) complexes as a theranostic nanoplatform for the radiotherapy-enhanced magnetic resonance imaging and chemotherapy of tumors and tumor metastasis. Nano Lett 19(2):1216–1226

Chapter 11
Polymer-Based Tumor-targeted Nanosystems

Teoman Benli-Hoppe and Ernst Wagner

Abstract Polymer-based delivery systems for tumor targeting may revolutionize cancer therapeutic strategies by effective delivery of drugs or diagnostic agents by virtue of passive or ligand-based active targeting mechanisms. Thus polymeric carriers will hopefully surmount the drawbacks of conventional cancer therapeutics, such as undesirable biodistribution, drug resistance of cancer cells, and systemic side effects. For this purpose, intense work is underway to develop polymer-based platforms, composed of polymeric nanoparticles (NPs), polymer micelles, polymersomes, polyelectrolyte polyplexes, polymer-lipid hybrid systems, and polymer-drug/protein conjugates as a suitable option for cancer treatment. By versatile polymer chemistry, the conversion of a variety of polymers into functional carriers is possible. Although first polymeric NP systems are already approved for clinical use, the gap between the bench and the bedside remains to be enormous, mainly because of two reasons. Pharmaceutical grade production of macromolecular drug formulations sets challenging demands on precision chemistry, purification, and high-end analytics. Molecular and pharmacological heterogeneity of tumor within a patient, between different patients and between different types of cancer, have resulted in a disparity between preclinical and clinical studies. This review aims to provide a detailed insight in tumor-targeting aspects of various polymer-based nanocarrier systems, to state current trends and introduce novel concepts, presenting examples of drug delivery and bioimaging of cancer.

Keywords Bioimaging · Drug conjugates · Nanomedicine · Polymer micelles · Polyplexes

T. Benli-Hoppe · E. Wagner (✉)
Pharmaceutical Biotechnology, Department of Pharmacy, Ludwig-Maximilians-Universität, Munich, Germany

Nanosystems Initiative Munich (NIM), Munich, Germany
e-mail: teoman.benli-hoppe@cup.uni-muenchen.de; ernst.wagner@cup.uni-muenchen.de

R. Huang, Y. Wang (eds.), *New Nanomaterials and Techniques for Tumor-targeted Systems*, https://doi.org/10.1007/978-981-15-5159-8_11

11.1 Introduction

At present, numerous polymeric nanosystems are being developed as cancer therapeutics. A broad spectrum of possible chemical modifications facilitates dynamic designs, where tumor-specific pharmacology [1, 2] and microenvironment can modulate improved delivery and release at the tumor site [3, 4]. Applied polymeric systems span the full nanomaterial size scale, from single polymer chains and drug conjugates up to large nanoparticle assemblies [5–12]. They can be classified by their unique physicochemical structures and properties, including solid polymeric NPs, polymeric micelles, polymer conjugates, polymersomes, polyplexes, and polymer-lipid hybrid systems [10–15]. Polymer NPs can be defined by their composition and morphology in a central core and a homogeneously or heterogeneously monolayer or multilayer shell. The therapeutic agent may either be conjugated to the NP, encapsulated between core and shell, or incorporated within the polymeric core. For a schematic overview of polymer-based platforms, see Fig. 11.1.

Polymers employed for fabrication of these nanocarrier platforms may be either of natural origin, such as albumin [16], hyaluronic acid (HA) [17], chitosan (CS) [18], and sodium alginate, or of synthetic origin, such as polyacrylic acid (PAA), polyglycolic acid (PGA), poly(lactide-co-glycolide) (PLGA), polylactic acid (PLA), dendrimers, and hyperbranched polymers [19] (for chemical structures, see Fig. 11.2). The design of polymeric nanosystems can be tailored for transport of a variety of drugs, proteins, nucleic acids, and bioimaging agents.

A prolonged blood circulation time is considered as a critical requirement for a preferred accumulation at the tumor site via the enhanced permeability and retention effect (EPR) [1, 2]. Therefore, nanosystems have to be designed to protect drugs from clearance by the kidney, liver, and reticuloendothelial system from unfavorably

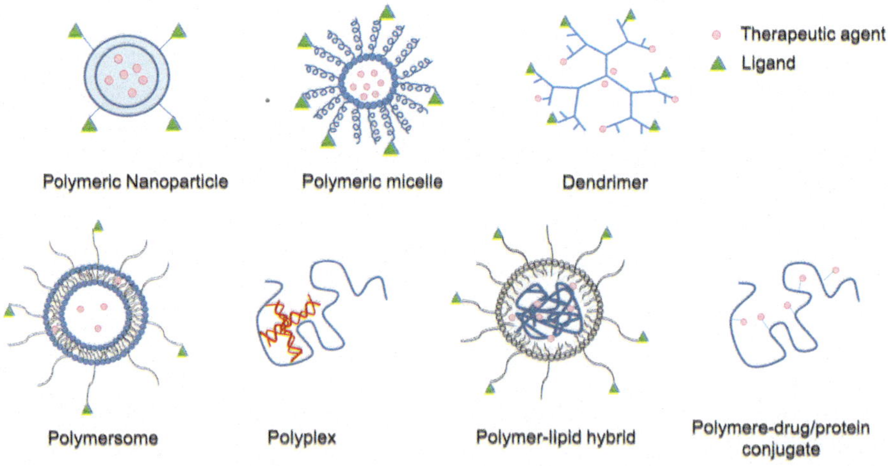

Fig. 11.1 Schematic overview of polymeric nanosystems. The blue color represents the polymeric platform. (Reprinted with permission from Ref. [14]. Copyright © 2019; DOVE Medical Press)

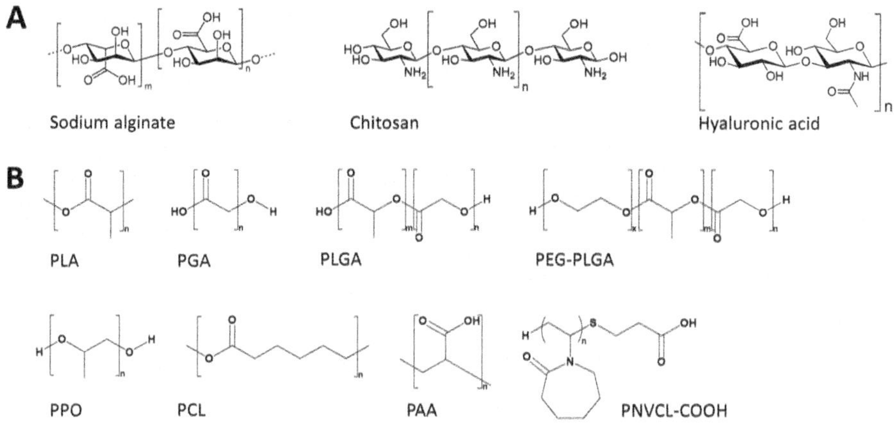

Fig. 11.2 Carrier polymers. (**a**) Natural polymers: sodium alginate, chitosan (CS), hyaluronic acid (HA). (**b**) Synthetic polymers: poly(lactic acid) (PLA), polyglycolic acid (PGA), poly(lactide-co-glycolide)(PLGA), poly(ethylene glycol)-poly(lactide-co-glycolide), (PEG-PLGA), poly(propylene oxide) (PPO), poly-ε-caprolactone (PCL), poly(acrylic acid) (PAA), poly(N-vinylcaprolactam) (pNVCL)

fast metabolism during systemic circulation. For this purpose, functionalization of the polymeric framework with shielding moieties like polyethylene glycol (PEG) [20–22], N-(2-hydroxypropyl) methacrylamide (pHPMA) [23, 24], hydroxyethyl starch (HES) [25], hyaluronic acid (HA) [26], poly(2-oxazoline) [27], or polysarcosine [28] can help to reduce nonspecific distribution [29], to achieve longer blood circulation times, and to reach tumor target tissue. In addition to passive targeting processes, active drug targeting by targeting ligands (peptide, aptamer, antibody, antibody fragment, small molecule) for specific tumor cell surface antigens [14, 30–36] and tumor microenvironment (TME)-triggered programmed drug release and activation mechanisms [3, 4] can be utilized. Polymer-based intracellular drug delivery may also overcome multidrug resistance (MDR) processes. Altogether, such measures should result in an increased antitumor efficacy with minimal systemic side effects [37–40].

However, despite their numerous advantages, polymeric nanosystems also have disadvantages. Depending on the application and the type of drug, various complex synthetic procedures and nanosystem assemblies may have to be applied. Such methods include spontaneous self-assembly by electrostatic and other noncovalent interactions, solvent evaporation, nanoprecipitation, emulsion diffusion, and salting out [41–43]. In some cases the acidity of their degradation products, the large-scale production, and the reproducibility of batch to batch in their synthesis are still a challenge for many polymer-based nanosystems [44]. In addition, there are still many unanswered questions about their toxicity profile and their long-term biological effects. Reviewing published work, Wilhelm et al. highlighted the real efficacy of only 0.7% of the dose of nanoparticles accumulating at the target tumor site on average [45]. This suggests that there definitely is room for further optimization.

The excellent therapeutic potential at both preclinical and clinical development stages led to diverse applications and perspectives of polymer-based nanosystems in bioimaging and nanomedicine. The fact that some formulations are already in clinical use further validates the efficiency of polymeric platforms for delivery of anticancer agents. This review details the tumor-targeting aspects of various polymer-based nanocarrier systems, focusing on polymeric nanoparticles, micelles, polymersomes, polyplexes, and polymer conjugates, using examples from recent research related to this field.

11.2 Polymeric Nanoparticles (NPs)

Polymeric NPs are solid colloidal systems in which the therapeutic agent is encapsulated, entrapped, adsorbed onto the polymeric matrix, or dissolved. Nanoparticles can be subdivided into different categories depending on their process of formation. The structure of the arising polymeric NPs may differ from nanocapsules (vesicular reservoir systems which are formed in a core-shell arrangement, in which the drug is entrapped in oily or aqueous cavity surrounded by a polymer membrane) to nanospheres (matrix systems in which the active substance is dispersed in the particles) [46–48].

The polymers used to create NPs are mainly biodegradable polyester, which are very suitable for biological systems [49]. Typical polymeric systems for passive and ligand-targeted delivery of therapeutics are poly(lactide) (PLA), poly(lactide-co-glycolide) (PLGA) (approved by the FDA for its biodistribution, biocompatibility, and biodegradability [50, 51]), poly-ε-caprolactone (PCL), and the evolution of PLGA, PEGylated PLGA (PEG-PLGA) as synthetic developed polymers [52]. Furthermore, there are also naturally occurring polymers used in drug delivery such as chitosan (CS), gelatin, albumin, and sodium alginate. However, synthetically polymers have proven to be more pure and reproducible when compared to their natural counterparts, which led to their preference in terms of therapeutic application [53]. Table 11.1 provides an overview of selected recent examples of NPs.

PLGA NPs can be employed for gene therapy purposes and for the delivery of bioactive agents such as proteins or drugs [44, 54]. For this reason, PLGA is one of the most important polymers for the development of anticancer drugs [55]. Chittasupho et al., for example, documented a NP system targeting an immunologically active receptor, intercellular adhesion molecule-1 (ICAM-1, CD54), by conjugating the cyclo-(1,12)-PenITDGEATDSGC (cLABL) peptide to the surface of the NP. As an anticancer drug, doxorubicin (DOX) was delivered. The PLGA-cLABL-DOX-NPs showed more rapid cellular uptake compared to NPs without shell functionalization in an A549 cancer model. The cytotoxicity evaluation of cLABL-targeted NPs compared to free drug showed similar IC50 values, indicating that the activity of DOX released from NPs was retained [56].

Table 11.1 Polymeric nanoparticles developed for delivery of drugs to treat various cancers

Carrier system	Targeting ligand	Drug	Cancer cell line	In vitro and in vivo studies	References
PLGA	cLABL	DOX	Adenocarcinomic human alveolar basal epithelial cells (A549)	In vitro	[56]
	–	DTX, TPGS	Human cervical carcinoma cells (HeLa)	In vitro, in vivo	[57]
	HA	SLM PTX	Breast cancer cell line (MCF-7)	In vitro	[58]
	OX26 mAb	TMZ	Human primary glioblastoma cell line (U215, U87)	In vitro	[59]
PEG-PLGA	A10 aptamer	DTX, PTX	PSMA-overexpressed LNCaP	In vitro, in vivo	[62]
	–	PTX	Human cervical carcinoma cells (HeLA)	In vitro, in vivo	[63]
	ITEM4	–	Glioblastoma (KR158)	In vitro, in vivo	[64, 65]
PCL	HA	CAP	Urethane-induced lung cancer (A549)	In vitro, in vivo	[66]
PEG-PCL	–	PTX TMX	Human ovarian adenocarci-noma cells (SK-OV-3)	In vitro, in vivo	[67]
	PD-L1	DOX	Gastric adenocarcinoma (MKN45 MGC803, HGC27)	In vitro	[68]
Poly (Cys-PCL)	–	PTX	Breast cancer cell line (4 T1)	In vitro, in vivo	[69]
PEG-CS	MTX	MTX, MMC	Hepatocellular carcinoma (H22)	In vitro, in vivo	[70]

Abbrevations:
CAP capsaicin; *cLABL* cyclo-(1,12)-PenITDGEATDSGC; *DOX* doxorubicin; *DTX* docetaxel; *HA* hyaluronic acid; *ITEM4* CD266 (TWEAK receptor) monoclonal antibody; *LNCaP* androgen-sensitive human prostate adenocarcinoma cells; *MMC* mitomycin C; *MTX* methotrexate; *OX26 mAb* CD71 TfR (transferrin R) monoclonal antibody; *PCL* poly-ε-caprolactone; *PD-L1* programmed death-ligand monoclonal antibody; *PEG* poly(ethylene glycol); *PEG-CS* PEGylated chitosan; *PLGA* poly(lactide-co-glycolide); *poly(Cys-PCL)* poly-dimethyl L-cysteine-poly-ε-caprolactone; *PSMA* prostate-specific membrane antigen; *PTX* paclitaxel; *SLM* salinomycin; *TMX* tamoxifen; *TMZ* temozolomide; *TPGS* D-α-tocopheryl polyethylene glycol succinate

Zhu et al. reported a PLGA-based NP system delivering docetaxel (DTX) together with vitamin E, D-α-tocopheryl polyethylene glycol 1000 succinate (TPGS), as a co-delivery system for HeLa cells in vivo and in vitro. The results showed that TPGS can act as a pore-forming agent with dual function for the production of porous NPs. This leads to favorable pharmacokinetic properties (higher drug encapsulation efficiency and faster drug release) and an active matrix component that provides (P-gp) ATPase and drug efflux inhibition to overcome the MDR of cancer cells. In sum, the strategy of co-delivery of anticancer drugs with TPGS by PLGA NPs was found as advantageous [57].

A PLGA NP co-delivery system, based on encapsulation of salinomycin (SLM), a drug against cancer stem cells (CSCs), and paclitaxel (PTX), a drug against cancer cells, was developed by the emulsion solvent diffusion method. Hyaluronic acid (HA) was coated onto the surface of the NPs as targeting domain to target CD44 receptors, which are overexpressed on CSCs. The results showed the specificity of simultaneous application of both drugs toward the tested MCF-7 cells. Furthermore, a longer circulation time was achieved by the coating, indicating the improved bioavailability of drugs from this nanoparticulate system [58].

The group of Pereira reported the encapsulation of temozolomide (TMZ) in PLGA NPs, which were functionalized with an OX26 type monoclonal antibody for transferrin receptor (TfR) targeting for treating glioblastoma tumor cells, since the TfR is highly overexpressed in this tumor type. U215 and U87 were used to evaluate the in vitro cytotoxicity of the NPs loaded with TMZ, showing an enhancement in anticancer activity of TMZ in comparison to free TMZ. The functionalization with the monoclonal antibody for TfR proved to be advantageous in enhancing the cellular internalization of the NPs into glioblastoma cells, suggesting a selective uptake mechanism [59].

PEG was incorporated into PLGA NPs with the purpose of providing a functional site for surface conjugation of targeting agents, for improving surface properties, and to overcome the limitation of PLGA to provide a very slowly drug release, due to its high crystallinity and low degradation rate [55, 60]. However, a small drawback to PEGylation was the accumulation of high molecular weight PEG in tissues such as the liver [61]. Cheng et al. presented functionalized PEG-PLGA NPs with A10 aptamer as a ligand against prostate-specific membrane antigen (PSMA)-overexpressing LNCaP cancer cells. The PEG-PLGAA10 aptamer NPs reported a nearly fourfold increased delivery of NPs to tumors as compared to equivalent NPs lacking this targeting ligand [62].

Moreno et al. employed PEG-PLGA NPs as a carrier for PTX to improve its therapeutic index. PTX-loaded NPs on HeLa cells were found to be three times more cytotoxic than taxol. In vivo, PTX-loaded NPs in transplantable liver tumor-bearing mice showed marked tumor growth inhibition and increased survival compared to taxol. This resulted from EPR phenomena with PTX-loaded NPs and their ability to maintain drug levels in the blood for a longer time [63].

The group of Kim designed and characterized a nanoformulation with decreased nonspecific adhesiveness and receptor targeting (DART) based on PEG-PLGA targeted with ITEM4 monoclonal antibody against the fibroblast growth factor-inducible 14 receptor (Fn14), which is upregulated in invasive glioblastoma. Tests were done in vitro (KR158 cells, generated from NF1 and p53 mutant mice) and in vivo (orthotopic murine malignant glioma), resulting in improved brain tissue dispersion and tumor cell uptake. In addition, a longer tumor retention was observed in vivo compared to the non-targeted version of the NPs [64, 65].

Parashar et al. developed a NP system containing PCL with conjugated HA as CD44-targeting unit for the delivery of capsaicin (CAP) against urethane-induced lung cancer. A superior performance of HA-PCL-CAP NPs in terms of A549 cell killing was reported when compared with CAP and PCL-CAP. The therapeutic in vivo efficacy in urethane-induced lung carcinoma was tested in a rat model, revealing a reduced tumor volume, restoration of oxidative stress marker, and improved animal survival rate, when compared to plain CAP and PCL-CAP NPs without surface functionalization [66].

Biodegradable PEG-PCL NPs loaded with PTX and tamoxifen (TMX) were found to be effective in overcoming multidrug resistance in ovarian adenocarcinoma. Such a NP system could lead to tenfold and twofold lower IC50 cytotoxicity values in SKOV3 cells, respectively, in comparison to free drug solution. Upon intravenous administration of PTX-TMX combination in PEG-PCL NP formulations, significant enhancement in antitumor efficacy and negligible treatment-associated toxicity were observed, as measured by body weight changes, blood cell counts, and hepatotoxicity [67].

Another PEG-PCL NP approach was described by Xu et al. They constructed PEG-PCL NPs with a surface-coated programmed death-ligand (PD-L1) monoclonal antibody for targeting the delivery of DOX as anticancer drug against gastric cancer (GC). The cellular uptake of MKN45 overexpressing PD-L1 showed that the functionalized NPs achieved significantly higher cellular uptake in comparison to non-targeted NPs, which was additionally underlined by the results of the in vitro cytotoxicity experiment (MGC803, MKN45, and HGC27). PEG-PCL-PD-L1-DOX NPs were significantly more toxic than their untargeted counterpart or free aqueous DOX. Furthermore, PEG-PCL-PD-L1-DOX NPs induced cell apoptosis and enhanced G2-M arrest in cancer cells, indicating the inhibition of microtubule synthesis [68].

Zhang et al. generated a novel pH and redox dual-sensitive polymeric NP for anticancer PTX delivery, by using PCL as polymeric main structure and modifying it with poly-dimethyl L-cysteine (Cys) via a pH-responsive imine bond. Poly(Cys–PCL)-PTX NPs released PTX in significant amounts only at mildly acidic pH and high concentration of GSH, but with almost no burst release under physiological conditions of plasma. Poly(Cys–PCL)-PTX were more cytotoxic to 4T1 cancer cells than pure PTX alone, showing efficient delivery to the tumor cells. In addition, in vivo results revealed a strong tumor inhibiting ability, good drug tolerability, and biosafety [69].

Jia et al. processed PEGylated chitosan (CS) NPs for the co-delivery of loaded methotrexate (MTX) and mitomycin C (MMC) as drug delivery system, in which MTX is also employed as a tumor-targeting ligand. The cellular uptake in a HeLa cell model illustrated that MTX modification improved the uptake compared to MMC-PEG-CS NPs, which confirmed to be internalized into the cells by FA receptor-mediated endocytosis. Moreover, MTX + MMC-PEG-CS NPs achieved high accumulation at the tumor site of mice bearing H22 tumors and a more efficient suppression of tumor cell growth than by the delivery of either drug alone, indicating a synergistic effect of the co-delivery [70].

In sum, a series of new polymeric NPs have been recently developed. Conjugation with tumor-specific ligands like antibodies, peptides, small molecules, and vitamins was found as a potent approach to enhance antitumoral efficacy. Polymeric NP formulations are the most progressed research field in polymer-based medicine, resulting in many clinical trials and already FDA-approved products. Indeed, in 2013, two of the top ten best-selling drugs in the USA were polymeric drugs, Copaxone® (glatiramer acetate, a immunomodulatory currently used to treat multiple sclerosis) and Neulasta®(pegfilgrastim, a stimulator of the production of white blood cells) [71]. A prominent example for cancer treatments is Abraxane® (FDA approved in 2005 for breast cancer, 2012 for non-small-cell lung carcinoma (NSCLC), and 2013 for pancreatic cancer), which consists of nanoparticle albumin-bound PTX. In addition, further clinical trials with Abraxane® are currently underway. Currently, a huge variety of trials is ongoing for cancer treating NP formulations. BIND-014, for example, a PEG-PLGA NP delivering docetaxel for the treatment of metastatic prostate cancer and NSCLC, was evaluated in a clinical phase II trial [72, 73]. Another promising candidate is CT-2103 (Opaxio, Xyotax). This NP is formulated out of poly(L-glutamic acid). It delivers PTX for treating ovarian cancer in a phase III trial. Additionally, CT-2103 is tested in a phase III trial, in combination with gemcitabine, to treat advanced NSCLC [15].Moreover, NKTR-102, a four-arm PEG polymer structure for delivering irinotecan, is under clinical evaluation against metastatic breast cancer and colorectal cancer (in combination with cetuximab) [15].

11.3 Polymeric Micelles

Polymeric micelles are amphiphilic self-assembled nanostructures consisting of a core with hydrophobic properties, which allows the incorporation of lipophilic drugs. Furthermore they possess a hydrophilic exterior surface, to protect the drug in the aqueous environment [74, 75] and for stabilizing the polymeric micelle against in vivo recognition by the reticuloendothelial system. Through such stealth-like properties, a longer circulation time in the bloodstream can be obtained [76, 77]. The size and morphology of the assembled micelles of normally 10–100 nm can be controlled through careful design of the hydrophobic/hydrophilic balance. A proper size in the sub-100 nm range may help the accumulation in cancerous tissue [78]. Other properties would be the reduced nonspecific interactions with biological components because of the biocompatible polymer shell. In addition, polymeric micelles have the capability of controlling drug release and can be structurally modified to improve their characteristics [79]. Because of their high molecular mass, the use of di- or tri-block copolymers has led to lower critical micelle concentration (CMC) guiding to a slower dissociation rate upon dilution, resulting in a higher stability in comparison to surfactant-based micelles (e.g., polyethoxylated castor oil or Cremophor® EL, polysorbate 80 or Tween® 80) [80], which extends blood circulation of nanomicelles. Table 11.2 provides an overview of selected recent examples of polymeric micelles evaluated as antitumoral agents.

Genexol-PM is an approved drug based on the development of polymeric micelle formulation of PTX, which consists of poly(ethylene glycol)-*block*-poly(D,L-lactic

Table 11.2 Polymeric micelles developed for delivery of drugs to treat various cancers

Carrier system	Targeting ligand	Drug	Cancer cell line	In vitro and in vivo studies	References
PEG-b-PLA	–	PTX	Breast, ovarian, lung, and head cancers	Phase I	[82]
	–	(o(LA (8-PTX)	Breast cancer cell line (4 T1)	In vitro, in vivo	[83]
	–	AZD2811	Colon adenocarcinoma (SW620)	In vitro, in vivo	[84]
PLGA		DOX	Human cholangiocarcinoma (HuCC-T1)	In vitro	[85]
PEG-PLGA	FA	DOX	Human nasopharyngeal epidermal carcinoma (KB)	In vitro, in vivo	[86]
	–	Pd(II), Alloc-DOX	Fibrosarcoma (HT1080)	In vivo	[112]
pNP-PEG-pNP-DOPE	GLUT1 scFv	DOX, CUR	Human primary glio-blastoma cell line (U87)	In vitro	[87]
PEG-DSPE+ mPEG-DSPE	FA	9-NC	Pancreatic carcinoma, human cervical cancer, human gastric carcinoma cell line	In vitro	[88]
PEG-DSPE	GE11	PTX	Human laryngeal cancer cell line (Hep-2)	In vitro	[89]
PEG-DSPE-PH	PDGF	TMZ	Human primary glio-blastoma cell line (U87)	In vitro, in vivo	[90]
H40-P (LG-Hyd-DOX)-b-PEG	cRGD	DOX	Human primary glio-blastoma cell line (U87)	In vitro	[91]
PEG-PAC+ PEG-PUC	–	THZ, DOX	Human breast cancer cell line (BT-474, MCF-7)	In vitro, in vivo	[92]
DACHPt-PEG-b-P(Glu)	cRGD	Oxaliplatin	Human primary glio-blastoma cell line (U87)	In vitro, in vivo	[93–97]
NOSC	–	PTX	Hepatocellular liver car-cinoma (HepG2)	In vitro, in vivo	[103]
p(MDS-co-CES)	–	PTX, herceptin	Human breast cancer cell line (MCF-7, T47D, BT474)	In vitro	[104]
P123-PAE	–	CUR	Human breast cancer cell line, hepatocellular liver carcinoma (MCF-7, HepG2)	In vitro, in vivo	[105]
PEG-pAsp (EDA-DM)	–	PAMAM-Pt(IV)	Lung adenocarcinoma (A549R)	In vitro	[106]
PEG-pAsp (AED)-CA	–	HRP, EIA	Lung adenocarcinoma (A549R)	In vitro	[107]

(continued)

Table 11.2 (continued)

Carrier system	Targeting ligand	Drug	Cancer cell line	In vitro and in vivo studies	References
pNVCL-b-PEG	FA	5-FU	Human endothelial cell line (EA.hy926)	In vitro	[108]
PEG-b-pTyr-LA	cRGD	DOX		In vitro, in vivo	[109, 110]
PEG-PPO-PEG	–	DOX	Adenocarcinoma of the esophagus and gastro-esophageal junctions	Phase II	[111]

Abbreviations:
5-FU 5-fluorouracil; *9-NC* 9-nitro-camptothecin; *AZD2811* Aurora B kinase inhibitor; *cRGD* cyclo (Arg-Gly-Asp-D-Phe-Cys); *CUR* curcumin; *DACHPt-PEG-b-P(Glu)* (1,2diaminocyclohexane) platinum(II)-poly(ethylene glycol)-b-poly(l-glutamic acid); *DOX* doxorubicin; *EIA* ethyl 3-indole acetate; *FA* folic acid; *GE11* EGFR peptide ligand; *GLUT1 scFv* GLUT1 single-chain variable fragment antibody; *H40-P(LG-Hyd-DOX)-b-PEG* H40-poly(L-glutamate-hydrazone-doxorubicin)-b-poly(ethylene glycol); *HRP* horseradish peroxidase; *NOSC* N-octyl-O-sulfate chitosan; *o(LA)* oligo(lactic acid); *P123-PAE* pluronic P123-poly(β-amino ester); *PAMAM-Pt(IV)* platinum(IV)-conjugated cationic poly(amidoamine); *PAsp(AED)-CA* poly (N-(2,2′-dithiobis(ethylamine)) aspartamide)-cholic acid; *Pd(II)* palladium; *PDGF* platelet derived growth factor peptide; *PEG-b-PLA* poly(ethylene glycol)-*block*-poly(D,L-lactic acid); *PEG-b-PTyr-LA* poly(ethylene glycol)-b-poly(L-tyrosine)-lipoic acid; *PEG-DSPE* poly(ethylene glycol)-distearoylphosphatidylethanolamine; *PEG-DSPE + mPEG-DSPE* poly(ethylene glycol) – distearoylphosphatidylethanolamine + methoxy-poly(ethylene glycol)-distearoylphosphatidylethanolamine; *PEG-DSPE-PH* poly(ethylene glycol) distearoylphosphatidylethanolamine-N-palmitoyl-homocysteine; *PEG-PAC + PEG-PUC* poly(ethylene glycol)-poly(carbonate) + poly(ethylene glycol) urea-functionalized poly(carbonate); *PEG-pAsp(EDA-DM)* poly(ethylene glycol)-poly[(*N*′-dimethylmaleoyl-2-aminoethyl)aspartamide]; *PLGA* poly(lactide-co-glycolide); *PEG-PLGA* poly(ethylene glycol)-poly(lactide-co-glycolide); *p(MDS-co-CES)* poly[(N-methyldiethylenaminesebacate)-co-[(cholesteryl oxocarbonylamido ethyl) methyl-bis-(ethylene) ammonium bromide] sebacate]; *pNP-PEG-pNP-DOPE* nitrophenylcarbonyl-PEG3400-nitrophenylcarbonyl-1,2-dioleoyl-*sn*-glycero-3-phosphoethanolamine; *pNVCL* poly(N-vinylcaprolactam); *PPO* poly(propylene oxide); *PTX* paclitaxel; *THZ* thioridazine; *TMZ* temozolomide

acid) (**PEG-*b*-PLA**) [81]. The increased MTD and lethal dose of 50% (LD50) of Genexol-PM compared to the classic PTX formulation displayed that it was less toxic. Genexol-PM was administered to animals at the same dose as lipid-based PTX; this resulted in the same concentration in the plasma with an increased accumulation of 2–3-fold in the heart, lung, kidney, and spleen. Prominently, Genexol-PM had twice as high PTX levels in tumors [82]. Nevertheless, Genexol-PM is unstable in the blood, is cleared rapidly, and causes dose-limiting toxicity. Therefore Tam et al. synthesized a prodrug for PTX (7-OH), using oligo (lactic acid) as a novel pro-moiety (o(LA)8-PTX) specifically for PEG-b-PLA micelles, gaining higher loading and slower release of o(LA)8-PTX over PTX, showing a better compatibility and stability with PEG-b-PLA in comparison with

PTX-PEG-b-PLA (Genexol-PM) [83]. Another approach in the field of PEG-PLA NP platforms was developed by Asthon et al. They worked with the clinical candidate AZD2811, an Aurora B kinase inhibitor. AZD2811 was successfully encapsulated as organic acid salt. In vitro, a continuous drug release for up to 1 week was observed. A corresponding extended pharmacodynamic reduction of tumor phosphorylated histone H3 levels was achieved in vivo for up to 4 days after a single administration, with a promising accumulation in tumor tissues. This resulted in an improvement in efficacy and tolerability in preclinical models with less frequent dosing [84].

Jeong et al. designed polymeric micelles composed of dextran and PLGA block copolymer for DOX delivery. They used DOX-resistant HuCC-T1 cells as in vitro cancer model. The micelles exhibited approximately fourfold higher cytotoxicity to DOX-resistant HuCC-T1 cells as compared to treatment with free DOX, reflecting the effective uptake of polymeric micelles in the tumor cells for overcoming drug resistance. Free DOX barely gained access into the tumor cells [85].

Yoo and Park took advantage of the folate receptor by conjugating folic acid (FA) onto a DOX-loaded PEG diblock copolymer of PLGA by coupling the ligand covalently through its γ-carboxyl group. The in vitro cytotoxicity study showed an increase in cell uptake and cytotoxicity in a KB cell model. The result for folate-conjugated micelles showed a marked improvement in in vivo antitumor activity compared to non-targeted micelles, with a twofold decrease in tumor growth rate [86].

Sarisozen et al. established a polymeric micelle system, with nitrophenylcarbonyl-PEG3400-nitrophenylcarbonyl (pNP-PEG-pNP) conjugated to 1,2-dioleoyl-*sn*-glycero-3-phosphoethanolamine (DOPE). The topoisomerase II inhibitor DOX and NF-κB inhibitor curcumin were co-incorporated for treating U87MG cells. In addition, the micelles were decorated with GLUT1 antibody single-chain variable (scFv) as a ligand, which promotes blood-brain barrier transport and glioblastoma targeting. Compared to non-targeted micelles, GLUT1 scFv surface modification increased the association of micelles and the nuclear localization of DOX threefold, which translated into higher cytotoxicity [87].

Folate targeting was also used by Han et al. [88]. They created folate-conjugated polymer micelles composed of a mixture of folate-polyethylene glycol distearoylphosphatidylethanolamine (FA-PEG-DSPE) and methoxy-polyethylene glycol-distearoylphosphatidylethanolamine (PEG-DSPE). Micelles loaded with the anticancer agent 9-nitrocamptothecin (9-NC) were evaluated using a pancreatic carcinoma cell line, a human cervical cancer cell line, and a human gastric carcinoma cell line. These experiments demonstrated a greater ability of the folate-conjugated polymeric micelles to actively attack tumor cells with overexpressed folate cell surface receptors as compared to analogous folate-free micelles or the free anticancer agent [88].

Ren et al. developed PEG-DSPE micelles containing PTX with modification of GE11 as peptide ligand for specific targeting of EGFR overexpressing cancer cell line Hep-2. Hep-2 cell proliferation was significantly inhibited by the GE11-PEG-DSPE-PTX system in comparison to untargeted PEG-DSPE-PTX and free PTX in vitro. Their results suggest a promising targeting strategy for enhancing PTX's tumor therapeutic effects on EGFR overexpressing cancer cell lines [89]. Likewise, Miller et al. worked with PEG-DSPE. They surface-functionalized the polymer system with a novel PDGF (platelet-derived growth factor) peptide. TMZ was used as cargo. PDGF-micelles encapsulated with TMZ showed specific uptake and increased killing in glial cells compared with untargeted micelles. In vivo studies prove selective accumulation of PDGF-micelles containing TMZ in orthotopic gliomas implanted in mice, and the pH-dependent release of TMZ preferably in the tumor tissue reduced the systemic toxicity [90].

Xiao et al. designed a hyperbranched amphiphilic block copolymer H40-poly (L-glutamate-hydrazone-doxorubicin)-b-poly(ethylene glycol) (i.e., H40-P (LG-Hyd-DOX)-b-PEG) and conjugated it with the peptide ligand (cRGD) cyclo (Arg-Gly-Asp-D-Phe-Cys) for integrin αvβ3 targeting. Unlike the previous approaches, DOX was not encapsulated, but covalently bound onto the hydrophobic segments of the amphiphilic block copolymer. Covalent binding proceeded via acid-labile hydrazone bonds, to enable pH-controlled drug release. H40-DOX-cRGD showed a much higher cellular uptake in U87MG cells due to integrin αvβ3-mediated endocytosis compared to non-targeted micelles, suggesting higher cytotoxicity and a much higher level of tumor accumulation [91].

Ke et al. investigated in their study the potency of cancer stem cell killing with a combination therapy of thioridazine (THZ) and DOX in a polymeric micelle-based system using a mixture of acid-functionalized poly(carbonate) and PEG diblock copolymer (PEG-PAC) and urea-functionalized poly(carbonate) (PUC) and PEG diblock copolymer (PEG-PUC). BT-474 and MCF-7 cells were used as tumor cell culture models. In vivo evaluation was performed using BT-474 xenografts in nude mice. Encapsulated THZ + DOX in the described micellar system showed a significant decrease of cancer volume in comparison to free THZ + DOX in vivo, resulting in a more suppression of effective tumor growth. The combination of co-delivering THZ + DOX achieved stronger antitumor efficacy when compared to single DOX or THZ [92].

Miura et al. developed polymeric micelles incorporating (1,2-diaminocyclohexane)platinum(II) (DACHPt), the parent complex of the potent anticancer drug oxaliplatin, through the metal complex formation-driven self-assembly of poly(ethylene glycol) (PEG)-b-poly(L-glutamic acid) and DACHPt in an aqueous milieu (DACHPt/m). They also used the cyclic peptide cRGD (Arg-Ala-Asp) as a ligand and compared it to cRAD (Arg-Ala-Asp) as a non-targeting ligand control. In contrast to the cRAD micelles, the micelles with cRGD accumulated with a high efficiency in the tumor cell line (U87MG), resulting in significant antitumor effects in an orthotopic mouse model of U87MG glioblastoma [93]. The same system of DACHPt-[PEG-*b*-p(Glu)] is used in several other cancer models by the same research group [94–97]. This type of nanomicelles is under clinical evaluation, for example, against spontaneous pancreatic cancer as a combination therapy with gemcitabine [98, 99], followed by a phase II study against NSCLC, biliary tract, and bladder cancer [100]. Most recently, the group extended their tumor-targeting efforts by incorporation of glucose into platinum-loaded polymeric micelles via the carbon

6 of glucose. By this measure, targeting to GLUT1 receptor which is overexpressed on many tumors as well expressed on vascular endothelium was possible, resulting in enhanced tumor accumulation (both in GLUT1-high OSC-19 and GLUT1-low U87MG tumors) antitumoral efficacy [101]. Analogous GLUT1 targeting into breast cancer stem–like cells was obtained for PLK1siRNA/polymer/gold sub-50 nm nanoparticles, with encouraging antitumoral effects [102].

Jin et al. synthesized an amphiphilic graft copolymer, N-octyl-O-sulfate chitosan (NOSC), for encapsulation of PTX for treating multidrug-resistant cancer. As a cell model, they used HepG2 cells and MDR HepG2 cells (HepG2-P). The micelles had a high cellular uptake, about twofold higher than taxol. A low efflux of PTX triggered optimal cytotoxicity in both cell lines. In a mouse model, the intravenously injected PTX-loaded micelles showed a higher tumor inhibition rate over that of taxol in tumor-bearing mice. This also resulted in longer survival time of mice [103].

Lee et al. reported micellar nanoparticles constructed of a biodegradable and amphiphilic copolymer, poly[(N-methyl bis(ethylene) amine sebacate)-co-[(cholesteryl oxocarbonylamidoethyl) methyl bis(ethylene) ammonium bromide] sebacate], p(MDS-co-CES). The micelles were used as carrier to co-deliver PTX and Herceptin for achieving targeted delivery of PTX to human epidermal growth factor receptor-2 (HER2/neu)-overexpressing human breast cancer cells (MCF7, T47D, and BT474) and an enhancement of cytotoxicity through synergistic activities. The targeting ability of their co-delivery system showed significantly higher cellular uptake in HER2-overexpressing BT474 cells as compared to HER2-negative HEK293 cells [104].

pH-responsive polymeric micelles were also used by the group of Cai et al. They created a pluronic P123-poly(β-amino ester) (P123-PAE). Curcumin served as a drug of choice which was efficiently encapsulated into the P123-PAE micelles. In vitro, curcumin P123 PAE showed similar antitumor activity against MCF-7 and HepG2 cells as compared to solubilized curcumin solution. In vivo, however, a longer circulation time of curcumin P123 PAE was observed, resulting in a longer retention time and therefore a superior pharmacokinetic outcome [105].

Li et al. [106] constructed anionic block copolymers, poly(ethylene glycol)-poly[(*N′*-dimethylmaleoyl-2-aminoethyl)aspartamide] (PEG-pAsp(EDA-DM)) with platinum(IV)-conjugated cationic poly(amidoamine) (PAMAM-Pt(IV)) prodrugs as drug delivery system with appreciable tumor tissue acidity-responsive charge conversion function. The efficient drug penetration together with elevated drug potency contributed to significant growth inhibition of A549R cells. Results demonstrated that this smart polymeric micelle drug delivery system managed to overcome the barriers of tumor tissue penetration and drug resistance. Efficient drug permeation coupled with increased drug potency contributed to a significant inhibition of A549R MCTS growth inhibition. The in vitro results confirmed that overcoming the barriers of tumor tissue penetration and drug resistance is successful with the described dual endogenous stimuli-responsive micelles [106].

Moreover, polymeric micelle systems can also be used as nanoreactors. Li et al., for example, designed one based on ternary block copolymer PEG-pAsp(AED)-CA, consisting of PEG, reduction-sensitive poly(N-(2,2′-dithiobis(ethylamine)) aspartamide) pAsp(AED), and cholic acid (CA) [107]. They applied this polymeric micelle system for co-delivering horseradish peroxidase (HRP) as an enzyme for prodrug activation and ethyl 3-indoleacetate (EIA) as the anticancer prodrug. HRP was integrated into a reduction-sensitive interlayer, whereas EIA was encapsulated in the hydrophobic core of the polymer. Cell survival tests on A549 cells showed high tumor cell killing rates in comparison to unloaded micelles [107].

Prabaharan et al. invented a thermosensitive poly(N-vinylcaprolactam)(pNVCL)-b-PEG block copolymer micelle system conjugated to FA (pNVCL-b-PEG-FA). Self-assembly into stable micelles worked in aqueous solution starting above 30 °C. At the physiological temperature of 37 °C, the release of the encapsulated anticancer agent 5-FU was slow, steady, and therefore more controlled than at a temperature of 25 °C. The 5-FU-loaded pNVCL-b-PEG-FA did not induce remarkable cytotoxicity against EA.hy926 cells but showed cytotoxicity effects against 4 T1 cells, due to the availability of the target cell folate receptor [108].

The group of Zhong generated biodegradable, reduction-responsive polytyrosine micelles (rPTM) based on poly(ethylene glycol)-b-poly(L-tyrosine)-lipoic acid (PEG-b-pTyr-LA). cRGD was used as targeting domain. The created NPs had a high drug loading content of DOX. cRGD-rPTM-Dox NP were tested in vitro and in vivo in an MDA-MB-231 cell line. A prolonged circulation time compared with the non-cross-linked cRGD-PTM-Dox control and significantly better accumulation in tumor tissues as compared to the non-targeted version were reported, outperforming even clinical liposomal DOX-NPs [109]. This NP platform was also successfully tested in vivo against HCT-116 with a nearly three times higher tumor accumulation compared to rPTM-Dox [110].

Juan et al. produced triblock copolymers of PEG and poly(propylene oxide) (PPO) with PEG-PPO-PEG structure for encapsulation of DOX. The so formed drug delivery system was named SP1049C and used in a phase II clinical trial. Patients with inoperable metastatic adenocarcinoma of the esophagus and gastro-esophageal junction, when treated with SP1049C, had high objective response rates and increased median survival, demonstrating superior antitumor activity of SP1049C compared with a standard formulation of DOX [111].

The group of Weissleder recently made use of palladium catalysts to catalyze intracellular reactions by converting prodrugs in their active forms. Palladium can cleave allyloxycarbonyl (alloc) groups under biological conditions. For this purpose, they used PEG-PLGA micelles with Pd(II) encapsulated in the core. The polymeric micelles were intravenously injected in cancerous mice and accumulated in the tumors by passive targeting based on the EPR effect. After 5 h, a prodrug of DOX, protected with an alloc ligand (alloc-DOX) at its active daunosamine amino group, was systemically delivered to the tumors. In the tumor, the accumulated Pd (II)-micelles could catalyze the release of DOX by cleaving off the alloc groups, which lead successfully to tumor regression and apoptosis [112].

In sum, many chemotherapeutic drugs have been formulated into polymeric micelles. Also plenty of polymeric micelle-based systems for tumor-targeting cancer therapy are under current clinical evaluation. Up to now, only Genexol-PM© has been approved, for the treatment of MBC and NSCLC, by the Ministry for Food and

Drug Safety of South Korea. There are ongoing phase III and IV clinical trials for FDA approval [15].

11.4 Polymersomes

Polymersomes are polymeric vesicles, which are fabricated through self-assembly of synthetic amphiphilic block copolymers, consisting of discrete hydrophilic and hydrophobic blocks, in aqueous solutions to form an architecture analogous to liposomes (vesicles derived from phospholipids) [113–115]. Compared to liposomes, polymersomes may have greater stability, storage capacity, and prolonged circulation time [116, 117] and can be easily functionalized by numerous functional groups, e.g., by targeting groups and stimuli-responsive segments [118]. Polymersomes have gained a growing interest in biomedical applications as nanocarrier systems for a variety of payloads. They achieve a programmable drug delivery and increased selective accumulation at tumor sites.

Table 11.3 provides an overview of selected recent examples of polymersomes evaluated in cancer models.

Polyphosphazene-based polymers (PEPs) have been studied as a delivery system of hydrophilic DOX hydrochloride (DOX·HCl) or hydrophobic DOX base (DOX). Xu et al. could show a high encapsulation rate of DOX·HCl and DOX due to strong intermolecular interaction with the PEP polymersomes. In a MCF-7 xenograft model in nude mice, a similar tumor growth inhibition was observed as with free DOX·HCl and DOX. However, a better life span safety and a strong reversal of drug resistance were noted due to improved pharmacokinetics [119].

Another approach of encapsulating DOX in a polymersome system was developed by Upadhyay et al. They used poly(γ-benzyl-L-glutamate)-block-hyaluronan (pBLG(23)-b-HYA(10)) as their polymersome system and evaluated the antitumor efficacy of DOX in an Ehrlich ascites carcinoma (EAC) murine mouse model. Due to the availability of HA on the surface of the polymersomes, CD44 receptor-mediated endocytosis is achieved. This resulted in a prolongation in tumor doubling time and an increased survival of mice [120].

The group of Voit also used DOX as their payload of choice. They developed photo-cross-linking stabilized polymersomes with a pH-adjustable membrane permeability and FA ligands as targeting domain. Poly(ethylene glycol)-block-poly [2-(diethylamino) ethyl methacrylate-stat-2-hydroxy-4-(methacryloyloxy) benzophenone] [PEG-b-p(DEAEMA-stat-BMA)] was used as the building block of their polymersomes. SBC-2 cells (with high FA receptor density on surface) were used for tumor targeting. The treatment of SBC-2 cells with FA-conjugated, Dox-loaded polymersomes arose in higher cell death as compared to SBC-2 cells treated with polymersomes without FA, showing the specific targeting efficiency of the FA modified polymersomes. In addition, when blocking folic receptors of SBC-2 by free FA, the toxicity potential of the polymersome system was significantly reduced, verifying the targeting specificity [121].

Table 11.3 Polymersomes developed for delivery of drugs to treat various cancers

Carrier system	Targeting ligand	Drug	Cancer cell line	In vitro and in vivo studies	References
PEPs	–	DOX	Human breast cancer cell line (MCF-7)	In vitro, in vivo	[119]
PBLG(23)-b-HYA(10)	–	DOX	Ehrlich ascites carcinoma (EAC)	In vivo	[120]
PEG-b-p (DEAEMA-stat-BMA)	FA	DOX	Small-cell lung carcinoma (SBC-2)	In vitro	[121]
PEG-b-cap	PR_b	Cisplatin	Human colon cancer cell line (DLD-1)	In vitro	[122]
PEG-b-P-CPTKMA–co-PEMA	–	Camptothecin	Lung adenocarcinoma (A549), hepatocellular carcinoma (H22)	In vivo	[123, 124]
PEG-Chol-10	–	DOX, 5-FU, LV	Human pancreatic cancer cell line (BxPC-3)	In vitro, in vivo	[127]
POEGMA-PDPA	cRGD	PTX	Peritoneal gastric carcinoma (MKN-45P), colon carcinoma (CT26)	In vitro, in vivo	[129]

Abbreviations:
5-FU 5-fluorouracil; *b-cap* block-poly(γ-methyl-ε-caprolactone); *cRGD* cyclo(Arg-Gly-Asp-D-Phe-Cys); *DOX* doxorubicin; *FA* folic acid; *LV* leucovorin; *PBLG(23)-b-HYA(10)* poly(γ-benzyl-l-glutamate)-block-hyaluronan; *PEG-b-P-CPTKMA–co-PEMA* poly(ethylene glycol)-block-poly-(camptothecin thioketal-linked methacrylate monomer)-co-(2-(pentamethyleneimino) ethyl methacrylate); *PEG-Chol-10* poly(ethylene glycol)-cholesteryl-decyl chain; *PEG-b-P(DEAEMA-stat-BMA)* poly(ethylene glycol)-block-poly[2-(diethylamino) ethyl methacrylate-stat-2-hydroxy-4-(methacryloyloxy) benzophenone]; *PEPs* polyphosphazene-based polymers; *POEGMA-PDPA* poly(oligoethylene glycol methacrylate)-poly(2-(diisopropylamino)ethyl methacrylate P[(OEG) 10MA]20-PDPA90; *PR_b* α(5)β(1) integrin-specific targeting peptide; *PTX* paclitaxel

Kokkoli et al. developed bioresorbable polymersomes for efficient and site-specific delivery of cisplatin to human colon cancer cells (DLD-1) overexpressing α(5)β(1) integrin. As a polymeric basis, PEG-block-poly(γ-methyl-ε-caprolactone) was used. The formed polymersomes were then functionalized with an α(5)β(1) integrin-specific targeting peptide named PR_b, which is highly specific to α(5)β(1) integrin than the conventionally RGD-targeting peptide that on the other hand is capable to bind a broad variety of integrins. Cisplatin-loaded PR_b-functionalized polymersomes displayed enhanced cytotoxicity toward DLD-1 cells compared to untargeted polymersomes. In addition, targeted polymersomes were less toxic to CACO-2 model human epithelial with a low expressed α(5)β(1) integrin rate, verifying the specific targeting of PR_b [122].

Polymersome systems can also be used as nanoreactors, which could be demonstrated by Ke et al. They developed therapeutic nanoreactor with selectively tumor acidity-responsive membrane permeability to activate a cascade reaction to in situ produce hydroxyl radicals (ROS). Their polymersomes are constructed out of

PEG-b-P-(CPTKMA (camptothecin thioketal-linked methacrylate monomer)-co-PEMA (2-(pentamethyleneimino) ethyl methacrylate)) as polymeric basis. Iron oxide nanoparticles were incorporated in the polymer membrane and glucose oxidase into the aqueous core, to trigger the ROS production by Fenton reaction. ROS would then release camptothecin by cleaving the thioketal linker. In an in vivo murine model of A549 and aggressive H22 mouse hepatoma, the nanoreactors proved to own highly effective tumor suppression abilities and at the same time low systemic toxicity [123, 124]. A very similar polymersome nanoreactor based on ROS production, which bursts the vesicle and unloads its cargo, quinone methide for specific tumor killing, was described by Li et al. [125].

Aibani et al. reported a polymersome system consisting of the electroneutral co-polymer constructed with PEG, cholesteryl chains, and decyl chains [126]. This system was used for drug delivery of DOX, 5-FU, and leucovorin (LV). The polymersomes were examined in pancreatic BxPC-3 cells in vitro and in vivo (ectopic BxPC-3 mouse model). The same concentration of the three free drugs served as comparison. In all experiments, the polymersomes had better efficacy and a higher maximum tolerated dose. When injected intratumorally, the tumor volume was reduced significantly as compared to the free drug combination. Furthermore, the polymersomes were less cardiotoxic [127].

Another approach of controlling the stability and cargo release of polymersomes was described by Simón-Gracia et al. They invented pH-sensitive polymersomes for PTX delivery to MKN-45P and CT26 carcinoma in a murine model. The polymersomes were assembled from diblock copolymer poly(oligoethylene glycol methacrylate)-poly(2-(diisopropylamino)ethyl methacrylate, p[(OEG)10MA]20-pDPA90 or pOEGMA–pDPA). The pOEGMA established a stealth-like character against the immune system, and the pDPA block served as a pH sensitizer. The polymersomes were highly stable in aqueous solution at neutral pH for up to 4 months, whereas exposure of pH <6 leads to quick disassembly. The PTX-pOEGMA-PDPA polymersomes showed robust accumulation and penetration in the in vivo tumor model and inhibited the tumor cell growth more efficiently than Abraxane®(albumin-paclitaxel nanoparticles), suggesting more efficient tumor cell uptake and/or cytoplasmic release of PTX [128]. Simón-Gracia et al. also worked on a more specific tumor targeting system in addition to the pH sensibility for their established system by conjugating RGD, for integrin $\alpha v\beta 3$ targeting, onto the polymersome shell. This led to higher tumor-selective accumulation and penetration than untargeted polymersomes, resulting in improved tumor growth inhibition and suppression compared to their other model and Abraxane® [129].

In sum, polymersomes present an encouraging novel alternative to classical liposomal formulations for encapsulation of drugs. For example, DOX-loaded polymeric vesicles made of PEG-block-PCL were able to retard tumor growth in a live animal on a par with the commercially available agent DOXIL® (a clinically administered liposomal formulation of DOX) [116] and therefore may possibly represent an alternative to liposomes like DOXIL® in the future. Furthermore, the polymeric basis of the vesicle membrane provides a broader basis for programmed disassembly than conventional lipid vesicles.

11.5 Polyplexes

Among synthetic nucleic acid carriers, cationic polymer-based nanosystems are widely investigated. An overview of the first five decades in developing nucleic acid – polycation polyelectrolyte complexes termed polyplexes [13] – can be found in references [31, 130]. Polyplexes offer various benefits, such as lower immunogenicity compared to viral vectors and, importantly, the capability of carrying either natural or chemically modified nucleic acid material. However, due to the presence of numerous positive charges on the surface of polyplexes, binding with serum complement proteins and activation of innate immune system, aggregation and nonspecific interaction remain the major issues in their application [29, 131–133]. Consequently, more stable packaging of polyplexes (e.g., by introducing hydrophobic elements or covalent cross-links within the particle core) and shielding the positive surface charges against undesired specific interactions with the bio-macromolecules, such as serum proteins, and avoiding an undesirable immune response are important measures to overcome these problems [134, 135]. These considerations resulted in the design of PEGylated polyplexes, which are polyelectrolyte complexes formulated by the self-assembly of oppositely charged anionic nucleic acids with PEG-polycation block copolymers. Alternatively to the use of diblock polymers, polyplex cores were formed with polycations and subsequently modified with shielding PEG or other shielding polymers [20, 136, 137]. The outer hydrophilic PEG layer of the core-shell architecture provides a stealth effect, minimizing nonspecific interactions and prolonging circulation time in vivo [31, 138]. The plasmid DNA (pDNA) or siRNA is complexed and condensed through electrostatic interactions, hydrogen bonding, coordinative interactions, and hydrophobic interactions [139]. However, entropy-driven electrostatic interactions between the cationic groups of the polymer and the negatively charged nucleic acids are the most prominent way of condensation [13, 31]. Condensation is necessary to prevent degradation and the release of cargo at off-target sites, neutralize negative charges, and reduce the size of larger nucleic acids such as pDNA [140, 141] with the aim of liberating the cargo in the cytosol, in the case of siRNA [142], or the nucleus, in the case of DNA [143, 144]. In both cases, cell uptake by endocytosis into cellular vesicles and endosomal escape mechanisms are involved [145–149]. See Fig. 11.3 for the nucleic acid delivery pathway of polyplexes. Therefore binding to the nucleic acid should be reversible, resulting in a well available therapeutic cargo once the specific target site is reached [150]. Another important factor of influencing the biological activity of polyplexes is the modulation of their shape (folded rod or collapsed sphere) and size, which can be precisely controlled by the length of the polycationic segments of the block copolymers [22, 151, 152] for additionally controlling the packaging of nucleic acid into an appropriate structure to achieve effective gene expression [153].

The most studied cationic polymers for the preparation of polyplex systems have been linear (LPEI) or branched (BPEI) polyethylenimine (PEI) and poly(L-Lysine) (PLL) [31, 154, 155]. PEI displays a high concentration of positively charged amino

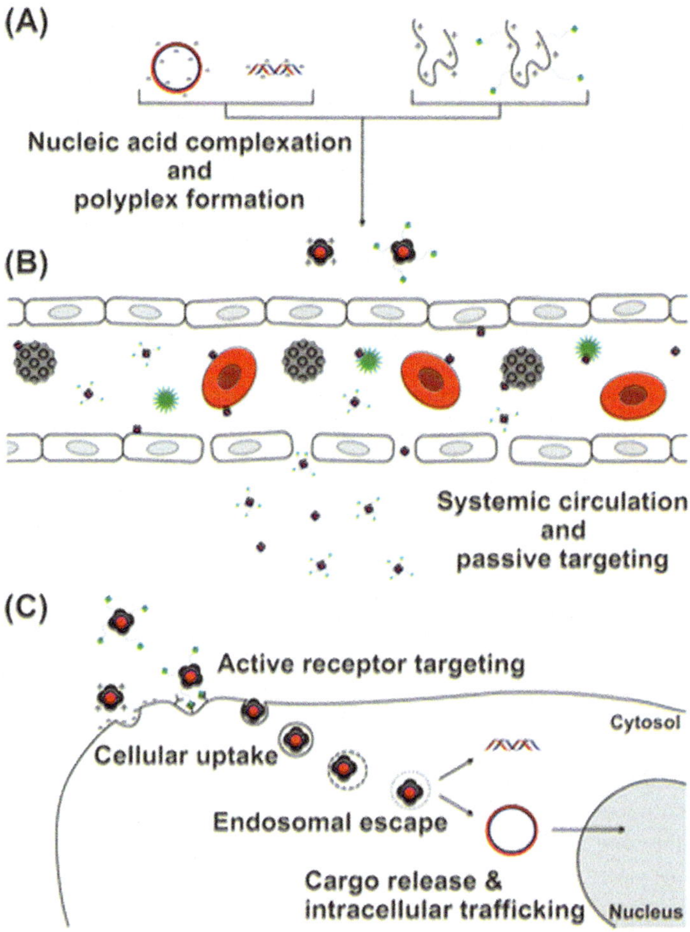

Fig. 11.3 Barriers in the nucleic acid delivery pathway of polyplexes. (**a**) Formation of stable polyplexes, (**b**) avoidance of rapid clearance and unspecific interactions with blood components, and (**c**) cellular barriers. (Reprinted with permission from Ref. [31]. Copyright © 2019; American Chemical Society)

groups (primary to tertiary), which enables effective electrostatic binding and condensation of negatively charged DNA [156]. Additionally, it possesses buffering capacity and polymer swelling at the acidic pH of the endosomes [157]. Optimized versions of PEI have already been applied in human clinical trials for cancer therapy. Nevertheless, PEI has also drawbacks like cytotoxicity and non-degradability. PLL, on the other hand, consists of only primary amines in its side chain, which interact with the DNA for the condensation. However, PLL has no intrinsic buffer-based endosomal escape mechanism, making it inferior as a gene delivery system when compared to PEI [158, 159]. Functionalization with endosomolytic agents such as virus-derived peptides is required to overcome the drawback of polymers such as

PLL [160–163]. Non-degradability of PEI is associated with medium- to long-term cytotoxicity [164]. Therefore the development of biodegradable analogs has been one line of optimization [165–167]. Both PEI and PLL are suitable polymers for polyplexes [168–175] but also pose risks such as triggering the complement activation of the innate immune system, which can lead to anaphylactic shocks [176]. Moreover, their polydisperse chemical nature as well as difficulties in precise modifications hampers their broader development. Therefore, the design of new more defined platforms, such as sequence-defined polymers [177, 178] as carrier systems to tackle these issues, has been a major research focus in pDNA and siRNA delivery. Table 11.4 provides a small selection of recent examples of polyelectrolyte complexes (polyplexes) as antitumoral agents. A wider overview can be found in references [31, 130].

Cationic (oligoethanamino)amide-based polymers (OAAs, see Fig. 11.4) synthesized by using Fmoc-/Boc-protected oligo(ethane amino)acids as building blocks for solid-phase-supported assembly represents one novel promising approach toward fully controlled syntheses of effective gene carriers, as firstly described by [179–184]. In addition to natural amino acids, artificial oligoamino acids (OAAs) such as succinoyltetraethylenepentamine (Stp) were introduced, providing a moderate positive charge density responsible for nucleic acid binding, intracellular endosomal release behavior within tumor cells based on the PEI-like proton sponge effect, and high biocompatibility due to decreased electropositivity and medium-small molecular weights [185]. Dohmen et al. conjugated them with FA for targeted delivery of siRNA in human cervix carcinoma cells. This polyplex system achieved folate receptor-specific cell targeting and silencing of the EG5 gene in receptor-positive tumors. The in vivo administration arose in silencing of the reporter gene and the lack of accumulation in non-target tissues [182].

For optimizing the bioactivity of anOAA polyplex system by folate receptor targeting in a cancer combination therapy, Lächelt et al. used MTX as a dual-functional ligand, inducing a folate receptor-mediated cellular uptake and an intracellular cytotoxic action. Molecularly precise OAAs were synthesized as carriers to encapsulate dsRNA polyinosinic–polycytidylic acid poly(I:C). The bioactivity, concerning dihydrofolate reductase (DHFR) inhibition, cytotoxicity, nucleic acid binding potency, cellular uptake of poly(I:C) polyplexes, and combined antifolate/poly(I:C) toxicity in KB cells, was investigated. More closely, the effect of different degrees of synthetic polyglutamylation on the conjugated MTX ligand was evaluated. The results demonstrated a correlation between polyglutamylation, the efficiency to deliver poly(I:C), and the MTX triggered cytotoxicity. With higher glutamylation degree, the transfection efficiency increased, leading to enhanced DHFR inhibition and higher cytotoxicity [186]. Optimized polyglutamylated MTX was used by Lee et al. [187] as a dual targeting ligand for siRNA delivery. A series of sequence-defined OAAs were prepared for eglin 5 (EG5) siRNA condensation into nanoplexes, targeted with different polyglutamylated MTX ligands. In vivo studies with KB tumor-bearing mice displayed an increase of intratumoral retention (168 h) of the siRNA, when compared to non-targeted control polyplexes (48 h). For this combination of MTX-conjugated polyplexes and eglin 5 (EG5) siRNA, the

Table 11.4 Polyelectrolyte complexes (polyplexes) developed for delivery of nucleic acids to treat various cancers

Carrier system	Targeting ligand	Drug	Cancer cell line	In vitro and in vivo studies	References
Cationizable (oligoethan amino) amide	FA	siRNA	Human cervix carcinoma (KB)	In vitro, in vivo	[182]
	cMBP2, GE11	pDNA	Hepatocellular carcinoma (Huh7), breast cancer cell line (MCF-7)	In vitro, in vivo	[193, 194, 196–198]
	cMBP2	pDNA	Hepatocellular carcinoma (Huh7)	In vitro, in vivo	[192]
	FA	pDNA	Human cervix carcinoma (KB)	In vitro	[222]
	FA	siRNA	Human cervix carcinoma (KB)	In vitro, in vivo	[190]
	FA	siRNA	Lymphocytic leukemia cell line (L1210)	In vitro, in vivo	[35]
	MTX	dsRNA	Human cervix carcinoma (KB)	In vitro	[186]
	MTX	siRNA	Human cervix carcinoma (KB)	In vitro, in vivo	[187]
	I6P7	pDNA	Human primary glioblastoma cell line (U87)	In vitro, in vivo	[199]
	GE11	miRNA-200c, siRNA	Bladder cancer (T24), breast cancer (MDA-MB 231), liver cancer (Huh7)	In vitro	[34]
	Tf, INF7	siRNA	Myelogenous leukemia (K562), human prostate cancer (D145), neuroblastoma (N2a)	In vitro, in vivo	[191]
	Angiopep-2	siRNA	Human glioma cell line (U87)	In vitro, in vivo	[188]
PEG-pAsp(TEP)-Chol	–	mRNA	Pancreatic cancer cell line (BxPC3)	In vitro, in vivo	[212]
PEG-pAsp(DET)-Chol	cRGD	pDNA	Human cervical carcinoma cells (HeLa)	In vitro, in vivo	[209]

(continued)

Table 11.4 (continued)

Carrier system	Targeting ligand	Drug	Cancer cell line	In vitro and in vivo studies	References
PEG-pAsp(DET)	–	pDNA	Colorectal carcinoma (CT26)	In vivo	[213]
VIPER (OEGMA-DMAEMA&DIPAMA-PDSEMA)	–	pDNA	Human cervix carcinoma (KB), lung adenocarcinoma (A549R	In vitro, in vivo	[214]

Abbreviations:
cMBP2 cMET-binding peptide; *cRGD* cyclo(Arg-Gly-Asp-D-Phe-Cys); *DIPAMA-PDSEMA* poly (2-diisopropylaminoethyl methacrylate)-co-poly(pyridyl disulfide ethyl methacrylate); *FA* folic acid; *GE11* EGFR peptide ligand; *I6P7* interleukin-6 receptor binding peptide; *INF7* influenza virus hemagglutinin-derived fusogenic peptide ligand to enable endosomal escape; *MTX* methotrexate; *OEGMA-DMAEMA* poly(oligo(ethylene glycol) monomethyl ether methacrylate)-co-poly (2-(dimethylamino)ethyl methacrylate); *PEG-PAsp(DET)* poly(ethylene glycol)-polyaspartamide-aminoethylene-diethylenetriamine; *PEG-PAsp(DET)-Chol* poly(ethylene glycol)-polyaspartamide-aminoethylene-diethylenetriamine-cholesterol; *PEG-PAsp(TEP)-Chol* poly(ethylene glycol)-polyaspartamide-aminoethylene-tetraethylenetriamine-cholesterol; *Tf* transferrin; *VIPER* virus-inspired polymer for endosomal release

enhanced antitumoral potency resulted in 50% of recurrence-free survival of mice [187].

The group of Jiang tested an OAA-based oligomer for siRNA delivery into glioma cells with tumor-cell-triggered siRNA release and high biocompatibility. As a targeting modification for enhancing the cellular uptake, angiopep-2 was chosen since it has high affinity to low-density-lipoprotein-receptor-related protein (LRP), which is overexpressed both on the surface of brain capillary endothelial cells (BCECs) and glioma cells. The angiopep-polyethylene glycol (PEG)/siRNA polyplexes indicated effective glioma-targeting siRNA delivery, and intracellular siRNA release, arising in a notable gene downregulation of over 70% of BAG3 expression, both in U87 cells and upon intravenous delivery in glioma model nude mice without significant toxicity [188].

Lee et al. also used folate receptor targeting in their polyplex system, optimized by co-formulating a folate-PEG-OAA with different lipo-OAAs (optionally tyrosine-modified, for optimizing stability and size) to generate targeted lipopolyplexes (TLPs) with the function of self-stabilizing by cysteine disulfide cross-links [189]. Intracellular distribution and siRNA release kinetics were analyzed. The best performer, TLP1, which was tyrosine tripeptide stabilized, unpacked siRNA after cellular uptake in a sustained manner. An up to fivefold higher intracellular siRNA stability after 4 h was noted compared to other TLPs. Furthermore, IV administration of TLP1 in a subcutaneous FR-positive leukemia (L1210) mouse

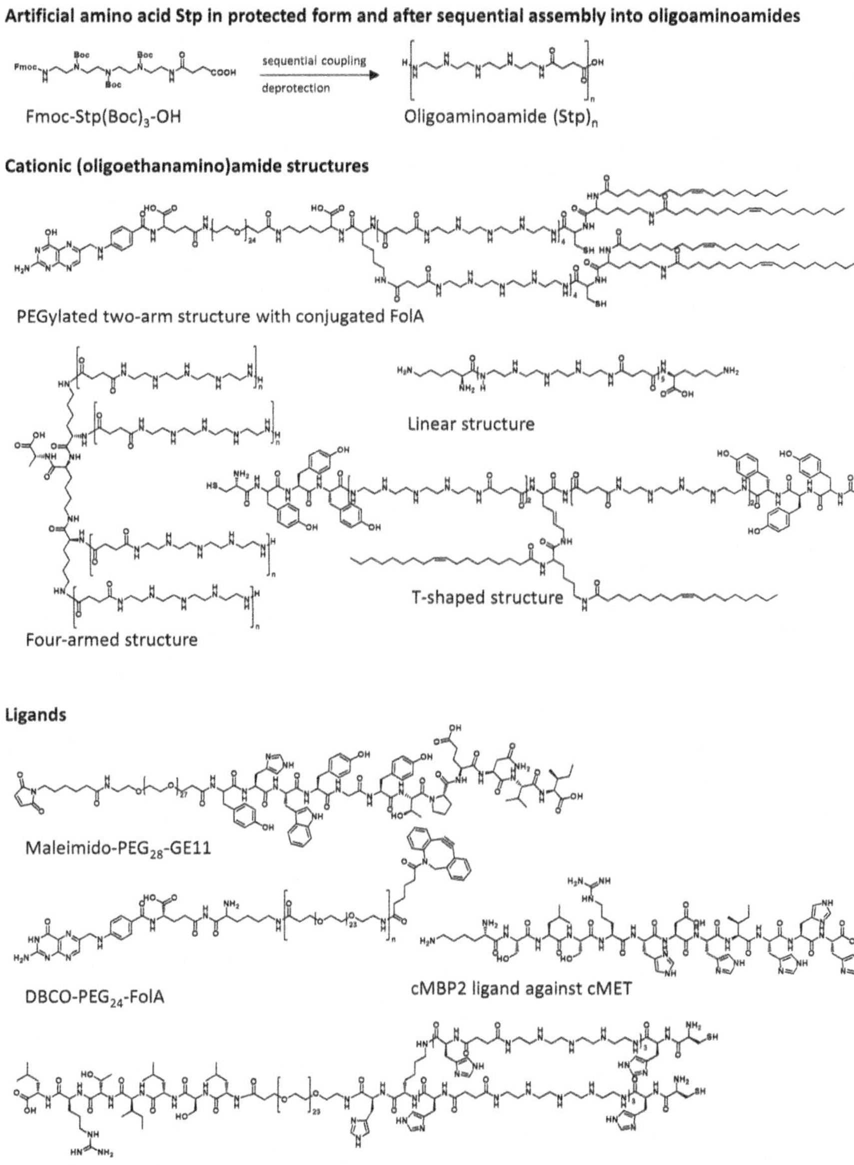

Fig. 11.4 Sequence-defined oligoaminoamides. Chemical structures of the building block Stp, cationic (oligoethanamino)amides, and targeting ligands

model could demonstrate a 65% EG5 gene silencing, measured at the tumoral mRNA level, without detectable adverse effects. These results demonstrated effective tumor-targeted delivery and superior protection of siRNA, through stabilizing the polyplex by introducing disulfide cross-links and tyrosine motifs [190].

Müller et al. advanced a polyplex system based on T-shaped cationizable lipo-OAA containing tyrosine motifs for stabilizing pDNA and siRNA [190, 191] and functionalized the formed polyplex with the targeting peptide ligand GE11 via a PEG linker. The resulting particles demonstrated receptor-mediated uptake into EGFR-positive T24 cells, MDA-MB 231 cells, and Huh7 cells. Furthermore, these formulations led to ligand-dependent gene silencing. RNA interference (RNAi) triggered antitumoral effects were observed for microRNA-200c mimic and EG5 siRNA. The treatment of T24 or MDA-MB 231 cancer cells with the polyplex delivering miR-200c led to decreased proliferation and migration, changes in cell cycle, and enhanced sensitivity toward DOX. In addition, the delivery of EG5 siRNA into Huh7 cells resulted in antitumoral activity, triggered by a loss of mitotic spindle separation and formation of mono-astral spindles [34]. Furthermore, Zhang et al. likewise used the T-shaped sequence-defined polycationic lipo-OAA described in [191] as a multifunctional transferrin receptor (TfR)-targeted siRNA delivery system. The created siRNA lipopolyplex cores were surface modifying with PEG-linked transferrin (Tf) for receptor targeting and the endosomolytic peptide INF7 for an efficient cytosolic release of the siRNA. The effect of Tf-INF7 polyplex internalization and target gene silencing was displayed for TfR overexpressing tumor cell lines (K562, D145, and N2a). As cargo EG5 siRNA was chosen, which led to a block of tumor cell growth and triggered apoptosis. Competition experiments with excess of Tf demonstrated TfR target specificity, showing the effectiveness of this targeting approach. Moreover, in vivo distribution studies illustrated enhanced tumor residence of siRNA in N2a tumor-bearing mice with the Tf-INF7 as compared to the untargeted version of this polyplex, but also reduced stability, which limited the in vivo performance of those polyplexes.

Since cMET is overexpressed on the cell surface of a variety of tumors, Kos et al. used cMET binding peptides (cMBP) as a pDNA polyplex targeting units. Sequence-defined oligomers, with a cationic (oligoethanamino) amide core containing terminal cysteines for redox-sensitive polyplex stabilization, and a monodisperse PEG layer for surface shielding were used as carriers. In addition, histidine units were implemented in the cationic core for promoting the endosomal escape. A cMET-overexpressing DU145 and Huh7 cell model was used. The resulting oligomers using cMBP2 as targeting ligand showed greatly increased cellular uptake and gene transfer over non-targeted control sequences, confirming the efficacy and target specificity of the pDNA polyplexes. The polyplex platform created in this way successfully demonstrated pDNA condensation, serum stability, and receptor-specific luciferase marker gene transfer in vivo in a Huh7 mouse model, both after local intratumoral or intravenous systemic injection. For systemic delivery improved compaction of pDNA by inclusion of a second three-arm oligomer was critical [192].

Urnauer et al. have successfully applied the same system to compact pDNA encoding sodium iodide symporter (NIS), a well-defined theranostic gene, which can target different cancer types, allowing noninvasive imaging of functional NIS expression and therapeutic radioiodide application [193–196] in a sequence-defined polymer made of PEG and a cationizable OAA. As targeting ligand, cMBP2 (cMET-targeting) or GE11 (EGFR-targeting) was used [197, 198]. In the in vivo mouse model, three cycles of polyplex and radioiodide (^{131}I) application significantly reduced tumor growth and prolonged survival of mice [197]. These data demonstrated the potential of cMET-targeted sequence-defined polymers combined with the unique theranostic function of NIS.

The group of Huang [199] reported the use of interleukin-6 receptor (IL6R) binding I6P7 peptide as a cascade-targeting ligand for non-viral gene delivery. Also here the polymeric foundation consisted of two sequence-defined OAAs, an untargeted three-arm oligomer structure mixed with an PEGylated, I6P7 ligand conjugated oligomer. Both oligomers again contained (Stp)-histidine units and cross-linking cysteines [185, 200]. The resulting polyplexes (I6P7-Stp-His/pDNA) were able to mediate higher gene expression in U87 cells than in healthy human astrocytes due to I6P7, which provides BBB-crossing functions, subsequent glioma targeting, and tumor inhibition effects. The transport across the BBB and specific glioma targeting could be demonstrated in vitro in a transwell bEnd.3 cell model, resulting in transfection of underlying U87 cells and in vivo in glioma-bearing mice. Moreover, targeted delivery of pDNA encoding inhibitor of growth 4 (pING4) significantly prolonged the survival time of orthotopic U87 glioma-bearing mice [199].

Klein et al. [35] established polyplexes based on the complexation of siRNA with the azido-functionalized sequence-defined cationizable lipo-oligomer, which contained two cholanic acids attached to an OAA backbone in T-shape configuration. The formed nanoparticles got additionally surface-functionalized by click chemistry with various folate-conjugated DBCO-PEG agents to provide a targeting and stealth effect. Testing targeting efficiency, cellular uptake, and gene silencing in vitro revealed bivalent DBCO-PEG24-FolA as the lead surface-functionalizing structure. Additionally, in vivo mouse experiments verified an extended biodistribution and intratumoral delivery in the FR-positive L1210 cell model. Furthermore, intravenous injection of analogous therapeutic EG5 siRNA (directed against the kinesin spindle motor protein eglin-5) lipopolyplexes mediated tumoral EG5 mRNA knockdown by ~60%. In combination with nature-derived tubulin inhibitor pretubulysin, the survival rate of aggressive leukemia-bearing mice was prolonged without apparent side effects. Analogous T-shaped lipo-oligomer polyplexes were successfully formulated for EG5 siRNA/methotrexate co-delivery [201], pretubulysin/methotrexate co-delivery [202], EG5 siRNA/pretubulysin co-delivery [203], and the delivery of a dual antitumoral conjugate of siRNA with the pro-apoptotic peptide KLK [204] into tumors. For this purpose, also novel targeting ligand conjugates, including EGFR targeting DBCO-PEG-GE11 and IL4R targeting DBCO-PEG-AP1 [204], were designed.

The group of Kataoka developed a polymeric carrier system based on self-assembly of a PEG-polyamino acid block copolymer for the delivery of pDNA, siRNA, and mRNA [205–208]. The functional polyamino acids possess the capacity of pH-responsive membrane destabilization, allowing endosomal escape of the polyplex into the cytoplasm. Ge et al. advanced this concept by installing the cyclic RGD peptide as ligand, in order to facilitate targeted pDNA delivery to the tumor site as well as promoting cellular uptake and intracellular trafficking behaviors. The created cRGD-PEG-pAsp(DET)-cholesteryl polyplexes were tested in an subcutaneous fibrotic BxPC3 human pancreatic cancer mouse model, achieving improved accumulation at the tumor site and potent tumor growth suppression by efficient gene expression of antiangiogenic protein (sFlt-1) at the tumor site [209]. Furthermore, Chol conjugation displayed a favorable stabilizing effect [210].

In regard to mRNA delivery, in vitro analyses using specific TLR-expressing HEK293 cells confirmed that the polyplex inclusion prevents mRNA from the recognition by Toll-like receptors (TLRs), showing the capability of this system to prevent triggering the immune response, normally caused by naked mRNA [211]. On this basis Uchida et al. created a mRNA delivery system, based on PEG-polycation block copolymers with conjugated cholesterol (Chol) moieties for an increased polyplex stability by hydrophobic interactions. As a cationic segment, polyaspartamide with four aminoethylene repeats in its side chain (PAsp(TEP)) was selected [205–208], since it helps to increase the nuclease resistance and high protein expression from the mRNA. The IV administration of PEG-pAsp(TEP)-Chol polyplexes for the delivery of mRNA encoding an anti-angiogenic protein (sFlt-1) in the BxPC3 pancreatic cancer subcutaneous inoculation mouse model led to enhanced blood retention of mRNA in comparison to polyplexes without Chol. Moreover, efficient sFlt-1 protein expression in tumor tissue resulted in remarkable inhibition of the tumor growth. In contrast, polyplexes without Chol failed to show a detectable therapeutic effect [212].

Another approach to use the PEG-pAsp(DET) polymeric system as cancer therapeutic has been described by Furugaki et al. [213]. They investigated the delivery of genes encoding the tumor-associated antigen squamous cell carcinoma antigen recognized by T cells-3 (SART3), adjuvant CD40L, and granulocyte macrophage colony-stimulating factor (GM-CSF) as a DNA vaccine platform in mouse tumor models with different types of major histocompatibility antigen complex (MHC).The triple gene vaccine (SART3/CD40L + GM-CSF gene-loaded polyplex) significantly extended the survival of mice with peritoneal dissemination of CT26 cells. Moreover, long-term surviving mice showed complete rejection when re-challenged with CT26 tumors, showing a successful inhibition of growth and metastasis of tumor cells [213].

The group of Pun reported a synthetic polymer designed to mimic viral mechanism of delivery named VIPER (virus-inspired polymer for endosomal release) [214]. The polymer is composed of a hydrophilic cationic block and a pH-sensitive block. The hydrophilic block is synthesized out of poly(oligo(ethylene glycol) monomethyl ether methacrylate)-co-poly(2-(dimethylamino)ethyl methacrylate) (p(OEGMA-DMAEMA)). Moreover, the pH-sensitive block is formed out of poly(2-diisopropylaminoethyl methacrylate)-co-poly(pyridyl disulfide ethyl methacrylate) (p(DIPAMA-PDSEMA)). The VIPER polyplex system was tested in vitro and in vivo on KB and A549 tumor models. The results demonstrated that VIPER can efficiently mediate both in vitro and in vivo gene transfer, showing superior efficiencies compared to the commonly used BPEI agent [214].

In sum, polyelectrolyte complexes (PECs) of therapeutic polyanionic nucleic acids with cationic polymers polyplexes are a most suitable delivery form, because the carrier also can condense the nucleic acid cargo into smaller NPs and protect it against biodegradation. Polyplexes have been evaluated in ex vivo and in vivo clinical cancer studies [215–217], but up to now not yet approved for therapy.

11.6 Polymer Conjugates

In contrast to polyanionic nucleic acids, most drug and protein cargos cannot directly build polyplexes. Direct covalent conjugation to polymers appears as more suitable delivery form. Polymer conjugates have been intensively reviewed in other reports [10–12]; therefore this section only briefly reviews the subject of polymer-protein delivery.

PEG-protein conjugates present one of the first and most important class of polymer therapeutics [218]. In comparison to small molecule drugs, site-specific conjugation with therapeutic proteins requires refined technologies. Recently, the group of Hua Lu [219] developed a very elegant protein-poly(amino acid) (P-AA) conjugation method by installation of two "chemical handles," including a thioester for native chemical ligation and a polyglycine nucleophile for sortase A-mediated ligation, at both ends of the P-AAs. P-AAs were based on $p(EG_3Glu)_{20}$, i.e., triethylene glycol ester of polyglutamate. In the following, they demonstrated that macrocyclization of interferon-P-AA conjugates improved the tumor retention, penetration, and antitumor efficacy of interferon [220].

Furthermore, the researcher demonstrated that this kind of site-specific polypeptide conjugation which they termed "PEPylation" reduces the risk of anti-conjugate immune responses and resultant antibody drug clearance (ABC) as was observed for the standard PEGylation proteins (interferon or human growth hormone) [221].

Zhang et al. [222] evaluated a library of 16 preselected sequence-defined oligoaminoamides for the delivery of the intracellularly active protein ribonuclease A, which elicits tumor cell killing via intracellular RNA digestion. The protein cargos were attached via bioreducible disulfide bonds. All oligomers contained PEG as a shielding unit and FA as targeting ligand. Folate-receptor-positive KB carcinoma cells were chosen as cancer model, validating both effective endolysosomal escape and subsequent intracellular transport, which was additionally verified using nls-EGFP. The investigation of structure-activity relationships elucidated that the incorporation of oleic acids play a vital role in the enhanced intracellular protein delivery, by facilitating membrane active characteristics for cytosolic delivery and efficient protein transduction, triggering potent tumor cell killing [222, 223]. Liu et al. [224] developed an analogous system with a different, traceless conjugation technique; an AzMMMan click linker [225] was applied for bioreversible protein linkage, which is cleaved under intracellular endosomal acidic conditions, releasing the intact unmodified protein.

11.7 Bioimaging and Theranostic Nanosystems

Improved pharmacokinetics and tumor accumulation will be a prime task in the optimization of the mentioned polymeric drug delivery system. Far less standardization will be possible with regard to the individual patient. Here, the heterogeneity of tumors, tumor stroma, and size dependence of vascularization are an open question with high variability within the same patient and between different patients [2, 226–228]. This uncertainty poses a formidable delivery challenge, which requires novel approaches such as performing tumor imaging before administration of the therapeutic nanoagent. In the ideal situation, functional bioimaging would detect and confirm the molecular target (such as a tumor-specific target receptor) and the accessibility of the tumor (across vasculature, stroma) with a diagnostic nanoagent, which would be harmless and would generate no side effects in non-target organs. The functional results from such a bioimaging would indicate the suitability of nanoagents. In case of alternative options, the most suitable nanoagent (size, surface modification) might be selected. For example, Cabral and colleagues [78] compared in preclinical mouse models the accumulation and antitumoral activity of long-circulating, drug-loaded polymeric micelles with diameters of 30, 50, 70, and 100 nm. All polymer micelles penetrated highly permeable tumors, but only the 30 nm micelles could penetrate poorly permeable pancreatic tumors and showed antitumor effects in a mouse model of this type of cancer. The researchers also showed that the delivery of the larger micelles could be enhanced by using a TGF-beta inhibitor for increased tumor permeability. Analogous treatments for reducing tumor stroma and improved vascular access had been developed by Jain and coworkers [229–231].

Theranostic nanoagents present the special case where the same nanosystem can be used as nontoxic diagnostic and subsequently as antitumoral therapeutic tool. Spitzweg and collaborators applied tumor-targeted pDNA polyplexes encoding sodium iodide symporter (NIS) as a well-defined theranostic gene [193, 194]. Functional NIS expression can be detected using the diagnostic radioisotope iodide ^{123}I by scintigraphy, or using ^{124}I-/^{18}Ftetrafluoroborate for positron emission tomography. Using three cycles of polyplex and therapeutic radioiodide^{131}I application significantly reduced tumor growth and prolonged survival of mice in several models [195–197]. Urnauer et al. also performed a dual targeting approach with LPEI-PEG2kDa-based polymer backbones and cMBP and GE11 as targeting ligands for c-MET and EGFR, respectively. Delivering NIS to hepatocellular cancer (orthotopic Huh7 cells), an increase of tumor-specific transduction efficiency of dual-targeted polyplexes compared to single-targeted polyplexes was demonstrated, resulting in a significant reduction of tumor growth [232].

Numerous polymeric nanoagents have been designed for multifunctional imaging and theranostics of cancer; the current section provides only a snapshot of activities [233, 234]. For pre-selection of cancer patients, integration of imaging properties into nanotheranostics should facilitate clinical translation and personalized nanomedicine administration. Imaging modalities include magnetic resonance imaging, incorporating superparamagnetic iron oxide nanoparticles (SPION) as MRI contrast agent into drug formulations [235, 236]. Ultrasound irradiation has been applied, with low-power diagnostic or high-power therapeutic irradiation, for example, using theranostic polymer microcapsules composed of hydrogen-bonded multilayers of tannic acid and poly(N-vinylpyrrolidone) that produce high imaging contrast and deliver the anticancer drug doxorubicin upon irradiation [237]. Optical imaging with near-infrared (NIR) light enables deep tissue penetration and the design of NIR-activatable polymeric nanoformulations for combined imaging and therapy of cancer [238]. NIR agents include standard small-molecule dyes but also polymer-coated quantum dots [239] or carbon nanodots for cancer-targeted photo-thermo-chemotherapy [240, 241].

Tian and colleagues designed tumor-targeted polypyrrole-bovine serum albumin-indocyanine green nanoparticles (SP94-PPy-BSA-ICG) for the detection by photoacoustic and near-infrared (NIR) fluorescence imaging and photothermal ablation of hepatocellular carcinoma (HCC) [242]. The NPs contained ICG for NIR and photoacoustic imaging, and PPy as photothermal and photoacoustic contrast agent. Based on the HCC targeting peptide SP94, a higher tumor accumulation was observed, and the NPs could effectively kill the tumor through photothermal therapy upon a single laser irradiation event. In further work, the same research group generated tantalum oxide (TaOx)-based core/shell NPs for triple-modality image-guided chemo-thermal therapy of esophageal carcinoma [243]. Hollow TaOx NPs encapsulating PPy and doxorubicin in the core and NIR dye800 conjugated on the shell enabled multimodal imaging including computed tomography (for the preliminary location of the tumor), photoacoustic (for the anatomical localization of the tumor), and fluorescence imaging (for real-time monitoring of the tumor margin) as well as pH- and thermal-sensitive drug release. Comparable chemo-thermal therapeutic efficacy of intravenous and aerosol administration was noted, with the latter probably more suitable for clinical therapy of esophageal carcinoma.

11.8 Conclusion

In the field of polymeric NP systems for cancer therapy, extensive studies have been conducted for effective and tumor-specific drug delivery. Studies demonstrate that polymer nanoparticles may become an important antitumoral treatment modality, also confirmed by the fact that some polymer-based systems are already in clinical use for delivering anticancer agents. This was made possible by important advances in this field including controlled stability, high carrier capacity, the ability to load the NP with both hydrophilic and hydrophobic anticancer agents, controlled drug release, and the possibility to deliver multiple drugs at the same time. One of the current main goals is to further improve their pharmacokinetics and rates in tumor accumulation. Another aim is the development of novel bioimaging concepts, for classifying patients in regard to the heterogeneity and vascular penetrability of tumors. Theranostic strategies, combining a diagnostic nanosystem for bioimaging as selection basis for a subsequent therapeutic nanosystem, as well as modulation of tumor stroma and vasculature, are encouraging directions for medical use.

Surface modification enables control of the transport kinetics and biodistribution of cargo and therefore enhances specific drug delivery at the tumor level while reducing toxicity to normal tissues. On the one hand, several ligands (e.g., TfR ligands, CPPs, LDLR ligands, integrins, carbohydrates (lectin ligands), EGFR ligands, folate) against various tumor-specific or associated markers with high binding affinity are extensively studied as efficient drug delivery tools in tumor targeting. On the other hand, the versatile polymer chemistry allows to modulate the properties of the polymeric matrix in a dynamic fashion. Thus, high therapeutic load and environment-triggered release [3, 244, 245] can be realized.

Nevertheless, the safety of polymeric nanocarriers is an additional important consideration, which needs to be assessed before proceeding to clinical study. In addition, complicated production designs may represent a hindrance for the translation into clinical use. In this regard, future considerations must demonstrate reliability and appropriateness to achieve the advantageous invention and evolution of these polymeric nano-platforms, by substantially improving reproducibility of production, enabling upscaling, clear mechanisms involved in the preparation and loading of anticancer agents, generation and control of membrane permeability, targeted delivery, and release of products on demand. This development of the optimization will be another step forward to explore the full potential of polymeric NP systems designed for tumor targeting.

Acknowledgment We greatly appreciate funding of our research on this subject by DFG SFB824 C8, SFB1032 B4, and SFB1066 B5, the DFG-NSFC Joint Sino-German research project WA 1648/7-1, and the Sino-German research project grant GZ995.

References

1. Fang J, Nakamura H, Maeda H (2011) The EPR effect: unique features of tumor blood vessels for drug delivery, factors involved, and limitations and augmentation of the effect. Adv Drug Deliv Rev 63(3):136–151
2. Golombek SK et al (2018) Tumor targeting via EPR: strategies to enhance patient responses. Adv Drug Deliv Rev 130:17–38
3. Wagner E (2007) Programmed drug delivery: nanosystems for tumor targeting. Expert Opin Biol Ther 7(5):587–593
4. Wu J et al (2019) Tumor microenvironment as the "regulator" and "target" for gene therapy. J Gene Med 21:e3088
5. Haley B, Frenkel E (2008) Nanoparticles for drug delivery in cancer treatment. Urol Oncol 26 (1):57–64
6. Kamaly N et al (2012) Targeted polymeric therapeutic nanoparticles: design, development and clinical translation. Chem Soc Rev 41(7):2971–3010
7. He C et al (2010) Effects of particle size and surface charge on cellular uptake and biodistribution of polymeric nanoparticles. Biomaterials 31(13):3657–3666
8. Zhao F et al (2011) Cellular uptake, intracellular trafficking, and cytotoxicity of nanomaterials. Small 7(10):1322–1337

9. Hu J et al (2018) Long circulating polymeric nanoparticles for gene/drug delivery. Curr Drug Metab 19(9):723–738
10. Duncan R (2003) The dawning era of polymer therapeutics. Nat Rev Drug Discov 2 (5):347–360
11. Li J et al (2015) Polymeric drugs: advances in the development of pharmacologically active polymers. J Control Release 219:369–382
12. Ekladious I, Colson YL, Grinstaff MW (2019) Polymer-drug conjugate therapeutics: advances, insights and prospects. Nat Rev Drug Discov 18(4):273–294
13. Felgner PL et al (1997) Nomenclature for synthetic gene delivery systems. Hum Gene Ther 8 (5):511–512
14. Prabhu RH, Patravale VB, Joshi MD (2015) Polymeric nanoparticles for targeted treatment in oncology: current insights. Int J Nanomedicine 10:1001–1018
15. Palazzolo S et al (2018) The clinical translation of organic nanomaterials for cancer therapy: a focus on polymeric nanoparticles, micelles, liposomes and exosomes. Curr Med Chem 25 (34):4224–4268
16. Elzoghby AO, Samy WM, Elgindy NA (2012) Albumin-based nanoparticles as potential controlled release drug delivery systems. J Control Release 157(2):168–182
17. Choi KY et al (2010) Self-assembled hyaluronic acid nanoparticles for active tumor targeting. Biomaterials 31(1):106–114
18. Agnihotri SA, Mallikarjuna NN, Aminabhavi TM (2004) Recent advances on chitosan-based micro- and nanoparticles in drug delivery. J Control Release 100(1):5–28
19. Panyam J, Labhasetwar V (2003) Biodegradable nanoparticles for drug and gene delivery to cells and tissue. Adv Drug Deliv Rev 55(3):329–347
20. Ogris M et al (1999) PEGylated DNA/transferrin-PEI complexes: reduced interaction with blood components, extended circulation in blood and potential for systemic gene delivery. Gene Ther 6(4):595–605
21. Meyer M et al (2009) Synthesis and biological evaluation of a bioresponsive and endosomolytic siRNA-polymer conjugate. Mol Pharm 6(3):752–762
22. Dirisala A et al (2014) Optimized rod length of polyplex micelles for maximizing transfection efficiency and their performance in systemic gene therapy against stroma-rich pancreatic tumors. Biomaterials 35(20):5359–5368
23. Johnson RN et al (2011) HPMA-oligolysine copolymers for gene delivery: optimization of peptide length and polymer molecular weight. J Control Release 155(2):303–311
24. Laga R et al (2012) Polymer coatings for delivery of nucleic acid therapeutics. J Control Release 161(2):537–553
25. Noga M et al (2012) Controlled shielding and deshielding of gene delivery polyplexes using hydroxyethyl starch (HES) and alpha-amylase. J Control Release 159(1):92–103
26. Guter M, Breunig M (2017) Hyaluronan as a promising excipient for ocular drug delivery. Eur J Pharm Biopharm 113:34–49
27. Manzenrieder F et al (2011) Stabilization of virus-like particles with poly(2-oxazoline)s. Angew Chem Int Ed Eng 50(11):2601–2605
28. Klein PM et al (2018) Efficient shielding of polyplexes using heterotelechelic polysarcosines. Polymers 10(6):689
29. Chollet P et al (2002) Side-effects of a systemic injection of linear polyethylenimine-DNA complexes. J Gene Med 4(1):84–91
30. Alexis F et al (2008) Factors affecting the clearance and biodistribution of polymeric nanoparticles. Mol Pharm 5(4):505–515
31. Lachelt U, Wagner E (2015) Nucleic acid therapeutics using polyplexes: a journey of 50 years (and beyond). Chem Rev 115(19):11043–11078
32. Liechty WB et al (2010) Polymers for drug delivery systems. Annu Rev Chem Biomol Eng 1:149–173
33. Zhu L, Torchilin VP (2012) Stimulus-responsive nanopreparations for tumor targeting. Integr Biol 5(1):96–107

34. Muller K et al (2016) EGF receptor targeted lipo-oligocation polyplexes for antitumoral siRNA and miRNA delivery. Nanotechnology 27(46):464001
35. Klein PM et al (2018) Folate receptor-directed orthogonal click-functionalization of siRNA lipopolyplexes for tumor cell killing in vivo. Biomaterials 178:630–642
36. Zhang W et al (2016) Targeted siRNA delivery using a Lipo-Oligoaminoamide nanocore with an influenza peptide and transferrin shell. Adv Healthc Mater 5(12):1493–1504
37. Sinha R et al (2006) Nanotechnology in cancer therapeutics: bioconjugated nanoparticles for drug delivery. Mol Cancer Ther 5(8):1909–1917
38. Chan JM et al (2010) Polymeric nanoparticles for drug delivery. Methods Mol Biol 624:163–175
39. Davis ME, Chen Z, Shin DM (2008) Nanoparticle therapeutics: an emerging treatment modality for cancer. Nat Rev Drug Discov 7:771
40. Brannon-Peppas L, Blanchette JO (2004) Nanoparticle and targeted systems for cancer therapy. Adv Drug Deliv Rev 56(11):1649–1659
41. Vauthier C, Bouchemal K (2009) Methods for the preparation and manufacture of polymeric nanoparticles. Pharm Res 26(5):1025–1058
42. Jain RA (2000) The manufacturing techniques of various drug loaded biodegradable poly (lactide-co-glycolide) (PLGA) devices. Biomaterials 21(23):2475–2490
43. Drogoz A et al (2007) Polyelectrolyte complexes from polysaccharides: formation and stoichiometry monitoring. Langmuir 23(22):10950–10958
44. Fernandez-Fernandez A, Manchanda R, McGoron AJ (2011) Theranostic applications of nanomaterials in cancer: drug delivery, image-guided therapy, and multifunctional platforms. Appl Biochem Biotechnol 165(7–8):1628–1651
45. Wilhelm S et al (2016) Analysis of nanoparticle delivery to tumours. Nat Rev Mater 1:16014
46. Parveen S, Sahoo SK (2008) Polymeric nanoparticles for cancer therapy. J Drug Target 16 (2):108–123
47. Letchford K, Burt H (2007) A review of the formation and classification of amphiphilic block copolymer nanoparticulate structures: micelles, nanospheres, nanocapsules and polymersomes. Eur J Pharm Biopharm 65(3):259–269
48. Guterres SS, Alves MP, Pohlmann AR (2007) Polymeric nanoparticles, nanospheres and nanocapsules, for cutaneous applications. Drug Target Insights 2:147–157
49. Thakur VK, Thakur MK (2003) Handbook of polymers for pharmaceutical technologies, processing and applications, vol 2. Wiley, Hoboken, p 496
50. Langroodi FA et al (2016) Evaluation of the effect of crocetin on antitumor activity of doxorubicin encapsulated in PLGA nanoparticles. Nanomed J 3(1):23–34
51. Avgoustakis K (2004) Pegylated poly(lactide) and poly(lactide-co-glycolide) nanoparticles: preparation, properties and possible applications in drug delivery. Curr Drug Deliv 1 (4):321–333
52. Mora-Huertas CE, Fessi H, Elaissari A (2010) Polymer-based nanocapsules for drug delivery. Int J Pharm 385(1):113–142
53. Nagavarma BVN et al (2012) Different techniques for preparation of polymeric nanoparticles-a review. Asian J Pharm Clin Res 5:16–23
54. Mishra RK et al (2018) Recent progress in selected bio-nanomaterials and their engineering applications: an overview. J Sci Adv Mater Devices 3(3):263–288
55. Kumari A, Yadav SK, Yadav SC (2010) Biodegradable polymeric nanoparticles based drug delivery systems. Colloids Surf B: Biointerfaces 75(1):1–18
56. Chittasupho C et al (2009) ICAM-1 targeting of doxorubicin-loaded PLGA nanoparticles to lung epithelial cells. Eur J Pharm Sci 37(2):141–150
57. Zhu H et al (2014) Co-delivery of chemotherapeutic drugs with vitamin E TPGS by porous PLGA nanoparticles for enhanced chemotherapy against multi-drug resistance. Biomaterials 35(7):2391–2400

58. Muntimadugu E et al (2016) CD44 targeted chemotherapy for co-eradication of breast cancer stem cells and cancer cells using polymeric nanoparticles of salinomycin and paclitaxel. Colloids Surf B: Biointerfaces 143:532–546
59. Ramalho MJ et al (2018) Receptor-mediated PLGA nanoparticles for glioblastoma multiforme treatment. Int J Pharm 545(1):84–92
60. Karavelidis V et al (2011) Evaluating the effects of crystallinity in new biocompatible polyester nanocarriers on drug release behavior. Int J Nanomedicine 6:3021–3032
61. Pasut G, Veronese FM (2007) Polymer–drug conjugation, recent achievements and general strategies. Prog Polym Sci 32(8):933–961
62. Cheng J et al (2007) Formulation of functionalized PLGA-PEG nanoparticles for in vivo targeted drug delivery. Biomaterials 28(5):869–876
63. Moreno D et al (2008) Characterization of cisplatin cytotoxicity delivered from PLGA-systems. Eur J Pharm Biopharm 68(3):503–512
64. Wadajkar AS et al (2017) Decreased non-specific adhesivity, receptor targeted (DART) nanoparticles exhibit improved dispersion, cellular uptake, and tumor retention in invasive gliomas. J Control Release 267:144–153
65. Schneider CS et al (2015) Minimizing the non-specific binding of nanoparticles to the brain enables active targeting of Fn14-positive glioblastoma cells. Biomaterials 42:42–51
66. Parashar P et al (2019) A facile approach for fabricating CD44-targeted delivery of hyaluronic acid-functionalized PCL nanoparticles in urethane-induced lung cancer: Bcl-2, MMP-9, caspase-9, and BAX as potential markers. Drug Deliv Transl Res 9(1):37–52
67. Devalapally H et al (2008) Modulation of drug resistance in ovarian adenocarcinoma by enhancing intracellular ceramide using tamoxifen-loaded biodegradable polymeric nanoparticles. Clin Cancer Res 14(10):3193–3203
68. Xu S et al (2019) PD-L1 monoclonal antibody-conjugated nanoparticles enhance drug delivery level and chemotherapy efficacy in gastric cancer cells. Int J Nanomedicine 14:17–32
69. Zhang X et al (2019) Poly(cystine–PCL) based pH/redox dual-responsive nanocarriers for enhanced tumor therapy. Biomater Sci 7(5):1962–1972
70. Jia M et al (2014) Development of both methotrexate and mitomycin C loaded PEGylated chitosan nanoparticles for targeted drug codelivery and synergistic anticancer effect. ACS Appl Mater Interfaces 6(14):11413–11423
71. Duncan R (2014) Polymer therapeutics: top 10 selling pharmaceuticals – what next? J Control Release 190:371–380
72. Hrkach J et al (2012) Preclinical development and clinical translation of a PSMA-targeted docetaxel nanoparticle with a differentiated pharmacological profile. Sci Transl Med 4 (128):128ra39–128ra39
73. Autio KA et al (2018) Safety and efficacy of BIND-014, a docetaxel nanoparticle targeting prostate-specific membrane antigen for patients with metastatic castration-resistant prostate cancer: a phase 2 clinical trial. JAMA Oncol 4(10):1344–1351
74. Lalatsa A et al (2012) Amphiphilic poly(L-amino acids) – new materials for drug delivery. J Control Release 161(2):523–536
75. Veronese FM et al (2005) PEG-doxorubicin conjugates: influence of polymer structure on drug release, in vitro cytotoxicity, biodistribution, and antitumor activity. Bioconjug Chem 16 (4):775–784
76. Xu W, Ling P, Zhang T (2013) Polymeric micelles, a promising drug delivery system to enhance bioavailability of poorly water-soluble drugs. J Drug Deliv 2013:340315
77. Soltani F et al (2015) Synthetic and biological vesicular nano-carriers designed for gene delivery. Curr Pharm Des 21(42):6214–6235
78. Cabral H et al (2011) Accumulation of sub-100 nm polymeric micelles in poorly permeable tumours depends on size. Nat Nanotechnol 6(12):815–823
79. Miyata K, Christie RJ, Kataoka K (2011) Polymeric micelles for nano-scale drug delivery. React Funct Polym 71(3):227–234

80. Oerlemans C et al (2010) Polymeric micelles in anticancer therapy: targeting, imaging and triggered release. Pharm Res 27(12):2569–2589
81. Kim SC et al (2001) In vivo evaluation of polymeric micellar paclitaxel formulation: toxicity and efficacy. J Control Release 72(1):191–202
82. Kim TY et al (2004) Phase I and pharmacokinetic study of Genexol-PM, a cremophor-free, polymeric micelle-formulated paclitaxel, in patients with advanced malignancies. Clin Cancer Res 10(11):3708–3716
83. Tam YT et al (2019) Poly(ethylene glycol)-block-poly(d,l-lactic acid) micelles containing oligo(lactic acid)8-paclitaxel prodrug: in vivo conversion and antitumor efficacy. J Control Release 298:186–193
84. Ashton S et al (2016) Aurora kinase inhibitor nanoparticles target tumors with favorable therapeutic index in vivo. Sci Transl Med 8(325):325ra17
85. Jeong Y-I et al (2011) Doxorubicin-incorporated polymeric micelles composed of dextran-b-poly(DL-lactide-co-glycolide) copolymer. Int J Nanomedicine 6:1415–1427
86. Yoo HS, Park TG (2004) Folate receptor targeted biodegradable polymeric doxorubicin micelles. J Control Release 96(2):273–283
87. Sarisozen C et al (2016) Nanomedicine based curcumin and doxorubicin combination treatment of glioblastoma with scFv-targeted micelles: in vitro evaluation on 2D and 3D tumor models. Eur J Pharm Biopharm 108:54–67
88. Han X et al (2009) 9-NC-loaded folate-conjugated polymer micelles as tumor targeted drug delivery system: preparation and evaluation in vitro. Int J Pharm 372(1–2):125–131
89. Ren H et al (2015) EGFR-targeted poly(ethylene glycol)-distearoylphosphatidylethanolamine micelle loaded with paclitaxel for laryngeal cancer: preparation, characterization and in vitro evaluation. Drug Deliv 22(6):785–794
90. Miller K et al (2016) Delivery of a drug cache to glioma cells overexpressing platelet-derived growth factor receptor using lipid nanocarriers. Nanomedicine (London) 11(6):581–595
91. Xiao Y et al (2012) Multifunctional unimolecular micelles for cancer-targeted drug delivery and positron emission tomography imaging. Biomaterials 33(11):3071–3082
92. Ke XY et al (2014) Co-delivery of thioridazine and doxorubicin using polymeric micelles for targeting both cancer cells and cancer stem cells. Biomaterials 35(3):1096–1108
93. Miura Y et al (2013) Cyclic RGD-linked polymeric micelles for targeted delivery of platinum anticancer drugs to glioblastoma through the blood-brain tumor barrier. ACS Nano 7 (10):8583–8592
94. Wu H et al (2014) Polymeric micelles loaded with platinum anticancer drugs target preangiogenic micrometastatic niches associated with inflammation. J Control Release 189:1–10
95. Rafi M et al (2012) Polymeric micelles incorporating (1,2-diaminocyclohexane)platinum (II) suppress the growth of orthotopic scirrhous gastric tumors and their lymph node metastasis. J Control Release 159(2):189–196
96. Mochida Y, Cabral H, Kataoka K (2017) Polymeric micelles for targeted tumor therapy of platinum anticancer drugs. Expert Opin Drug Deliv 14(12):1423–1438
97. Endo K et al (2013) Tumor-targeted chemotherapy with the nanopolymer-based drug NC-6004 for oral squamous cell carcinoma. Cancer Sci 104(3):369–374
98. Cabral H et al (2013) Targeted therapy of spontaneous murine pancreatic tumors by polymeric micelles prolongs survival and prevents peritoneal metastasis. Proc Natl Acad Sci U S A 110 (28):11397–11402
99. Doi T et al (2017) NC-6004 phase I study in combination with gemcitabine for advanced solid tumors and population PK/PD analysis. Cancer Chemother Pharmacol 79(3):569–578
100. Subbiah V et al (2018) Phase Ib/II trial of NC-6004 (nanoparticle cisplatin) plus gemcitabine in patients with advanced solid tumors. Clin Cancer Res 24(1):43–51
101. Suzuki K et al (2019) Glucose transporter 1-mediated vascular translocation of nanomedicines enhances accumulation and efficacy in solid tumors. J Control Release 301:28–41

102. Yi Y et al (2019) Glucose-linked sub-50-nm unimer polyion complex-assembled gold nanoparticles for targeted siRNA delivery to glucose transporter 1-overexpressing breast cancer stem-like cells. J Control Release 295:268–277
103. Jin X et al (2014) Paclitaxel-loaded N-octyl-O-sulfate chitosan micelles for superior cancer therapeutic efficacy and overcoming drug resistance. Mol Pharm 11(1):145–157
104. Lee AL et al (2009) The co-delivery of paclitaxel and Herceptin using cationic micellar nanoparticles. Biomaterials 30(5):919–927
105. Cai X et al (2016) pH-responsive copolymers based on pluronic P123-poly(β-amino ester): synthesis, characterization and application of copolymer micelles. Colloids Surf B: Biointerfaces 142:114–122
106. Li J et al (2014) Dual endogenous stimuli-responsive polyplex micelles as smart two-step delivery nanocarriers for deep tumor tissue penetration and combating drug resistance of cisplatin. J Mater Chem B 2(13):1813–1824
107. Li H et al (2017) Design of block copolymer micellar aggregates for co-delivery of enzyme and anticancer prodrug. Chem Asian J 12(2):176–180
108. Prabaharan M et al (2009) Thermosensitive micelles based on folate-conjugated poly (N-vinylcaprolactam)-block-poly(ethylene glycol) for tumor-targeted drug delivery. Macromol Biosci 9(8):744–753
109. Xue S et al (2018) Construction of small-sized, robust, and reduction-responsive polypeptide micelles for high loading and targeted delivery of chemotherapeutics. Biomacromolecules 19 (8):3586–3593
110. Gu X et al (2019) cRGD-decorated biodegradable polytyrosine nanoparticles for robust encapsulation and targeted delivery of doxorubicin to colorectal cancer in vivo. J Control Release 301:110–118
111. Valle JW et al (2011) A phase 2 study of SP1049C, doxorubicin in P-glycoprotein-targeting pluronics, in patients with advanced adenocarcinoma of the esophagus and gastroesophageal junction. Investig New Drugs 29(5):1029–1037
112. Miller MA et al (2017) Nano-palladium is a cellular catalyst for in vivo chemistry. Nat Commun 8:15906
113. Lee JS, Feijen J (2012) Polymersomes for drug delivery: design, formation and characterization. J Control Release 161(2):473–483
114. Discher DE, Eisenberg A (2002) Polymer vesicles. Science 297(5583):967–973
115. Mai Y, Eisenberg A (2012) Self-assembly of block copolymers. Chem Soc Rev 41 (18):5969–5985
116. Levine DH et al (2008) Polymersomes: a new multi-functional tool for cancer diagnosis and therapy. Methods 46(1):25–32
117. Rideau E et al (2018) Liposomes and polymersomes: a comparative review towards cell mimicking. Chem Soc Rev 47(23):8572–8610
118. Chang H-Y, Sheng Y-J, Tsao H-K (2014) Structural and mechanical characteristics of polymersomes. Soft Matter 10(34):6373–6381
119. Xu J et al (2014) High loading of hydrophilic/hydrophobic doxorubicin into polyphosphazene polymersome for breast cancer therapy. Nanomedicine 10(2):349–358
120. Upadhyay KK et al (2012) The in vivo behavior and antitumor activity of doxorubicin-loaded poly(gamma-benzyl l-glutamate)-block-hyaluronan polymersomes in Ehrlich ascites tumor-bearing BalB/c mice. Nanomedicine 8(1):71–80
121. Yassin MA et al (2015) Overcoming concealment effects of targeting moieties in the PEG corona: controlled permeable polymersomes decorated with folate-antennae for selective targeting of tumor cells. Small 11(13):1580–1591
122. Petersen MA, Hillmyer MA, Kokkoli E (2013) Bioresorbable polymersomes for targeted delivery of cisplatin. Bioconjug Chem 24(4):533–543
123. Ke W et al (2019) Therapeutic polymersome nanoreactors with tumor-specific activable cascade reactions for cooperative cancer therapy. ACS Nano 13(2):2357–2369

124. Li J et al (2017) Polymer prodrug-based nanoreactors activated by tumor acidity for orchestrated oxidation/chemotherapy. Nano Lett 17(11):6983–6990
125. Li J et al (2017) Therapeutic vesicular nanoreactors with tumor-specific activation and self-destruction for synergistic tumor ablation. Angew Chem Int Ed 56(45):14025–14030
126. Martin C et al (2016) Cholesteryl to improve the cellular uptake of polymersomes within HeLa cells. Int J Pharm 511(1):570–578
127. Aibani N et al (2018) Electroneutral polymersomes for combined cancer chemotherapy. Acta Biomater 80:327–340
128. Simón-Gracia L et al (2016) Paclitaxel-loaded polymersomes for enhanced intraperitoneal chemotherapy. Mol Cancer Ther 15(4):670–679
129. Simón-Gracia L et al (2016) iRGD peptide conjugation potentiates intraperitoneal tumor delivery of paclitaxel with polymersomes. Biomaterials 104:247–257
130. Zhang P, Wagner E (2017) History of polymeric gene delivery systems. Top Curr Chem (Cham) 375(2):26
131. Plank C et al (1996) Activation of the complement system by synthetic DNA complexes: a potential barrier for intravenous gene delivery. Hum Gene Ther 7(12):1437–1446
132. Xiaoli S, Na Z (2010) Cationic polymer optimization for efficient gene delivery. Mini-Rev Med Chem 10(2):108–125
133. Jin L et al (2014) Current progress in gene delivery technology based on chemical methods and nano-carriers. Theranostics 4(3):240–255
134. Nomoto T et al (2011) In situ quantitative monitoring of polyplexes and polyplex micelles in the blood circulation using intravital real-time confocal laser scanning microscopy. J Control Release 151(2):104–109
135. Li SD, Huang L (2006) Gene therapy progress and prospects: non-viral gene therapy by systemic delivery. Gene Ther 13:1313
136. Blessing T et al (2001) Different strategies for formation of pegylated EGF-conjugated PEI/DNA complexes for targeted gene delivery. Bioconjug Chem 12(4):529–537
137. Ogris M et al (2003) Tumor-targeted gene therapy: strategies for the preparation of ligand-polyethylene glycol-polyethylenimine/DNA complexes. J Control Release 91(1–2):173–181
138. Keiji I, Kazunori K (2011) Progress and prospects of polyplex nanomicelles for plasmid DNA delivery. Curr Gene Ther 11(6):457–465
139. Nie X et al (2019) Interactions in DNA condensation: an important factor for improving the efficacy of gene transfection. Bioconjug Chem 30(2):284–292
140. Bloomfield VA (1991) Condensation of DNA by multivalent cations: considerations on mechanism. Biopolymers 31(13):1471–1481
141. Wagner E et al (1991) Transferrin-polycation-DNA complexes: the effect of polycations on the structure of the complex and DNA delivery to cells. Proc Natl Acad Sci U S A 88 (10):4255–4259
142. Reinhard S, Wagner E (2017) How to tackle the challenge of siRNA delivery with sequence-defined Oligoamino amides. Macromol Biosci 17(1):1600152
143. Brunner S et al (2000) Cell cycle dependence of gene transfer by lipoplex, polyplex and recombinant adenovirus. Gene Ther 7(5):401–407
144. Levacic AK et al (2017) Minicircle versus plasmid DNA delivery by receptor-targeted polyplexes. Hum Gene Ther 28(10):862–874
145. Boeckle S et al (2006) Melittin analogs with high lytic activity at endosomal pH enhance transfection with purified targeted PEI polyplexes. J Control Release 112(2):240–248
146. Kloeckner J et al (2006) DNA polyplexes based on degradable oligoethylenimine-derivatives: combination with EGF receptor targeting and endosomal release functions. J Control Release 116(2):115–122
147. Walker GF et al (2005) Toward synthetic viruses: endosomal pH-triggered deshielding of targeted polyplexes greatly enhances gene transfer in vitro and in vivo. Mol Ther 11 (3):418–425

148. Behr J-P (1997) The proton sponge: a trick to enter cells the viruses did not exploit. CHIMIA Int J Chem 51(1–2):34–36
149. Lachelt U et al (2014) Fine-tuning of proton sponges by precise diaminoethanes and histidines in pDNA polyplexes. Nanomedicine 10(1):35–44
150. Christie RJ, Kataoka K, Nishiyama N (2010) Minireview: delivering the code: polyplex carriers for deoxyribonucleic acid and ribonucleic acid interference therapies. Endocrinology 151(2):466–473
151. Tockary TA et al (2013) Tethered PEG crowdedness determining shape and blood circulation profile of polyplex micelle gene carriers. Macromolecules 46(16):6585–6592
152. Chen Q et al (2015) A tadpole-shaped gene carrier with distinct phase segregation in a ternary polymeric micelle. Soft Matter 11(14):2718–2722
153. Osada K et al (2012) Enhanced gene expression promoted by the quantized folding of pDNA within polyplex micelles. Biomaterials 33(1):325–332
154. Mintzer MA, Simanek EE (2009) Nonviral vectors for gene delivery. Chem Rev 109 (2):259–302
155. Nicolas J et al (2013) Design, functionalization strategies and biomedical applications of targeted biodegradable/biocompatible polymer-based nanocarriers for drug delivery. Chem Soc Rev 42(3):1147–1235
156. Dunlap DD et al (1997) Nanoscopic structure of DNA condensed for gene delivery. Nucleic Acids Res 25(15):3095–3101
157. Tang MX, Szoka FC (1997) The influence of polymer structure on the interactions of cationic polymers with DNA and morphology of the resulting complexes. Gene Ther 4(8):823–832
158. Liu G et al (2001) Biological properties of poly-L-lysine-DNA complexes generated by cooperative binding of the polycation. J Biol Chem 276(37):34379–34387
159. Godbey WT, Wu KK, Mikos AG (1999) Tracking the intracellular path of poly(ethylenimine)/ DNA complexes for gene delivery. Proc Natl Acad Sci U S A 96(9):5177–5181
160. Wagner E et al (1992) Influenza virus hemagglutinin HA-2 N-terminal fusogenic peptides augment gene transfer by transferrin-polylysine-DNA complexes: toward a synthetic virus-like gene-transfer vehicle. Proc Natl Acad Sci U S A 89(17):7934–7938
161. Cotten M et al (1994) Psoralen treatment of adenovirus particles eliminates virus replication and transcription while maintaining the endosomolytic activity of the virus capsid. Virology 205(1):254–261
162. Wagner E (1998) Effects of membrane-active agents in gene delivery. J Control Release 53 (1–3):155–158
163. Meyer M et al (2008) Breathing life into polycations: functionalization with pH-responsive endosomolytic peptides and polyethylene glycol enables siRNA delivery. J Am Chem Soc 130 (11):3272–3273
164. Hall A et al (2017) Polyplex evolution: understanding biology, optimizing performance. Mol Ther 25(7):1476–1490
165. Kloeckner J et al (2006) Gene carriers based on hexanediol diacrylate linked oligoethylenimine: effect of chemical structure of polymer on biological properties. Bioconjug Chem 17(5):1339–1345
166. Knorr V et al (2008) Acetal linked oligoethylenimines for use as pH-sensitive gene carriers. Bioconjug Chem 19(8):1625–1634
167. Russ V et al (2008) Novel degradable oligoethylenimine acrylate ester-based pseudodendrimers for in vitro and in vivo gene transfer. Gene Ther 15(1):18–29
168. Abourbeh G et al (2012) PolyIC GE11 polyplex inhibits EGFR-overexpressing tumors. IUBMB Life 64(4):324–330
169. Nie Y et al (2011) Dual-targeted polyplexes: one step towards a synthetic virus for cancer gene therapy. J Control Release 152(1):127–134
170. Wang J et al (2014) Retro-inverso CendR peptide-mediated polyethyleneimine for intracranial glioblastoma-targeting gene therapy. Bioconjug Chem 25(2):414–423

171. Jorge AF et al (2015) Combining polyethylenimine and Fe(III) for mediating pDNA transfection. Biochim Biophys Acta 1850(6):1325–1335
172. Noga M et al (2014) Characterization and compatibility of hydroxyethyl starch-polyethylenimine copolymers for DNA delivery. J Biomater Sci Polym Ed 25(9):855–871
173. Rodl W et al (2019) Synthesis of polyethylenimine-based nanocarriers for systemic tumor targeting of nucleic acids. Methods Mol Biol 1943:83–99
174. Haase R et al (2013) Generation of a tumor- and tissue-specific episomal non-viral vector system. BMC Biotechnol 13:49
175. Ogris M, Wagner E (2012) Synthesis of linear polyethylenimine and use in transfection. Cold Spring Harb Protoc 2012(2):246–250
176. Merkel OM et al (2011) In vitro and in vivo complement activation and related anaphylactic effects associated with polyethylenimine and polyethylenimine-graft-poly(ethylene glycol) block copolymers. Biomaterials 32(21):4936–4942
177. He D, Wagner E (2015) Defined polymeric materials for gene delivery. Macromol Biosci 15 (5):600–612
178. Lehto T, Wagner E (2014) Sequence-defined polymers for the delivery of oligonucleotides. Nanomedicine 9(18):2843–2859
179. Hartmann L et al (2006) Solid-phase supported polymer synthesis of sequence-defined, multifunctional poly(amidoamines). Biomacromolecules 7(4):1239–1244
180. Martin I et al (2012) Solid-phase-assisted synthesis of targeting peptide-PEG-oligo(ethane amino)amides for receptor-mediated gene delivery. Org Biomol Chem 10(16):3258–3268
181. Frohlich T et al (2012) Structure-activity relationships of siRNA carriers based on sequence-defined oligo (ethane amino) amides. J Control Release 160(3):532–541
182. Dohmen C et al (2012) Nanosized multifunctional polyplexes for receptor-mediated siRNA delivery. ACS Nano 6(6):5198–5208
183. Schaffert D et al (2011) Solid-phase synthesis of sequence-defined T-, i-, and U-shape polymers for pDNA and siRNA delivery. Angew Chem Int Ed 50(38):8986–8989
184. Salcher EE et al (2012) Sequence-defined four-arm oligo(ethanamino)amides for pDNA and siRNA delivery: impact of building blocks on efficacy. J Control Release 164(3):380–386
185. Schaffert D, Badgujar N, Wagner E (2011) Novel Fmoc-polyamino acids for solid-phase synthesis of defined polyamidoamines. Org Lett 13(7):1586–1589
186. Lachelt U et al (2014) Synthetic polyglutamylation of dual-functional MTX ligands for enhanced combined cytotoxicity of poly(I:C) nanoplexes. Mol Pharm 11(8):2631–2639
187. Lee DJ et al (2016) Dual antitumoral potency of EG5 siRNA nanoplexes armed with cytotoxic bifunctional glutamyl-methotrexate targeting ligand. Biomaterials 77:98–110
188. An S et al (2015) Peptide-like polymers exerting effective glioma-targeted siRNA delivery and release for therapeutic application. Small 11(38):5142–5150
189. Lee DJ et al (2016) Tumoral gene silencing by receptor-targeted combinatorial siRNA polyplexes. J Control Release 244(Pt B):280–291
190. Lee DJ et al (2017) Systemic delivery of folate-PEG siRNA lipopolyplexes with enhanced intracellular stability for in vivo gene silencing in leukemia. Bioconjug Chem 28 (9):2393–2409
191. Troiber C et al (2013) Stabilizing effect of tyrosine trimers on pDNA and siRNA polyplexes. Biomaterials 34(5):1624–1633
192. Kos P et al (2015) Histidine-rich stabilized polyplexes for cMet-directed tumor-targeted gene transfer. Nanoscale 7(12):5350–5362
193. Ahn BC (2012) Sodium iodide symporter for nuclear molecular imaging and gene therapy: from bedside to bench and back. Theranostics 2(4):392–402
194. Klutz K et al (2011) Epidermal growth factor receptor-targeted 131I-therapy of liver cancer following systemic delivery of the sodium iodide symporter gene. Mol Ther 19(4):676–685
195. Schmohl KA et al (2017) Imaging and targeted therapy of pancreatic ductal adenocarcinoma using the theranostic sodium iodide symporter (NIS) gene. Oncotarget 8(20):33393–33404

196. Urnauer S et al (2017) Systemic tumor-targeted sodium iodide symporter (NIS) gene therapy of hepatocellular carcinoma mediated by B6 peptide polyplexes. J Gene Med 19(5)
197. Urnauer S et al (2016) Sequence-defined cMET/HGFR-targeted polymers as gene delivery vehicles for the theranostic sodium iodide symporter (NIS) gene. Mol Ther 24(8):1395–1404
198. Morys S et al (2018) EGFR targeting and shielding of pDNA lipopolyplexes via bivalent attachment of a sequence-defined PEG agent. Macromol Biosci 18(1):1700203
199. Wang S et al (2017) Antitumoral cascade-targeting ligand for IL-6 receptor-mediated gene delivery to glioma. Mol Ther 25(7):1556–1566
200. Scholz C et al (2014) Correlation of length of linear oligo(ethanamino) amides with gene transfer and cytotoxicity. ChemMedChem 9(9):2104–2110
201. Steinborn B et al (2018) Epidermal growth factor receptor targeted methotrexate and small interfering RNA co-delivery. J Gene Med 20(7–8):e3041
202. Truebenbach I et al (2019) Combination chemotherapy of L1210 tumors in mice with pretubulysin and methotrexate lipo-oligomer nanoparticles. Mol Pharm 16(6):2405–2417
203. Truebenbach I et al (2019) Co-delivery of pretubulysin and siEG5 to EGFR overexpressing carcinoma cells. Int J Pharm 569:118570
204. Luo J et al (2019) IL4-receptor-targeted dual antitumoral apoptotic peptide—siRNA conjugate lipoplexes. Adv Funct Mater 29:1900697
205. Kanayama N et al (2006) A PEG-based biocompatible block catiomer with high buffering capacity for the construction of polyplex micelles showing efficient gene transfer toward primary cells. ChemMedChem 1(4):439–444
206. Miyata K et al (2008) Polyplexes from poly(aspartamide) bearing 1,2-diaminoethane side chains induce pH-selective, endosomal membrane destabilization with amplified transfection and negligible cytotoxicity. J Am Chem Soc 130(48):16287–16294
207. Itaka K et al (2010) Biodegradable polyamino acid-based polycations as safe and effective gene carrier minimizing cumulative toxicity. Biomaterials 31(13):3707–3714
208. Kim HJ et al (2010) Introduction of stearoyl moieties into a biocompatible cationic polyaspartamide derivative, PAsp(DET), with endosomal escaping function for enhanced siRNA-mediated gene knockdown. J Control Release 145(2):141–148
209. Ge Z et al (2014) Targeted gene delivery by polyplex micelles with crowded PEG palisade and cRGD moiety for systemic treatment of pancreatic tumors. Biomaterials 35(10):3416–3426
210. Chen Q et al (2017) Polyplex micelle installing intracellular self-processing functionalities without free catiomers for safe and efficient systemic gene therapy through tumor vasculature targeting. Biomaterials 113:253–265
211. Uchida S et al (2013) In vivo messenger RNA introduction into the central nervous system using polyplex nanomicelle. PLoS One 8(2):e56220
212. Uchida S et al (2016) Systemic delivery of messenger RNA for the treatment of pancreatic cancer using polyplex nanomicelles with a cholesterol moiety. Biomaterials 82:221–228
213. Furugaki K et al (2014) Intraperitoneal administration of a tumor-associated antigen SART3, CD40L, and GM-CSF gene-loaded polyplex micelle elicits a vaccine effect in mouse tumor models. PLoS One 9(7):e101854
214. Cheng Y, Yumul RC, Pun SH (2016) Virus-inspired polymer for efficient in vitro and in vivo gene delivery. Angew Chem Int Ed Eng 55(39):12013–12017
215. Schreiber S et al (1999) Immunotherapy of metastatic malignant melanoma by a vaccine consisting of autologous interleukin 2-transfected cancer cells: outcome of a phase I study. Hum Gene Ther 10(6):983–993
216. Sidi AA et al (2008) Phase I/II marker lesion study of intravesical BC-819 DNA plasmid in H19 over expressing superficial bladder cancer refractory to bacillus Calmette-Guerin. J Urol 180(6):2379–2383
217. Davis ME et al (2010) Evidence of RNAi in humans from systemically administered siRNA via targeted nanoparticles. Nature 464:1067–1070
218. Veronese FM, Pasut G (2005) PEGylation, successful approach to drug delivery. Drug Discov Today 10(21):1451–1458

219. Hou Y et al (2016) A concise approach to site-specific topological protein-poly(amino acid) conjugates enabled by in situ-generated functionalities. J Am Chem Soc 138 (34):10995–11000
220. Hou Y et al (2018) Macrocyclization of interferon-poly(alpha-amino acid) conjugates significantly improves the tumor retention, penetration, and antitumor efficacy. J Am Chem Soc 140 (3):1170–1178
221. Hou Y et al (2019) Therapeutic protein PEPylation: the helix of nonfouling synthetic polypeptides minimizes antidrug antibody generation. ACS Cent Sci 5(2):229–236
222. Zhang P et al (2015) Enhanced intracellular protein transduction by sequence defined tetraoleoyl oligoaminoamides targeted for cancer therapy. Adv Funct Mater 25(42):6627–6636
223. Zhang P et al (2017) Lipo-oligomer nanoformulations for targeted intracellular protein delivery. Biomacromolecules 18:2509–2520
224. Liu X et al (2016) pH-reversible cationic RNase A conjugates for enhanced cellular delivery and tumor cell killing. Biomacromolecules 17(1):173–182
225. Maier K, Wagner E (2012) Acid-labile traceless click linker for protein transduction. J Am Chem Soc 134(24):10169–10173
226. Lammers T et al (2016) Cancer nanomedicine: is targeting our target? Nat Rev Mater 1(9):pii: 16069
227. Yuan F et al (1995) Vascular permeability in a human tumor xenograft: molecular size dependence and cutoff size. Cancer Res 55(17):3752–3756
228. Ojha T et al (2017) Pharmacological and physical vessel modulation strategies to improve EPR-mediated drug targeting to tumors. Adv Drug Deliv Rev 119:44–60
229. Diop-Frimpong B et al (2011) Losartan inhibits collagen I synthesis and improves the distribution and efficacy of nanotherapeutics in tumors. Proc Natl Acad Sci U S A 108 (7):2909–2914
230. Chauhan VP et al (2013) Angiotensin inhibition enhances drug delivery and potentiates chemotherapy by decompressing tumour blood vessels. Nat Commun 4:2516
231. Zhao Y et al (2019) Losartan treatment enhances chemotherapy efficacy and reduces ascites in ovarian cancer models by normalizing the tumor stroma. Proc Natl Acad Sci U S A 116 (6):2210–2219
232. Urnauer S et al (2019) Dual-targeted NIS polyplexes-a theranostic strategy toward tumors with heterogeneous receptor expression. Gene Ther 26:93–108
233. Mi P et al (2017) Molecular cancer imaging with polymeric nanoassemblies: from tumor detection to theranostics. Macromol Biosci 17(1):1600305
234. Arranja AG et al (2017) Tumor-targeted nanomedicines for cancer theranostics. Pharmacol Res 115:87–95
235. Upponi JR et al (2018) Polymeric micelles: theranostic co-delivery system for poorly water-soluble drugs and contrast agents. Biomaterials 170:26–36
236. Li K, Nejadnik H, Daldrup-Link HE (2017) Next-generation superparamagnetic iron oxide nanoparticles for cancer theranostics. Drug Discov Today 22(9):1421–1429
237. Chen J et al (2017) Theranostic multilayer capsules for ultrasound imaging and guided drug delivery. ACS Nano 11(3):3135–3146
238. Yue X, Zhang Q, Dai Z (2017) Near-infrared light-activatable polymeric nanoformulations for combined therapy and imaging of cancer. Adv Drug Deliv Rev 115:155–170
239. Zintchenko A et al (2009) Drug nanocarriers labeled with near-infrared-emitting quantum dots (quantoplexes): imaging fast dynamics of distribution in living animals. Mol Ther 17 (11):1849–1856
240. Wang S et al (2017) Augmented glioma-targeted theranostics using multifunctional polymer-coated carbon nanodots. Biomaterials 141:29–39
241. Wang Y et al (2014) Synthesis of core-shell graphitic carbon@silica nanospheres with dual-ordered mesopores for cancer-targeted photothermochemotherapy. ACS Nano 8 (8):7870–7879

242. Jin Y, Yang X, Tian J (2018) Targeted polypyrrole nanoparticles for the identification and treatment of hepatocellular carcinoma. Nanoscale 10(20):9594–9601
243. Jin Y et al (2017) A tantalum oxide-based core/shell nanoparticle for triple-modality image-guided chemo-thermal synergetic therapy of esophageal carcinoma. Cancer Lett 397:61–71
244. Howard KA et al (2007) Nanocarrier stimuli-activated gene delivery. Small 3(1):54–57
245. Hager S, Wagner E (2018) Bioresponsive polyplexes – chemically programmed for nucleic acid delivery. Expert Opin Drug Deliv 15(11):1067–1083

Chapter 12
Other New Tumor-targeted Systems

Yibo Xie, Min Qian, Xiaoyi Zhang, and Rongqin Huang

Abstract Nowadays there are more and more rare material-based tumor-targeted systems such as black phosphorus-based tumor-targeted systems, upconversion nanoparticle-based tumor-targeted systems, reactive oxygen and nitrogen species (RONS) scavenger-based tumor-targeted systems, perfluorocarbon nanoparticles, sulfide-based tumor-targeted systems, oxide-based tumor-targeted systems, and biomimetic material-based tumor-targeted systems. The properties of the systems are mostly dependent on the specialties of materials. This chapter will introduce both the materials and systems and focus on the targeting efficiency, bioactivity, and the relating mechanisms of killing tumor cells and inhibiting the progress of tumor growth.

Keywords Black phosphorus · Upconversion · RONS scavenger · Perfluorocarbon · Biomimetic materials

12.1 Introduction

Increased generation of reactive oxygen species (ROS) and other material-based tumor-targeted systems like black phosphorus (BP), upconversion nanoparticles, and so on has been developed for targeting tumor sites. The properties of the system are mostly dependent on the specialties of materials, and some are related to the anatomical and pathophysiological condition like the enhanced permeability and retention (EPR) effect. BP have attracted considerable attention for applications in optoelectronics, energy storage, and biomedicine due to its exceptional properties. The properties of the system are mostly dependent on the specialties of materials. Rare-earth doped upconversion nanoparticles (UCNPs) have been widely applied in

Y. Xie · M. Qian · X. Zhang · R. Huang (✉)
Department of Pharmaceutics, School of Pharmacy, Key Laboratory of Smart Drug Delivery, Ministry of Education, Fudan University, Shanghai, China
e-mail: rqhuang@fudan.edu.cn

R. Huang, Y. Wang (eds.), *New Nanomaterials and Techniques for Tumor-targeted Systems*, https://doi.org/10.1007/978-981-15-5159-8_12

bioimaging like cell labeling, racking, and multicolor upconversion luminescent (UCL) imaging, due to their superior physicochemical features such as high photostability, deep tissue penetration capability, multi-wavelength emission, and extremely weak autofluorescence compared to other conventional molecular probes like organic fluorophores. As increased production of reactive oxygen species (ROS) and changes in redox status have been observed in cancer cells, recent studies have shown that this biochemical property of cancer cells can be treated with new materials such as perfluorocarbon (PFO) with high oxygen capacity and low boiling point. These specialties can endow perfluorocarbon (PFO) some specific medical functions. Oxide-based tumor-targeted systems also have robust therapy effect, while biomimetic material-based tumor-targeted systems have low side effect. This chapter will introduce both the materials and systems, focusing on the targeting efficiency, bioactivity, and the mechanism of killing tumor cells and inhibiting the progress of tumor growth.

12.2 Black Phosphorus-Based Tumor-targeted Systems

Recently, as a new 2D material, BP have attracted considerable attention for applications in optoelectronics, energy storage, and biomedicine due to its exceptional properties [1–3]. Either from bulk BP through top-down methods (e.g., mechanical cleavage and liquid-phase exfoliation) or P source via bottom-up methods (e.g., chemical vapor deposition (CVD) and wet chemistry), BP nanomaterials, including BP nanosheets (BP NSs) and BP quantum dots (BP QDs), of different thickness can be produced with a layer-dependent band gap that spans from 0.3 eV (a bulk value) to $\approx$2.0 eV (a monolayer value), leading to the excellent optical properties from ultraviolet (UV) to near-infrared (NIR) regions for bioimaging [4]. Furthermore, the large NIR extinction coefficient and high photothermal conversion efficiency also qualified BP as a promising nanomaterial for photothermal therapy (PTT) and photodynamic therapy (PDT) [5]. Meanwhile, the puckered honeycomb 2D structure of BP is held together by van der Waals forces and provides an ultralarge surface, where many drugs, bioactive molecules, fluorescent molecules, and metal atoms can be loaded via nondestructive non-covalent bonds. Furthermore, the structure has the advantages of maintaining the bioactivity of the loaded agents and allowing a number of biological applications, e.g., targeted drug delivery, biomolecule detection, cell imaging, and cancer therapy [6, 7]. Most importantly, BP has inappreciable cytotoxicity and excellent biodegradability in vivo owing to the final degradation products which are nontoxic phosphate and phosphonate, guaranteeing BP as an especially attractive and omnipotent nanoplatform for multidisciplinary biomedical applications compared to other inorganic nanomaterials [8]. However, BP has been found to be highly reactive to oxygen and water, which leads to changes in the composition and physical properties of BP. The lack of water and air stability under ambient conditions hinders its potential biomedical applications [9].

Herein, abundant efforts and strategies such as ligand surface coordination, covalent aryl diazonium functionalization, and capping layer protection have been developed to prevent BP from rapid ambient degradation. In 2016, after Lee et al. first investigated BP QDs for biological imaging applications, Shao and coworkers developed PLGA encapsulated BP QDs (BPQDs/PLGA) nanospheres with excellent biodegradability and biocompatibility for highly efficient tumor ablation under NIR laser illumination in vivo where the hydrophobic PLGA not only isolated the interior BP QDs from oxygen and water to enhance the photothermal stability but also controlled the degradation rate of BP QDs (Fig. 12.1) [8]. Analogously, for the targeted chemo, gene, and photothermal therapy against multidrug-resistant cancer, Zeng et al. improved both stability and photothermal performance of bare BP NSs via a simple polydopamine (PDA) modification where PDA itself has a strong NIR absorbility and high photothermal conversion efficiency and can also biodegrade at low pH values which is rather suitable for the enhancement of unitary photothermal effect and manipulation of intratumor degradation [11].

Recently, Chen et al. prepared an ordered mesoporous silica-sandwiched black phosphorus nanosheet (BP@MS) with the vertical pore coating, which could not only enhance the dispersibility of BP and increase the loading efficiency of its doxorubicin (DOX) but also promote post-modifications, such as PEGylation and conjugation of targeting ligand, TKD peptides, thereby generating BSPT [12].

Although the vast biomedical application prospects of BP are very exciting and promising, biomedical exploration of BP nanomaterials is still relatively new compared to that of other 2D counterparts. Many critical issues like low yield, easy aggregation, and lack of biosafety and biocompatibility information still need to be systematically investigated for the future pervasive applications.

12.3 Upconversion Nanoparticle-Based Tumor-targeted Systems

As a representative for inorganic nanomedicines, rare-earth doped upconversion nanoparticles (UCNPs) have been widely applied in bioimaging like cell labeling, racking, and multicolor upconversion luminescent (UCL) imaging, due to their superior physicochemical features such as high photostability, deep tissue penetration capability, multi-wavelength emission, and extremely weak autofluorescence compared to other conventional molecular probes like organic fluorophores [13–16]. Generally, UCNPs consist of a crystalline host matrix (e.g., $NaYF_4$, LaF_3, Y_2O_3, etc.) doped with diverse lanthanide ions (e.g., Yb^{3+}, Er^{3+}, Tm^{3+}, etc.) where the sensitizer ion can absorb and transfer NIR energy to the activator ion leading to the conversion of low-energy near-infrared (NIR) light into high-energy ultraviolet/visible (UV/vis) light [13, 17]. In order to make up for the shortcomings of ceach-imaging modality, several kinds of functional ions have been co-doped into the matrix of the UCNPs to achieve multimodal imaging with high detection sensitivity and spatial resolution. Furthermore, following the "detect-to-treat" strategy, their biomedical applications have gradually shifted from single tumor imaging to

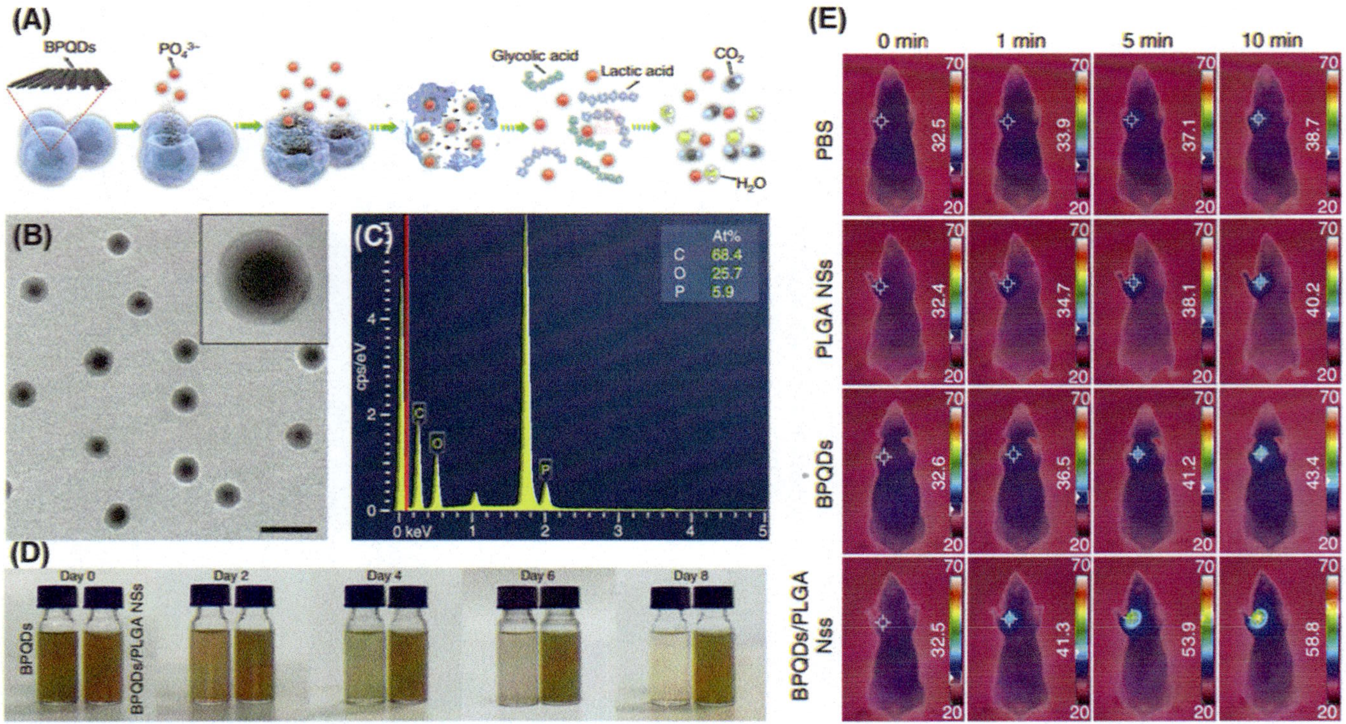

Fig. 12.1 (**a**) Schematic representation of the degradation process of the BPQDs/PLGA nanospheres in the physiological environment. (**b**) TEM image of the BPQDs/PLGA nanospheres (scale bar, 200 nm) with the inset displaying the magnified TEM image of a BPQDs/PLGA nanosphere. (**c**) Energy dispersive X-ray spectroscopy analysis of the BPQDs/PLGA nanospheres. (**d**) Photographs of the BP QDs and BPQDs/PLGA nanospheres with the same amount of BP QDs (20 p.p.m.) after storing in water for different periods of time. (**e**) Infrared thermographic maps in the MCF7 breast tumor-bearing nude mice irradiated by the 808 nm laser (1 W cm^{-2}) at 24 h after separate intravenous injection with 100 ml of PBS, PLGA nanospheres, BPQDs (1 mg ml^{-1}), and BPQDs/PLGA nanospheres (1 mg BP ml^{-1}) with the color bar referring to the relative temperature. (Reproduced with permission from Ref. [8]. Copyright©2016, Springer Nature)

multimodal imaging-guided light/heat-triggered cancer therapy which is leap-forward development of UCNP-based nanomedicines and cancer theranostic techniques. However, the relatively low NIR-UV/vis conversion efficiency of UCNPs remains a major challenge to its future clinical translational applications, although NIR light is more penetrating than conventional visible or ultraviolet light in various types of photo-induced therapies. Low power density and short duration on NIR radiation will be used for negligible heating damage, which extrudes virtual improvements in upconversion efficiency [18].

The Zhao's group adopted this highly efficient and improved method. They synthesized a Janus mesoporous silica composite (UCNPs@ SiO_2@mSiO_2&PMO, PMO: periodic mesoporous organosilica) modified with heat-sensitive phase-change materials (PCM) 1-tetradecanol and light-sensitive azo molecules. By loading hydrophobic PTX and hydrophilic DOX into the PMO and UCNPs@SiO_2@mSiO_2 domains, respectively, they can combine the use of NIR irradiation and heat, which could trigger the simultaneous release of both PTX and DOX to achieve a significantly higher cell-killing efficiency (over 50%), while single heat could only cause the release of PTX for a killing efficiency of 25% [19].

Besides, the conversion of NIR light into UV/vis made UCNPs such suitable for assembling with those UV-triggered photodynamic sensors for the improvement of tissue-penetration depth excitation and this method would greatly expand the clinical application of those excellent UV-triggered photosensors like TiO_2. Thus, with a uniform TiO_2 layer coating stably on UCNPs, Zhang and coworkers developed a core-shell-structured PDT agent (TiO_2-UCN). Thanks to the NIR-to-UV upconversion of the UCNPs, different ROS ($^1O_2$, OH·, and $O_2·^-$) were produced by the TiO_2 layer causing cell death upon NIR irradiation [20]. In the following study, they have conducted detailed in vivo and in vitro experiments on TiO_2-UCN whose surface was modified by Mal-PEG to explore its application in NIR-excited deep PDT (Fig. 12.2). The results showed that Mal-PEG-TiO_2-UCN had significant PDT effect on NIR radiation, and the survival period of mice was extended to more than 60 days [21].

12.4 RONS Scavenger-Based Tumor-targeted Systems

12.4.1 Ceria-Based Tumor-targeted Systems

Ceria nanoparticles (CeNPs) are well-known catalysts that show remarkable pharmacological potential due to their antioxidant properties. Such property is derived from the mixed valence state of Ce^{3+} and Ce^{4+}, while Ce^{3+} is responsible for eliminating superoxide anions ($O_2·^-$) and hydroxyl radicals (·OH), and Ce^{4+} eradicates hydrogen peroxide (H_2O_2) [22]. CeNPs scavenge free radicals by reversibly binding oxygen and shifting between the Ce^{3+} (reduced) and Ce^{4+} (oxidized) forms at the particle surface (Fig. 12.3). This ability is comparable to biological antioxidants. And attributed to this, we name SOD-mimetic activity for the ability to

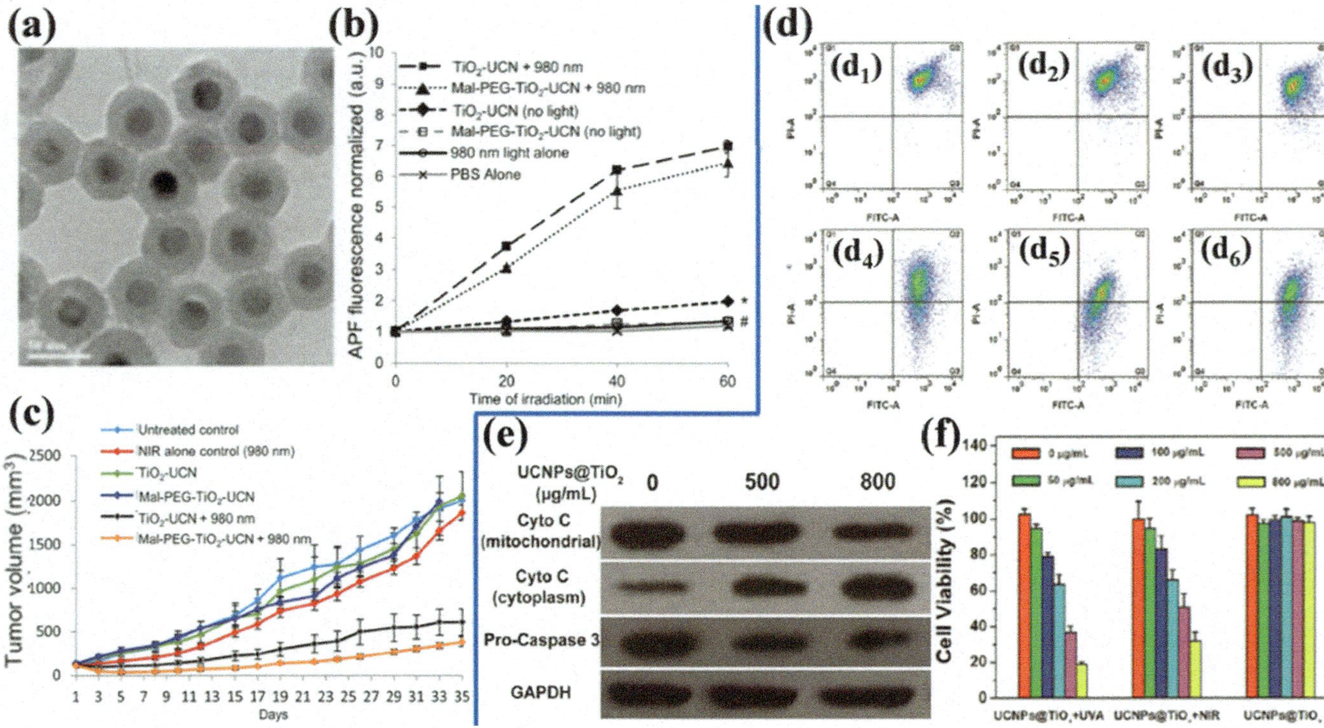

Fig. 12.2 (**a**) TEM images of Mal-PEG-TiO_2-UCN. (**b**) Comparison of ROS production from nonirradiated and irradiated nanoparticles in PBS. (**c**) In vivo tumor growth up to 35 d on different groups of mice after various treatments indicated. (**a–c**) Reproduced with permission from Ref. [10]. (**d**) The change of the mitochondria membrane potential was objectively measured using flow cytometry: (d1) control, (d2) UCNPs@TiO_2 alone, (d3) 980 nm NIR laser irradiation alone, d4–d6) UCNPs@TiO_2 + 980 nm NIR laser irradiation for different times (10, 20, and 30 min). (**e**) Expression of Cyto C and pro-caspase 3 proteins by Western blotting analysis after 30 min irradiation using a 980 nm laser; GAPDH was used as an internal control. (**f**) In vitro viabilities of HeLa cells treated with UCNPs@TiO_2 under 365 nm UV light irradiation, 980 nm NIR laser irradiation, and no irradiation. ((**d–f**) Reproduced with permission from Ref. [21]. Copyright© 2015, American Chemical Society)

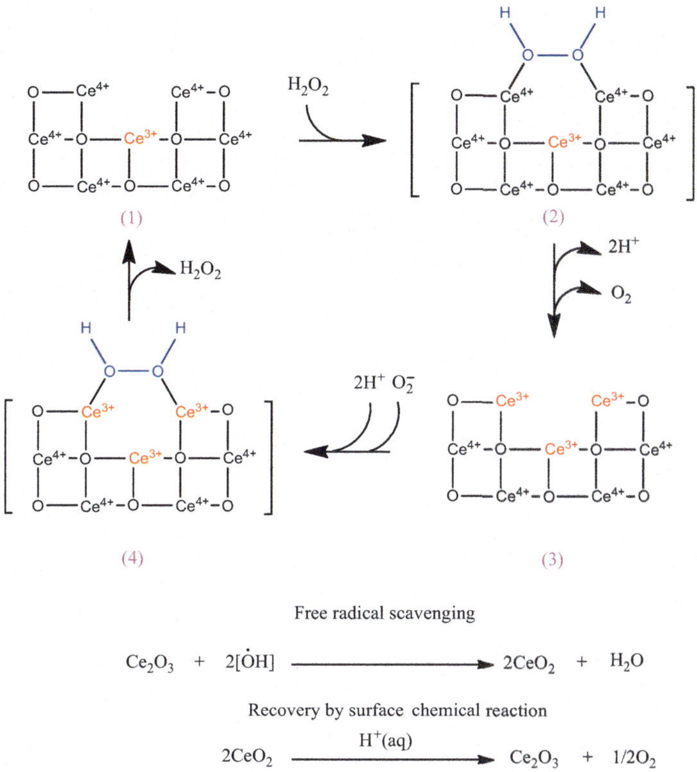

Fig. 12.3 A model of the reaction mechanism for CeNPs to eliminate $O_2\cdot^-$, OH and H_2O_2

eliminate superoxide anions ($O_2\cdot^-$) and hydroxyl radicals (·OH) efficiently and catalase (CAT)-mimetics for eradicating hydrogen peroxide (H_2O_2) efficiently.

Lee et al. showed that the antioxidant power of CeNPs was nine times greater than that of the benchmark antioxidant. What's more, even in high concentrations of hydrogen peroxide, the CeNPs were reused more than 20 times. It should be noted that the subsequent studies have shown that the properties of CeNP depend not only on the diameter of the nanocrystals but also the things that coated on the surface of nanocrystals. The researchers found that the smallest nanosized ceria had the greatest antioxidant activity. Lee et al. attributed this to the improved stability of cerium (IV) and the increased concentration of cerium (III) on the surface of the nanocrystal as lattice expansion increases and oxygen atoms are released [23]. As mentioned earlier, the things that coat on the surface of nanomaterials also play a significant role on their antioxidant capacity, while thicker polymers can reduce their antioxidant capacity. The CeNPs quenched the H_2O_2 molecule by a Fenton-like reaction. The antioxidant capacity of cerium depends on the concentration of cerium (III) and the thickness of the polymer shell. Cerium oxide nanocrystals coated with poly (acrylie acid) or oleic acid (3.8 nm, incorporating 44% cerium (III)), quench more H_2O_2 than

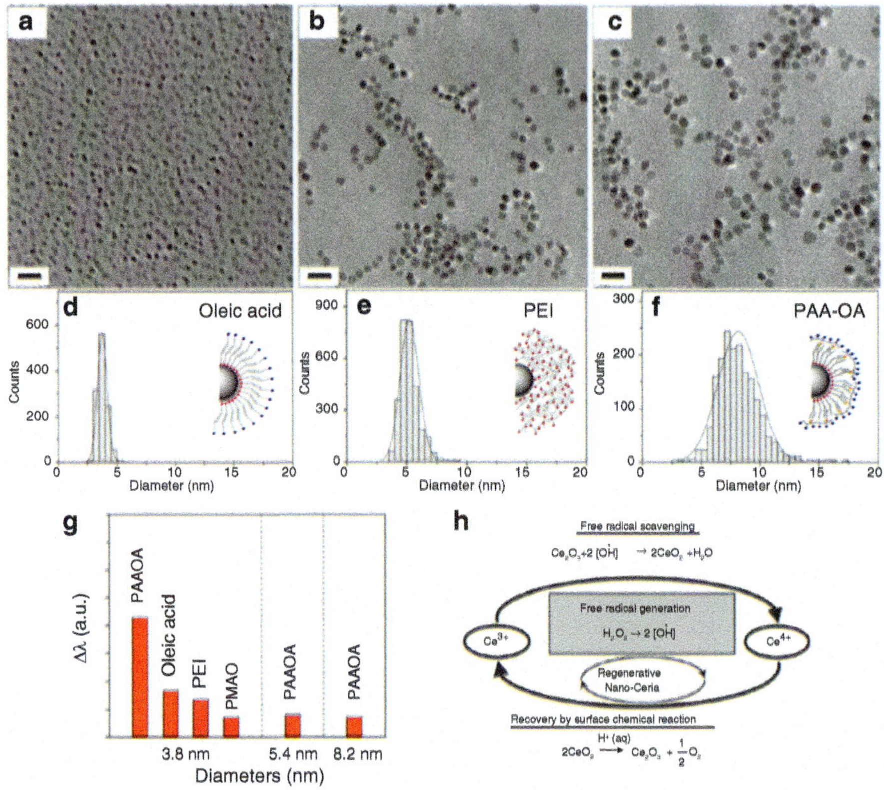

Fig. 12.4 TEM micrographs (**a**)–(**c**) of water-soluble nanoceria. All scale bars are 20 nm from (**a**) to (**c**). (**a**) Oleic acid-coated CeO_2 nanoparticles (3.8 ± 0.4 nm); (**b**) PEI-coated CeO_2 nanoparticles (5.4 ± 1.0 nm); (**c**) PAAOA-coated CeO_2 nanoparticles (8.2 ± 1.7 nm). The size distribution histograms (**d**)–(**f**) are placed at the bottom of the corresponding images together with a schematic depiction of the coated nanomaterial. (**g**) The extent of H_2O_2 quenching capacity depends on the diameter and surface stabilizers of nanoceria. To compare the surface polymer-dependent H_2O_2 quenching efficiency of nanoceria, the extent of the band shift ($\Delta\lambda$) was measured at 0.30 absorbance after the injection of H_2O_2 from the control. For diameter-dependent H_2O_2 quenching, three CeO_2 suspensions with different diameters were utilized (d = 3.8, 5.4, and 8.2 nm; PAAOA-coated CeO_2). Surface coating-dependent H_2O_2 quenching was shown by 3.8 nm CeO_2 covered with four different polymers (PAAOA, oleic acid, PEI, and PMAO). Reproduced with permission from Ref. [23]. Copyright©2013, American Chemical Society. (**h**) Schematic detailing of the proposed regenerative properties of nanoceria, and probable mechanism of free radical scavenging and auto-catalytic behavior of cerium oxide nanoparticles. (Reproduced with permission from Ref. [24]. Copyright©2007, Elsevier)

cerium oxide nanocrystals covered with thicker polymer (poluethylene imine or polymaleicanhydride-alt-1-octadacccne) stablizers (8.2 nm, incorporating 30% cerium (III)) (Fig. 12.4). Das et al. demonstrated the ability of CNPs to protect neurons in adult spinal cord oxidative injury induced by H_2O_2 in vitro, which is significant to treatment for disease in CNS (Fig. 12.4h) [24].

12.4.2 Melanin-Based Tumor-targeted Systems

Melanin is a well-known biopolymer that is widely distributed in almost all living organisms. With many distinct properties of antibiotic function, thermoregulation, free radical quenching, some nervous system involvement and NIR absorption, and melanin-based nanoparticles have attracted widespread attention. Liu YL et al. developed a novel photothermal therapeutic agent based on biopolymer dopamine-melanin colloidal nanospheres. Due to its excellent pharmacokinetic properties, it can be widely distributed in the human body, showing strong biocompatibility and biodegradability. In terms of pharmacodynamics, the drug can effectively damage the tumor tissue without damaging the healthy tissue in the low power density and short irradiation time after administration, which is attributed from up to 40% photothermal conversion efficiency [25].

Melanin nanoparticles, especially modified by PEG(PEG-MeNPs), have a SOD-mimetic catalytic mechanism toward $O_2{\cdot}^-$, and such properties were illustrated at Fig. 12.5. PEG-MeNPs had a wide range of antioxidant activities against a variety of toxic electrons ($O_2{\cdot}^-$,H_2O_2,·OH, $ONOO^-$, and·NO), which showed they had more potential as a potent electron scavenger, than that of natural antioxidant enzymes (such as SOD) and currently studied nano-antioxidants targeting specific electrons [23]. In addition, Liu et al. further investigated the possibility of using these NPs as antioxidants in the body. In this study, the antioxidant activity of MeNPs modified by polyethylene glycol (PEG) against $O_2{\cdot}^-$ was detected by electron paramagnetic resonance (EPR) spectroscopy and DEPMPO (as free radical producer). PET-MeNPs reduces the EPR amplitude of DEPMPO-OOH (nearly equal to the background level). X-ray photoelectron spectroscopy (XPS) analysis showed that these NPs did not covalently react with $O_2{\cdot}^-$ but had catalytic activity similar to SOD. This was attributed to the free radicals in the polymer matrix, which act as catalytic centers for the removal of electrons from the superoxide. What's more, this antioxidant activity was so stable that it could remain constant at different pH values, even after being stored at 4 °C for 1 year. In further experiments, the group verified that MeNPs can also react with $O_2{\cdot}^-$ and form H_2O_2. Thus, the SOD-mimic catalysis of $O_2{\cdot}^-$ can be represented by the reactions. Furthermore, it can be inferred from the EPR results that PEG-MeNPs can also clear ·OH, which may be due to blocking the Fenton reaction; finally, the ·NO and $ONOO^-$ generated by materials would be nitrated and nitrosated by PEG-MeNPs [26].

$$\text{melanin}\cdot + O_2^{\cdot-} \longrightarrow \text{melanin}^- + O_2$$

$$\text{melanin}^- + O_2^{\cdot-} + H_2O \longrightarrow \text{melanin}\cdot + H_2O_2 + 2OH^-$$

Fig. 12.5 A model of the reaction mechanism for MeNPs to eliminate $O_2{\cdot}^-$

12.4.3 Selenium-Based Tumor-targeted Systems

Selenium (Se) is an important component of the liver's antioxidant defense system and plays a crucial role in antioxidant stress. So far, many studies have shown that supplementing selenium can upregulate the activity or expression levels of related enzymes, such as GPx, thus preventing the accumulation of free radicals and reducing cell damage.

Zhai et al. further evaluated the antioxidant capacity of these nanoparticles by synthesizing stable SeNPs using chitosan (CS) of different molecular weights. Evaluation using models of DPPH (2, 2-diphenyl-1-picryl-hydrazine hydrate), ABTS (2,2′ -azo (3-ethylbenzothiazolin-6-sulfonic acid), and lipid peroxides showed that these CS-SeNPs scavenge free radicals at different rates. The scavenging ability of ABTS on low molecular weight CS-SeNPs (CS(l) -SeNPs) and high molecular weight CS-SeNPs (CS(h) -SeNPs) was $87.45\% \pm 7.63\%$ and $89.44\% \pm 5.03\%$, respectively [27].

In vitro, using BABLC-3 T3 or Caco-2 model showed that selenium could effectively inhibit the generation of electrons in cells in a Se concentration-dependent manner. Further tests on the stability of topical or oral CS-SeNPs indicated that the stability of CS(l)-SeNPs could effectively protect the glutathione peroxidase activity in mice and prevent the formation of lipochrome induced by ultraviolet light or *d*-galactose.

12.4.4 Manganese-Based Tumor-targeted Systems

Manganese dioxide (MnO_2) nanoparticles (MnNPs) possess inherent high peroxidase-, oxidase-, and catalase-like activities; several studies demonstrated that MnO_2 nanoparticles have the catalase (CAT)-mimetic activities which can directly oxidize hydrogen peroxide to produce oxygen and Mn^{2+} according to the following reactions which were shown in Fig. 12.6.

Additionally, MnNPs have the SOD-mimetic activity by reacting with superoxide anions ($O_2\cdot^-$) and producing hydrogen peroxide. Li et al. used the inherent enzymatic properties of MnNPs (catalytic properties and as an intelligent cell protective shell) to package living cells separately. The authors independently evaluated the

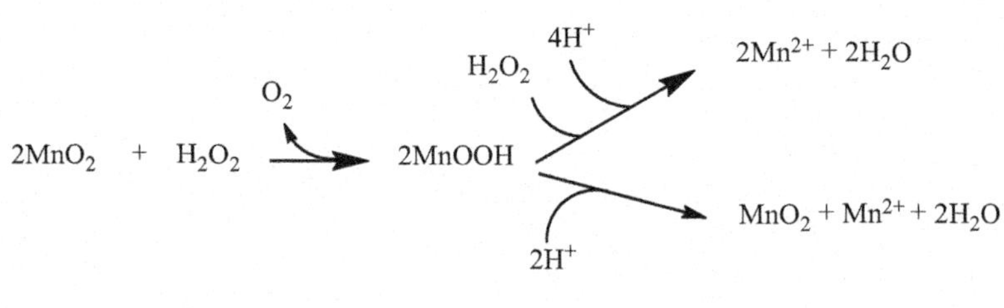

Fig. 12.6 A model of the reaction mechanism for MnNPs to eliminate H_2O_2

multi-enzyme activity of MnNPs on yeast cells. The SOD-like activity was determined by improved NBT assay. The results showed that there was a dose-dependent correlation between high concentration of $O_2\cdot^-$ removal capacity and catalytic activity, with the IC_{50} value of 3.5 $\mu g \cdot mL^{-1}$. The catalase-like activity of these MnO_2 shells was studied by terephthalic acid (TA) assay. The results showed that the removal efficiency of OH· free radical is 70%, corresponding to 64 $\mu g \cdot mL^{-1}$. The determination of H_2O_2 concentration showed that H_2O_2 was decomposed by MnNPs in a dose-dependent manner. The experiments showed that the intrinsic multi-enzyme activity and high stability of nanozyme shells effectively protected the encapsulated cells from long-term physical and chemical stimulation [28].

12.4.5 Platinum-Based Tumor-targeted Systems

Studies have shown that platinum nanoparticles (PtNPs) also outperformed on scavenging RONS; PtNPs showed SOD-/catalase-mimetic activity and the property that was similar to mitochondrial electron transfer complex I. In vitro, Watanabe et al. showed that polyacrylic acid (PAA)-protected platinum nanoparticle species (PAA-Pt) cleared AOO· generated by AAPH thermal decomposition in a dose-dependent manner, demonstrating that PAA-protected platinum nanoparticle species (PAA-Pt) had antioxidant effects. When AOO· was half cleared, the corresponding NPs concentration was 58 ± 4 μM, while the control group showed no antioxidant activity. Tests showed that PAA-Pt is at least six times better at removing RONS than other metal nanoparticles. To further evaluate PAA-Pt inhibition of AOO-·-induced linoleic acid peroxidation, the authors determined oxygen consumption and thiobarbituric acid reactive substance (TBARS). According to the experimental results, the oxygen consumption decreased significantly after adding NPs, but the oxygen consumption rate was not statistically different from that of the control group. From the above results, the authors inferred that the peroxides inhibition mechanism of PAA-Pt may be mainly attributed to clear LOO·, thus inhibiting the spread of linarenate peroxides [29].

Malondialdehyde (MDA) produced in the process of lipid peroxidation showed that PAA-Pt reduced the production of fat peroxides by inhibiting the proliferation of nitrite peroxidation chain induced by AOO·. In vivo, PTNPs are also used as a scavenger for RONS. Katsumi et al. demonstrated for the first time in mouse model that these NPs prevent liver ischemia/reperfusion injury. In their study, two PTNPs of different sizes were injected intravenously into mice with hepatic injury caused by occlusion of the portal vein and then reperfusion for 6 h. The results indicated that both types of NPs accumulated in hepatic non-parenchymal cells post-injection. The above experimental results indicated that the distribution behavior of the two NPs after injection was approximately the same, and both of them were accumulated in liver non-parenchymal cells. In terms of pharmacodynamics, the smaller the particle size of Pt-NPs, the more obvious the decrease of the plasma alanine aminotransferase (ALT) and aspartic aminotransferase (AST) activities. In addition, smaller NPs can effectively reduce the increase of lipid peroxides in ischemic liver and inhibit the increase of the ratio of oxidized GSH to reduced glutathione [30].

12.4.6 *Carbon Nanotubes-Based Tumor-targeted Systems*

Fenoglio et al. discovered for the first time that carbon nanotubes (CNTs) can be used to scavenge electrons (mainly for hydroxyl radicals (·OH) and superoxide anions ($O_2\cdot^-$)). Carbon nanotubes have an excellent scavenging capacity of hydroxyl radicals on the surface of 0.27 μmol·m^{-2}. This quenching effect is attributed to reactions that occur on the carbon skeleton, such as the fullerene reaction [28].

Recently, some studies showed that fullerenes (C60), another type of crystalline carbon, can also act as free radical scavengers. The authors suggest that this is due to the high electron affinity of the fullerene (C60) carbon skeleton. Because of this, free radicals undergo carbon-centered radical addition. Galano used density functional theory to calculate and simulate the ability of single-walled carbon nanotubes (SWNTs) as free radical scavengers, and the results suggested that SWNTs can be used as free radical sponges [31].

In order to facilitate the readers to further understand the characteristics and advantages of each materials, we attributed the characteristics of each materials to Table 12.1.

Table 12.1 Comparison of different metal-based RONS scavenger tumor-targeted systems

Name	Size (nm)	Function	Activity
Ceria NPs (CeNPs) [20]	1 ~ 10	$\cdot OH/O_2\cdot^-$ H_2O_2	SOD-mimetics Catalase (CAT)-mimetics
Carbon nanotubes (CNTs) [28]		$\cdot OH/O_2\cdot^-$	0.27 μmol·m^{-2} (OH·)
Manganese NPs (MnNPs) [25]	15 ~ 50	$\cdot OH/O_2\cdot^-$	SOD-mimetic catalase (CAT)-mimetics
Platinum NPs (PtNPs) [26]	1 ~ 5	AOO·	SOD-mimetics Catalase mimetics Mitochondrial electron transfer complex I -like
Melanin nanoparticles (MeNPs) [23]	>100	$O_2\cdot^-$ ·NO; $ONOO^-$	SOD-mimetics
Selenium nanoparticles (SeNPs) [24]	~50	Se supplementation can increase enzyme levels, such as those of GPx, which prevents the accumulation of free radical species, thereby reducing cellular damage	

12.5 Perfluorocarbon-Based Tumor-targeted Systems

Perfluorocarbon (PFC) has some specialties like high oxygen capacity and low boiling point. These specialties can endow perfluorocarbon (PFC) some medical functions.

Since the discovery of PTT and PDT therapy, it has been regarded as a less invasive but effective alternative to tumor treatment. Through (passive-enhanced permeability and retention) EPR effect or active-targeting effect between the material ligand and tumor cells, the materials gathered in the tumor site absorbed near-infrared (NIR) laser light and convert it into heat to produce singlet oxygen ($\cdot O_2$) with cytotoxicity, so as to achieve the treatment of tumor.

Many studies have shown that anoxic environments in tumors can reduce the effectiveness of PDT. According to clinical results, the internal environment of most solid tumors is anoxic due to the reduction of oxygen supply which is caused by microcirculation and deterioration of diffusion. In addition, a certain amount of oxygen is consumed by the PDT process, and the vascular closure effect further aggravates the hypoxia in the tumor site. This condition will reduce the photodynamic effect of PDT, which leads to poor treatment effect of PDT. To enhance the efficacy of PDT, Cheng et al. developed a novel self-oxygen-enriched photodynamic therapy (Oxy-PDT) using perfluorocarbon nanoparticles loaded with photosensitizer. Due to its high oxygen capacity, PFC can maintain a higher oxygen content than the tumor matrix at a given oxygen partial pressure. In the process of PDT, with the help of the high oxygen content of PFC, although the oxygen content in the tumor was still limited, enough O_2 could always be enriched in the PFC nanodroplets for the photodynamic consumption of PS, so as to obtain better efficacy. The authors showed that this strategy applies to many photosensitizers [32]. Besides, Jia et al. made a formulation for activated ultrasound imaging and photothermal therapy of cancer. N-perfluoropentane (PFP) coated hollow mesoporous Prussian blue (HPB) nanotubes (HPB-PFP) were synthesized for in vivo tumor diagnosis and regression. The experiment showed that the materials had good stability, and at the same time, the shell of the materials has good photothermal conversion efficiency, which can absorb the near-infrared laser and convert it into heat. The generated heat, on the one hand, causes tumor cell death and tumor ablation by increasing the temperature of tumor tissue and, on the other hand, promotes the continuous gasification and bubbling of encapsulation solution PFP (low boiling point). These PFP bubbles can cause tissue impedance mismatch, so that significantly enhanced B-mode ultrasound imaging signals can be detected in vitro and obvious echo signals can be generated for tumor tissues of nude mice in vivo [33]. The subsequent experimental results indicated that the biocompatible HPB-PFP nanotheranostics designed with high colloidal stability and photothermal efficiency are expected to find a variety of biomedical applications in the detection and treatment of tumors under the guidance of ultrasound imaging in the absence of obvious in vivo and in vitro cytotoxicity.

12.6 Oxide-Based Tumor-targeted Systems

12.6.1 Iron Oxide-Based Tumor-targeted Systems

Magnetic iron oxide (IO) nanoparticles have the characteristics of long blood retention time, biodegradability, and low toxicity, making them one of the main nanomaterials for in vivo and in vitro biomedical applications. In addition, IO nanoparticles, due to their large surface area, can be designed to coated with a large number of functional groups to cross-link to tumor-targeted ligands (such as monoclonal antibodies, peptides, or small molecules) for diagnosis, imaging, or delivery of drugs. In 2018, Park JH et al. synthesized synthetic particles consisting of iron oxide (IO) nuclei in chains clustered in dextran coatings for tumor targeting and imaging. When coupled with tumor-targeting peptides, the interaction between the material and tumor-targeting compounds is more effective than that of spherical nanoparticles. Compared to the untargeted spherical IO nanoparticles, the untargeted IO nanoworms showed similar in vivo circulation and enhanced passive accumulation in xenograft tumors of mice [34]. IO nanoparticles have unique paramagnetism compared to other nanoparticles, which can produce significant magnetic susceptibility effect, enhance the contrast between T2 and T∗2, and generate T1 effect at very low concentration. This technique is widely used in clinical tumor imaging [35]. Antibody-based targeted IO nanoparticles for in vitro or in vivo imaging have been studied in several laboratories and were found to maintain both the properties of the antibody and the magnetic particles. Peptides targeting tumor cell surface-related receptors can be internalized through receptor-mediated endocytosis, thereby increasing the cell uptake of IO nanoparticles and enhancing the MRI effect. Therefore, such peptides are ideal ligands for constructing targeted IO nanoparticles for tumor imaging.

Toxic and side effects on normal tissues are urgent challenges for IO and other nanomaterials. Chengyi Li et al. have developed a drug delivery system without side effects on normal physiological tissues. In this paper, a multifunctional drug delivery system with ordered mesoporous resin as polymer core and uniform Fe nanodot-doped silica as biodegradable shell is proposed. In this core-shell structure, the Fe-doped silica shell was used as a tight inorganic cap that seals doxorubicin in the mesoporous polymer core and also was used as superparamagnetism agent for magnetic targeting and magnetic resonance imaging (MRI). Importantly, in the acidic tumor environment, Fe extraction would induce the degradation of the cap layer to slowly release the loaded drug and achieve ultralow drug leakage in the normal in vivo blood circulation (physiological environment).

Due to this unique core-shell structure, the material has an ultralow drug leakage characteristic, thereby avoiding side effects on normal tissues. This strategy has been shown to improve the outcome of targeted chemotherapy under the guidance of MRI. It showed the great potential of materials as an alternative to cancer treatment in clinical practice [36].

12.6.2 Calcium Carbonate-Based Tumor-targeted Systems

Calcium carbonate ($CaCO_3$) has broad biomedical utilizations owing to its availability, low cost, safety, biocompatibility, pH-sensitivity, and slow biodegradability. Due to the pH-dependent properties of such materials (as shown in Fig.12.7) and the potential for targeted modification, they can be used in targeted delivery systems for anticancer drugs. In addition, the nanoparticles can be used as sustained-release system to retain the drug in cancerous tissue for a longer period of time after administration due to the slow degradation of $CaCO_3$ matrices [37].

The scheme for $CaCO_3$ nanomaterials to delivery drugs (DOX) is shown in Fig.12.8. Shafiu Kamba et al. prepared $CaCO_3$ nanocrystal carriers with pH sensitivity and high DOX loading capacity and then tested the relevant properties. According to their results, the rate of DOX release from the prepared nanocrystals was significantly increased under acidic pH conditions that simulated the tumor environment, compared to the normal physiological pH value (7.4) [38].

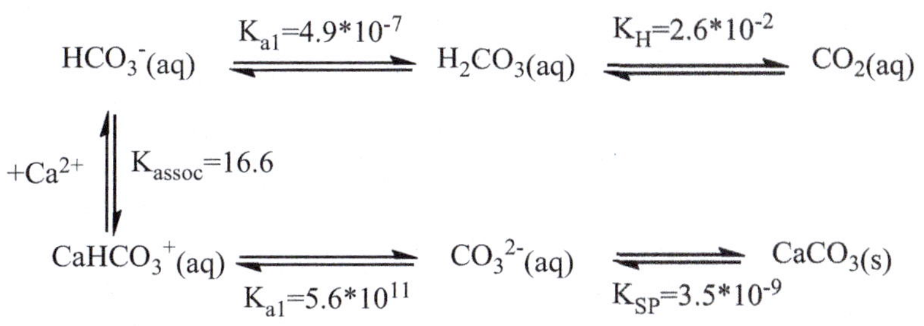

Fig. 12.7 The pH-dependent solubility of $CaCO_3$; equilibrium relationship between carbon dioxide (CO_2) and $CaCO_3$

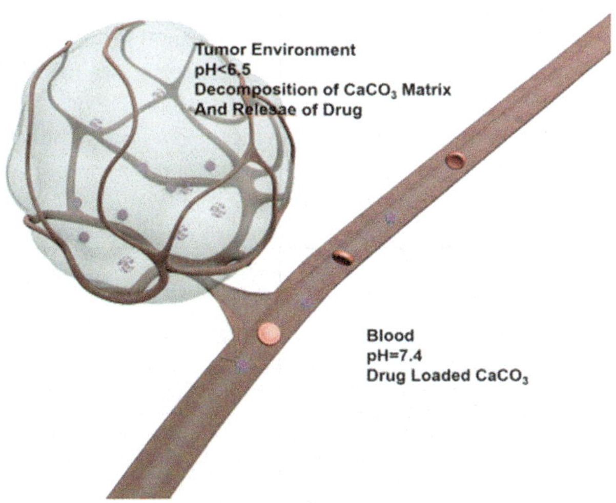

Fig. 12.8 $CaCO_3$ nanoparticles as drug carrier; slow decomposition at normal physiological pH (7.4) and a faster decomposition and release of drug at acidic pH (<6.5) tumor environment

Svenskaya et al. used a porous $CaCO_3$ carrier (polycrystalline vaterite) to co-transport the photosensitizer. The results showed that the carrier degrades rapidly in acid medium. According to the authors, pH-dependent release of the system in the acidic microenvironment might lead to intrinsic cancer sensitivity [39].

Zhou et al. prepared a comprehensive and easily prepared anticancer theranostic platform with high drug loading, which is named an aptamer-functionalized calcium carbonate nanostructure (apt-CCN). They found that the formulation could specifically bind to targeted cancer cells and rapidly internalized by cells and then selectively reached lysosomes through receptor-mediated endocytosis, where DOX is easily released at a relatively low lysosomal pH (4.5–5.5) [40].

12.6.3 Manganese Dioxide-Based Tumor-targeted Systems

Tumor hypoxia is a major mechanism of resistance to radiation therapy (RT), which is associated with poor prognosis in affected cancer patients. Hypoxia at tumor site not only promotes the invasion of tumor cells but also leads to chemotherapy resistance of tumor cells. Tumor-associated macrophages (TAMs), which are located in the tumor hypoxia area, have been reported to cooperate with tumor cells to play a significant role in promoting tumor cell proliferation and drug resistance.

Shin et al. developed fucoidan-coated manganese dioxide nanoparticles (Fuco-MnO_2-NPs) and tested the therapeutic potential with RT using pancreatic cancer models. In vitro experiments showed that Fuco-MnO_2-NPs could effectively generate oxygen in the presence of H_2O_2, thus improving the hypoxic condition of tumor microenvironment and significantly inhibiting the expression of HIF-1 in human pancreatic cancer cells. Fuco-MnO_2-NPs successfully reversed the radiation tolerance induced by hypoxia by reducing the survival rate of clonogenic, increasing the damage of RT to DNA and the death of apoptotic cells. Compared to RT alone, Fuco-MnO_2-NPs combined with RT could significantly delay tumor growth in $BxPC_3$ xenograft mice. In the experimental group, the expression of phosphorylated vascular endothelial growth factor receptor 2 (VEGFR2) and CD31 was decreased, and fucoidan-coated NPs (rather than bare NPs) further inhibited tumor angiogenesis. It is suggested that Fuco-MnO_2-NPs might enhance RT through dual targeting of tumor hypoxia and angiogenesis. This has important clinical value in the treatment of hypoxic and radioresistant pancreatic cancer [41].

Tumor-associated macrophages (TAMs) residing at the site of hypoxic region of tumors have been known to cooperate with tumor cells and promote proliferation and chemoresistance. Song et al. found another way that could effectively alleviate tumor hypoxia by targeted delivery of manganese dioxide nanoparticles (MnO_2 NPs) to hypoxic areas. Due to the high accumulation of TAMs in hypoxia areas and high reactivity of MnO_2 NPs, the materials could stimulate the production of O_2 and upregulate the pH, which are conduce to alleviate tumor hypoxia. In addition, the author reprogrammed TAMs through modified by hyaluronic acid (HA). Due to this scheme, the TAMs would have the ability to anti-inflammatory, pro-tumoral M2

TAMs to pro-inflammatory, antitumor M1 macrophages, which would enhance the ability of MnO_2 NPs to lessen tumor hypoxia and modulate chemoresistance. HA-coated mannan combined with MnO_2 particles (Man-HA-MnO_2) significantly decreased the expression of hypoxia-inducible factor-1 α (HIF-1α) and vascular endothelial growth factor (VEGF) in tumor and increased tumor oxygenation. Compared to chemotherapy alone, the combination of Man-HA-MnO_2 NPs and doxorubicin significantly enhanced the therapeutic effect of breast cancer, significantly increased the apparent diffusion coefficient (ADC) of breast cancer, and inhibited the growth and proliferation of tumor cells. In terms of imaging, the response of Man-HA-MnO_2 NPs to endogenous H_2O_2 enhances the T1- and T2-MRI manifestations of tumor imaging and detection [42].

12.7 Sulfide-Based Tumor-targeted Systems

12.7.1 MoS_2.

Molybdenum disulfide (MoS_2) has been widely explored for biomedical applications due to its brilliant photothermal conversion ability. Zhang et al. established a new HA-modified multifunctional MoS_2-based drug delivery system (MoS_2-SS-HA). Modified by hyaluronic acid (HA), MoS_2 nanosheets could deliver drugs under the near-infrared (NIR) photothermal trigger, thus realizing a cancer treatment system with synergism of chemotherapy and photothermal therapy. MoS_2-SS-HA has the characteristics of uniform diameter (125 nm), good biocompatibility, and good stability in physiological solution. Due to these, the materials could be combined with erlotinib (Er), an insoluble anticancer drug. The release rate of Er triggered by near-infrared laser (NIR) was significantly increased, indicating that the composite has potential as a light-response system. The results showed that Er emission could be controlled by controlling the irradiation time and power of near-infrared laser. In addition, MTT and confocal imaging experiments were carried out. The results showed that MoS_2 could selectively target CD44-positive lung cancer cells, especially for drug-resistant cells such as A549 and h1975. Through the study of tumor ablation in vivo, it has been proved that the combined therapy had better synergistic effect than chemotherapy or photothermal therapy alone [43].

Liu et al. developed another hyaluronic acid (HA)-modified MoS_2-based functional nanoplatform, which can realize the targeted drug delivery of camptothecin (CPT) and dual stimulation-response drug release. Specifically, HA was connected with MoS_2 through disulfide bond to form detachable HA shells on the surface of MoS_2. On the one hand, this unique design could effectively prevent the drug leakage in the blood circulation process; on the other hand, it could significantly accelerate the release of the drug when system was in touch with tumor-associated glutathione (GSH). In addition, MoS_2 could transform the energy from radiated by NIR into heat, which would further improve the drug release rate and induce the photothermal ablation of tumor cells. According to the experimental results in vitro and in vivo, MoS_2-SS-HA-CPT which combined photothermal and chemotherapy could significantly inhibit the cell proliferation and tumor growth of lung cancer cells under near-infrared irradiation [44].

12.7.2 TiS_2

How to improve the efficacy of drugs and reduce the side effects is the key for targeted delivery of drugs. Recently, it has been pointed out that subcellular-targeted drug delivery can maximize the anticancer effect and reduce adverse reactions. Sen et al. developed a mitochondrial-targeted drug delivery nanoplatform based on IR780 iodide (IR780) and titanium disulfide (TiS_2) nanosheets. With the large specific surface area of TiS_2 nanosheets, resveratrol (RV) was coloaded. The prepared nanocomposites ((IR780-TiS_2/RV) had good stability and biocompatibility, which were used in photothermal-triggered tumor chemotherapy. The results showed that the loading rates of RV and IR780 were 112% and 56%, respectively. Under near-infrared (NIR) irradiation, the heat generated by IR780-TiS_2/RV was enough to trigger and promote RV release. Because of the specific conjugation of mitochondria and IR780, IR780-TiS_2/RV could target and further accumulate in mitochondria, which would release RV under near-infrared excitation, reduce mitochondrial membrane potential, rapidly induce the upregulation of key endogenous apoptosis factors such as cytochrome c, and initiate caspase cascade reaction, so as to achieve the effect of chemotherapy. The experimental results showed that IR780-TiS_2/RV nanocomposites had high antitumor activity both in vitro and in vivo without obvious tissue toxicity. The authors believed that the near-infrared triggered IR780-TiS_2/RV nanoplatform was a promising chemotherapeutic drug in clinic [45].

12.8 Biomimetic Material-Based Tumor-targeted Systems

12.8.1 Cell Membrane

Inspired by the nature, researchers put forward the concept of bionics. This concept has been applied to the drug delivery system of cancer treatment. By wrapping nanoparticles with cell membrane, nanoparticles can be endowed with various functions of natural cells. With the help of cell membrane coating technology, nanoparticles could more effectively circulate in vivo to overcome the fast elimination in circulation. In addition, due to the diversity of surface functional molecules, cell membrane nanoparticles (CMBNPs) can specifically interact with the complex biological microenvironment of tumors. The camouflage effect of various cell membranes on CMBNPs was studied, and different tumor-targeting strategies were developed to enhance the delivery of anti-tumor drugs [46].

Compared to blood cells, cancer cells have unlimited replication potential, immune escape, and homologous-targeting ability [47]. In order to make the cells had a better proliferation ability, the cancer cells used in the preparation process were not obtained from patients' self-plasma or donors but are easily obtained by cell culture in vitro [48]. Studies have shown that in the process of metastasis, the aggregation of homotypic cancer cells is essential for the establishment of secondary lesions in distant tissues and organs [49, 50]. It has been shown that the process of tumor aggregation depends on the surface adhesion molecules (such as N-cadherin, galectin-3, EpCAM) on the cancer cell membranes [51]. Using natural cell membrane for carrier surface functionalization could make the diversity of membrane surface proteins completely replicate from the source cells to the engineered nanoparticles, so that the engineered nanoparticles have a variety of functions (targeting, adhesion). Tumor cells have inherent characteristics of immune escape and homologous adhesion. Inspired by this, the design strategy based on various tumor cell membrane-encapsulated nanoparticles has been used for tumor-targeted diagnosis and treatment. Rao L et al. developed a novel cancer cell membrane-cloaked upconversion nanoprobe with low immunogenicity and homologous-targeting effect [52]. The synthesis strategy was to hybridize the cancer cell membrane with DSPE-PEG to form camouflaged nanoparticles (ICNP), while ICG was wrapped in the PLGA nucleus as a probe. Because of the near-infrared fluorescence emission characteristics of ICG, the core-shell nanoparticles could be used for high-specific in vivo tumor imaging. The author chose different core materials and constructed a multifunctional delivery system. Doxorubicin (DOX) was incorporated into the gold nanocage to form the core, which was further encapsulated by 4 T1 breast cancer cell membrane. Related experiments showed that the system combined the advantages of photothermal therapy and chemotherapy.

The results showed that the nanoparticles had good targeting efficiency, high accumulation, and high thermal reactivity to 4T1 cells. In addition, a cancer cell membrane-coated probe (indocyanine green, ICG)-loaded lipid polymer NPs has been proved to be an excellent nanosystem for homologous target dual-modal imaging and imaging-guided photothermal therapy.

12.8.2 Erythrocyte/Platelet

Platelets, the smallest circulating blood cells, are fragments of cytoplasm produced from mature megakaryocytes in bone marrow. Under normal physiological conditions, there are about 150,000–350,000 platelets per microliter in the blood. The function of these platelets is to maintain the integrity of the vascular system, which plays an important role in the process of hemostasis, wound healing, inflammatory response, and thrombosis after vascular injury. Their average life span is 8–9 days. Traditionally viewed as major cellular components in hemostasis and thrombosis, the contribution of platelets to the progression of cancer is an emerging area of research interest. Recently, many scholars have shown that platelets play an important role in promoting tumor angiogenesis, helping tumor survive in the blood, and promoting the interaction between tumor cells and blood vessels [53]. The interaction between circulating tumor cells (CTCs) and platelets is of great significance. The results showed that activated platelets could change shape; release granules containing chemokines, growth factors, and protease; and then increase their adhesiveness, forming heterogeneous aggregation with CTCs and leukocytes [54].

Recently, Hu et al. inserted tumor necrosis factor (TNF)-related apoptosis-inducing ligand (TRAIL) into the outer membrane and loaded DOX into the inner nanoparticles to prepare platelet membrane-coated nanocarriers. The results showed that the platelet membrane-encapsulated nanovehicle (PM-NV) had a strong antitumor effect on subcutaneous tumor and metastasis animal models. The authors suggested that because of the overexpression of P-selectin on the platelet membrane of this system, it could closely bind to the upregulated CD44 receptor on cancer cells and then actively target CTCs and deliver TRAIL to induce apoptosis of CTCs [55]. However, there was no relevant experiment to directly prove its mechanism. In addition, platelet membrane-coated nanoparticles could also be used for tumor imaging. For example, a magnetic nanoparticle encapsulated in a platelet membrane has been reported to enhance the imaging and treatment of cancer. Fe_3O_4 magnetic nanoparticles encapsulated in platelet membrane (derived from mouse blood) still had the ability of long circulation of platelets and tumor targeting. The results showed that the system can be used to enhance MR imaging and photothermal treatment of tumor, so as to achieve personalized diagnosis and treatment of tumor.

12.9 Conclusion

Although the degree of the researches for these materials are not as deep as other types of drug delivery systems mentioned above, the drug delivery systems described in this chapter often show some advantages in lethality: low side effects or using for synergistic treatment. For example, with the development of these researches, PDT- or ROS-related materials not only have great advantages in killing tumor in situ but also can enhance immunity, slow down, or even eliminate tumor metastasis and recurrence, which makes PDT therapy has incomparable advantages compared to traditional therapy. We believe that the study of these materials can be enlightening to the traditional targeting system, such as the bionics of biomaterials for targeting liposomes. With the development of these researches, these materials will certainly contribute to the development and application of drugs in the future.

References

1. Zhang S et al (2014) Extraordinary photoluminescence and strong temperature/angle-dependent Raman responses in few-layer phosphorene. ACS Nano 8:9590–9596
2. Xia F et al (2014) Two-dimensional material nanophotonics. Nat Photonics 8:899–907
3. Xi L et al (2015) The renaissance of black phosphorus. Proc Natl Acad Sci USA 112:4523–4530
4. Liu H et al (2014) Phosphorene: an unexplored 2D semiconductor with a high hole mobility. ACS Nano 8:4033–4041
5. Sun Z et al (2015) Ultrasmall black phosphorus quantum dots: synthesis and use as photothermal agents. Angew Chem Int Ed 127:11688–11692
6. Tao W et al (2017) Black phosphorus nanosheets as a robust delivery platform for cancer theranostics. Adv Mater 29:1603276
7. Zhou WH et al (2018) Enhanced cytosolic delivery and release of CRISPR/Cas9 by black phosphorus nanosheets for genome editing. Angew Chem Int Ed 57:10268–10272
8. Shao J et al (2016) Biodegradable black phosphorus-based nanospheres for in vivo photothermal cancer therapy. Nat Commun 7:12967
9. Favron A et al (2015) Photooxidation and quantum confinement effects in exfoliated black phosphorus. Nat Mater 14:826–832
10. Lee HU et al (**2016**) Black Phosphorus(BP) nanodots for potential biomedical applications. Samll 2:214–219
11. Zeng X et al (2018) Polydopamine-modified black phosphorous nanocapsule with enhanced stability and photothermal performance for tumor multimodal treatments. Adv Sci 5:1800510
12. Chen L et al (2020) Multifunctional mesoporous black phosphorous-based nanosheet for enhanced tumor-targeted combined therapy with biodegradation-mediated metastasis inhibition. Biomaterials 236:11970
13. Zhou J et al (2012) Upconversion nanophosphors for small-animal imaging. Chem Soc Rev 41:1323–1349
14. Liu Y et al (2013) Lanthanide-doped luminescent nanoprobes: controlled synthesis, optical spectroscopy, and bioapplications. Chem Soc Rev 42:6924–6958
15. Chen H et al (2013) Synthesis of brightly PEGylated luminescent magnetic upconversion nanophosphors for deep tissue and dual MRI imaging. Small 10:160–168
16. Gu Z et al (2013) Recent advances in design and fabrication of upconversion nanoparticles and their safe theranostic applications. Adv Mater 25:3758–3779
17. Gai S et al (2014) Recent progress in rare earth micro/nanocrystals: soft chemical synthesis, luminescent properties, and biomedical applications. Chem Rev 114:2343–2389
18. Fan W et al (2016) On the latest three-stage development of nanomedicines based on upconversion nanoparticles. Adv Mater 28:3987–4011
19. Li X et al (2014) Anisotropic growth-induced synthesis of dual-compartment janus mesoporous silica nanoparticles for bimodal triggered drugs delivery. J Am Chem Soc 136:15086–15092
20. Idris NM et al (2014) Photoactivation of core–shell titania coated upconversion nanoparticles and their effect on cell death. J Mater Chem B 2:7017–7026
21. Lucky SS et al (2015) Titania coated upconversion nanoparticles for near-infrared light triggered photodynamic therapy. ACS Nano 9:191–205
22. Yuan P et al (2019) Mitochondria-targeting, intracellular delivery of native proteins using biodegradable silica nanoparticles. Angew Chem Int Ed 58:7657–7661
23. Lee SS et al (2013) Antioxidant properties of cerium oxide nanocrystals as a function of nanocrystal diameter and surface coating. ACS Nano 7:9693–9703
24. Das M et al (2007) Auto-catalytic ceria nanoparticles offer neuroprotection to adult rat spinal cord neurons. Biomaterials 28:1918–1925
25. Liu YL et al (2013) Dopamine-melanin colloidal nanospheres: an efficient near-infrared photothermal therapeutic agent for in vivo cancer therapy. Adv Mater 25:1353–1359

26. Liu Y et al (2017) Comprehensive insights into the multi-Antioxidative mechanisms of melanin nanoparticles and their application to protect brain from injury in ischemic stroke. J Am Chem Soc 139:856–862
27. Zhai X et al (2017) Antioxidant capacities of the selenium nanoparticles stabilized by chitosan. J Nano 15:4
28. Li W et al (2017) Manganese dioxide nanozymes as responsive cytoprotective shells for individual living cell encapsulation. Angew Chem Int Ed 56:13661–13665
29. Watanabe A et al (2009) In vitro free radical scavenging activity of platinum nanoparticles. Nanotechnology 20:455105
30. Katsumi H et al (2014) Pharmacokinetics and preventive effects of platinum nanoparticles as reactive oxygen species scavengers on hepatic ischemia/reperfusion injury in mice. Metallomics 6:1050–1056
31. Fenoglio I et al (2006) Reactivity of carbon nanotubes: free radical generation or scavenging activity. Free Radic Biol Med 40:1227–1233
32. Cheng Y et al (2015) Perfluorocarbon nanoparticles enhance reactive oxygen levels and tumour growth inhibition in photodynamic therapy. Nat Commun 6:9785–9788
33. Jia X et al (2015) Perfluoropentane-encapsulated hollow mesoporous prussian blue Nanocubes for activated ultrasound imaging and photothermal therapy of cancer. ACS Appl Mater Interfaces 7:4579–4588
34. Park JH et al (2008) Magnetic iron oxide nanoworms for tumor targeting and imaging. Adv Mater 20:1630
35. Peng XH et al (2008) Targeted magnetic iron oxide nanoparticles for tumor imaging and therapy. Int J Nanomedicine 3:311–321
36. Li C et al (2018) Side effects-avoided theranostics achieved by biodegradable magnetic silica-sealed mesoporous polymer-drug with ultralow leakage. Biomaterials 186:1–7
37. Shafiu Kamba A et al (2013) A pH-sensitive, biobased calcium carbonate aragonite nanocrystal as a novel anticancer delivery system. BioMed Res Int 2013:1–10
38. Svenskaya Y et al (2013) Anticancer drug delivery system based on calcium carbonate particles loaded with a photosensitizer. Biophys Chem 182:11–15
39. Zhou C et al (2015) Aptamer $CaCO_3$ Nanostructures: A Facile, pH-Responsive, specific platform for targeted anticancer theranostics. Chem Asian J 10:166–171
40. Shin S et al (2018) Fucoidan-manganese dioxide nanoparticles potentiate radiation therapy by co-targeting tumor hypoxia and angiogenesis. Mar Drugs 16:510
41. Song M et al (2015) Bioconjugated manganese dioxide nanoparticles enhance chemotherapy response by priming tumor-associated macrophages toward M1-like phenotype and attenuating tumor hypoxia. ACS Nano 10:633–647
42. Dizaj SM et al (2015) Calcium carbonate nanoparticles as cancer drug delivery system. Expert Opin Drug Deliv 12:1649–1660
43. Zhang C et al (2019) Functionalized MoS_2-erlotinib produces hyperthermia under NIR. J Nanobiotech 17:76
44. Liu J et al (2019) Redox/NIR dual-responsive MoS_2 for synergetic chemo-photothermal therapy of cancer. J Nanobiotechnol 17:78–93
45. Xiang S et al (2019) Mitochondria-targeted and resveratrol-loaded dual-function titanium disulfide nanosheets for photothermal-triggered tumor chemotherapy. Nanoscale Res Lett 14:211–220
46. Tan SW et al (2015) Cell or cell membrane-based drug delivery systems. Theranostics 8:863–881
47. Hanahan D et al (2011) Hallmarks of cancer: the next generation. Cell 144:646–674
48. Sherr CJ (1996) Cancer cell cycles. Science 274:1672–1677
49. Glinsky VV et al (2003) Intravascular metastatic cancer cell homotypic aggregation at the sites of primary attachment to the endothelium. Cancer Res 63:3805–3811

50. Khaldoyanidi SK et al (2003) MDA-MB-435 human breast carcinoma cell homo- and hetero-typic adhesion under flow conditions is mediated in part by Thomsen-Friedenreich antigen-galectin-3 interactions. J Biol Chem 278:4127–4134
51. Chen Z et al (2016) Cancer cell membrane-biomimetic nanoparticles for homologous-targeting dual-modal imaging and photothermal therapy. ACS Nano 10:10049–10057
52. Rao L et al (2016) Cancer cell membrane-coated upconversion nanoprobes for highly specific tumor imaging. Adv Mater 28:3460–3466
53. Bambace NM et al (2011) The platelet contribution to cancer progression. J Thromb Haemost 9:237–249
54. Nash GF et al (2002) Platelets and cancer. Oncology 3:425–430
55. Hu Q et al (2015) Anticancer platelet-mimicking Nanovehicles. Adv Mater 27:7043–7050

Chapter 13
Clinical Applications of Tumor-targeted Systems

Xinxin Zhang

Abstract Cancer therapies are currently limited to surgery, radiation, and chemotherapy. All these methods risk damage to normal tissues or incomplete eradication of the cancer. The limits of conventional cancer therapies promoted the development and application of various nanotechnologies. Recent advances in nanotechnology have established the importance in the detection, diagnosis, and treatment of tumors. Although nanotechnology has attracted much attention, the actual use of nanotechnology to clinical cancer treatment is just in its start-up. This chapter reviews the technologies being investigated clinically and their therapeutic efficacy and safety. In addition, we discuss the challenges that must be overcome to realize more effective nanotherapeutics for cancer patients.

Keywords Tumor targeted · Nanoparticle · Drug delivery system · Tumor diagnosis and therapy · Clinical application

13.1 Introduction

Despite the continuous improvement of cancer-fighting strategies, malignancies are one of the leading causes of death worldwide. The combination of surgery, radiation, and chemotherapy is still the standard anticancer therapeutic strategy. However, traditional anticancer drugs display poor selectivity to targeting site, leading to systemic toxicity, drug-resistant cancer cells, and tumor recurrence. The central promise of targeted drug delivery technologies is improved efficacy by increasing the drug concentration at a desired site while simultaneously minimizing toxicity by reducing off-target accumulation. The US National Cancer Institute (NCI) established the Alliance for Nanotechnology in Cancer to promote the development of cancer nanotechnology in 2004. Only several years later, nano-based targeting

X. Zhang (✉)
Shanghai Institute of Materia Medica, Chinese Academy of Sciences, Shanghai, China
e-mail: xinxinzhang@simm.ac.cn

R. Huang, Y. Wang (eds.), *New Nanomaterials and Techniques for Tumor-targeted Systems*, https://doi.org/10.1007/978-981-15-5159-8_13

delivery systems showed breakthrough in the burgeoning field and applied in cancer treatment and tumor earlier diagnosis.

Nanotechnology is a union of several fields of science, including not only medicine but also areas such as physics, chemistry, and molecular biology. It has the potential to overcome the disadvantages of conventional drug delivery by adjusting pharmacokinetics and delivery, resulting in diminishing side effects and therefore enhancing efficiency. Nowadays, nanotherapeutics with features of improving blood circulation and reducing toxicity have been authorized for cancer therapy, such as liposomes, albumin nanoparticles, and polymeric micelles. Many other nanotechnology-enable treatments have shown great promise in clinical development, including radiation therapy, gene or RNA interference therapy, and chemotherapy, which are expected to achieve definitive results in the near future [1].

Most therapeutic nanoparticles for solid tumor treatment are administered systemically and accumulate in the tumor through the enhanced permeation and retention (EPR) effect due to the poor lymphatic drainage and vascular leakage in tumor [2–6]. However, the realization of EPR effect is obviously oversimplified, as EPR can be affected by many biological factors, such as blood circulation, tumor extracellular microenvironment, tumor tissue penetration, and cancer cell internalization. From another perspective, the physicochemical properties of nanoparticles, such as particle size, geometry, and surface targeting ligand, can also influence these biological processes, thus determining the therapeutic outcomes [7–9]. Nevertheless, it is important to point out that most of our current understanding of nanoparticle behavior in vivo is based on animal data, and its translation to humans remains largely unexplored.

This chapter aims to offer an overview of technologies being investigated clinically and their therapeutics on efficacy and safety. We present recent progress in exploring EPR effect and identifying markers to predict responses to nanotherapies and in developing new strategies to enhance systemic nanoparticle delivery for more pronounced therapeutic and diagnostic benefit [10, 11].

13.2 The EPR Effect-Based Nanomedicines

Tumors are characterized by rapid cell proliferation, strong metabolism, abnormal neovascularization in the tumor, loose connection of endothelial cells in neovascularization, and lack of lymphatic drainage pathway, resulting in EPR effect in tumor tissues [2]. Passive targeting therapy plays an important role in tumor therapy by using this EPR effect. Hydrodynamics shows that the nanodelivery system with size of 10–500 nm can escape the lymphatic drainage pathway and enter the tumor tissue through the vascular endothelial cell gap, thereby achieving specific distribution and accumulation in tumor tissues (Fig. 13.1) [12]. The EPR effect has become the foundation of nanoparticle delivery to solid tumors [5, 6]. Stratifying the subpopulations of cancer patients based on their possibility of EPR effect

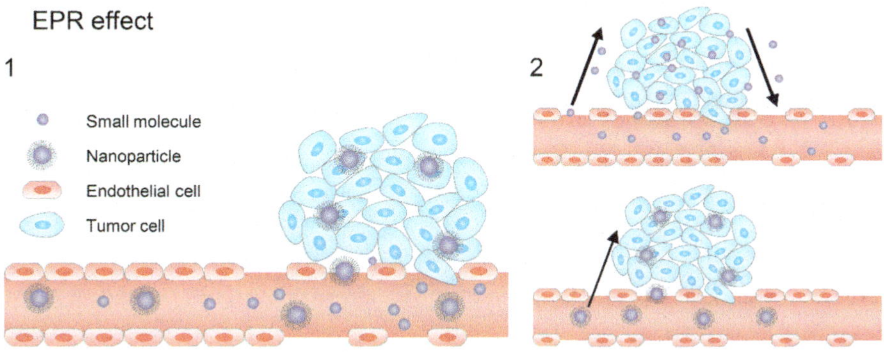

Fig. 13.1 Passive targeting of nanoparticles based on EPR effect [15]. (1) Nanoparticles can selectively accumulate in tumor through leaky vasculature. (2) The effect of size on tumor retention. Drug molecules spread freely inside and outside the tumor blood vessels, leading to low drug concentration in the tumor. In contrast, drug-loaded nanoparticles with larger size cannot return to the blood circulation, resulting in a higher effective drug accumulation

has been confirmed value in the preliminary clinical studies [13, 14]. It indicates that the predictive markers for EPR may play a significant role in clinical translation of tumor nanotherapy.

13.2.1 Passive Targeting Liposomes for Cancer Therapy

Passively targeted liposomes (Table 13.1) were the first class of therapeutic nanoparticles to receive clinical approval for cancer treatment and still represent a large proportion of clinical-stage nanotherapeutics. For example, Myocet is a novel liposome-encapsulated formulation of doxorubicin, which aims to improve the safety profile of doxorubicin while at least maintaining its proven antitumor efficacy, which has been used to treat metastatic breast cancer and ovarian cancer, myeloma, and HIV-related Kaposi's sarcoma. Comparative phase III trials in patients with metastatic breast cancer showed that Myocet regimens are significantly less cardiotoxic than conventional doxorubicin [16]. Doxil, the first FDA-approved long circulation nanodrug (1995), can passively target to tumors, and its doxorubicin is released and becomes available to tumor cells by as yet unknown means. Doxil exhibits prolonged drug circulation time and avoidance of the RES due to the use of pegylated nano-liposomes [17]. Trials comparing Doxil with traditional chemotherapy agents have generally found that the pegylated liposomal formulation reduces many adverse events. The FDA has approved Doxil for patients with recurrent ovarian cancer, AIDS-related Kaposi sarcoma, and multiple myeloma [18]. In addition, Onivyde (irinotecan liposome injection) was approved by FDA in 2015 for the treatment of post-gemcitabine metastatic adenocarcinoma of the pancreas [19].

Table 13.1 Summary of clinical-stage passive targeting liposomes for cancer therapy

Nanomedicines	Research phase	Clinical application	Refs.
Liposomal doxorubicin (Myocet)	Approved in Europe and Canada	Metastatic breast cancer	[16]
Liposomal doxorubicin (Doxil)	FDA approved	HIV-related Kaposi sarcoma, ovarian cancer, and multiple myeloma	[17, 18]
Liposomal irinotecan (Onivyde or MM-398)	FDA approved	Post-gemcitabine metastatic pancreatic cancer	[19]
Liposomal cytarabine (Depocyt)	FDA approved	Lymphoma meningitis	[27]
Liposomal cytarabine-daunorubicin (CPX-351 or Vyxeos)	FDA approved	High-risk acute myeloid leukemia	[27]
Mifamurtide (Mepact)	Approved in Europe	Nonmetastatic, resectable osteosarcoma	[16–18]
Liposomal cisplatin (Lipoplatin)	Phase III	NSCLC	[20]
Liposomal paclitaxel (EndoTAG-1)	Phase III	Pancreatic cancer liver metastases, HER2- negative and triple-negative breast cancer	[23–26]

Several liposomal nanomedicines also showed great potential for tumor treatment in clinical trials. Liposomal cisplatin (Lipoplatin) is pegylated liposome, which is developed to enhance the drug targeting to tumors and decrease systemic toxicity of cisplatin by prolonging circulation time [20]. Lipoplatin in combination with paclitaxel has been shown to be much less toxic than the original cisplatin combined with paclitaxel. Nephrotoxicity was negligible after liposomal cisplatin administration [21]. Liposomal paclitaxel (EndoTAG-1) have reached the phase III clinical trials [22] for treatment of pancreatic cancer, liver metastases, and HER2-negative and triple-negative breast cancer [23–26]. EndoTAG-1 have good performances in reducing allergic reactions and drug side effects, improving tumor-targeted and tolerated doses.

It is well-known that drug-loaded liposomes can improve pharmacokinetics and biodistribution; however, compared with traditional parent drugs, few marketed liposomes have prominent overall survival (OS) advantages. In patients with acute myeloid leukemia, the liposomal cytarabine-daunorubicin (Vyxeos) showed improved OS of 9.56 months in phase III clinical trial compared with that of 5.95 months with the standard treatment of cytarabine and daunorubicin. Vyxeos was approved by FDA in 2017, which is encouraging for the field of cancer nanomedicine [27].

13.2.2 Passive Targeting Nanoparticles for Cancer Therapy

Many other types of nanoparticles can act as delivery systems for antitumor drugs [28], including lipid carriers (solid lipid nanoparticles and micelles); polymer carriers [29] such as hydrogels, polymers, dendritic macromolecules, and nanofibers; metallic nanoparticles [30] (gold, silver, titanium); carbon structure (nanotubes, nanodiamond, and graphene); and inorganic particles, such as silica. Different types of materials hold different optimal delivery modes. For example, albumin [31] is an effective solutionizing platform enabling formulation of hydrophobic chemotherapy drugs, such as paclitaxel [32]. Metal particles are promising photothermal therapeutic agents. Carbon nanoparticles have mediated an unprecedented increase in magnetic resonance (MR) efficiency to improve image-based diagnosis. Polymer nanoparticles can stabilize small interfering RNA (siRNA) therapy through their reasonable and scalable design without the need for toxic multi-cationic agents. The corresponding characteristics of different types of nanoparticles promoted their clinical application [32] and improved their curative effect. Table 13.2 lists some nanoparticles that have been approved by the FDA for marketing or clinical trials.

Albumin-bound paclitaxel (Abraxane) was the second generation of commercial nanomedicines. The albumin can solubilize the hydrophobic drugs without using solvent Cremophor EL, which can avoid allergic reactions and administrate at higher doses. A phase III study of Abraxane in 454 metastatic breast cancer patients showed a higher response rate (33%) and a lower neutropenia rate (10%) compared to Taxol (19 and 21%, respectively). Abraxane dissociates rapidly into albumin and paclitaxel after intravenous injection, which will not change the biodistribution and pharmacokinetics of paclitaxel. Although the every-3-weeks dosing schedule of Abraxane is better than paclitaxel in patients with breast cancer in terms of response rate, the progression-free survival (PFS) reveal similar tendencies compared with once-per-week dosing schedule [33].

Table 13.2 Summary of clinical and market research on passive targeting nanoparticles

Nanomedicines	Research phase	Clinical application	Refs.
Albumin-bound paclitaxel (Abraxane)	FDA approved	Metastatic breast cancer	[31, 32]
Zinostatin stimalamer (SMANCS)	Approved in Japan	HCC	[31]
Polymeric micelle paclitaxel (Genexol-PM)	Approved in Korea	Breast cancer and NSCLC	[33]
Doxorubicin transdrug (Livatag)	Phase II/III	Liver cancer (HCC)	[42]
Au nanoparticles (AuNPs)	Phase I	Mammary, head, and neck or glioma tumors	[37]
CYT-6091	Phase I	Solid tumor	[36]
CRLX101	Phase II	Metastatic renal cell carcinoma	[34]

Polymer nanoparticles, including polymer micelles, synthetic polymer nanoparticles, or dendritic molecules, have achieved considerable commercial conversion capabilities, primarily due to their ability to improve stability, biocompatibility, and drug release and circulation. Zinostatin stimalamer (SMANCS) is a styrene-maleic acid polymer coupled with neocarzinostatin, an antibiotic with antitumor activity. It was approved in Japan in 1994 for the treatment of advanced and recurrent HCC, with SMANCS in the HCC artery. This treatment leads to the deposition of SMANCS in hepatocellular carcinoma and the gradual release of SMANCS into tumor tissues. Preclinical studies have also shown that SMANCS have higher tumor blood ratio compared to non-nanometer SMANCS due to EPR. Clinical trials were conducted on HCC patients. SMANCS showed strong activity in patients who took 3–4 mg of the drug every 3–4 weeks. In fact, 95% of patients had tumors that shrank. Livatag, a doxorubicin-loaded polyisohexylcyanoacrylate (PIHCA) nanoparticle, has been developed by Onxeo for the treatment of the most common form of liver cancer (HCC). Livatag can avoid drug resistance by short-circuiting the mechanisms of tumor multidrug resistance through the masking of the anticancer agent. The phase II study in liver cancer patients appears to have resulted in an average survival time of 32 months for patients treated with Livatag, compared with 15 months for those receiving the current best treatment. A phase III clinical study of Livatag is currently under way, and preliminary results appear to indicate that the treatment is well tolerated. Currently, clinical results of BIND-014 and CRLX101 were frustrating [34], emphasizing that development strategies need to be reconsidered, including selection of potential patients who may respond to nanotherapeutics.

Inorganic nanomaterials, such as iron oxide nanoparticles and gold nanoshells, have also been evaluated in cancer patients. NanoTherm is made up of 15 nm nano iron oxide nanoparticles coated with aminosilane and produced by Magforce. They are placed inside tumors and heated by alternating magnetic fields, a technique known as magnetohyperthermia. A clinical study of NanoTherm was conducted on 60 patients with glioblastoma. The application of alternating magnetic field increased the internal temperature of the tumor to 43–45 °C without obvious side effects. The mean survival rate after the first recurrence of glioma was about 13.2 months, which was longer than the 6 months of conventional treatment [35]. CYT-6091, the 27 nm colloidal gold particles combined with recombinant human tumor necrosis factor alpha and thiolated polyethylene glycol on the surface, was tested in advanced stage cancer patients (phase I) [36]. Doses from 50 to 600 μg/m^2 were well tolerated, and no maximum tolerated dose (MTD) was reached, exceeding the previous MTD for native rhTNF by threefold. After 24 h treatment, nanoparticles of gold were found to target tumors in patient biopsies, which was visualized using electron microscopy. Au nanoparticles (AuNPs), under development by nanoprobe, are 11 nm gold nanoparticles conjugated with antiepidermal growth factor receptor (EGFR) antibodies and PEG. In preclinical study, mice bearing mammary, head, and neck or glioma tumors were treated by intravenous administration of AuNPs and by exposing the mice to x-ray radiotherapy. The tumor was found to disappear after 20–30 days treatment, and the 1-year survival rate increased to 50–86%, which is higher than 0–20% observed for receiving radiation therapy alone [37].

13.3 Enhancing Drug Targeting to the Tumor

The core of a drug-targeting system is to improve the efficacy by increasing the concentration of the drug at the desired location while minimizing toxicity by reducing nontarget accumulation. In the past few years, a large number of nanosystems have been developed to achieve this drug-targeting promise [1]. Previous generations of tumor-targeted systems were designed to maximize passive targeting through EPR effect for enhancing tissue penetration and tumor site accumulation. The surface of these particles is always covered by inert materials (such as polyethylene glycol) to reduce the phagocytic elimination of protein, prolong the circulation time, and thereby enhance the accumulation of tumors [38]. To further improve targeting efficiency, many current generation carrier systems rely on the prospect of enhanced active (or molecular) targeting. In active targeting technology, modification of target molecules, such as antibodies (ab), antibody fragments, peptides, and small nucleic acid molecules, on the surface of nanoparticles can selectively increase their adhesion and absorption to target cells by binding these target molecules to receptor-binding molecules [39].

At present, many tumor-active targeting nanodelivery systems have entered clinical trials (Table 13.3), but most of them fail in clinical phase I and phase II, and fewer have reached further trials or even come to market. Therefore, the following is a summary of the current clinical research on tumor-active targeting systems and the clinical results of clinical trials, as well as the potential reasons for the low conversion success rate, in the hope of learning from these clinical failures.

Liposomes can be designed to target tumor antigens, such as HER2, and change their microdistribution within tumor sites. This approach can direct liposomes into tumor cells rather than macrophages. MM-302 is a HER2-targeted liposome encapsulating doxorubicin in its core and anti-HER2 antibodies conjugated to its surface [40], which is under development by Merrimack for patients with HER2-positive metastatic breast cancer. As an estimated 25% of breast cancer patients are defined as HER2-positive, MM-302 was specifically designed to maximize doxorubicin uptake into tumor cells while minimizing uptake into nontarget cells and tissues and to enable combination with trastuzumab [41]. Preclinical studies of MM-302 showed the extended pharmacokinetics and EPR-mediated deposition in tumors. The anti-HER2 antibodies on the surface of MM-302 specifically increase targeting to HER2-overexpressing tumor cells with a resultant increase in antitumor activity relative to pegylated liposomal doxorubicin [40]. Phase I clinical study of MM302 was carried out on 47 patients with advanced HER2-positive breast cancers, did not show any decline in cardiac functions [42], and thus had a manageable safety profile. MM310 is an Ephrin A2-targeted liposome that binds to a pH-sensitive taxane prodrug, developed by Merrimack Pharmaceuticals. Preclinical studies of MM310

Table 13.3 Summary of clinical research on active targeting systems

Name	Target head	Research phase	Clinical application	Sponsor	Refs.
MM-302	HER2 targeting	Phase II/III	HER2-positive breast cancer	Merrimack pharmaceuticals	[40–42]
MM-310	EphA2 targeting	Phase I	Solid tumors	Merrimack pharmaceuticals	[43]
Anti-EGFR-IL-dox	EGFR targeting	Phase II	Breast cancer	Swiss group for clinical cancer research	[44]
C225-ILs-dox	EGFR targeting	Phase I	Glioblastoma	University hospital, Switzerland	[1]
SGT-53	TFR1 targeting	Phase II	Recurrent glioblastoma; neoplasm; metastatic pancreatic cancer	SynerGene Therapeutics, Inc.	[47, 48]
SGT-94	TFR1 targeting	Phase I	Solid tumors expressing high levels of TFR1	SynerGene Therapeutics, Inc.	[47, 49–50]
BIND-014	Prostate-specific ligand (PSMA) targeting	Phase II	Urothelial carcinoma; cholangiocarcinoma; squamous cell carcinoma	BIND therapeutics	[7, 45]
CALAA-01	Transferrin targeting	Phase I	Solid tumors	Calando pharmaceuticals	[9]
TKM-080301	PLK1	Phase I/II	Hepatocellular carcinoma; liver cancer	Arbutus biopharma corporation	[59, 60]

demonstrated strong antitumor activity against the human lung cell A549 and the triple negative breast cancer xenograft models with a long circulating half-life (8–12 h) and good stability. Then, MM310 has advanced to a phase I clinical trial in patients with solid tumors [43].

Nanoparticles such as immunoliposomes are generated by combining a liposome or a stable and robust drug carrier with a ligand, such as a monoclonal antibody or antibody fragment. Immunoliposomes directly target tumor cells and provide a so-called bystander killing effect by diffusion of small molecule drugs to neighboring tumor cells. Anti-EGFR immunoliposomes (anti-EGFR-IL-dox) loaded with doxorubicin is made by combing polyethylene glycol liposomes with EGFP, and it was sponsored by Swiss Group for Clinical Cancer Research. In preclinical in vitro and in vivo studies, anti-EGFR immunoliposomes showed substantial antitumor effects and were significantly more effective than all other treatments, including the corresponding free or liposomal drug [44]. Data from a phase I trial, in 26 patients with different solid tumors, show very little toxicity and signs of efficacy of anti-EGFR-IL-dox. C225-ILs-dox, another doxorubicin-loaded anti-EGFR immunoliposomes, are intravenously administrated to patients with relapsed high-grade gliomas in phase I clinical trial. Then, the pharmacokinetics of drug in peripheral blood, cerebrospinal fluid, and tumor tissue resection was investigated.

Since the discovery and characterization of prostate-specific membrane antigen (PSMA), investigators have exploited the unique diagnostic and therapeutic potential provided by this cell-surface biomarker in prostate cancer management. BIND-014 is a docetaxel-encapsulating nanoparticle and a hydrophilic polyethylene glycol corona decorated with small-molecule PSMA-targeting ligands. Prostate-specific membrane antigen (PSMA) is expressed in most prostate cancer and many other malignancies vasculature, but not in the vasculature of healthy tissues. The modified nanoparticles can be internalized by binding to PSMA-expressing cells and increase the accumulation of docetaxel in the tumor sites. In preclinical studies, BIND-014 exhibited increased intratumoral drug concentration, enhanced antitumor activity, and decreased distribution and clearance in xenograft models of prostate, breast, and non-small cell lung cancer as compared with traditional docetaxel [7]. BIND-014 confirmed safety at the dosage of 60 mg/m^2 every 3 weeks in phase II clinical trial [45].

Beyond their widely reported use as carriers for chemotherapeutics, nanoparticles have shown potential for the delivery of various new anticancer therapeutic agents, including molecularly targeted agents, antisense oligonucleotides, and small interfering RNA and mRNA. The cationic liposomes SGT-53 containing wild-type p53 gene achieve tumor targeted by binding to the anti-transferrin receptor single-chain antibody fragment. Preclinical studies of SGT-53 indicated that it could enhance radiation/chemical sensitivity to tumors [46]. Tumor-specific targeting and great antitumor effect in patients with advanced solid tumor were demonstrated in phase I clinical trial of SGT-53[47]. Phase II clinical trial of the tumor-targeted SGT-53 nanocomplex in combination with chemotherapeutic agent temozolomide is undergoing investigation in recurrent glioblastoma. SGT-94 is a liposome containing RB94 gene (plasmid DNA), which can target tumor through binding the anti-transferrin receptor single-chain antibody fragment. RB94, a tumor-suppressor gene, is a modified form of the retinoblastoma gene RB110 and has shown enhanced tumor-suppressor activity compared to RB110 in all tumor cell types. SGT-94 was administered intravenously in a first-in-man study in metastatic genitourinary cancer [46, 48], demonstrating selective tumor targeted and well tolerance [49]. CALAA-01, a targeted nanocomplex that contains anti-R2 siRNA, is the first RNA interference (RNAi)-based, experimental therapeutic to be administered to cancer patients. The nanocomplex formulation consists of nonchemically modified siRNA (C05C), cyclodextrin-containing polymer (CAL101), stabilizing agent (AD-PEG), and transferrin protein targeting agent (AD-PEG-Tf). The cationic polymer interacts electrostatically with anionic siRNA to assemble into nanocomplexes that prevent degradation of siRNA in serum. The siRNA-encapsulated nanocarriers can specifically bind to cells that overexpress the transferrin receptor (TfR) and enter the cells through endocytosis. Intracellularly, siRNA can be released from the nanocomplex through the chemical component in the polymer to exert RNA interference function. CALAA-01 was well tolerated during the initial dose-escalation portion of the phase I study. The delivery system used in CALAA-01 was demonstrated to target delivery of functional siRNA. However, the full potential of CALAA-01 was not evaluated in this phase Ia/Ib clinical study [9]. TKM-080301 is a siRNA-loaded lipid nanoparticle for targeting polo-like kinase 1 (PLK1), a protein overexpressed in cancer cells promoting an uncontrolled cell proliferation. The preclinical evaluation of TKM-080301 includes in vitro and in vivo pharmacological activity studies, effects on immune system, and toxicity. TKM-080301 exhibited inhibitory effect on cancer cell proliferation with PLK-1 silencing. It also showed antitumor effect in xenograft models of human cancer [50]. The first phase I/II clinical trial of TKM-080301 evaluated safety and toxicity. The result showed that siRNA was generally well tolerated and TKM-080301 presented a preliminary antitumor efficacy. In phase 2 and 3, regarding adverse effects, patients showed mild to moderate infusion-related reactions with cytokines upregulated [51].

13.4 Nanoparticles for Targeted Tumor Diagnostics and Imaging

In recent years, as personalized cancer treatments have gained much attention, the need for more advanced imaging techniques has grown dramatically. Nanoparticles can be used not only as imaging contrast agents but also as auxiliary diagnostic agents [1]. Metal and semiconductor nanoparticles real unique electronic, optical, and catalytic properties, making them hold great potential in imaging technology. Among them, gold, semiconductor metals, and iron oxide are of particular interest [52, 53]. They are promising to improve imaging techniques such as magnetic resonance imaging (MRI), near-infrared (NIR) fluorescence imaging, X-ray imaging, computed tomography (CT), and positron emission spectroscopy (PET) [54]. Nanoparticle probes can enhance the signal sensitivity of imaging technology and provide better spatial resolution at the molecular and cellular levels. Table 13.4 lists some nanoparticles that have entered clinical trials from phase I to phase IV, and the following is a summary of the current research on nanoparticles for targeted tumor diagnostics and imaging.

13.4.1 Magnetic Resonance Imaging

Magnetic resonance imaging uses NPs as a contrast agent to diagnose diseases and study biological processes such as cancer metastasis and inflammation. NPs can provide positive contrast in spin-lattice relaxation images and negative contrast in transverse relaxation images. Introduced in the mid-1980s, superparamagnetic iron oxide nanoparticles (SPIONs) are effective contrast agent with high sensitivity in MRI [53, 55]. They are typically composed of ferrite cores (magnetite or other insoluble ferrites) with the size of 5–10 nm and a water-soluble coating layer. In the

Table 13.4 Summary of clinical research on nanoparticles for targeted tumor diagnostics and imaging

Name	Research phase	Clinical application	ClinicalTrials. gov identifier
Superparamagnetic iron oxide	Phase IV	Pancreatic cancer	NCT00920023
Superparamagnetic iron oxide (SENTINAC-01)	Phase III	Breast cancer	NCT02249208
Resovist (BAY86-4884, SH U 555 A)	Phase III	Hepatic neoplasms	NCT00307866
Sinerem (USPIO)-enhanced MRI	Phase III	Prostate cancer bladder cancer	NCT00622973
Fluorescent cRGDY-PEG-Cy5.5-C dots	Phase I, phase II	Head and neck melanoma, breast cancer, colorectal cancer	NCT02106598
(64Cu)-labeled PSMA-targeting particle tracer or 64Cu-NOTA-PSMAi-PEG-Cy5.5-C′ dots	Phase I	Prostate cancer	NCT04167969

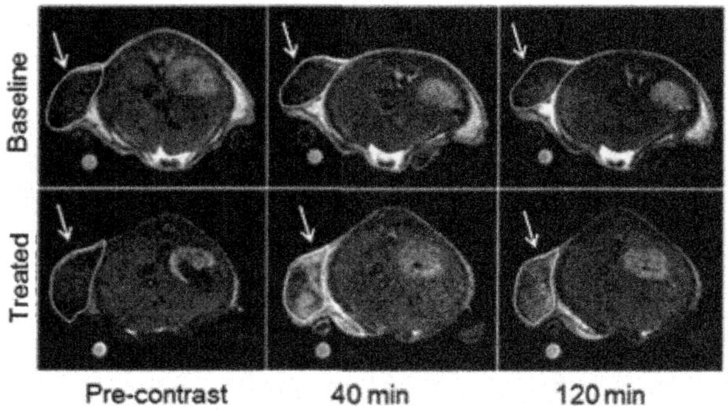

Fig. 13.2 Gadolinium NPs self-assemble in the presence of caspase 3/7 enzymes ("treated" group), increasing MRI contrast. (Reproduced from Ref. [67])

presence of a magnetic field, these magnetic domains have large magnetic moments and generate local disturbances in magnetic field homogeneity. Due to these magnetic disturbances, large difference in susceptibility between iron oxide crystals and nearby protons leads to rapid spin de-phasing, resulting in short transverse relaxation time and darkening of the image relative to the unaffected area. It is noteworthy that SPIONs have been used in animal research since the 1990s; extensive studies have also been carried out in the past decade to make them suitable for biomedical applications in patients [56]. SPIONs are mainly used as transverse relaxation contrast agents and create dark regions in the MR images [57]. Several SPIONs have undergone clinical trials in cancer patients (Fig. 13.2), including ferumoxytol [58], ferumoxtran, and ferucarbotran NPs [59]. However, some magnetic nanoparticle-based nanocarriers loaded with drugs were evaluated in patients and did not produce convincing results, leading to termination of clinical development in phase II/III [60]. Different forms of NPs have been used in contrast-enhanced MRI, such as molecular targeting SPIONs, to detect mRNA levels after chronic drug exposure [61] or track gene expression tracking [62], more complex magnetic iron oxides for theranostic applications in cancer [63], gadolinium-doped NPs for locating radiosensitizers (Figs. 13.2) [64, 65], manganese oxide NPs as alternatives to Gd-based agents [66], and other metals for cell tracking.

13.4.2 Optical Imaging

Optical imaging technology has high spatial resolution, but the detection depth of the signal in the tissue is limited due to the inherent attenuation of signals in a special wavelength range. Many NPs propose inherent fluorescent and are suited for optical or near-infrared (NIR) imaging. The first-in-human trial of fluorescent core shell silica NPs (Cornell dots or C dots) received FDA investigational new drug (IND) approval as a drug for targeted molecular imaging [59]. For positron emission tomography (PET) imaging, these renally excreted silica particles were labeled with (124) I and modified with cRGDY peptides for molecular targeting. For example, (124) I-cRGDY-PEG-C dot particles containing the dye Cy5 are inherently fluorescent, which may be used as hybrid PET-optical imaging agents for lesion detection, cancer staging, and treatment management. The safety, pharmacokinetics, clearance properties, and radiation dosimetry of (124) I-cRGDY-PEG-C dots were assessed by serial PET and computerized tomography after intravenous administration in patients [68]. The results are consistent with the well-tolerated tracer of inorganic granules, exhibiting stability and distinct in vivo, reproducible pharmacokinetic signatures defined by renal excretion. No toxic or adverse events caused by the particles were observed [69].

In addition, the application of "The use of nanoparticles to guide the surgical treatment of prostate cancer" is being recruited in clinical trials. The purpose of this study is to evaluate the safety of 64Cu-NOTA-PSMA-PEG-Cy5.5-C′ dot tracer in tumor cell identification before and during surgery for prostate cancer. The researchers want to find out whether PET/MRI scans performed after the injection of this tracer are more accurate than the imaging scans commonly used to locate deposits of prostate tumor cells [70]. It is the first time that this tracer will be used in people who are undergoing surgery for prostate cancer.

Photoacoustic imaging, also called optoacoustic tomography (OAT), is a new medical imaging method that uses optical illumination and ultrasonic detection to overcome this limitation [71]. Because ultrasound waves can penetrate through tissues with minimal scattering and attenuation, OAT can locate optical absorption objects deep in the tissue, including abnormal angiogenesis in advanced tumors. However, angiogenesis is not sufficient to distinguish tumors from healthy tissue in the early stages of the disease, making exogenous contrast agents necessary [72]. For this purpose, gold nanoparticles as exogenous contrast agents have gained attention in photoacoustic imaging. Due to the phenomenon of plasma resonance, gold nanoparticles exhibit a narrow and strong absorption and scattering band. The photoacoustic effect produced by the accumulated gold nanoparticles makes them have higher sensitivity and better image resolution [57].

13.4.3 Multimodal Imaging

A primary benefit of utilizing a NP platform for imaging is the relative ease of incorporating contrast agents for multiple imaging modalities. Clinically, C dots which incorporate Cy5 are used for fluorescent imaging, cRGD peptide is capable of targeting tumor through integrin, and 124I is available in PET imaging. Preliminary studies of these NPs revealed no toxicity to late-stage melanoma patients and preferential accumulation in tumor sites, visualized by serial PET scans [59].

Compared to conventional agents, nanoparticles have showed targeting ability, long circulation time, and improved signal intensity in cancer diagnosis and therapy. With the help of nanotechnology, imaging modalities have become more powerful than before, and about 50 nanomedicines have been approved for clinical trials by the FDA. Major obstacles for translation to clinical prative involve the safety and biodistribution of developed nanoparticles. Despite the impressive progress that has been made, most nano-contrast agents are still in the experimental stage; very few of them were evaluated in humans. Ongoing research should focus on enhancing targeting selectivity while minimizing toxicity [73].

It is noteworthy that nanotheranostics provide a promising strategy to monitor the drug biodistribution by integrating imaging and drug delivery functions into a single NP formulation, which offer vital insights for tumor identification and predicting efficacy of nanomedicines [13].

13.5 Challenges in Clinical Translation

Nanotechnology has transformed by crafting promising avenues in therapeutics and diagnosis, showing enormous progress in the field of cancer nanomedicine (Fig. 13.3). Considering the heterogeneity of the tumor, the extent of hypoxia and expression of enzyme required for drug release may be different, which may lead to unpredictable drug release. There is clearly room for further improvement. One strategy is using dual-stimulus reactive triggers to increase tumor specificity, but attention should be given to further characterizing these systems and improving the scalability of the formulation [74, 75]. For multi-loading nanoparticles, the optimized cargo loading, the biocompatibility, and efficacy are important aspects to be considered. Collectively, several factors must be considered to improve the translation of nanomedicines from bench to bedside [76]. Here, we discuss key issues that affect the development of clinically feasible nanosystems.

Most nanoparticles are thought to rely on the EPR effect to accumulate in tumors. It has been known that EPR effect varies between patients and tumor types, even with the same patient and tumor type in different stages. However, few studies have been carried out to address the effect of EPR on the therapeutic efficacy of nanomedicines. Some preliminary clinical trials have demonstrated the value of subpopulation of patients based on their possibility of response to EPR, suggesting

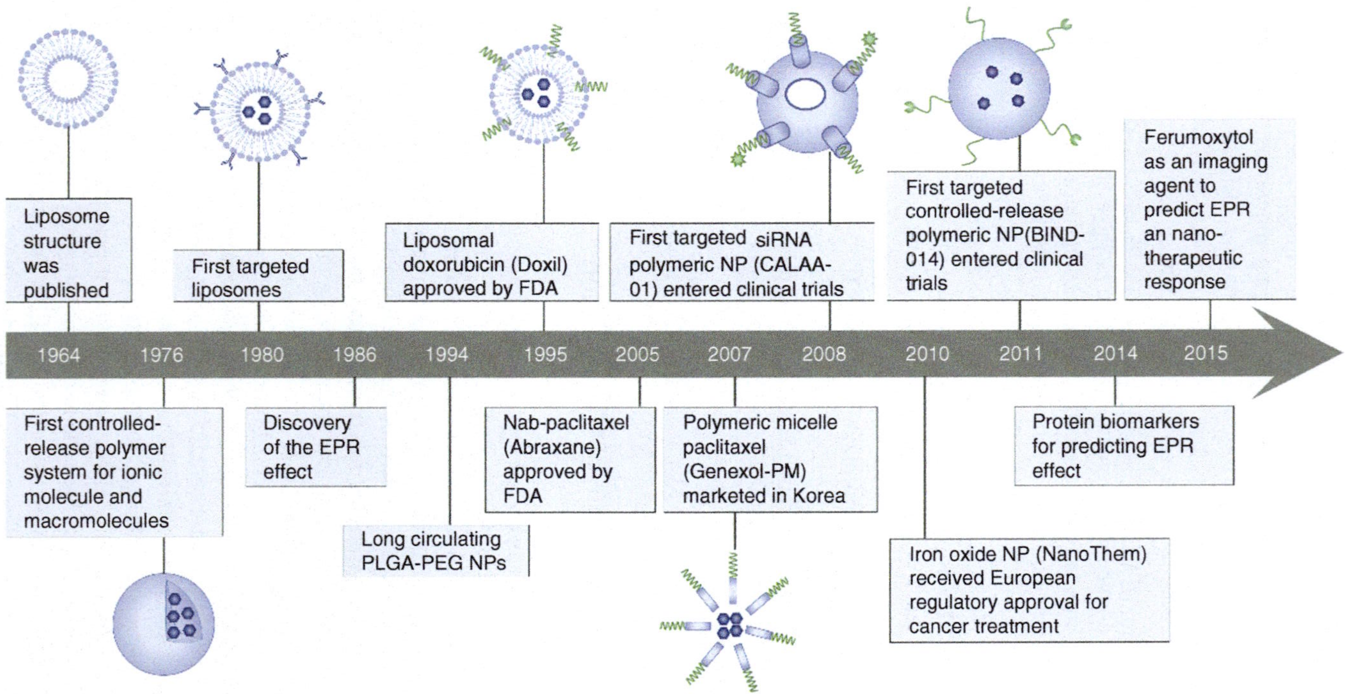

Fig. 13.3 Timeline of major developments in the field of nanomedicine for tumor therapy

markers for EPR effect may play an important role in the clinical success of cancer nanotherapies. However, more reliable methods for assessing the effect on nanoparticle delivery should be explored further in relevant preclinical and clinical models.

Recently, we have also gradually realized the challenges and opportunities that lie ahead. It is now recognized that in vitro models may not reliable to predict the utility of nanoparticles, because the tumor cells cultured in multi-well plates lack the complexity of biological tissue and fluid flow. The complex interactions between NPs and the physiological barrier may not be simulated on such a platform. For example, a dual-targeted (Tf and mAb 2C5) nanoparticle system failed to reproduce the in vitro effectiveness when tested in mouse models [77]. Recent efforts to develop biomimetic 3D tumor models may help bridge this gap to some extent. Current 3D model can be further improved by incorporating stromal cells, extracellular matrix proteins, or even related mechanical forces [78], producing a more reliable platform for nanomedicine evaluating and preclinical study designing.

Design of preclinical trials is crucial for predicting the efficacy and safety of the nanoparticles. The translation of nanotherapeutics may be promoted by developing animal models that highly simulate the heterogeneity and anatomy of human tumors, such as the humanized mouse model, the high-fidelity patient-derived xenografts (PDXs) model, and the genetically engineered mouse model (GEMMs) with metastasis [79]. In addition, it is important to assess changes in immunological parameters, such as cytokine levels or immune cell numbers in preclinical studies. These comprehensive analyses will help to predict the efficacy and toxicity of nanotherapeutics in patients. Also, the clinical translatability of nanoparticles may be benefit from phase 0 studies, which is useful in timely evaluation of the PK and tumor localization potential of the nanotherapeutics in humans. They are much cheaper to implement than phase I trials, and researchers can get feedback more quickly on the clinical feasibility of a given nanosystem.

The heterogeneity and complexity of the tumors indicate the need for patient selection with specific biomarker, for example, to identify those most likely to benefit from the nanotherapies. Ideally, nanodrugs should be administrated the same way in preclinical models as they are expected to use in patients. For example, intratumoral injection is commonly used in preclinical studies of oncolytic viral therapy and nonviral nanocarriers [80–82]. However, it may show limited utility in patients with metastatic tumor. Intravenous injections expose nanoparticles to a variety of biological barriers and thus will be less effective than intratumoral administration.

The transition of NP technology from preclinical development to clinical development, subsequent commercialization, and wider application will encounter several challenges, including the escalating complexity of chemical, good manufacturing practice (GMP), and manufacturing and control (CMC) requirements. GMP and CMC involve different but overlapping methods and regulations to ensure that products always meet predetermined quality standards. The scale-up of simple NPs, such as polymeric and liposomal systems, can be achieved through off-the-shelf production unit operations that are widely used in the pharmaceutical industry. The scale-up of more complex nanomaterials, such as nanomedicines that integrate biological targeting ligands or components, may require the development of new manufacturing processes or modifications to existing unit operations, which may induce additional CMC and GMP challenges [83, 84]. Although the current mainstream of NP manufacturing is still synthesis, it is possible to produce NPs suitable for industrial production through technologies such as PRINT and turbulent jet mixing, which may accelerate their clinical translation.

13.6 Conclusion

Although there have been some intriguing successes, current technologies for tumor-targeted nanoparticles do not produce predictable outcomes; it is unclear how much closer these innovations have led the field to fulfilling the promise of drug targeting and where exactly the field should now focus its attention. To make nanotherapy safer and more effective, full understanding is necessary about the nano-biosystem interactions, the targeting of nanoparticles to the tumor microenvironment, and the transport of nanoparticles to tumor cells. Addressing the challenges of manufacture and scale-up of nanoparticles, as well as evaluating nanoparticles in vitro and in vivo, may facilitate clinical translation. As more and more nanomedicines containing existing drugs are now approved, the new generation of nanomedicines are expected to use novel molecular entities and therapeutic agents, such as kinase inhibitors, siRNA, and gene editing.

This chapter has explored the importance of the convergence of nanotechnology and tumor biology for more successful development and clinical translation of nanotherapeutics. We believe that a more nuanced view of targeting will facilitate the design of clinically useful systems to better deliver on the potential of nanoparticle drug delivery.

References

1. Shi J et al (2017) Cancer nanomedicine: progress, challenges and opportunities. Nat Rev Cancer 17(1):20–37
2. Tanaka T et al (2004) Tumor-targeted based on the effect of enhanced permeability and retention (EPR) and the mechanism of receptor-mediated endocytosis (RME). Int J Pharm 277(1-2):39–61
3. Matsumura Y, Maeda H (1986) A new concept for macromolecular therapeutics in cancer chemotherapy: mechanism of tumoritropic accumulation of proteins and the antitumor agent smancs. Cancer Res 46(12 Pt 1):6387–6392
4. Gerlowski LE, Jain RK (1986) Microvascular permeability of normal and neoplastic tissues. Microvasc Res 31(3):288–305
5. Bertrand N et al (2014) Cancer nanotechnology: the impact of passive and active targeting in the era of modern cancer biology. Adv Drug Deliv Rev 66:2–25
6. Maeda H (2015) Toward a full understanding of the EPR effect in primary and metastatic tumors as well as issues related to its heterogeneity. Adv Drug Deliv Rev 91:3–6

7. Hrkach J, Von Hoff D, Mukkaram Ali M, Andrianova E, Auer J, Campbell T, De Witt D, Figa M, Figueiredo M, Horhota A, Low S, McDonnell K, Peeke E, Retnarajan B, Sabnis A, Schnipper E, Song JJ, Song YH, Summa J, Tompsett D, Troiano G, Van Geen Hoven T, Wright J, LoRusso P, Kantoff PW, Bander NH, Sweeney C, Farokhzad OC, Langer R, Zale S (2012) Preclinical development and clinical translation of a PSMA-targeted docetaxel nanoparticle with a differentiated pharmacological profile. Sci Transl Med 4(128):128ra39
8. Eliasof S et al (2013) Correlating preclinical animal studies and human clinical trials of a multifunctional, polymeric nanoparticle. Proc Natl Acad Sci USA 110(37):15127–15132
9. Zuckerman JE et al (2014) Correlating animal and human phase Ia/Ib clinical data with CALAA-01, a targeted, polymer-based nanoparticle containing siRNA. Proc Natl Acad Sci USA 111(31):11449–11454
10. Joyce JA (2005) Therapeutic targeting of the tumor microenvironment. Cancer Cell 7 (6):513–520
11. Meads MB, Gatenby RA, Dalton WS (2009) Environment-mediated drug resistance: a major contributor to minimal residual disease. Nat Rev Cancer 9(9):665–674
12. Prabhakar U et al (2013) Challenges and key considerations of the enhanced permeability and retention effect for nanomedicine drug delivery in oncology. Cancer Res 73(8):2412–2417
13. Arrieta O et al (2014) High liposomal doxorubicin tumour tissue distribution, as determined by radiopharmaceutical labelling with (99m)Tc-LD, is associated with the response and survival of patients with unresectable pleural mesothelioma treated with a combination of liposomal doxorubicin and cisplatin. Cancer Chemother Pharmacol 74(1):211–215
14. Ramanathan R et al (2014) Abstract CT224: pilot study in patients with advanced solid tumors to evaluate feasibility of ferumoxytol (FMX) as tumor imaging agent prior to MM-398, a nanoliposomal irinotecan (nal-IRI). Cancer Res 74:CT224–CT224
15. Danhier F, Feron O, Preat V (2010) To exploit the tumor microenvironment: passive and active tumor-targeted of nanocarriers for anti-cancer drug delivery. J Control Release 148(2):135–146
16. Romeo C et al (2019) Non-pegylated liposomal doxorubicin (NPLD, Myocet(R))+carboplatin in patients with platinum sensitive ovarian cancers: a ARCAGY-GINECO phase IB-II trial. Gynecol Oncol 152(1):68–75
17. Barenholz Y (2012) Doxil(R)–the first FDA-approved nano-drug: lessons learned. J Control Release 160(2):117–134
18. Smith AD (2013) Big moment for nanotech: oncology therapeutics poised for a leap. OncLive. https://www.onclive.com/publications/Oncology-live/2013/June-2013/Big-Moment-for-Nanotech-Oncology-Therapeutics-Poised-for-a-Leap
19. Inman S (2015) FDA approves second-line MM-398 regimen for metastatic pancreatic cancer. OncLive. http://www.onclive.com/web-exclusives/fda-approves-mm-398-regimen-for-metastatic-pancreatic-cancer
20. Stathopoulos GP et al (2010) Liposomal cisplatin combined with paclitaxel versus cisplatin and paclitaxel in non-small-cell lung cancer: a randomized phase III multicenter trial. Ann Oncol 21 (11):2227–2232
21. Kelly K et al (2001) Randomized phase III trial of paclitaxel plus carboplatin versus vinorelbine plus cisplatin in the treatment of patients with advanced non--small-cell lung cancer: a Southwest Oncology Group trial. J Clin Oncol 19(13):3210–3218
22. Koudelka S, Turanek J (2012) Liposomal paclitaxel formulations. J Control Release 163 (3):322–334
23. US National Library of Medicine (2008) Clinical Trials.gov. https://clinicaltrials.gov/ct2/show/NCT00377936?term
24. US National Library of Medicine (2010) Clinical Trials.gov. https://clinicaltrials.gov/ct2/show/NCT00542048?term
25. US National Library of Medicine (2012) Clinical Trials.gov. https://clinicaltrials.gov/ct2/show/NCT00448305?term
26. US National Library of Medicine (2013) Clinical Trials.gov. https://clinicaltrials.gov/ct2/show/NCT01537536?term

27. Lancet JE et al (2018) CPX-351 (cytarabine and daunorubicin) Liposome for injection versus conventional cytarabine plus daunorubicin in older patients with newly diagnosed secondary acute myeloid leukemia. J Clin Oncol 36(26):2684–2692
28. Fais S et al (2016) Evidence-based clinical use of nanoscale extracellular vesicles in nanomedicine. ACS Nano 10(4):3886–3899
29. Lai WF, Wong WT (2018) Design of polymeric gene carriers for effective intracellular delivery. Trends Biotechnol 36(7):713–728
30. Das RK, Brar SK, Verma M (2016) Checking the biocompatibility of plant-derived metallic nanoparticles: molecular perspectives. Trends Biotechnol 34(6):440–449
31. Boyer TD et al (2016) Terlipressin plus albumin is more effective than albumin alone in improving renal function in patients with cirrhosis and hepatorenal syndrome type 1. Gastroenterology 150(7):1579–1589.e2
32. Shitara K et al (2018) Pembrolizumab versus paclitaxel for previously treated, advanced gastric or gastro-oesophageal junction cancer (KEYNOTE-061): a randomised, open-label, controlled, phase 3 trial. Lancet 392(10142):123–133
33. Kim HS et al (2015) A prospective phase II study of cisplatin and cremophor EL-free paclitaxel (Genexol-PM) in patients with unresectable thymic epithelial tumors. J Thorac Oncol 10 (12):1800–1806
34. Voss MH et al (2017) A randomized phase II trial of CRLX101 in combination with bevacizumab versus standard of care in patients with advanced renal cell carcinoma. Ann Oncol 28(11):2754–2760
35. Grauer O et al (2019) Combined intracavitary thermotherapy with iron oxide nanoparticles and radiotherapy as local treatment modality in recurrent glioblastoma patients. J Neuro-Oncol 141 (1):83–94
36. Libutti SK et al (2010) Phase I and pharmacokinetic studies of CYT-6091, a novel PEGylated colloidal gold-rhTNF nanomedicine. Clin Cancer Res 16(24):6139–6149
37. El-Sayed IH, Huang X, El-Sayed MA (2006) Selective laser photo-thermal therapy of epithelial carcinoma using anti-EGFR antibody conjugated gold nanoparticles. Cancer Lett 239 (1):129–135
38. Cheng CJ, Tietjen GT, Saucier-Sawyer JK, Saltzman WM (2015) A holistic approach to targeting disease with polymeric nanoparticles. Nat Rev Drug Discov 14:239–247
39. Accardo A, Tesauro D, Morelli G (2013) Peptide-based targeting strategies for simultaneous imaging and therapy with nanovectors. Polym J 45(5):481–493
40. Espelin CW et al (2016) Dual HER2 targeting with Trastuzumab and liposomal-encapsulated Doxorubicin (MM-302) demonstrates synergistic antitumor activity in breast and gastric Cancer. Cancer Res 76(6):1517–1527
41. Lee H et al (2017) (64)cu-MM-302 positron emission tomography quantifies variability of enhanced permeability and retention of nanoparticles in relation to treatment response in patients with metastatic breast cancer. Clin Cancer Res 23(15):4190–4202
42. Alphandery E et al (2015) Cancer therapy using nanoformulated substances: scientific, regulatory and financial aspects. Expert Rev Anticancer Ther 15(10):1233–1255
43. Huang ZR et al (2019) Formulation optimization of an ephrin A2 targeted immunoliposome encapsulating reversibly modified taxane prodrugs. J Control Release 310:47–57
44. Mamot C et al (2012) Tolerability, safety, pharmacokinetics, and efficacy of doxorubicin-loaded anti-EGFR immunoliposomes in advanced solid tumours: a phase 1 dose-escalation study. Lancet Oncol 13(12):1234–1241
45. Autio KA et al (2018) Safety and efficacy of BIND-014, a Docetaxel nanoparticle targeting prostate-specific membrane antigen for patients with metastatic castration-resistant prostate cancer: a phase 2 clinical trial. JAMA Oncol 4(10):1344–1351
46. Pirollo KF et al (2008) Tumor-targeted nanocomplex delivery of novel tumor suppressor RB94 chemosensitizes bladder carcinoma cells in vitro and in vivo. Clin Cancer Res 14(7):2190–2198
47. Senzer N et al (2013) Phase I study of a systemically delivered p53 nanoparticle in advanced solid tumors. Mol Ther 21(5):1096–1103

48. Zhou J et al (2009) Early RB94-produced cytotoxicity in cancer cells is independent of caspase activation or 50 kb DNA fragmentation. Cancer Gene Ther 16(1):13–19
49. Siefker-Radtke A et al (2016) A phase 1 study of a tumor-targeted systemic Nanodelivery system, SGT-94, in genitourinary cancers. Mol Ther 24(8):1484–1491
50. El Dika I et al (2019) An open-label, Multicenter, phase I, dose escalation study with phase II expansion cohort to determine the safety, pharmacokinetics, and preliminary antitumor activity of intravenous TKM-080301 in subjects with advanced hepatocellular carcinoma. Oncologist 24(6):747–e218
51. Titze-de-Almeida R, David C, Titze-de-Almeida SS (2017) The race of 10 synthetic RNAi-based drugs to the pharmaceutical market. Pharm Res 34(7):1339–1363
52. Alekseeva AV et al (2006) Gold nanorods: synthesis and optical properties. Colloid J 68 (6):661–678
53. Mornet S et al (2004) Magnetic nanoparticle design for medical diagnosis and therapy. J Mater Chem 14(14):2161–2175
54. Durr NJ et al (2007) Two-photon luminescence imaging of cancer cells using molecularly targeted gold nanorods. Nano Lett 7(4):941–945
55. Bulte JWM, Kraitchman DL (2004) Iron oxide MR contrast agents for molecular and cellular imaging. NMR Biomed 17(7):484–499
56. Zhu L et al (2017) Magnetic nanoparticles for precision oncology: theranostic magnetic iron oxide nanoparticles for image-guided and targeted cancer therapy. Nanomedicine (Lond) 12 (1):73–87
57. Gindy ME, Prud'homme RK (2009) Multifunctional nanoparticles for imaging, delivery and targeting in cancer therapy. Expert Opin Drug Deliv 6(8):865–878
58. Bashir MR et al (2015) Emerging applications for ferumoxytol as a contrast agent in MRI. J Magn Reson Imaging 41(4):884–898
59. Ehlerding EB et al (2018) Big potential from small agents: nanoparticles for imaging-based companion diagnostics. ACS Nano 12(3):2106–2121
60. Zhi D et al (2020) Targeting strategies for superparamagnetic iron oxide nanoparticles in cancer therapy. Acta Biomater 102:13–34
61. Liu CH et al (2009) DNA-based MRI probes for specific detection of chronic exposure to amphetamine in living brains. J Neurosci 29(34):10663–10670
62. Ren J et al (2016) Noninvasive tracking of gene transcript and neuroprotection after gene therapy. Gene Ther 23(1):1–9
63. Thomas R, Park IK, Jeong YY (2013) Magnetic iron oxide nanoparticles for multimodal imaging and therapy of cancer. Int J Mol Sci 14(8):15910–15930
64. Lux F et al (2015) Gadolinium-based nanoparticles for theranostic MRI-radiosensitization. Nanomedicine (Lond) 10(11):1801–1815
65. Coughlin AJ et al (2014) Gadolinium-conjugated gold nanoshells for multimodal diagnostic imaging and photothermal cancer therapy. Small 10(3):556–565
66. Pan D et al (2011) Revisiting an old friend: manganese-based MRI contrast agents. Wiley Interdiscip Rev Nanomed Nanobiotechnol 3(2):162–173
67. Ye D et al (2014) Caspase-responsive smart gadolinium-based contrast agent for magnetic resonance imaging of drug-induced apoptosis. Chem Sci 4(10):3845–3852
68. US National Library of Medicine (2019) Clinical Trials.gov. https://clinicaltrials.gov/ct2/show/NCT02106598
69. Phillips E et al (2014) Clinical translation of an ultrasmall inorganic optical-PET imaging nanoparticle probe. Sci Transl Med 6(260):260ra149
70. US National Library of Medicine (2019) Clinical Trials.gov https://clinicaltrials.gov/ct2/show/NCT04167969
71. Karabutov AA, Savateeva EV, Oraevsky AA (2003) Optoacoustic tomography: new modality of laser diagnostic systems. Laser Phys 13(5):711–723
72. Kolkman RGM, Steenbergen W, van Leeuwen TG (2006) In vivo photoacoustic imaging of blood vessels with a pulsed laser diode. Lasers Med Sci 21(3):134–139

73. Han X et al (2019) Applications of nanoparticles in biomedical imaging. Nanoscale 11 (3):799–819
74. Gao W et al (2013) Chemotherapeutic drug delivery to cancer cells using a combination of folate targeting and tumor microenvironment-sensitive polypeptides. Biomaterials 34 (16):4137–4149
75. Xiao D et al (2014) A dual-responsive mesoporous silica nanoparticle for tumor-triggered targeting drug delivery. Small 10(3):591–598
76. Gharpure KM et al (2015) Nanotechnology: future of oncotherapy. Clin Cancer Res 21 (14):3121–3130
77. Sawant RR et al (2013) The effect of dual ligand-targeted micelles on the delivery and efficacy of poorly soluble drug for cancer therapy. J Drug Target 21(7):630–638
78. Goodman TT, Ng CP, Pun SH (2008) 3-D tissue culture systems for the evaluation and optimization of nanoparticle-based drug carriers. Bioconjug Chem 19(10):1951–1959
79. Francia G et al (2011) Mouse models of advanced spontaneous metastasis for experimental therapeutics. Nat Rev Cancer 11(2):135–141
80. Bartlett DL et al (2013) Oncolytic viruses as therapeutic cancer vaccines. Mol Cancer 12(1):103
81. Al-Ghananeem AM et al (2009) Intratumoral delivery of paclitaxel in solid tumor from biodegradable hyaluronan nanoparticle formulations. AAPS Pharm Sci Tech 10(2):410–417
82. Chattopadhyay N et al (2012) Role of antibody-mediated tumor-targeted and route of administration in nanoparticle tumor accumulation in vivo. Mol Pharm 9(8):2168–2179
83. Xu Q et al (2012) Anti-tumor activity of paclitaxel through dual-targeting carrier of cyclic RGD and transferrin conjugated hyperbranched copolymer nanoparticles. Biomaterials 33 (5):1627–1639
84. Ko HY et al (2011) A multimodal nanoparticle-based cancer imaging probe simultaneously targeting nucleolin, integrin alphavbeta3 and tenascin-C proteins. Biomaterials 32 (4):1130–1138

Chapter 14
Challenges and Perspectives of Tumor-targeted Systems

Yi Wang

Since the first nano-drug (Doxil®, a kind of liposome formulation) was approved by the Food and Drug Administration (FDA), both the scale of clinical translation and the research depth of the basic theory of nanomedicine for tumor treatments have been greatly improved [1]. More than 800 clinical trials of nanomaterials have been undergoing evaluation [2]. Delivering cargos by nanocarriers demonstrates superior advantages over free drugs including controlled pharmacokinetics, targeted tissue distribution, limited adverse side effects, and inspiring diagnostic and therapeutic efficacy [3]. However, due to superficial knowledge of cancer, applications of clinical nanomedicines are still "in prison" due to the drug resistance, tumor relapse, and failure to elicit the enhanced anti-tumor efficacy [4]. Up to now, only about 170 nanomedicines and nano-imaging agents were approved for tumor treatments by the FDA and European Medicine Agency (EMA), most of which are organic nanomaterials such as liposomal and polymeric systems [5, 6]. Although inorganic nanocarriers and other novel nanomaterials have been developed, only a few of them were investigated in clinical trials and even fewer were approved by the FDA [7]. As the old saying goes, without a chill, there is no sweet plum blossom. For the development of nanomaterials in tumor diagnosis and therapy, more efforts are needed.

Y. Wang (✉)
Center for Advanced Low-dimension Materials, Donghua University, Shanghai, China
e-mail: ywang@dhu.edu.cn

R. Huang, Y. Wang (eds.), *New Nanomaterials and Techniques for Tumor-targeted Systems*, https://doi.org/10.1007/978-981-15-5159-8_14

14.1 Considering the Integrity of the Tumor Microenvironment

The tumor microenvironment always overarches problems limiting the overall therapeutic performance and clinical application of nanomedicine. The tumor microenvironment plays an important role in a wide variety of cancer therapeutics. However, the preparations of tumor-related nanoparticles are often considered by dividing the tumor microenvironment into independent individuals such as tumor vascular normalization [8], pH correction [9], inhibition of tumor-related factors [10], interference with cellular glucose metabolism [11], or targeting to overexpressed receptors on the surface of tumor cells [12], respectively. Because of consistency between the tumor microenvironment and tumor, if just focus on one or two factors, and lack the co-discussion of the interaction with cargos and tumor, the therapeutic outcome of nanomedicines will be comprised consequently. In addition, a recent report pointed out that the accumulation of the nanoparticles in tumor was only 0.7% of the injected dose. With active tumor-targeted ligand modification, the accumulation of nanocarriers would increase to 0.9%. And finally, about 0.0014% of ligand-modified nanoparticles directly interacted with tumor cells [13, 14]. Thus, a comprehensive understanding of the tumor microenvironment and its relationship with tumor growth and proliferation is of great importance for the design and application of tumor-targeted nanomaterials. Firstly, a full picture of the tumor microenvironment and cell composition characteristics including the biological factors, cell types, physiological characteristics of the tumor microenvironment, and organelle biological characteristics is all-important. On this basis, the network between biological factors and substances both inside and outside of tumor cells should be established to clearly understand the interaction of each cytokine and the corresponding biological characteristics, including organelles, microenvironment, and the relationship between tumor cells. And then, finding out the first domino of each network that promotes the growth and proliferation of tumor cells is essential for design and preparation of the nanoparticle systems to perform the greatest slaughter of cancer.

14.2 Rational Characterizations

Most of nanomedicines underwent inadequate characterizations in which the fluorescence imaging and pharmacodynamics of cargos rather than the properties of nanomaterials such as distribution, metabolism, and excretion in vivo were studied. And this counterfeit perfect result is the main reason why most nanomaterials fail to translate in clinical trials or cancer therapy. Therefore, the comprehensive characterization of nanomaterials is extremely important in the construction of tumor-targeted nanoparticles. Firstly, the characterization of nanoparticles (alone or carrying cargos) over time is needed to assess potential changes in their size, zeta potential, purity, geometry, surface properties, stability (both colloidal and chemical), aggregation, and agglomeration during mass production and storage. Secondly, the effects of extracellular matrix (ECM), matrix metalloproteinases (MMPs) and related enzymes, acidosis, and hypoxia within the tumor microenvironment of nanoparticles are needed to be evaluated. All shelf-life stability, degradation profiles in different body fluids, and possible potential changes in their physiochemical properties in the body after administration need to be assessed by some classical characterizations. Thirdly, the fate of nanomaterials with or without cargos in vivo needs to be further characterized, including the distribution in normal and tumor tissues, the metabolic and excretory characteristics. In addition, long-term biodistribution and rate of clearance of nanoparticles from different organs need to be evaluated. Most importantly, the above results could not only rely on a single characterization, but also need to be represented and verified by different methods. And the preclinical evaluation methods and models should be explored to decrease the gap between preclinical and clinical outcomes, and also accelerate the bench-to-bedside translation.

14.3 Biosafety Analysis

The biosafety of drug carriers to normal tissues and organs and the associated side effects are important considerations. However, the commercialization of nanomaterials is faster than the time it takes to assess their risk on health and environment. A limited number of studies focused on assessing the impact of nanoparticles on organ toxicity, inflammation, immunotoxicity, and genotoxicity in chronic exposure settings. Based on the standard in vivo methodology for subchronic and chronic safety test of materials, firstly, the mechanisms of their long-term toxicity in the accumulated organs should be included in the study design. And the evaluation of the toxic effect of nanoparticles on different organs as a function of dose, route, and frequency of administration is also required. Meanwhile, construction of a database of acute to chronic toxicity of nanoparticles classified by different parameters affecting their toxicity would be useful to calculate the relative maximum safe doses of nanoparticles that need to be co-delivered with biological agents for specific delivery applications. Understanding the underlying mechanisms of subchronic and chronic tissue toxicity, immunotoxicity, and genotoxicity might greatly accelerate the clinical translation of nanoparticles. Finally, development of predictive in vitro and animal models which are more relevant to humans is requested for long-term toxicology studies.

14.4 Development of Natural, Pollution-Free Nanomaterials

Natural materials, which possess low immunogenicity and toxicity, have been successfully exploited as biomaterials to prepare nanoparticles for tumor treatment. For example, alginate or chitosan belonging to ECM or ECM components have been used as nanocarriers [15]. Milk, fruit, and vegetables have also been used as raw materials for preparation of carbon quantum dots which owns features such as photothermal imaging [16–19]. Meanwhile, more researches are focusing on the preparation of nanomaterials by protein modification or cell reutilization. Viral nanoparticles (VNP) and self-assembling protein cages have shown great efficacy at tumor-targeted delivery, because they naturally use their native tropism for tumor cell-surface displayed vimentin and the EPR effect to extravasate through tumor neovasculature to efficiently penetrate into tumors and be internalized by tumor cells [20, 21]. And VNP can simultaneously display imaging, targeting, or homing moieties and load drugs as cargos for maximum anticancer activity [22]. In recent years, on the basis of liposomes, the nanoscaled systems prepared by fusion of one or more cell membranes have attracted more and more attentions. Cytomembrane-related liposomes (CRL) not only have good drug-loading capacity, but also can reduce the scavenging effect of the immune system in vivo [23, 24]. Meanwhile, via proper membrane selection, CRL could be endowed with strong homing effect, which can efficiently enhance their accumulation in target tissues and organs, releasing cargos into cells through membrane fusion. Thus, more natural, pollution-free nanomaterials should be developed for tumor diagnosis and therapy.

14.5 Combination Therapy

The complexity of the tumor microenvironment and heterogeneity of tumor cells lead to compromised therapeutic efficacy of monotherapy [25]. And to cross this mountain, synergistic combinations of different therapeutics based on nanotechnology have exhibited a promising potential in cancer therapy. Nanomaterials with or without photothermal, photodynamic, magnetic, and upconversion properties could load two or more antitumoral therapeutic agents which sometimes show good effects themselves or sometimes on increasing the sensitivity of tumor cells to another one, and/or integrate different therapeutic modes in one single platform to achieve the combination therapy for tumors. However, when these combined delivery systems arrive at the tumor site, nanomaterials with specific properties will respond to external or internal stimuli in a short time to carry out their missions such as photothermal therapy and upconversion. Meanwhile, the cargos release in large quantities and most of them will be in the form of a free state which is easily expelled by the organism [26]. This multidimensional combination therapy based on the monomers of commercially available preparations and other inhibitors of signaling proteins or pathways are not as good as those of commercially available preparations [27]. In this case, a combined drug delivery strategy for nanoscaled delivery systems and commercially available preparations is needed. The monomers of commercially available preparations contained in nanomaterials are replaced with more comprehensive inhibitors or drugs, which provide more possibilities for maximizing the effectiveness of commercially available preparations in tumor therapy. And this treatment strategy is particularly obvious in the research of tumor immunotherapy. A tumor microenvironment-activatable prodrug vesicle engineered by integrating an oxaliplatin prodrug and PEGylated photosensitizer into a single nanoplatform was prepared, showing tumor-specific accumulation, activation, and deep penetration in response to the tumoral acidic and enzymatic microenvironment. When combination with αCD47-mediated CD47 blockade following a designed dosing regimen, CD47 blockade and immunogenic cell death induction efficiently inhibited the growth of both primary and abscopal tumors, suppressed tumor metastasis, and prevented tumor recurrence [28]. On the other hand, based on upconversion-polymer hybrid nanoparticles, a synergistic strategy combining photodynamic therapy with photothermal therapy was established in which polydopamine nanoparticle was used as the thermal agent and chlorin e6 as photosensitizer, providing a powerful toolbox to control antitumor. When combined with immune checkpoint blockades following a designed dosing regimen, this system could inhibit tumor relapse and metastasis as well as prolong the survival of tumor-bearing mice in two types of tumor metastasis models [29]. In addition, more and more regulators of multiple signaling pathways which play key roles in oncogenic transformation and drug resistance are identified. And co-delivery nanotechnology which releases all the cargoes at the same time or place should transform to the procedural delivery system in which the release of the first drug will open the door for the release and action of the next one. This kind of drug delivery platforms with procedural drug release capacity will be one of the potential trends in preparations of tumor-targeted drug delivery systems in the future.

As substantial progress has been made in tumor biology, it is very much expected to achieve major breakthroughs in fields of nanotechnology and targeted drug delivery in the near future. By suitable combination, each therapeutic manner can be strengthened. Based on the background and analysis above, it is rational to believe that the nanotechnology-based drug delivery techniques provide great promises for cancer treatment and are expected to offer superior therapeutic outcomes to the current drug cocktail therapy.

References

1. Peng JR, Yang Q, Shi K, Xiao X, Wei XW, Qian ZY et al (2019) Intratumoral fate of functional nanoparticles in response to microenvironment factor: implications on cancer diagnosis and therapy. Adv Drug Deliv Rev 143:37–67
2. Search results from ClinicalTrials.gov.

3. Majumder J, Taratula O, Minko T et al (2019) Nanocarrier-based systems for targeted and site specific therapeutic delivery. Adv Drug Deliv Rev 144:57–77
4. Miao L, Guo ST, Lin CM, Liu Q, Huang L et al (2017) Nano-formulations for combination or cascade anticancer therapy. Adv Drug Deliv Rev 115:3–22
5. Search results from http://www.fda.gov/Drugs/InformationOnDrugs/ucm142438.htm
6. Search results from https://www.accessdata.fda.gov/scripts/cder/daf/#opennewwindow
7. Choi YH, Han HK et al (2018) Nanomedicines: current status and future perspectives in aspect of drug delivery and pharmacokinetics. J Pharm Investig 48:43–60
8. Peng Q, Li M, Wang Z, Jiang M, Yan X, Lei S, Zhang H, Zhang W, Liu YY, Luo F et al (2013) Polarization of tumor-associated macrophage is associated with tumor vascular normalization by endostatin. Thorac Cancer 4:295–3057
9. Thews O, Riemann A et al (2019) Tumor pH and metastasis: a malignant process beyond hypoxia. Cancer Metast Rev 38(1–2):113–129
10. Wang SS, Li CY, Qian M, Jiang HL, Shi W, Chen J, Lachelt U, Wagner E, Lu WY, Wang Y, Huang RQ et al (2017) Augmented glioma-targeted theranostics using multifunctional polymer-coated carbon nanodots. Biomaterials 141:29–39
11. Gonzalez PS et al (2018) Mannose impairs tumour growth and enhances chemotherapy. Nature 563(7733):719–723
12. Jiang Z, Guan J, Qian J, Zhan C et al (2019) Peptide ligand-mediated targeted drug delivery of nanomedicines. Biomater Sci 7:461–471
13. Wilhelm S, Tavares AJ, Dai Q, Ohta S, Audet J, Dvorak HF, Chan WC et al (2016) Analysis of nanoparticle delivery to tumours. Nat Rev Mater 1:16014
14. Dai Q, Wilhelm S, Ding D, Syed AM, Sindhwani S, Zhang Y, Chen YY, MacMillan P, Chan WC et al (2018) Quantifying the ligand-coated nanoparticle delivery to cancer cells in solid tumors. ACS Nano 12:8423–8435
15. Hinderer S, Layland SL, Schenke-Layland K et al (2016) ECM and ECM-like materials-biomaterials for applications in regenerative medicine and cancer therapy. Adv Drug Deliv Rev 97:260–269
16. Wang L, Zhou HS et al (2014) Green synthesis of luminescent nitrogen-doped carbon dots from milk and its imaging application. Anal Chem 86(18):8902–8905
17. Zhu L, Yin Y, Wang CF, Chen S et al (2013) Plant leaf-derived fluorescent carbon dots for sensing, patterning and coding. J Mater Chem C 1(32):4925
18. Sachdev A, Gopinath P et al (2015) Green synthesis of multifunctional carbon dots from coriander leaves and their potential application as antioxidants, sensors and bioimaging agents. Analyst 140:4260–4269
19. Zhai H, Zheng B, Yang F, Wang M, Xiao D et al (2018) Synthesis of water-soluble fluorescent carbon dots from Setcreasea purpurea boom and its application for Br2 detection. Anal Methods 10(1):151–157
20. Ma Y, Nolte RJ, Cornelissen JJ et al (2012) Virus-based nanocarriers for drug delivery. Adv Drug Deliv Rev 64:811–825
21. Beatty PH, Lewis JD et al (2019) Cowpea mosaic virus nanoparticles for cancer imaging and therapy. Adv Drug Deliv Rev 145:130–144
22. Srinivasan M, Rajabi M, Mousa SA et al (2015) Multifunctional nanomaterials and their applications in drug delivery and cancer therapy. Nano 5(4):1690–1703
23. Zheng BB, Xu JJ, Chen GK, Zhang SH, Xiao ZY, Lu W et al (2019) Bacterium-mimicking vector with enhanced adjuvanticity for cancer immunotherapy and minimized toxicity. Adv Funct Mater:1901437
24. Chai Z, Ran D, Lu L, Zhan C, Ruan H, Hu X, Lu WY et al (2019) Ligand-modified cell membrane enables targeted delivery of drug nanocrystals to glioma. ACS Nano 13 (5):5591–5601
25. Jessie LSA, Bertrand ZY, Michael GW, Ze L, Wientjes GM et al (2016) Delivery of cancer therapeutics to extracellular and intracellular targets: determinants, barriers, challenges and opportunities. Adv Drug Deliv Rev 97:280–301

26. Dai WB, Wang XY, Song G, Liu TZ, He B, Zhang H, Wang XQ, Zhang Q et al (2017) Combination antitumor therapy with targeted dual-nanomedicines. Adv Drug Deliv Rev 115:23–45
27. Chen W, Guo Z, Zhu Y, Qiao N, Zhang Z, Sun X et al (2019) Combination of bacterial-photothermal therapy with an anti-PD-1 peptide depot for enhanced immunity against advanced cancer. Adv Funct Mater:1906623
28. Zhou FY, Feng B, Yu HJ, Wang DG, Wang TT, Ma YT, Wang SL, Li YP et al (2019) Tumor microenvironment-activatable prodrug vesicles for nanoenabled cancer chemoimmunotherapy combining immunogenic cell death induction and CD47 blockade. Adv Mater 31:1805888
29. Yan SQ, Zeng XM, Tang YA, Liu BF, Wang Y, Liu XG et al (2019) Activating antitumor immunity and anti metastatic effect through polydopamine-encapsulated core-shell upconversion nanoparticles. Adv Mater 31:1905825

The manufacturer's authorised representative in the EU is Springer Nature Customer Service Centre GmbH, Europaplatz 3, 69115 Heidelberg, Germany. If you have any concerns regarding our products, please contact ProductSafety@springernature.com

Printed and bound by CPI Group (UK) Ltd, Croydon, CR0 4YY
28/07/2026
02179195-0006